Differential Geometry from a Singularity Theory Viewpoint

Publishers' page

Differential Geometry from a Singularity Theory Viewpoint

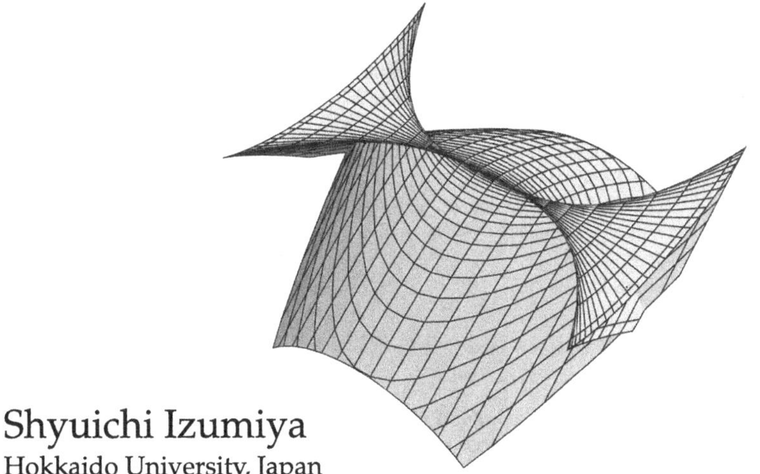

Shyuichi Izumiya
Hokkaido University, Japan

Maria del Carmen Romero Fuster
Universitat de València, Spain

Maria Aparecida Soares Ruas
University of São Paulo, Brazil

Farid Tari
University of São Paulo, Brazil

NEW JERSEY • LONDON • SINGAPORE • BEIJING • SHANGHAI • HONG KONG • TAIPEI • CHENNAI • TOKYO

Published by

World Scientific Publishing Co. Pte. Ltd.
5 Toh Tuck Link, Singapore 596224
USA office: 27 Warren Street, Suite 401-402, Hackensack, NJ 07601
UK office: 57 Shelton Street, Covent Garden, London WC2H 9HE

Library of Congress Cataloging-in-Publication Data
Differential geometry from a singularity theory viewpoint / by Shyuichi Izumiya
(Hokkaido University, Japan) [and three others].
pages cm
Includes bibliographical references and index.
ISBN 978-9814590440 (hardcover : alk. paper)
1. Surfaces--Areas and volumes. 2. Singularities (Mathematics) 3. Geometry, Differential.
4. Curvature. I. Izumiya, Shyuichi.
QA645.D54 2015
516.3'6--dc23

2015033184

British Library Cataloguing-in-Publication Data
A catalogue record for this book is available from the British Library.

Printed in Singapore

Preface

The geometry of surfaces is a subject that has fascinated many mathematicians and users of mathematics. This book offers a new look at this classical subject, namely from the point of view of singularity theory. Robust geometric features on a surface in the Euclidean 3-space, some of which are detectable by the naked eye, can be captured by certain types of singularities of some functions and mappings on the surface. In fact, the mappings in question come as members of some natural families of mappings on the surface. The singularities of the individual members of these families of mappings measure the contact of the surface with model objects such as lines, circles, planes and spheres.

This book gives a detailed account of the theory of contact between manifolds and its link with the theory of caustics and wavefronts. It then uses the powerful techniques of these theories to deduce geometric information about surfaces immersed in the Euclidean 3, 4 and 5-spaces as well as spacelike surfaces in the Minkowski space-time.

In Chapter 1 we argue the case for using singularity theory to study the extrinsic geometry of submanifolds of Euclidean spaces (or of other spaces). To make the book self-contained, we devote Chapter 2 to introducing basic facts about the extrinsic geometry of submanifolds of Euclidean spaces. Chapter 3 deals with singularities of smooth mappings. We state the results on finite determinacy and versal unfoldings which are fundamental in the study of the geometric families of mappings on surfaces treated in the book. Chapter 4 is about the theory of contact introduced by Mather and developed by Montaldi. In Chapter 5 we recall some basic concepts in symplectic and contact geometry and establish the link between the theory of contact and that of Lagrangian and Legendrian singularities. We apply in Chapters 6, 7 and 8 the singularity theory framework exposed in the previous chapters to the study of the extrinsic differential geometry of surfaces in the Euclidean 3, 4 and 5-spaces respectively. The codimension of the surface in the ambient space is 1, 2 or 3 and this book shows how some aspects of the geometry of the surface change with its codimension. In Chapter 9 we chose spacelike surfaces in the Minkowski space-time to illustrate how to approach the study of submanifolds in Minkowski spaces using singularity theory. Most of the results in the previous chapters are local in nature. Chapter 10 gives a flavour of global results on closed surfaces using local invariants obtained from the local study of the surfaces in the previous chapters.

The emphasis in this book is on how to apply singularity theory to the study of the extrinsic geometry of surfaces. The methods apply to any smooth submanifolds of higher dimensional Euclidean space as well as to other settings, such as affine, hyperbolic or Minkowski spaces. However, as it is shown in Chapters 6, 7 and 8, each pair (m, n) with m the dimension of the submanifold and n of the ambient space needs to be considered separately.

This book is unapologetically biased as it focuses on research results and interests of the authors and their collaborators. We tried to remedy this by including, in the Notes of each chapter, other aspect and studies on the topics in question and as many references as we can. Omissions are inevitable, and we apologise to anyone whose work is unintentionally left out.

Currently, there is a growing and justified interest in the study of the differential geometry of singular submanifolds (such as caustics, wavefronts, images of singular mappings etc) of Euclidean or Minkowski spaces, and of submanifolds with induced (pseudo) metrics changing signature on some subsets of the submanifolds. We hope that this book can be used as a guide to anyone embarking on the study of such objects.

This book has been used (twice so far!) by the last-named author as lecture notes for a post-graduate course at the University of São Paulo, in São Carlos. We thank the following students for their thorough reading of the final draft of the book: Alex Paulo Francisco, Leandro Nery de Oliveira, Lito Edinson Bocanegra Rodríguez, Martin Barajas Sichaca, Mostafa Salarinoghabi and Patricia Tempesta. Thanks are also due to Catarina Mendes de Jesus for her help with a couple of the book's figures and to Asahi Tsuchida, Shunichi Honda and Yutaro Kabata for correcting some typos. Most of the results in Chapter 4 are due to James Montaldi. We thank him for allowing us to reproduce some of his proofs in this book.

We are also very grateful to Masatomo Takahashi for reading the final draft of the book and for his invaluable comments and corrections.

S. Izumiya, M. C. Romero Fuster, M. A. S. Ruas and F. Tari
August, 2015

Contents

Preface v

1. The case for the singularity theory approach 1

1.1 Plane curves 2
1.1.1 The evolute of a plane curve 3
1.1.2 Parallels of a plane curve 5
1.1.3 The evolute from the singularity theory viewpoint 7
1.1.4 Parallels from the singularity theory viewpoint . . 10
1.2 Surfaces in the Euclidean 3-space 11
1.2.1 The focal set 13
1.3 Special surfaces in the Euclidean 3-space 14
1.3.1 Ruled surfaces 15
1.3.2 Developable surfaces 19
1.4 Notes 21

2. Submanifolds of the Euclidean space 23

2.1 Hypersurfaces in $\mathbb{R}^{n+1}$ 23
2.1.1 The first fundamental form 24
2.1.2 The shape operator 25
2.1.3 Totally umbilic hypersurfaces 29
2.1.4 Parabolic and umbilic points 32
2.2 Higher codimension submanifolds of $\mathbb{R}^{n+r}$ 36
2.2.1 Totally ν-umbilic submanifolds 40
2.2.2 ν-parabolic and ν-umbilic points 41
2.2.3 The canal hypersurface 41

3. Singularities of germs of smooth mappings 45

3.1 Germs of smooth mappings 45
3.2 Multi-germs of smooth mappings 47
3.3 Singularities of germs of smooth mappings 47
3.4 The Thom-Boardman symbols 51
3.5 Mather's groups 51
3.6 Tangent spaces to the $\mathcal{G}$-orbits 52
3.7 Finite determinacy 55
3.8 Versal unfoldings 56
3.9 Classification of singularities 59
3.9.1 Germs of functions 60
3.9.2 Discriminants and bifurcation sets 64
3.10 Damon's geometric subgroups 69
3.11 Notes 70

4. Contact between submanifolds of $\mathbb{R}^n$ 73

4.1 Contact between submanifolds 74
4.2 Genericity 80
4.3 The meaning of generic immersions 85
4.4 Contact with hyperplanes 88
4.5 The family of distance squared functions 90
4.6 The family of projections into hyperplanes 92
4.7 Notes 95

5. Lagrangian and Legendrian Singularities 97

5.1 Symplectic manifolds 98
5.1.1 Lagrangian submanifolds and Langrangian maps . 100
5.1.2 Lagrangian singularities 101
5.2 Contact manifolds 105
5.2.1 Legendrian submanifolds and Legendrian maps . . 107
5.2.2 Legendrian singularities 108
5.3 Graph-like Legendrian submanifolds 113
5.4 Versal unfoldings and Morse families of functions 117
5.5 Families of functions on hypersurfaces in $\mathbb{R}^n$ 119
5.5.1 The family of height functions 122
5.5.2 The extended family of height functions 124
5.5.3 The family of distance squared functions 127
5.6 Contact from the viewpoint of Lagrangian and Legendrian singularities 128

5.6.1 Contact of hypersurfaces with hyperplanes 128
5.6.2 Contact of hypersurfaces with hyperspheres 132
5.6.3 Contact of submanifolds with hyperplanes 134

6. Surfaces in the Euclidean 3-space 139

6.1 First and second fundamental forms 139
6.2 Surfaces in Monge form 145
6.3 Contact with planes 146
6.4 Contact with lines 159
6.4.1 Contour generators and apparent contours 160
6.4.2 The generic singularities of orthogonal projections 165
6.5 Contact with spheres 178
6.6 Robust features of surfaces 183
6.6.1 The parabolic curve 184
6.6.2 The flecnodal curve 186
6.6.3 The ridge curve 189
6.6.4 The sub-parabolic curve 194
6.7 Notes 198

7. Surfaces in the Euclidean 4-space 201

7.1 The curvature ellipse 202
7.2 Second order affine properties 207
7.2.1 Pencils of quadratic forms 211
7.3 Asymptotic directions 213
7.4 Surfaces in Monge form 218
7.5 Examples of surfaces in $\mathbb{R}^4$ 219
7.6 Contact with hyperplanes 221
7.6.1 The canal hypersurface 225
7.6.2 Characterisation of the singularities of the height function 229
7.7 Contact with lines 232
7.7.1 The geometry of the projections 237
7.8 Contact with planes 242
7.9 Contact with hyperspheres 246
7.10 Notes 249

8. Surfaces in the Euclidean 5-space 251

8.1 The second order geometry of surfaces in $\mathbb{R}^5$ 252

8.2 Contacts with hyperplanes 259
8.3 Orthogonal projections onto hyperplanes, 3-spaces and planes 267
8.3.1 Contact with lines 267
8.3.2 Contact with planes 268
8.3.3 Contact with 3-spaces 270
8.4 Contacts with hyperspheres 272
8.5 Notes 277

9. Spacelike surfaces in the Minkowski space-time 281

9.1 Minkowski space-time 283
9.1.1 The hyperbolic space and the Poincaré ball model 284
9.2 The lightcone Gauss maps 285
9.3 The normalised lightcone Gauss map 291
9.4 Marginally trapped surfaces 292
9.5 The family of lightcone height functions 293
9.6 The Lagrangian viewpoint 296
9.7 The lightcone pedal and the extended lightcone height function: the Legendrian viewpoint 300
9.8 Special cases of spacelike surfaces 304
9.8.1 Surfaces in Euclidean 3-space 305
9.8.2 Spacelike surfaces in de Sitter 3-space 305
9.8.3 Spacelike surfaces in Minkowski 3-space 306
9.8.4 Surfaces in hyperbolic 3-space 307
9.9 Lorentzian distance squared functions 309
9.9.1 Lightlike hypersurfaces 311
9.9.2 Contact of spacelike surfaces with lightcones . . . 313
9.10 Legendrian dualities between pseudo-spheres 315
9.11 Spacelike surfaces in the lightcone 317
9.11.1 The Lightcone Theorema Egregium 320
9.12 Notes 325

10. Global viewpoint 329

10.1 Submanifolds of Euclidean space 330
10.1.1 Surfaces in $\mathbb{R}^3$ 330
10.1.2 Wavefronts 333
10.1.3 Surfaces in $\mathbb{R}^4$ 335
10.1.4 Semiumbilicity 337

10.2 Spacelike submanifolds of Minkowski space-time 339
10.3 Notes 343

Bibliography 347

Index 363

Chapter 1

The case for the singularity theory approach

The study of curves and surfaces in the Euclidean space is a fascinating and important subject in differential geometry. We highlight in this chapter how singularity theory can be used not only to recover classical results on curves and surfaces in a simpler and more elegant way but also how it reveals the rich and deep underlying concepts involved.

We start with the evolute and parallels of a plane curve. We first use classical differential geometry techniques to obtain the shape of the evolute and parallels. We then define the family of distance squared functions on the plane curve and recover from the singularities type of the members of this family geometric information about the curve itself. We outline how to use the Lagrangian and Legendrian singularity theory framework to deduce properties of the evolute that are invariant under diffeomorphisms. We proceed similarly for surfaces in the Euclidean 3-space and consider the singularities of their focal sets. We deal in the last section with the singularities of ruled and developable surfaces.

We refer to [do Carmo (1976)] for a detailed study of the differential geometry of curves and surfaces.

Throughout this book, a given map is said to be *smooth* (or C^∞) if its partial derivatives of all order exist and are continuous.

The Euclidean n-space is the vector space $\mathbb{R}^n$ endowed with the scalar product

$$\langle \mathbf{u}, \mathbf{v} \rangle = u_1 v_1 + \cdots + u_n v_n$$

for any $\mathbf{u} = (u_1, \ldots, u_n)$ and $\mathbf{v} = (v_1, \ldots, v_n)$ in $\mathbb{R}^n$.

We also view the Euclidean n-space as a set of points. The vector space $\mathbb{R}^n$ comes with a standard orthogonal basis $\mathbf{e}_1 = (1, \ldots, 0)$, $\ldots$, $\mathbf{e}_n = (0, \ldots, 1)$. We choose a point $O = (0, \ldots, 0)$ to be the origin and denote by $\Sigma = (O, \mathbf{e}_1, \ldots, \mathbf{e}_n)$ the standard orthonormal coordinates system in $\mathbb{R}^n$. Then, a point p in the Euclidean n-space is the endpoint of the vector Op and its coordinates $(x_1, \dots, x_n)$ in the system Σ are the coordinates of the vector Op in the basis $\{\mathbf{e}_1, \dots, \mathbf{e}_n\}$. Curves, surfaces, submanifolds in $\mathbb{R}^n$ are considered as subsets of points in $\mathbb{R}^n$.

The vector product of $n-1$ vectors $\mathbf{u}_1, \dots, \mathbf{u}_{n-1}$ in $\mathbb{R}^n$, is defined by

$$\mathbf{u}_1 \times \cdots \times \mathbf{u}_{n-1} = \begin{vmatrix} e_1 & \cdots & e_n \\ u_1^1 & \cdots & u_n^1 \\ \vdots & \cdots & \vdots \\ u_1^{n-1} & \cdots & u_n^{n-1} \end{vmatrix},$$

where $\mathbf{u}_i = (u_1^i, \dots, u_n^i)$. By the property of the determinant, we have

$$\langle \mathbf{u}, \mathbf{u}_1 \times \cdots \times \mathbf{u}_{n-1} \rangle = \det(\mathbf{u}, \mathbf{u}_1, \dots, \mathbf{u}_{n-1}).$$

1.1 Plane curves

A smooth curve in the Euclidean n-space is a smooth map $\gamma : I \to \mathbb{R}^n$, where I is an open interval of $\mathbb{R}$. The *trace* of γ, which we still denote by γ, is the set of points $\gamma(I)$ in $\mathbb{R}^n$. The curve γ is said to be *regular* if $\gamma'(t)$ is not the zero vector for any t in I. Points where $\gamma'(t)$ is the zero vector are called *singular points* of γ.

We consider here smooth and regular plane curves ($n = 2$ above). We shall suppose that the curve $\gamma : I \to \mathbb{R}^2$ is parametrised by arc length and denote the arc length parameter by s. Then, $\mathbf{t}(s) = \gamma'(s)$ is a unit tangent vector to γ. We denote by $\mathbf{n}(s)$ the unit normal vector to γ obtained by rotating $\mathbf{t}(s)$ anti-clockwise by an angle of $\pi/2$. It follows from the fact that $\langle \mathbf{t}(s), \mathbf{t}(s) \rangle = 1$ that $\langle \mathbf{t}'(s), \mathbf{t}(s) \rangle = 0$, so

$$\mathbf{t}'(s) = \kappa(s)\mathbf{n}(s), \tag{1.1}$$

for some smooth function $\kappa(s)$, called *the curvature* of γ at s.

We have, similarly, $\langle \mathbf{n}'(s), \mathbf{n}(s) \rangle = 0$, so $\mathbf{n}'(s) = \alpha(s)\mathbf{t}(s)$ for some function $\alpha(s)$. Differentiating the identity $\langle \mathbf{t}(s), \mathbf{n}(s) \rangle = 0$ and using (1.1) gives $\alpha(s) = -\kappa(s)$, so that

$$\mathbf{n}'(s) = -\kappa(s)\mathbf{t}(s).$$

We can use (1.1) to deduce that

$$\kappa(s) = \langle \mathbf{t}'(s), \mathbf{n}(s) \rangle.$$

When the parameter t of the curve γ is not necessarily the arc length parameter, the curvature is given by the formula

$$\kappa(t) = \frac{\det(\gamma'(t), \gamma''(t))}{||\gamma'(t)||^3}.$$

(One can re-parametrise γ by arc length and use the chain rule to get the above formula, see for example [Bruce and Giblin (1992)].) If we write $\gamma(t) = (x(t), y(t))$, then

$$\kappa = \frac{x'y'' - x''y'}{(x'^2 + y'^2)^{\frac{3}{2}}},$$

where all the functions are evaluated at t.

The curvature function determines completely the curve up to rigid motions (i.e., up to translations and rotations about points in the plane). Indeed,

Theorem 1.1 (Fundamental Theorem of Plane Curves). *Given a smooth function $\kappa(s) : I \to \mathbb{R}$, there is a smooth and regular curve $\gamma : I \to \mathbb{R}^2$ parametrised by arc length s with curvature $\kappa(s)$. The curve γ is unique up to rigid motions of $\mathbb{R}^2$.*

An *inflection point* of γ is a point where $\kappa(t) = 0$. An inflection point is referred to as an *ordinary inflection* if $\kappa(t) = 0$ but $\kappa'(t) \neq 0$.

A *vertex* of γ is a point where $\kappa(t) \neq 0$ and $\kappa'(t) = 0$. A vertex is called an *ordinary vertex* if $\kappa(t) \neq 0$, $\kappa'(t) = 0$ and $\kappa''(t) \neq 0$.

We define the following types of singularities of plane curves.

Definition 1.1. (1) A smooth curve $\gamma : I \to \mathbb{R}^2$ has an ordinary cusp singularity at $t_0 \in I$ if t_0 is a singular point of γ and the vectors $\gamma''(t_0)$ and $\gamma'''(t_0)$ are linearly independent (Figure 1.1, left).

(2) A smooth curve $\gamma : I \to \mathbb{R}^2$ has a $(3,4)$-singularity at $t_0 \in I$ if $\gamma'(t_0) = \gamma''(t_0) = (0,0)$ and $\gamma'''(t_0)$ and the fourth derivative vector $\gamma^{(4)}(t_0)$ are linearly independent (Figure 1.1, right).

1.1.1 *The evolute of a plane curve*

The *evolute* of a curve $\gamma : I \to \mathbb{R}^2$ is the plane curve ε parametrised by

$$\varepsilon(t) = \gamma(t) + \frac{1}{\kappa(t)}\mathbf{n}(t),\ t \in I, \tag{1.2}$$

Fig. 1.1 A cusp singularity left and a (3, 4)-singularity right.

where $\mathbf{n}(t)$ is the unit normal vector obtained by rotating the unit tangent vector $\gamma'(t)/||\gamma'(t)||$ anti-clockwise by $\pi/2$.

The evolute is well defined and is a smooth curve away from the inflection points of γ. We can use classical differential geometry techniques to study its geometry.

Proposition 1.1. *The evolute of a smooth and regular curve $\gamma : I \to \mathbb{R}^2$ is a regular curve except at points corresponding to the vertices of γ. The evolute has an ordinary cusp singularity at points corresponding to ordinary vertices of γ.*

Proof. We take γ parametrised by arc length s. Differentiating (1.2) and dropping the argument s, we get

$$\varepsilon' = -\frac{\kappa'}{\kappa^2}\mathbf{n},$$

and this is the zero vector at $s_0 \in I$ if and only if $\kappa'(s_0) = 0$, that is, if and only if γ has a vertex at s_0.

We obtain by differentiating again

$$\begin{aligned} \varepsilon'' &= \frac{\kappa'}{\kappa}\mathbf{t} + \left(\frac{2\kappa'^2 - \kappa''\kappa}{\kappa^3}\right)\mathbf{n}, \\ \varepsilon''' &= \left(\frac{2\kappa''\kappa - 3\kappa'^2}{\kappa^2}\right)\mathbf{t} + \left(\frac{\kappa'\kappa^4 - \kappa'''\kappa^2 + 6\kappa''\kappa'\kappa - 6\kappa'^3}{\kappa^4}\right)\mathbf{n}. \end{aligned}$$

At a vertex s_0 of γ, $\kappa'(s_0) = 0$ and the expressions for ε'' and ε''' at s_0 simplify and become

$$\begin{aligned} \varepsilon''(s_0) &= -\frac{\kappa''(s_0)}{\kappa^2(s_0)}\mathbf{n}(s_0), \\ \varepsilon'''(s_0) &= \frac{2\kappa''(s_0)}{\kappa(s_0)}\mathbf{t}(s_0) - \frac{\kappa'''(s_0)}{\kappa^2(s_0)}\mathbf{n}(s_0). \end{aligned}$$

The vectors $\varepsilon''(s_0)$ and $\varepsilon'''(s_0)$ are linearly independent if and only if $\kappa''(s_0) \neq 0$. Thus, the evolute of γ has an ordinary cusp singularity if and only if the corresponding point on γ is an ordinary vertex of γ. □

An ellipse has four ordinary vertices and Figure 1.2, left, shows the evolute of an ellipse with its four ordinary cusps. The vertices of a curve are points where the curve is most or least curved and these can be detected (approximately) by the naked eye for the ellipse in Figure 1.2, left. It is not possible to do so for the ellipse in Figure 1.2, right, as its principal axes have almost the same length. The ellipse in Figure 1.2, right, looks like a circle but is not a circle as its evolute is not a point. We can find the vertices of the ellipse in Figure 1.2, right, by considering the limiting tangent lines to the evolute at its ordinary cusp. These lines intersect the ellipse at its vertices.

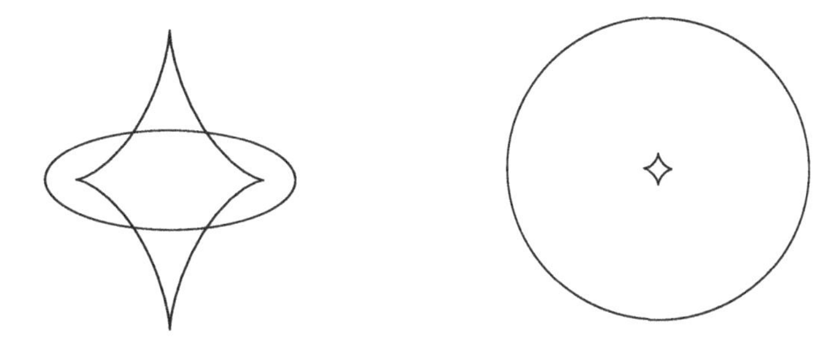

Fig. 1.2 Evolutes of ellipses: the difference between the lengths of the principal axes of the ellipse on the left is noticeable, whereas that of the ellipse on the right is negligible. The ellipse on the right looks like a circle but is not a circle.

1.1.2 *Parallels of a plane curve*

A *parallel* (or a wavefront) of a curve $\gamma : I \to \mathbb{R}^2$ is the curve obtained by moving each point on γ along its unit normal vector by a fixed distance d. When γ is parametrised by arc length, a parametrisation of a parallel is given by

$$\rho_d(s) = \gamma(s) + d\mathbf{n}(s),\ s \in I.$$

We have $\rho_d'(s) = (1 - d\kappa(s))\mathbf{t}(s)$, so a parallel is singular at points where $d = 1/\kappa(s)$. This means that the singular points of a parallel are located on the evolute of γ. As d varies, the singular points of the parallels of γ

trace the evolute of γ. Figure 1.3 shows the parallels of an ellipse with their singular points tracing the evolute of the ellipse.

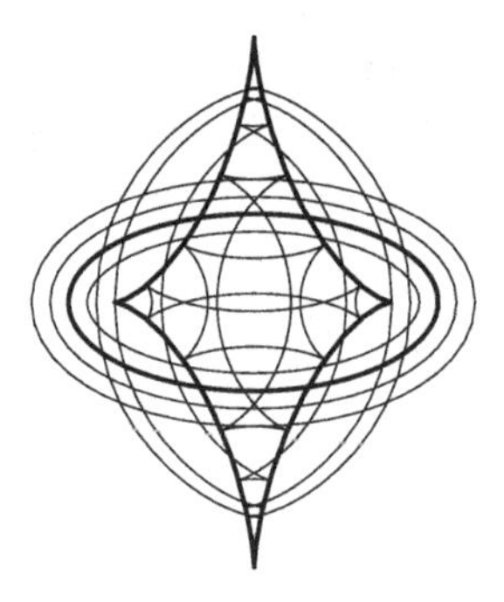

Fig. 1.3 The parallels of an ellipse. The ellipse and its evolute are drawn in thick.

Proposition 1.2. *The s of a smooth and regular curve γ have an ordinary cusp singularity at regular points on the evolute. The parallel through an ordinary singularity of the evolute has a $(3,4)$-singularity at that point.*

Proof. We have the following successive derivatives of the parametrisation of a parallel:

$$\begin{aligned} \rho_d' &= (1-d\kappa)\mathbf{t}, \\ \rho_d'' &= -d\kappa'\mathbf{t} + \kappa(1-d\kappa)\mathbf{n}, \\ \rho_d''' &= -(d\kappa'' + \kappa^2(1-d\kappa))\mathbf{t} + \kappa'(1-3d\kappa)\mathbf{n}, \\ \rho_d^{(4)} &= -(d\kappa''' + 3\kappa'\kappa(1-2d\kappa))\mathbf{t} \\ &\quad +(-\kappa(d\kappa'' + \kappa^2(1-d\kappa)) + \kappa''(1-3d\kappa) - 3d\kappa'^2)\mathbf{n}. \end{aligned}$$

At a singularity s_0 of the parallel $d = 1/\kappa(s_0)$, so

$$\begin{aligned} \rho_d''(s_0) &= -\frac{\kappa'(s_0)}{\kappa(s_0)}\mathbf{t}(s_0), \\ \rho_d'''(s_0) &= -\frac{\kappa''(s_0)}{\kappa(s_0)}\mathbf{t}(s_0) - 2\kappa'(s_0)\mathbf{n}(s_0). \end{aligned}$$

The vectors $\rho_d''(s_0)$ and $\rho_d'''(s_0)$ are linearly independent if and only if $\kappa'(s_0) \neq 0$, equivalently, if and only if $\gamma(s_0)$ is not a vertex of γ. If this is the case, the parallel ρ_d with $d = 1/\kappa(s_0)$ has an ordinary cusp singularity at $s = s_0$ (see Definition 1.1).

Suppose that $d = 1/\kappa(s_0)$ and $\kappa'(s_0) = 0$. Then $\rho_d'(s_0) = \rho_d''(s_0) = 0$ and

$$\begin{aligned}\rho_d'''(s_0) &= -\frac{\kappa''(s_0)}{\kappa(s_0)}\mathbf{t}(s_0),\\ \rho_d^{(4)}(s_0) &= -\frac{\kappa'''(s_0)}{\kappa(s_0)}\mathbf{t}(s_0) - 3\kappa''(s_0)\mathbf{n}(s_0).\end{aligned}$$

The vectors $\rho_d'''(s_0)$ and $\rho_d^{(4)}(s_0)$ are linearly independent if and only if $\kappa''(s_0) \neq 0$, equivalently, if and only if $\gamma(s_0)$ is an ordinary vertex of γ. If this is the case, the parallel ρ_d with $d = 1/\kappa(s_0)$ has a $(3,4)$-singularity at s_0 (see Definition 1.1). $\square$

1.1.3 *The evolute from the singularity theory viewpoint*

In sections 1.1.1 and 1.1.2 we obtained geometric information about the evolute and parallels of a plane curve by direct computation of the successive derivatives of a parametrisation of the curve. This method has several limitations. For instance, it does not explain which singularities could appear in the evolute and parallels and how these bifurcate as the original curve is deformed. It also misses to capture the deep concepts involved. We outline these concepts in this section.

We consider the *contact* (see Chapter 4) of a smooth and regular plane curve $\gamma : I \to \mathbb{R}^2$ with circles. A circle of centre $\boldsymbol{a}$ and radius r is the level set $D_a(p) = r^2$ of the *distance squared function* $D_a : \mathbb{R}^2 \to \mathbb{R}$, given by

$$D_a(p) = ||p - a||^2 = \langle p - a, p - a\rangle.$$

The contact of γ with the level sets of D_a can be measured by the vanishing of successive derivatives of the function

$$g(s) = D_a(\gamma(s)) = \langle\gamma(s) - a, \gamma(s) - a\rangle.$$

A point $\gamma(s_0)$ is on a circle C of centre a and radius r if and only if $g(s_0) = r^2$. The curve γ and the circle C have an ordinary tangency at $\gamma(s_0)$ if and only if $g(s_0) = r^2$, $g'(s_0) = 0$ and $g''(s_0) \neq 0$. Higher orders of tangency between γ and C are captured by the vanishing of the successive derivatives of g at s_0.

Definition 1.2. We say that γ and C have $k+1$*-point contact* at s_0 if $g^{(i)}(s_0) = 0$ for $i = 1, \dots, k$ but $g^{(k+1)}(s_0) \neq 0$. Then s_0 is said to be a

Table 1.1 Geometric conditions for the singularities of g.

g	Conditions	Geometric interpretation
A_1	$a = \gamma(s_0) + \lambda \mathbf{n}(s_0)$, $\lambda \neq \frac{1}{\kappa(s_0)}$	The centre of the circle C lies on the normal line to γ at s_0 but is not on the evolute of γ.
A_2	$a = \gamma(s_0) + \frac{1}{\kappa(s_0)}\mathbf{n}(s_0)$, $\kappa'(s_0) \neq 0$	The centre of the circle C lies on the evolute of γ but s_0 is not a vertex of γ.
A_3	$a = \gamma(s_0) + \frac{1}{\kappa(s_0)}\mathbf{n}(s_0)$, $\kappa'(s_0) = 0$, $\kappa''(s_0) \neq 0$	The centre of the circle C lies on the evolute of γ and s_0 is an ordinary vertex of γ.

singularity of g of type A_k. We say that γ and the circle C have $\geq k$*-point contact* at s_0 if $g^{(i)}(s_0) = 0$ for $i = 1, \dots, k$ and call s_0 a singularity of g of type $A_{\geq k}$.

Suppose that $g(s_0) = r^2$. We can recover geometric information about the curve γ at s_0 from the singularity type of the function g at s_0.

Proposition 1.3. *Let $\gamma : I \to \mathbb{R}^2$ be a smooth and regular plane curve and let C be a circle of centre a and radius r. Suppose that $g(s_0) = r^2$ for some $s_0 \in I$. Then g has a singularity of type A_1, A_2 or A_3 at s_0 if and only if the geometric conditions in Table 1.1 are satisfied.*

Proof. We take γ parametrised by arc length. Then the result follows by observing that

$$\begin{aligned}
\tfrac{1}{2}g' &= \langle \mathbf{t}, \gamma - a \rangle, \\
\tfrac{1}{2}g'' &= \kappa \langle \mathbf{n}, \gamma - a \rangle + 1, \\
\tfrac{1}{2}g''' &= \kappa' \langle \mathbf{n}, \gamma - a \rangle - \kappa^2 \langle \mathbf{t}, \gamma - a \rangle, \\
\tfrac{1}{2}g^{(4)} &= (\kappa'' - \kappa^3)\langle \mathbf{n}, \gamma - a \rangle - 3\kappa\kappa' \langle \mathbf{t}, \gamma - a \rangle - \kappa^2,
\end{aligned}$$

where all the functions are evaluated at s. □

It is possible to carry on and identify geometrically the singularities of g of type A_k, with $k > 3$. However, in general, or to be more precise for *generic* curves, the function g has only singularities of type A_1, A_2 or A_3. (The concept of genericity is dealt with in Chapter 4. Intuitively, a property of an object is generic if it persists when the object is deformed.)

We consider the functions D_a, $a \in \mathbb{R}^2$, all together as members of the *family of distance squared functions* $D : \mathbb{R}^2 \times \mathbb{R}^2 \to \mathbb{R}$, given by

$$D(p, a) = D_a(p).$$

The restriction of D to a plane curve γ is the family $D : I \times \mathbb{R}^2 \to \mathbb{R}$, given by

$$D(s,a) = \langle \gamma(s) - a, \gamma(s) - a \rangle.$$

The *catastrophe set* of the family D is defined to be the set

$$C_D = \left\{ (s,a) \in I \times \mathbb{R}^2 \,:\, \frac{\partial D}{\partial s}(s,a) = 0 \right\},$$

and the (local) *bifurcation set* of D is defined as the set

$$\begin{aligned} B_D &= \left\{ a \in \mathbb{R}^2 \,:\, \exists (s,a) \in C_D \text{ such that } \tfrac{\partial^2 D}{\partial s^2}(s,a) = 0 \right\} \\ &= \left\{ a \in \mathbb{R}^2 \,:\, D_a \text{ has an } A_{\geq 2} - \text{singularity at some } s \in I \right\}. \end{aligned}$$

Proposition 1.4. (1) *The local bifurcation set of the family of distance squared functions on γ is the evolute of γ.*

(2) *The catastrophe set C_D is a regular surface in $I \times \mathbb{R}^2$. The set of critical values of the catastrophe map $\pi_{C_D} : C_D \to \mathbb{R}^2$, with $\pi_{C_D}(s,a) = a$, is the local bifurcation set of the family of distance squared functions.*

Proof. The proof of (1) follows from the definition of the bifurcation set and from Proposition 1.3. As for (2), we prove in Chapter 5 a more general result that shows that C_D is a regular surface. □

We can now outline the underlying singularity theory concepts involved in the study of the evolute. These are developed in subsequent chapters.

1. The cotangent bundle $T^*\mathbb{R}^2$, with the canonical projection $\pi : T^*\mathbb{R}^2 \to \mathbb{R}^2$ to the base space, has the canonical symplectic structure $\omega = \sum_{i=1}^{2} dq_i \wedge dp_i$.
2. There is a Lagrangian immersion $L(D) : C_D \to T^*\mathbb{R}^2$, that is, $L(D)$ is an immersion and $L(D)(C_D)$ is a Lagrangian surface in $T^*\mathbb{R}^2$.
3. The following diagramme commutes

$$\begin{array}{ccc} & & T^*\mathbb{R}^2 \\ & \overset{L(D)}{\nearrow} & \downarrow \pi \\ C_D & \xrightarrow[\pi_{C_D}]{} & \mathbb{R}^2 \end{array}$$

   so the catastrophe map $\pi_{C_D} = \pi \circ L(D)$ is a Lagrangian map.
4. It follows from Proposition 1.4 that the evolute is the set of critical values of the Lagrangian map π_{C_D}, i.e., it is a *caustic*. As a consequence, it has only Lagrangian singularities.

5. Lagrangian surfaces are given by generating families and the family D of distance squared functions is the generating family of $L(D)(C_D)$.
6. There is an equivalence relation between two generating families of functions such that if two families are equivalent, then their bifurcation sets are diffeomorphic. As a consequence, it is enough to compute the bifurcation sets of the model families (which have simple expressions) to deduce the diffeomorphism models of the evolute. We can then conclude that the evolute of γ has an ordinary cusp singularity at points corresponding to ordinary vertices of γ.

1.1.4 *Parallels from the singularity theory viewpoint*

Parallels of a plane curve γ can also be studied using the family of distance squared function D on γ. For each $r > 0$, consider the family of functions $\tilde{D}_r : I \times \mathbb{R}^2 \to \mathbb{R}$, given by

$$\tilde{D}_r(s, a) = D(s, a) - r^2.$$

The set

$$\begin{aligned}\Sigma_*^{\pm}(\tilde{D}_r) &= \{(s, a) \in I \times \mathbb{R}^2 \,:\, \tilde{D}_r(s, a) = \tfrac{\partial \tilde{D}_r}{\partial s}(s, a) = 0\} \\ &= \{(s, a) \in I \times \mathbb{R}^2 \,:\, a = \gamma(s) \pm r\mathbf{n}(s)\}\end{aligned}$$

is a smooth and regular surface in $I \times \mathbb{R}^2$. Consider the projection $\pi_2 : \Sigma_*^{\pm}(\tilde{D}) \to \mathbb{R}^2$, given by $\pi_2(s, a) = a$. Then $\pi_2(\Sigma_*^{\pm}(\tilde{D}_r))$ is the parallel $\rho_{\pm r}$ (i.e., the parallel $\rho_{\pm r}$ is *the discriminant* of the family $\tilde{D}_r$).

We outline below the underlying singularity theory concepts involved in the study of the parallels. These are also developed in subsequent chapters.

1. The projective cotangent bundle $PT^*\mathbb{R}^2$, with the canonical projection $\pi : PT^*\mathbb{R}^2 \to \mathbb{R}^2$ to the base space, has a canonical contact structure.
2. There is a Legendrian immersion $\mathscr{L}(\tilde{D}_r) : \Sigma_*^{\pm}(\tilde{D}_r) \to PT^*\mathbb{R}^2$, that is, $\mathscr{L}(\tilde{D}_r)$ is an immersion and $\mathscr{L}(\tilde{D}_r)(\Sigma_*^{\pm}(\tilde{D}_r))$ is a Legendrian surface in $PT^*\mathbb{R}^2$.
3. The following diagramme commutes

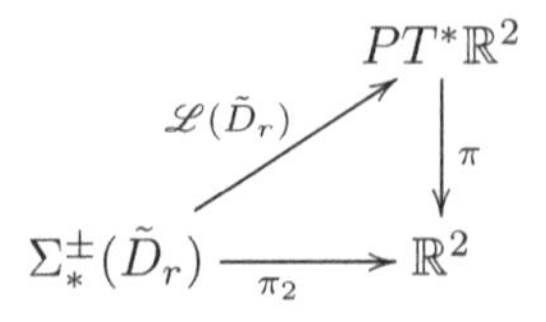

so the projection $\pi_2 = \pi \circ \mathscr{L}(D_r)$ is a Legendrian map.

4. It follows that the parallels $\rho_{\pm r}$ are wavefronts. As a consequence, they have only Legendrian singularities.
5. Legendrian surfaces are given by generating families and the family $\tilde{D}_r$ is the generating family of $\mathscr{L}(\tilde{D}_r)(\Sigma_*^{\pm}(\tilde{D}_r))$.
6. There is an equivalence relation between two generating families of functions such that if two such families are equivalent, then their discriminant sets are diffeomorphic. As a consequence, it is enough to compute the discriminant sets of the model families to deduce the diffeomorphism models of the parallels.
7. To study how the parallels evolve as r varies, we consider the big family $\tilde{D} : I \times \mathbb{R}^2 \times \mathbb{R}_+ \to \mathbb{R}$ given by $\tilde{D}(s, a, r) = \tilde{D}_r(s, a)$. The set $\Sigma_*^{\pm}(\tilde{D})$ is a regular 3-dimensional manifold in $I \times \mathbb{R}^2 \times \mathbb{R}_+$. We obtain a Legendrian immersion $\mathscr{L}(\tilde{D}) : \Sigma_*^{\pm}(\tilde{D}) \to PT^*(\mathbb{R}^2 \times \mathbb{R}_+)$ and a big wavefront $\pi \circ \mathscr{L}(D)(\Sigma_*^{\pm}(\tilde{D}))$. The individual fronts are recovered by slicing the big front by the planes $r = constant$.
8. We deduce that the singularities of the parallel ρ_r of the curve γ are ordinary cusps at points $\rho_r(s_0)$ where $r = 1/\kappa(s_0)$ when s_0 is not a vertex of γ. The parallels undergo the transitions given by the generic section of the swallowtail surface (big wavefront) at an ordinary vertex of γ (Figure 1.4). See also Figure 1.3 which shows how the parallels to an ellipse are stacked together at the vertices of the ellipse.

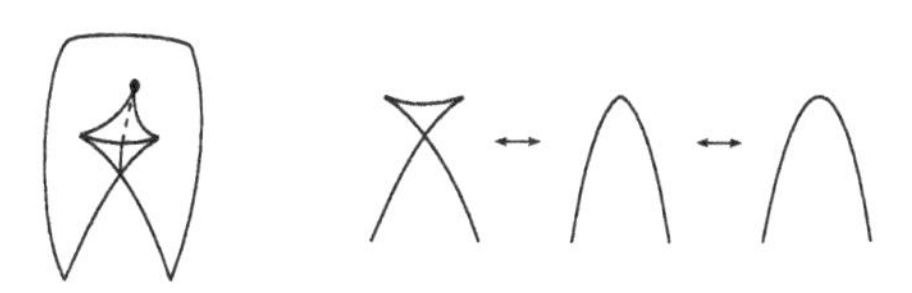

Fig. 1.4 A swallowtail surface (big wavefront) left and its generic sections (right).

1.2 Surfaces in the Euclidean 3-space

Let M be a smooth and regular surface in the Euclidean 3-space $\mathbb{R}^3$. We consider local properties of M and take at a point $p \in M$ a local parametrisation $\mathbf{x} : U \to \mathbb{R}^3$, where U is an open subset of $\mathbb{R}^2$ with $p \in \mathbf{x}(U)$. We denote by $u = (u_1, u_2)$ the parameters in U.

Let S^2 denote the unit sphere in $\mathbb{R}^3$. The map $N : M \to S^2$ which assigns to each point $p \in M$ a unit normal vector $N(p)$ to M at p is called the *Gauss map*. (We require M to be orientable if we want to define N on the whole surface.)

The normal vector $N(p)$ is also orthogonal to the tangent plane $T_{N(p)}S^2$ of the sphere S^2 at $N(p)$, so we can identify $T_{N(p)}S^2$ with the tangent plane T_pM of the surface M at p. With this identification, we can consider the differential map dN_p as a linear operator $T_pM \to T_pM$.

The *shape operator* or the *Weingarten map* at $p \in M$ is the linear map $W_p : T_pM \to T_pM$ given by

$$W_p(\mathbf{w}) = -dN_p(\mathbf{w}),$$

for any $\mathbf{w} \in T_pM$. The map W_p is a self-adjoint operator, that is, W_p is a linear map which satisfies

$$\langle W_p(\mathbf{w}_1), \mathbf{w}_2 \rangle = \langle \mathbf{w}_1, W_p(\mathbf{w}_2) \rangle$$

for any $\mathbf{w}_1, \mathbf{w}_2 \in T_pM$.

The scalar product $\langle , \rangle$ in $\mathbb{R}^3$ defines a scalar product in T_pM by restriction. Therefore, T_pM admits an orthonormal basis of eigenvectors $\mathbf{v}_1, \mathbf{v}_2$ of the shape operator W_p, see Theorem 2.1 in Chapter 2. We have $W_p(\mathbf{v}_1) = \kappa_1 \mathbf{v}_1$ and $W_p(\mathbf{v}_2) = \kappa_2 \mathbf{v}_2$.

The eigenvalues κ_1 and κ_2 of W_p, which are functions of p, are called *the principal curvatures* of M at p.

The directions given by the eigenvectors $\mathbf{v}_1$ and $\mathbf{v}_2$ depend on p and are called *the principal directions*.

The determinant $K(p) = \det(W_p) = \kappa_1(p)\kappa_2(p)$ of the shape operator is the *Gaussian curvature* of the surface M at the point p. A point $p \in M$ is called *elliptic* if $K(p) > 0$; *hyperbolic* if $K(p) < 0$; *parabolic* if $K(p) = 0$; *umbilic* if $\kappa_1(p) = \kappa_2(p)$ (i.e., if W_p is a multiple of the identity map).

The Gaussian curvature is an intrinsic invariant of the surface.

Theorem 1.2 (Theorema Eugregium of Gauss). *Two locally isometric surfaces in $\mathbb{R}^3$ have the same Gaussian curvature.*

Remark 1.1. There are non-isometric surfaces in $\mathbb{R}^3$ with the same Gaussian curvature. A theorem of Bonnet states that if some compatibility conditions (the Gauss and Mainardi-Codazi conditions) are imposed on the coefficients of the first and second fundamental forms of the surface, then these coefficients determine completely the surface up to isometries. See, for example, [do Carmo (1976)] for a proof.

1.2.1 *The focal set*

The focal set of a surface in $\mathbb{R}^3$ is the analogue of the evolute of a plane curve and is defined as follows.

Definition 1.3. Let M be a smooth and regular surface in $\mathbb{R}^3$ without parabolic points, and let $\mathbf{x} : U \to \mathbb{R}^3$ be a local parametrisation of M. The focal set (or the evolute) of M is the set $\varepsilon(M) = \varepsilon_1(M) \cup \varepsilon_2(M)$ where

$$\varepsilon_i(M) = \{p + \frac{1}{\kappa_i(p)} N(p),\, p \in M\}, i = 1, 2.$$

Given a non-umbilic point $p \in M$, there are two distinct focal points corresponding to p, one is on ε_1 and the other on ε_2. These focal points coincide when p is an umbilic point.

What does the focal set look like?

It is very hard to answer this question satisfactorily using classical differential geometry techniques. Note that away from umbilic points, the focal set is locally the union of the images of the smooth maps $\phi_i : U \to \mathbb{R}^3$, given by

$$\phi_i(u_1, u_2) = \mathbf{x}(u_1, u_2) + \frac{1}{\kappa_i(u_1, u_2)} N(u_1, u_2),\ i = 1, 2.$$

The maps ϕ_i, $i = 1, 2$, are from a 2-dimensional space to a 3-dimensional space. Whitney ([Whitney (1944)]) showed that the cross-cap is the only $\mathcal{A}$-stable local singularity for such maps ($\mathcal{A}$ means we are considering smooth changes of coordinates in the source and target, see Chapter 3). An $\mathcal{A}$-model for the cross-cap is $f(u_1, u_2) = (u_1, u_1u_2, u_2^2)$, see Figure 1.5. However, the cross-cap singularity can never appear on the focal set. This follows, for instance, from the fact that away from the umbilic points of the surface the focal set has a limiting normal direction at its singular points but the cross-cap does not have one at its singular point.

Also, at umbilic points, the two focal points coincide and the maps ϕ_i are not differentiable at such points. Thus, using the maps ϕ_i directly will not lead to a satisfactory understanding of the focal set.

We consider the contact of the surface M with spheres. This is measured by the singularities of the members of the family distance squared functions $D : U \times \mathbb{R}^3 \to \mathbb{R}$, with $D(u, a) = \langle \mathbf{x}(u) - a, \mathbf{x}(u) - a \rangle$. The singularity theory approach outlined in section 1.1.3 for the study of the evolute of a plane curve can also be used to study the focal set of a surface in $\mathbb{R}^3$. We give in Chapter 5 the proof that the focal set is a caustic. As a consequence, it has only Lagrangian singularities and these are as in Figure 1.6 for a generic surface in $\mathbb{R}^3$.

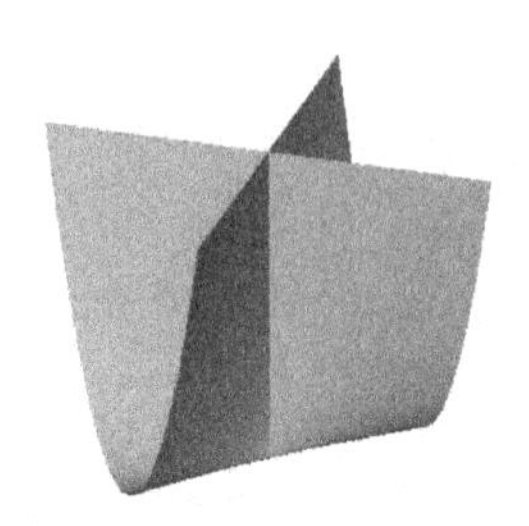

Fig. 1.5 A cross-cap.

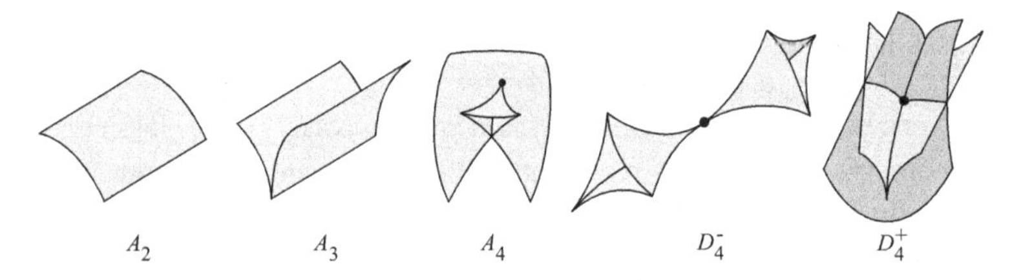

Fig. 1.6 Models of the focal set of a generic surface in $\mathbb{R}^3$. The fourth and fifth figures model the focal sets ε_1 and ε_2 joining at an umbilic point. The first three figures model the focal sets ε_1 or ε_2. One can have the following generic combinations for the pair $(\varepsilon_1, \varepsilon_2)$: (A_2, A_2), (A_2, A_3), (A_2, A_4), (A_3, A_3) or vice-versa.

1.3 Special surfaces in the Euclidean 3-space

Let $\gamma : I \to \mathbb{R}^3$ be a smooth and regular space curve parametrised by arc length s. Then $\mathbf{t}(s) = \gamma'(s)$ is a unit tangent vector to γ at s. The *curvature* of γ at s is defined as

$$\kappa(s) = ||\mathbf{t}'(s)||.$$

When $\kappa(s) \neq 0$ for all $s \in I$, the *unit principal normal vector* of γ at s is defined as the vector

$$\mathbf{n}(s) = \frac{\mathbf{t}'(s)}{||\mathbf{t}'(s)||}.$$

Then

$$\mathbf{t}'(s) = \kappa(s)\mathbf{n}(s).$$

The unit vector $\mathbf{b}(s) = \mathbf{t}(s) \times \mathbf{n}(s)$ is called the *unit binormal vector* of γ at s. It can be shown that $\mathbf{b}'(s)$ is parallel to $\mathbf{n}(s)$, so

$$\mathbf{b}'(s) = -\tau(s)\mathbf{n}(s),$$

for some smooth function $\tau(s)$. The scalar $\tau(s)$ is called the *torsion* of γ at s.

At each point on γ where $\kappa(s) \neq 0$, we have a positively oriented Serret-Frenet frame $\{\mathbf{t}(s), \mathbf{n}(s), \mathbf{b}(s)\}$ which moves along the curve. This motion is described by the Frenet-Serret formulae

$$\begin{cases} \mathbf{t}'(s) = \kappa(s)\mathbf{n}(s) \\ \mathbf{n}'(s) = -\kappa(s)\mathbf{t}(s) + \tau(s)\mathbf{b}(s) \\ \mathbf{b}'(s) = -\tau(s)\mathbf{n}(s). \end{cases}$$

The curvature and torsion functions determine completely the space curve up to rigid motions. Indeed,

Theorem 1.3 (Fundamental Theorem of Space Curves). *Given smooth functions $\kappa(s), \tau(s) : I \to \mathbb{R}$, with $\kappa(s) > 0$, there is a smooth and regular curve $\gamma : I \to \mathbb{R}^3$ parametrised by arc length s with curvature $\kappa(s)$ and torsion $\tau(s)$. The curve γ is unique up to rigid motions of $\mathbb{R}^3$.*

One can use singularity theory, as in the case of plane curves, to obtain a great deal of geometric information about space curves (see for example [Bruce and Giblin (1992)]). We consider below surfaces in $\mathbb{R}^3$ generated by space curves.

1.3.1 *Ruled surfaces*

A *ruled surface* in $\mathbb{R}^3$ is one which is swept out by a family of lines (the *rulings*) moving along a space curve (the *directrix curve* or the *base curve*). Ruled surfaces are extensively used in architecture. (The reader can surf the internet to see pictures of well known buildings parts of which have the shape of ruled surfaces.) Ruled surfaces are also used in abstract art (see for example some of Barbara Hepworth's work, available on the web).

Let $\gamma : I \to \mathbb{R}^3$ denote the base curve and assume that it is a smooth curve (it needs not be a regular curve) and let $\beta : I \to \mathbb{R}^3 \setminus \{(0,0,0)\}$ be a smooth map which parametrises the directions of the rulings. We take, without loss of generality, $||\beta(t)|| = 1$ for all $t \in I$. Then the ruled surface M determined by γ and β is the surface parametrised by

$$\mathbf{x}(t,u) = \gamma(t) + u\beta(t)$$

with $t \in I$ and $u \in \mathbb{R}$. The partial derivatives of $\mathbf{x}$ are given by

$$\begin{aligned} \mathbf{x}_t(t,u) &= \gamma'(t) + u\beta'(t), \\ \mathbf{x}_u(t,u) &= \beta(t), \end{aligned}$$

and we get

$$\mathbf{x}_t \times \mathbf{x}_u(t,u) = \gamma'(t) \times \beta(t) + u\beta'(t) \times \beta(t).$$

The surface M is singular at (t_0, u_0) if and only if $\mathbf{x}_t(t_0, u_0)$ and $\mathbf{x}_u(t_0, u_0)$ are linearly dependent, equivalently, if and only if $\mathbf{x}_t \times \mathbf{x}_u(t_0, u_0)$ is the zero vector. This occurs when

$$\gamma'(t_0) \times \beta(t_0) + u_0\beta'(t_0) \times \beta(t_0) = (0,0,0).$$

The surface M can be re-parametrised so as to make the singular points easier to detect. We shall assume that $\beta'(t) \neq 0$ for all $t \in I$; such surfaces are referred to as *non-cylindrical ruled surfaces.*

Given a non-cylindrical ruled surface M determined by γ and β, there is a curve $\sigma : I \to \mathbb{R}^3$ contained in M with

$$\langle \sigma'(t), \beta'(t) \rangle = 0 \tag{1.3}$$

for all $t \in I$. To show this, we observe that as the trace of σ lies on M, we have

$$\sigma(t) = \gamma(t) + u(t)\beta(t) \tag{1.4}$$

for some smooth function $u : I \to \mathbb{R}$, which we determine as follows. We have, by differentiating (1.4),

$$\sigma'(t) = \gamma'(t) + u'(t)\beta(t) + u(t)\beta'(t),$$

so equation (1.3) becomes

$$\langle \gamma'(t) + u'(t)\beta(t) + u(t)\beta'(t), \beta'(t) \rangle = 0. \tag{1.5}$$

The assumption $\|\beta(t)\| = 1$ implies $\langle \beta(t), \beta'(t) \rangle = 0$, so that equation (1.5) becomes

$$\langle \gamma'(t), \beta'(t) \rangle + u(t)\langle \beta'(t), \beta'(t) \rangle = 0.$$

Therefore,

$$u(t) = -\frac{\langle \gamma'(t), \beta'(t) \rangle}{\langle \beta'(t), \beta'(t) \rangle}. \tag{1.6}$$

The curve σ in (1.3) with $u(t)$ as in (1.6) is called the *striction curve* of the ruled surface M. It can be shown that the striction curve does not depend on the choice of the base curve γ on M (see [do Carmo (1976)]). Observe that the striction curve need not be a regular curve.

We can now re-parametrise M by taking the striction curve σ as the base curve. The new parametrisation of M is given by

$$\mathbf{y}(t,u) = \sigma(t) + u\beta(t)$$

with $t \in I$ and $u \in \mathbb{R}$. Then,

$$\begin{aligned}\mathbf{y}_t(t,u) &= \sigma'(t) + u\beta'(t),\\ \mathbf{y}_u(t,u) &= \beta(t),\end{aligned}$$

and

$$\mathbf{y}_t \times \mathbf{y}_u(t,u) = \sigma'(t) \times \beta(t) + u\beta'(t) \times \beta(t).$$

Since $\langle \sigma'(t), \beta'(t)\rangle = 0$ and $\langle \beta(t), \beta'(t)\rangle = 0$, it follows that

$$\sigma'(t) \times \beta(t) = \lambda(t)\beta'(t)$$

where the function $\lambda : I \to \mathbb{R}$ is given by

$$\lambda(t) = \frac{\langle \sigma'(t) \times \beta(t), \beta'(t)\rangle}{\langle \beta'(t), \beta'(t)\rangle}. \tag{1.7}$$

Now,

$$\|\mathbf{y}_t \times \mathbf{y}_u(t,u)\|^2 = (\lambda(t)^2 + u^2)\|\beta'(t)\|^2 \tag{1.8}$$

so the singularities of the surfaces M occur when $u = 0$ and $\lambda(t) = 0$. That is, the singularities of a ruled surface are located on its striction curve and at points where $\lambda(t) = 0$.

Example 1.1. The cross-cap parametrised by $f(u_1, u_2) = (u_1, u_1u_2, u_2^2)$ is a ruled surface. Indeed, we can rewrite f in the form

$$f(u_1, u_2) = (0, 0, u_2^2) + u_1(1, u_2, 0) = \gamma(u_2) + u_1\beta(u_2),$$

with $\gamma(u_2) = (0, 0, u_2^2)$ (a singular base curve) and $\beta(u_2) = (1, u_2, 0)$.

We leave it as an exercise to show that the striction curve is the double point curve of the cross-cap.

The Gaussian curvature of a ruled surface M is always non-positive, i.e. $K(p) \le 0$ for any $p \in M$. In fact, at the regular points of M, we have

$$K(t,u) = -\frac{\lambda(t)^2}{(\lambda(t)^2 + u^2)^2}, \tag{1.9}$$

and this vanishes along the rulings that pass through the singular points of the surface. It follows that the singular points of a ruled surface are located at the intersection of the striction curve with the rulings along which the Gaussian curvature is zero.

Now that we have the location of the singularities of a ruled surface using tools from classical differential geometry, we can go further and ask about the nature of such singularities and the shape of the ruled surface at its singular points. Here, as in the case of the focal set, a ruled surface is the image of a map from a 2-dimensional space to a 3-dimensional space. However, unlike the focal set, the generic singularities of ruled surfaces are the same as the generic singularities of maps $\mathbb{R}^2 \to \mathbb{R}^3$.

Theorem 1.4. *Generically, a ruled surface has only a cross-cap singularity* (Figure 1.5).

The proof of Theorem 1.4 relies on a special version of the transversality theorem in the space of ruled surfaces. The proof depends heavily on Mather's results (cf. [Mather (1969b,c); Martinet (1982)]), which is given in [Izumiya and Takeuchi (2001)].

Remark 1.2. More degenerate singularities of ruled surfaces are studied in [Martins and Nuño-Ballesteros (2009)].

An example of a ruled surface with two cross-cap singularities is the *Plücker conoid.* Its base curve is given by $\gamma(t) = (0, 0, \sin(2t))$ and its rulings are along $\beta(t) = (\cos(t), \sin(t), 0)$, $0 \le t \le 2\pi$; Figure 1.7.

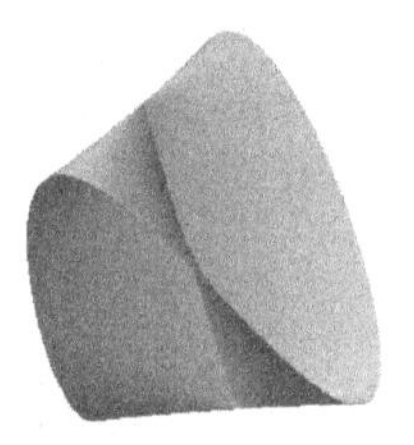

Fig. 1.7 Plücker conoid.

Another example of a ruled surface with a cross-cap singularity is the surface swept out by the principal normal lines to a regular space curve. Let $\gamma : I \to \mathbb{R}^3$ be a space curve parametrised by arc length and suppose that $\kappa(s) \neq 0$, for all $s \in I$. The principal normal surface is parametrised by

$$\mathbf{x}(s,u) = \gamma(s) + u\mathbf{n}(s),$$

with $(s,u) \in I \times \mathbb{R}^2$. We have

$$\mathbf{x}_s \times \mathbf{x}_u(s,u) = -u\tau(s)\mathbf{t}(s) + (1 - n\kappa(s))\mathbf{n}(s),$$

so the principal normal surface is singular at (s_0, u_0) if and only if $\tau(s_0) = 0$ and $u_0 = 1/\kappa(s_0)$. The singularity is of type cross-cap if $\tau'(s_0) \neq 0$ (see [Izumiya and Takeuchi (2003)]).

1.3.2 *Developable surfaces*

A ruled surface M is called a *developable surface* if its Gaussian curvature at all its regular points is zero. It follows from (1.7) that a ruled surface M is a developable surface if and only if

$$\langle \sigma'(t) \times \beta(t), \beta'(t) \rangle = \det(\sigma'(t), \beta(t), \beta'(t)) = 0$$

for all $t \in I$.

An example of a developable surface is the *tangent developable* of a space curve, whose rulings are along the tangent directions of the space curve ($\beta(t)$ is parallel to $\gamma'(t)$).

Classically, non-singular developable surfaces are classified as follows (see [Vaisman (1984)]).

Theorem 1.5. *Let* $\mathbf{x}(t,u) = \gamma(t) + u\beta(t)$ *be a parametrisation of non-singular developable surface* M *with* $\|\beta(t)\| = 1$*. Then* M *is one of the following surfaces:*

(i) *a subset of a plane.*
(ii) *a subset of a cylindrical surface* ($\beta(t)$ *is constant*).
(iii) *a subset of a conical surface* ($\gamma(t)$ *is constant*).
(iv) *a subset of a tangent developable* ($\beta(t)$ *is parallel to* $\gamma'(t)$).

A developable surface can have (stable) singularities which cannot be removed by deforming it within the space of developable surfaces. We deduce from (1.8) and (1.9) that the locus of singular points of a developable surface is precisely its striction curve. For a non-cylindrical developable surface, the condition $\det(\sigma'(t), \beta(t), \beta'(t)) = 0$ implies that there exist smooth functions λ and μ such that $\sigma'(t) = \lambda(t)\beta(t) + \mu(t)\beta'(t), t \in I$. The functions λ and μ determine σ, so the triple (λ, μ, β) determines the developable surface. Therefore, we can identify the space of non-cylindrical developable surfaces with the space of mappings $C^\infty(I, \mathbb{R} \times \mathbb{R} \times S^2)$. We endow the space $C^\infty(I, \mathbb{R} \times \mathbb{R} \times S^2)$ with the Whitney C^∞-topology (see Chapter 4).

We require the following definition in order to further our understanding of the singular points of developable surfaces.

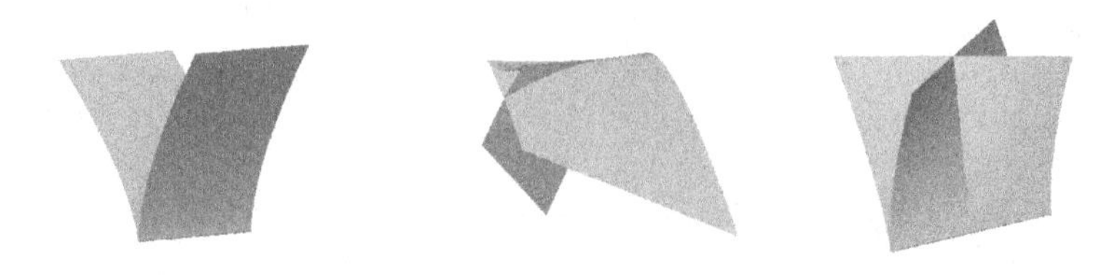

Fig. 1.8 Generic singularities of developable surfaces: cuspidal edge (left), swallowtail (centre) and cuspidal cross-cap (right).

Definition 1.4. Let U, U' be open subsets of $\mathbb{R}^2$ and V, V' be open subsets of $\mathbb{R}^3$. A surface $\mathbf{x} : U \to V$ is a cuspidal edge, swallowtail or cuspidal cross-cap if there exist diffeomorfisms $h : U \to U'$ and $k : V \to V'$ such that $k \circ \mathbf{x} \circ h^{-1}$ is given in the following form:

(i) cuspidal edge: (s, t^2, t^3) (Figure 1.8, left),
(ii) swallowtail: $(3s^4 + s^2t, 4s^3 + 2st, t)$ (Figure 1.8, centre),
(iii) cuspidal cross-cap: (s^3, s^3t^3, t^2) (Figure 1.8, right).

A non-cylindrical developable surface is the tangent developable surface of its striction curve (even when the striction curve is singular). One can then use the classification of the generic singularities of tangent developable surfaces in [Cleave (1980); Ishikawa (1993); Mond (1989); Shcherbak (1986)] to obtain the classification of the generic singularities of developable surface.

Theorem 1.6. *For a generic $(\lambda, \mu, \beta) \in C^\infty(I, \mathbb{R}\times\mathbb{R}\times S^2)$, the singularities of the corresponding non-cylindrical developable surface are cuspidal edges, swallowtails and cuspidal cross-caps* (Figure 1.8).

Remark 1.3. The result in Theorem 1.6 shows that the generic singularities of developable surfaces are distinct from those of general ruled surfaces (compare Theorem 1.4).

We turn now to tangent developable surfaces. Such surfaces have always singularities along their base curves γ. As an immediate consequence, the swallowtail singularity, which is a generic singularity of developable surface (Theorem 1.6), is not a generic singularity of a tangent developable of a regular space curve.

In fact, a tangent developable surface is a cuspidal edge surface along $\gamma(t)$ if $\tau(t) \neq 0$ (Figure 1.8, left). It is incredible that the classification of the generic singularities of tangent developable surfaces was only established recently. It is shown in [Cleave (1980)] that the tangent developable of a space curve has a cuspidal cross-cap singularity at $\gamma(t_0)$ if $\tau(t_0) = 0$ and $\tau'(t_0) \neq 0$. These conditions are generic for space curves, so this makes the cuspidal cross-cap singularity a generic singularity of tangent developable surfaces.

One can still define the tangent developable of certain singular space curves ([Ishikawa (1995)]). Consider for example the space curve $\gamma(t) = (t^2, t^3, t^4)$, which is singular at $t = 0$. The direction of $\gamma'(t)$ is the same as that of $\beta(t) = (2, 3t, 4t^2)$ for $t \neq 0$. The ruled surface

$$\mathbf{x}(t, u) = \gamma(t) + u\beta(t) = (t^2 + 2u, t^3 + 3tu, t^4 + 4tu)$$

is still called the tangent developable surface of the singular space curve $\gamma(t) = (t^2, t^3, t^4)$ ([Ishikawa (1995)]). This surface has a swallowtail singularity at the origin (this was first observed by Arnol'd in [Arnol'd (1981)]).

We deform the curve γ within the family of curves $\gamma_\epsilon(t) = (t^2, t^3 - \epsilon t, t^4)$, with $\gamma_0 = \gamma$. The tangent developable of the regular curve γ_ϵ, with $\epsilon \neq 0$, is

$$\mathbf{x}_\epsilon(t, u) = (t^2 + 2tu, t^3 - \epsilon t + u(3t^2 - \epsilon), t^4 + 4t^3 u),$$

and this has a cuspidal cross-cap singularity. Therefore, the swallowtail singularity is also not a generic singularity of the tangent developable of a singular space curve.

1.4 Notes

We considered briefly the contact of plane curves with circles and showed how this gives information about the evolute of the curve. This contact captures also some infinitesimal symmetry of the curve. The locus of centres of circles tangent to the curve at two or more points is called the *Symmetry Set* (SS) of the curve. The SS is used extensively in computer vision and shape recognition (see for example [Giblin and Brassett (1985); Siddiqi and Pizer (2008)] and [Damon (2003, 2004, 2006)]).

The contact of a space curve with lines is measured by the singularities of its projections to planes. The generic singularities of such projections are obtained in [David (1983)], and in [Dias and Nuño (2008); Oset Sinha and Tari (2010)] is considered the singularities as well as the inflections of the projected curves.

The contact of a space curve with planes is measured by the singularities of the height functions. These pick up the points of zero torsion of the curve. Such points can be counted on some curves (see for example [Nuño Ballesteros and Romero Fuster (1992, 1993); Sedykh (1992)]).

One question of interest is how to approximate a surface in $\mathbb{R}^3$ by a developable surface. In [Izumiya and Otani (2015)] is considered the approximation of a surface by the developable surface tangent to it along a curve.

Chapter 2

Submanifolds of the Euclidean space

In this chapter, we consider some aspects of the extrinsic geometry of a submanifold M of dimension n of the Euclidean space $\mathbb{R}^{n+r}$, with $r \geq 1$. When $r = 1$, M is a hypersurface and has locally a well defined unit normal vector field which can be extended to the whole hypersurface if it is orientable. This normal vector field defines a map, called the Gauss map, from M to the unit hypersphere in $\mathbb{R}^{n+1}$. A great deal of the extrinsic geometry of the hypersurface M can be derived from the Gauss map and its derivative map, the shape operator. The shape operator is, at each point of M, a self adjoint operator from the tangent space of M at p to itself. Its eigenvalues are called the principal curvatures and when they all are equal at a given point p, the point p is called an umbilic point. A hypersurface is totally umbilic if all its points are umbilic points. Totally umbilic hypersurfaces form our models of hypersurfaces and the extrinsic geometry of a general hypersurface in $\mathbb{R}^{n+1}$ is studied in subsequent chapters by looking at its contact with these model hypersurfaces.

At each point p of a submanifold M of codimension $r > 1$, there is an r-dimensional space of normal vectors to M at p. One can define a shape operator on M along a fixed unit normal vector field on M. All the concepts and results on the shape operator on a hypersurface can be carried over for the shape operator along the chosen normal vector field.

2.1 Hypersurfaces in $\mathbb{R}^{n+1}$

A hypersurface in $\mathbb{R}^{n+1}$ is a codimension 1 submanifold. As our study is local in nature, we consider patches of hypersurfaces given by embeddings $\mathbf{x} : U \to \mathbb{R}^n$ of an open subset U of $\mathbb{R}^n$ in $\mathbb{R}^{n+1}$. We write $M = \mathbf{x}(U)$ and identify M and U via the embedding $\mathbf{x}$.

Let $u = (u_1, \dots, u_n)$ denote the coordinates of a point in U with respect to the canonical basis of $\mathbb{R}^n$. The tangent space T_pM of M at $p = \mathbf{x}(u)$ is an n-dimensional vector space generated by the linearly independent vectors of the partial derivatives of $\mathbf{x}$ at u. We take these vectors as a basis $B(\mathbf{x})$ of T_pM, so

$$B(\mathbf{x}) = \{\mathbf{x}_{u_1}(u), \dots, \mathbf{x}_{u_n}(u)\}.$$

2.1.1 *The first fundamental form*

The scalar product $\langle , \rangle$ in $\mathbb{R}^{n+1}$ induces a scalar product $\langle , \rangle_p$ on T_pM by restriction. Any tangent vector in T_pM is a vector in $\mathbb{R}^{n+1}$, so for two tangent vectors $\mathbf{u}$ and $\mathbf{v}$ in T_pM, we have $\langle \mathbf{u}, \mathbf{v} \rangle_p = \langle \mathbf{u}, \mathbf{v} \rangle$. To simplify notation, we drop the subscript in the notation for induced scalar product on T_pM and indicate it by $\langle , \rangle$.

Definition 2.1. The first fundamental form on M is the quadratic form $\mathrm{I}_p : T_pM \to \mathbb{R}$, given by

$$\mathrm{I}_p(\mathbf{w}) = \langle \mathbf{w}, \mathbf{w} \rangle = ||\mathbf{w}||^2.$$

The first fundamental form, also called the induced metric on M, expresses the way the hypersurface M inherits the metric of the ambient space $\mathbb{R}^{n+1}$. It is a tool for taking measurements on M, such as the length of curves and angles between curves on M. For instance, the length of a curve $\gamma : [a, b] \to M$, is defined as

$$l(\gamma) = \int_a^b ||\gamma'(t)||dt.$$

We can express the first fundamental form with respect to the basis $B(\mathbf{x})$ of T_pM at $p = \mathbf{x}(u)$ as follows. For $\mathbf{w} = \sum_{i=1}^{i=n} w_i \mathbf{x}_{u_i}(u)$ in T_pM, we have

$$\mathrm{I}_p(\mathbf{w}) = \sum_{i,j=1}^{n} w_i w_j g_{ij}(u)$$

with

$$g_{ij}(u) = \langle \mathbf{x}_{u_i}(u), \mathbf{x}_{u_j}(u) \rangle.$$

The functions g_{ij} are clearly smooth functions on U and satisfy $g_{ij}(u) = g_{ji}(u)$ for all $i, j = 1, \dots, n$ and all u in U. These functions form an $n \times n$-symmetric matrix (g_{ij}) and the first fundamental form can be written in matrix form

$$\mathrm{I}_p(\mathbf{w}) = \mathbf{w}^T (g_{ij}(u)) \mathbf{w}.$$

Definition 2.2. The functions g_{ij}, $i, j = 1, \ldots, n$ are called the coefficients of the first fundamental form of M with respect to the parametrisation $\mathbf{x}$, and the $n \times n$-symmetric matrix (g_{ij}) is called the matrix of the first fundamental form.

It is worth observing that the matrix $(g_{ij}(u))$ is not singular for any u in U. Indeed, as it is a symmetric matrix, it is conjugate to a diagonal matrix D. The scalar product is positive definite, so the diagonal entries of D are all strictly positive numbers. The determinant of $(g_{ij}(u))$, being equal to the product of the diagonal entries of D, is therefore distinct from zero.

2.1.2 *The shape operator*

The way M sits in $\mathbb{R}^{n+1}$ can be described by the variation of its tangent spaces, or equivalently, by the variation of a normal vector to M. We fix an orientation of $\mathbb{R}^{n+1}$. Given a hypersurface patch M, we can choose a unique unit normal vector $N(p)$ on M (i.e., orthogonal to T_pM at all p in M) which induces a positive orientation of M. This can always be done for hypersurface patches. For a general hypersurface, it is necessary that it is orientable in order to be able to make a consistent choice of a normal vector field on the whole of M.

Let S^n be the unit hypersphere in $\mathbb{R}^{n+1}$,

$$S^n = \{x \in \mathbb{R}^{n+1} : \langle x, x \rangle = 1\}.$$

Definition 2.3. Let M be a hypersurface in $\mathbb{R}^{n+1}$. The map $N : M \to S^n$ which associates to each point p on M the normal vector $N(p)$ to M at p is called the *Gauss map*.

The tangent spaces T_pM and $T_{N(p)}S^n$ are n-vector spaces in $\mathbb{R}^{n+1}$ orthogonal to $N(p)$ and can be identified. With this identification, the derivative map dN_p of the Gauss map at p is considered as a linear transformation $T_pM \to T_pM$.

Definition 2.4. The linear transformation $W_p = -dN_p : T_pM \to T_pM$ is called the shape operator or the *Weingarten map* of M at p.

Theorem 2.1 (The Principal Axis Theorem).
The transformation $W_p : T_pM \to T_pM$ is a self-adjoint operator, that is, $\langle W_p(\mathbf{u}), \mathbf{v}\rangle_p = \langle \mathbf{u}, W_p(\mathbf{v})\rangle_p$ for any $\mathbf{u}$ and $\mathbf{v}$ in T_pM. As a consequence, it

has n real eigenvalues $\kappa_i(p), i = 1, \ldots, n$ and T_pM admits an orthonormal basis $\{\mathbf{v}_1(p), \ldots, \mathbf{v}_n(p)\}$ of eigenvectors of W_p.

Proof. We take a parametrisation $\mathbf{x} : U \to \mathbb{R}^{n+1}$ of M and write $p = \mathbf{x}(u)$. To prove that W_p is a self-adjoint operator it is enough to show that $\langle W_p(\mathbf{x}_{u_i}), \mathbf{x}_{u_j}\rangle = \langle \mathbf{x}_{u_i}, W_p(\mathbf{x}_{u_j})\rangle$ for the vectors in the basis $B(\mathbf{x})$ of T_pM. Differentiating the identity $\langle N, \mathbf{x}_{u_i}\rangle = 0$ with respect to u_j gives $\langle N_{u_j}, \mathbf{x}_{u_i}\rangle + \langle N, \mathbf{x}_{u_i u_j}\rangle = 0$. Therefore,

$$\langle N_{u_j}, \mathbf{x}_{u_i}\rangle = -\langle N, \mathbf{x}_{u_i u_j}\rangle = -\langle N, \mathbf{x}_{u_j u_i}\rangle = \langle N_{u_i}, \mathbf{x}_{u_j}\rangle,$$

which implies that W_p is a self-adjoint operator as $W_p(\mathbf{x}_{u_i}) = -N_{u_i}$.

For the remaining part of the proof see for example [Mac Lane and Birkhoff (1967)], p.403. □

Definition 2.5. The eigenvalues $\kappa_i(p), i = 1, \ldots, n$, of W_p are called the *principal curvature* of M at p and its eigenvectors are called the *principal directions* of M at p. The *Gauss-Kronecker curvature* of M at p is defined to be

$$K(p) = \det W_p = \prod_{i=1}^{n} \kappa_i(p),$$

and its *mean curvature* at p is defined to be

$$H(p) = \frac{1}{n}\operatorname{trace} W_p = \frac{1}{n}\sum_{i=1}^{n} \kappa_i(p).$$

Example 2.1. A hyperplane orthogonal to a non-zero vector $\mathbf{u}$ in $\mathbb{R}^{n+1}$ is the set of points

$$H(\mathbf{u}, r) = \{p \in \mathbb{R}^{n+1} \,:\, \langle p, \mathbf{u}\rangle = r\},$$

where c is some real number. The Gauss map of $H(\mathbf{u}, r)$ is the constant map $N(p) = \mathbf{u}/||\mathbf{u}||$. It follows that the principal curvatures at all points on the hyperplane are equal to zero. In particular, the Gauss-Kronecker curvature of a hyperplane is identically zero. We can then describe hyperplanes as *flat* objects in $\mathbb{R}^{n+1}$.

From the above, one can view the Gauss-Kronecker curvature of a hypersurface M as a measurement of the deviation of M from being a hyperplane.

Example 2.2. A hypersphere of centre c in $\mathbb{R}^{n+1}$ and radius r is the set of points

$$S^n(c, r) = \{p \in \mathbb{R}^{n+1} \,:\, \langle p - c, p - c\rangle = r^2\}.$$

The Gauss map of $S^n(c, r)$ is given by $N(p) = (p - c)/r$. Its derivative map is a multiple of the identity map of T_pM by the scalar $1/r$. Therefore, the principal curvatures at all points on the hypersphere are equal to $-1/r$. It follows that the Gauss-Kronecker curvature of the hypersphere is equal to $(-1)^n/r^n$ at all points. In particular, hyperspheres can be described as *round* objects.

We can use the fact that W_p is a self-adjoint operator to define a bilinear symmetric map $T_pM \times T_pM \to \mathbb{R}$ by $(\mathbf{u}, \mathbf{v}) \mapsto \langle W_p(\mathbf{u}), \mathbf{v}\rangle$, and associate to it a quadratic form $\mathrm{II}_p : T_pM \to \mathbb{R}$ given by $\mathrm{II}_p(\mathbf{w}) = \langle W_p(\mathbf{w}), \mathbf{w}\rangle$.

Definition 2.6. The quadratic form $\mathrm{II}_p : T_pM \to \mathbb{R}$, given by

$$\mathrm{II}_p(\mathbf{w}) = \langle W_p(\mathbf{w}), \mathbf{w}\rangle,$$

is called the *second fundamental form* of M at p.

The Gauss map of M parametrised by $\mathbf{x} : U \to \mathbb{R}^{n+1}$ is given by

$$N(u) = \frac{\mathbf{x}_{u_1}(u) \times \cdots \times \mathbf{x}_{u_n}(u)}{\|\mathbf{x}_{u_1}(u) \times \cdots \times \mathbf{x}_{u_n}(u)\|}, \tag{2.1}$$

with $u \in U$. The map N is infinitely differentiable. We write $dN_u(\mathbf{x}_{u_i}) = N_{u_i}(u)$.

The second fundamental form is also a smooth function on U and can be expressed in the basis $B(\mathbf{x})$ of T_pM at $p = \mathbf{x}(u)$ as follows. For $\mathbf{w} = \sum_{i=1}^n w_i \mathbf{x}_{u_i}(u)$ in T_pM, we have

$$\mathrm{II}_p(\mathbf{w}) = \sum_{i,j=1}^{n} w_i w_j h_{ij}(u)$$

with

$$h_{ij}(u) = -\langle N_{u_i}(u), \mathbf{x}_{u_j}(u)\rangle = \langle N(u), \mathbf{x}_{u_i u_j}(u)\rangle. \tag{2.2}$$

The functions h_{ij} are smooth functions on U for all $i, j = 1, \ldots, n$ and satisfy $h_{ij} = h_{ji}$. They form an $n \times n$-symmetric matrix (h_{ij}) and the second fundamental form can be written in matrix form

$$\mathrm{II}_p(\mathbf{w}) = \mathbf{w}^T(h_{ij}(u))\mathbf{w}.$$

Definition 2.7. The functions h_{ij}, $i, j = 1, \ldots, n$ are called the *coefficients of the second fundamental form* and the matrix (h_{ij}) is called the *matrix of the second fundamental form.*

Theorem 2.2 (The Weingarten formula). *The matrix of the shape operator W_p with respect to the basis $B(\mathbf{x})$ of T_pM at $p = \mathbf{x}(u)$ is given by $(h_i^j(u))$ with*

$$(h_i^j(u)) = (h_{ik}(u))(g^{kj}(u))$$

for $(g^{ij}(u)) = (g_{ij}(u))^{-1}$, where the matrices, $(g_{ij}(u))$ and $(h_{ij}(u))$ are, respectively, those of the first and second fundamental forms of M at p. That is,

$$N_{u_i}(u) = -\sum_{j=1}^{n} h_i^j(u)\mathbf{x}_{u_j}(u).$$

Proof. Since the vectors $\mathbf{x}_{u_1}(u), \ldots, \mathbf{x}_{u_n}(u), N(u)$ form a basis of $\mathbb{R}^{n+1}$, there exist scalar functions $\alpha_i^j(u)$ and $\beta_i(u)$, $i, j = 1, \ldots, n$ such that

$$N_{u_i}(u) = \sum_{j=1}^{n} \alpha_i^j(u)\mathbf{x}_{u_j}(u) + \beta_i(u)N(u).$$

Differentiating the identity $\langle N(u), N(u)\rangle = 1$, yields $\langle N(u), N_{u_i}(u)\rangle = 0$, so $\beta_i(u) = 0$ for $i = 1, \ldots, n$ and all for $u \in U$.

We have, by definition of the coefficients of the second fundamental form,

$$-h_{ik}(u) = \langle N_{u_i}(u), \mathbf{x}_{u_k}(u)\rangle = \sum_{j=1}^{n} \alpha_i^j(u)\langle \mathbf{x}_{u_j}(u), \mathbf{x}_{u_k}(u)\rangle = \sum_{j=1}^{n} \alpha_i^j(u) g_{jk}(u).$$

It follows that

$$-(h_{ik}(u)) = (\alpha_i^j(u))(g_{jk}(u)),$$

so

$$(h_i^j(u)) = -(\alpha_i^j(u)) = (h_{ik}(u))(g_{jk}(u))^{-1} = (h_{ik}(u))(g^{kj}(u)). \qquad \square$$

One consequence of Theorem 2.2 is the following formula for the Gauss-Kronecker curvature with respect to the parametrisation $\mathbf{x}$ of M.

Corollary 2.1. *The Gauss-Kronecker curvature at $p = \mathbf{x}(u)$ is given by*

$$K(u) = \frac{\det(h_{ij}(u))}{\det(g_{ij}(u))}.$$

Proof. The matrix of the shape operator in the basis $B(\mathbf{x})$ is (λ_{ij}) in Theorem 2.2. The Gauss-Kronecker curvature, is by definition, $K(u) = \det(\lambda_{ij}(u))$ and from Theorem 2.2

$$K(u) = \det(h_{ik}(u))\det(g_{kj}(u))^{-1} = \frac{\det(h_{ij}(u))}{\det(g_{ij}(u))}. \qquad \square$$

2.1.3 *Totally umbilic hypersurfaces*

René Thom and Ian Porteous ([Thom (1976); Porteous (1983a)]) suggested studying the extrinsic differential geometry of hypersurfaces (and submanifolds in general) in the Euclidean space by comparing them to model hypersurfaces. This comparison is done in terms of the contact between submanifolds and is detailed in Chapter 4. Our model hypersurfaces are the totally umbilic hypersurfaces which we define below.

Definition 2.8. A point p on a hypersurface M is an *umbilic* point if all the principal curvatures at p are equal. Equivalently, a point p is an umbilic point if the shape operator W_p is a scalar multiple of the identity map of T_pM. A hypersurface M is called *totally umbilic* if all its points are umbilic points.

Example 2.3. It follows from Example 2.1 and Example 2.2 that hyperplanes and hyperspheres in $\mathbb{R}^{n+1}$ are totally umbilic hypersurfaces.

Hyperplanes and hypersurfaces are not merely examples of totally umbilic hypersurfaces. In fact they are, together with their patches, the only totally umbilic hypersurfaces in $\mathbb{R}^{n+1}$.

Theorem 2.3. *Suppose that a hypersurface patch M is totally umbilic with the principal curvatures at p all equal to $\kappa(p)$. Then $\kappa(p)$ is independent of p and is equal to a constant κ. Furthermore,*
(i) *if $\kappa \neq 0$, then M is contained in a hypersphere.*
(ii) *if $\kappa = 0$, then M is contained in a hyperplane.*

Proof. We take a parametrisation $\mathbf{x} : U \to \mathbb{R}^{n+1}$ of the hypersurface and write $p = \mathbf{x}(u)$. Since $W_p = \kappa(u)1_{T_pM}$, the function $\kappa(u)$ is differentiable in U. We have $N_{u_i}(u) = -\kappa(u)\mathbf{x}_{u_i}(u), i = 1, \ldots, n$, so for $i, j = 1, \ldots, n$,

$$N_{u_iu_j}(u) = -\kappa_{u_j}(u)\mathbf{x}_{u_i}(u) - \kappa(u)\mathbf{x}_{u_iu_j}(u). \tag{2.3}$$

The maps $\mathbf{x}$ and N are C^∞-maps, therefore $N_{u_iu_j}(u) - N_{u_ju_i}(u) = 0$ and $\mathbf{x}_{u_iu_j}(u) - \mathbf{x}_{u_ju_i}(u) = 0$ for $i, j = 1, \ldots, n$. We deduce from this and from the relations (2.3) that

$$\kappa_{u_j}(u)\mathbf{x}_{u_i}(u) - \kappa_{u_i}(u)\mathbf{x}_{u_j}(u) = 0$$

for $i, j = 1, \ldots, n$. However, the vectors $\mathbf{x}_{u_1}(u), \ldots, \mathbf{x}_{u_n}(u)$ are linearly independent, so $\kappa_{u_i}(u) = 0$ for $i = 1, \ldots, n$, that is, $\kappa(u)$ is a constant function κ. We have two possibilities depending on whether κ is zero or not.

Suppose that $\kappa \neq 0$ and consider the map from $U \to \mathbb{R}^{n+1}$ given by

$$u \mapsto \mathbf{x}(u) + \frac{1}{\kappa}N(u).$$

Differentiating it gives

$$\mathbf{x}_{u_i}(u) + \frac{1}{\kappa}N_{u_i}(u) = \mathbf{x}_{u_i}(u) + \frac{1}{\kappa}(-\kappa\mathbf{x}_{u_i}(u)) = 0.$$

Therefore, $\mathbf{x}(u)+(1/\kappa)N(u)$ is equal to a constant, say, a. We have then

$$\|\mathbf{x}(u) - a\| = \frac{1}{|\kappa|},$$

which means that M is contained in the hypersphere $S^n(a, 1/|\kappa|)$.

If $\kappa = 0$, $N_{u_i}(u) = 0$ for all u in U, so N is a constant map, say $N(u) = \mathbf{n}$. We have $\langle \mathbf{x}_{u_i}(u), \mathbf{n}\rangle = \langle \mathbf{x}_{u_i}(u), N(u)\rangle = 0$ for all $i = 1, \dots, n$, so $\langle \mathbf{x}(u), \mathbf{n}\rangle$ is a constant function, that is,

$$\langle \mathbf{x}(u), \mathbf{n}\rangle = c,$$

for some real number c. This means that M is contained in the hyperplane $H(\mathbf{n}, c)$. □

Totally umbilic hypersurfaces with $\kappa \neq 0$ in Theorem 2.3 can also be identified via their evolutes. The concept of an evolute of a plane curve extends naturally to hypersurfaces in $\mathbb{R}^{n+1}$. Suppose that all the principal curvatures of M at all its points are not zero. Then we associate to the point p the n focal points

$$\varepsilon_i(p) = p + \frac{1}{\kappa_i(p)}N(p).$$

Definition 2.9. The *evolute* or *focal set* of a hypersurface M in $\mathbb{R}^{n+1}$ with nowhere vanishing principal curvatures is the set

$$\varepsilon(M) = \bigcup_{i=1}^{n} \varepsilon_i(M),$$

where

$$\varepsilon_i(M) = \{p + \frac{1}{\kappa_i(p)}N(p),\ p \in M\}.$$

Remark 2.1. When the principal curvatures are pairwise distinct at a point p, they are smooth functions near p and the evolute is locally the union of n disjoint patches of hypersurfaces, parametrised by the smooth maps ε_i. (Some components of the evolute can be singular!) If two principal curvatures κ_i and κ_j coincide at p, then the components of the evolute $\varepsilon_i(M)$ and $\varepsilon_j(M)$ have a non-empty intersection. Also, the maps ε_i and ε_j are no longer differentiable at p.

A parametrisation $\mathbf{x} : U \to \mathbb{R}^{n+1}$ of the hypersurface M induces parametrisations (not necessarily smooth, see Remark 2.1) $\varepsilon_i : U \to \mathbb{R}^{n+1}$ of the n components of the evolute, with

$$\varepsilon_i(u) = \mathbf{x}(u) + \frac{1}{\kappa_i(u)} N(u),$$

where we assume $\kappa_i(u) \neq 0$ for all $u \in U$ and $i = 1, \ldots, n$.

Proposition 2.1. *Let M be a hypersurface in $\mathbb{R}^{n+1}$ with nowhere vanishing principal curvatures. Then the following are equivalent.*

(i) *M is contained in a hypersphere.*

(ii) *The evolute $\varepsilon(M)$ of M is a point in $\mathbb{R}^{n+1}$.*

Proof. If M, parametrised by $\mathbf{x} : U \to \mathbb{R}^{n+1}$, is contained in a hypersphere, then by Theorem 2.3,

$$\varepsilon_i(u) = \mathbf{x}(u) + \frac{1}{\kappa} N(u).$$

In particular, the maps ε_i are identical for $i = 1 \ldots, n$. The partial derivatives of ε_i are identically zero, so ε_i is a constant map, which implies that the evolute of M is a point.

Conversely, if the evolute is a point, say a, then $\varepsilon_i(u) = a$ on U. Consequently,

$$\frac{1}{\kappa_i(u)} = -\langle \mathbf{x}(u) - a, N(u) \rangle,$$

is independent of i and is a smooth function. We denote $\kappa_i(u)$ by $\kappa(u)$. Differentiating the map ε_i and using Theorem 2.2 yields

$$\begin{aligned} \frac{\partial \varepsilon_i}{\partial u_j}(u) &= \mathbf{x}_{u_j}(u) + \frac{1}{\kappa(u)} N_{u_j}(u) - \frac{\kappa_{u_j}(u)}{\kappa(u)^2} N(u) \\ &= \frac{1}{\kappa(u)} \sum_{k=1}^{n} (\kappa(u)\delta_j^k - \lambda_{kj}(u))\mathbf{x}_{u_k}(u) - \frac{\kappa_{u_j}(u)}{\kappa(u)^2} N(u) = 0, \end{aligned}$$

where $\delta_j^k = 1$ if $k = j$ and 0 otherwise and the λ_{kj} are as in Theorem 2.2. Thus, κ is a constant function and $(\lambda_{kj}(u)) = \kappa(\delta_j^k)$. This means that M is a totally umbilic hypersurface with $\kappa \neq 0$. By Theorem 2.3, M is contained in a hypersphere.

We remark that for a totally umbilic hypersurface M, the point $\varepsilon(M)$ is the centre of the hypersphere which contains M. □

Another way to identify totally umbilic hypersurfaces with $\kappa = 0$ in Theorem 2.3 is via the dual hypersurface of M.

Definition 2.10. The *dual* of M is the hypersurface in the dual projective space $\mathbb{R}P^{n*}$ formed by all the tangent spaces of M.

We represent the dual hypersurface in the double cover $S^n \times \mathbb{R}$ of $\mathbb{R}P^{n*}$ as follows. Let $\mathbf{x} : U \to \mathbb{R}^{n+1}$ be a parametrisation of M, and define the map $\mathbf{x}^* : U \to S^n \times \mathbb{R}$ by

$$\mathbf{x}^*(u) = (N(u), \langle \mathbf{x}(u), N(u) \rangle).$$

Observe that $\mathbf{x}^*(u)$ represents the hyperplane $H(N(u), \langle \mathbf{x}(u), N(u) \rangle)$ which is precisely the tangent space T_pM of M at $p = \mathbf{x}(u)$.

Definition 2.11. The image of the map $\mathbf{x}^*$ is called the *cylindrical pedal of M* and represents the dual hypersurface of M in $S^n \times \mathbb{R}$ ([Bruce (1981); Romero Fuster (1983)]).

Proposition 2.2. *Let $M = \mathbf{x}(U)$ be a hypersurface in $\mathbb{R}^{n+1}$. Then the following are equivalent.*

(i) *M is contained in a hyperplane.*
(ii) *The Gauss map of M is a constant map.*
(iii) *The cylindrical pedal of M is a point.*

Proof. By the Weingarten formula in Theorem 2.2, the Gauss map of M is a constant map if and only if $\lambda_{ij}(u)$ are identically zero, for $i, j = 1, \ldots, n$. This occurs if and only if all the eigenvalues of W_p are zero at all points on M, that is, if and only if M is a totally umbilic hypersurface with κ in Theorem 2.3 equal to zero. Thus, (i) and (ii) are equivalent.

Now, if M is contained in a hyperplane $H(\mathbf{n}, c)$, then

$$\langle \mathbf{x}(u), \mathbf{n} \rangle = c,$$

for all u in U, so $N(u)$ is the constant vector $\mathbf{n}$ or $-\mathbf{n}$. It follows that $\mathbf{x}^*(u) = (N(u), \langle \mathbf{x}(u), N(u) \rangle) = (\pm \mathbf{n}, c)$ is a constant map, which implies that the cylindrical pedal of M is a point. Thus, (i) implies (iii).

If the image of $\mathbf{x}^*$ is a point, then the Gauss map is a constant map. Therefore, (iii) implies (ii). □

2.1.4 *Parabolic and umbilic points*

In the previous section we dealt with the property of the whole hypersurface patch M being totally umbilic (either flat or round). We restrict now to local properties of M at a given point and use Theorem 2.3 to classify, as follows, umbilic points on a general hypersurface.

Definition 2.12. An umbilic point p on a hypersurface M in $\mathbb{R}^{n+1}$ with principal curvatures $\kappa_i(p)$, $i = 1, \ldots, n$, all equal to κ is said to be a *flat umbilic point* if $\kappa = 0$.

Proposition 2.3. *A point $p = \mathbf{x}(u)$ on a hypersurface $M = \mathbf{x}(U)$ is a flat umbilic point if and only if all the coefficients of the second fundamental form vanish at u, that is, $h_{ij}(u) = 0$ for all $i, j = 1, \ldots, n$.*

Proof. The proof is an immediate consequence of Theorem 2.2. □

Points on a hypersurface where the Gauss-Kronecker curvature vanishes are also of interest. As the Gauss-Kronecker curvature is the product of the principal curvature, it vanishes at a given point when at least one of the principal curvatures is zero at that point.

Definition 2.13. A point p on a hypersurface M in $\mathbb{R}^{n+1}$ is a *parabolic point* if the Gauss-Kronecker curvature vanishes at p, that is, if $K(p) = 0$. The locus of all the parabolic points on M is called the *parabolic set* of M.

When M is a surface in $\mathbb{R}^3$, it is possible to identify the shape of M at a point p according to the sign of the Gaussian curvature $K(p)$. The point p is hyperbolic if $K(p) < 0$ and elliptic if $K(p) > 0$. The parabolic set separates these two regions on the surface. At an elliptic point p the two principal curvatures have the same sign. All the curves on M obtained by slicing M by planes passing through p and containing the normal vector $N(p)$ bend in the same direction, that is, all their curvatures have the same sign. Then the surface M is locally at p on one side of its tangent plane. At a hyperbolic point, the two principal curvatures have opposite signs and the surface looks like a horse saddle; it has parts on both sides of its tangent plane. When $n > 3$, we can have a positive Gauss-Kronecker curvature at p with an even number of principal curvatures having opposite signs. For this reason, the concept of a point being elliptic or hyperbolic is reserved for surfaces in $\mathbb{R}^3$.

The parabolic, umbilic and flat umbilic points on M can be detected via the families of height functions and distance squared functions on M, parametrised by $\mathbf{x} : U \to \mathbb{R}^{n+1}$. We start with the family of height functions $H : U \times S^n \to \mathbb{R}$, given by

$$H(u, \mathbf{v}) = \langle \mathbf{x}(u), \mathbf{v} \rangle.$$

We label $h_{\mathbf{v}}$ the function defined on U by $h_{\mathbf{v}}(u) = H(u, \mathbf{v})$ and call it the height function on M along $\mathbf{v}$.

Proposition 2.4. *Let M be a hypersurface parametrised by $\mathbf{x} : U \to \mathbb{R}^{n+1}$. Then, for a given u in U, $\partial h_{\mathbf{v}}/\partial u_i(u) = 0$ for $i = 1, \ldots, n$ if and only if $\mathbf{v} = \pm N(u)$.*

Proof. We have $\partial h_{\mathbf{v}}/\partial u_i(u) = \langle \mathbf{x}_{u_i}(u), \mathbf{v}\rangle$. Thus, $\partial h_{\mathbf{v}}/\partial u_i(u) = 0$ for $i = 1, \ldots, n$ if and only if v is a unit normal vector of M at $p = \mathbf{x}(u)$. Equivalently, if and only if $\mathbf{v} = \pm N(u)$. □

We fix now $\mathbf{v} = N(u)$ at a given point $p = \mathbf{x}(u)$ on M. The second order partial derivatives of $h_{\mathbf{v}}$ at u are, for $i, j = 1, \ldots, n$,

$$\frac{\partial^2 h_{\mathbf{v}}}{\partial u_i \partial u_j}(u) = \langle \mathbf{x}_{u_i u_j}(u), \mathbf{v}\rangle = \langle \mathbf{x}_{u_i u_j}(u), N(u)\rangle = h_{ij}(u), \qquad (2.4)$$

where h_{ij} are the coefficients of the second fundamental form as defined in Definition 2.7. The second order derivatives of $H_{\mathbf{v}}$ at u form an $n \times n$-symmetric matrix, called the *Hessian matrix* of $h_{\mathbf{v}}$ at u, which we denote by

$$\mathcal{H}(h_{\mathbf{v}})(u) = \left(\frac{\partial^2 h_{\mathbf{v}}}{\partial u_i \partial u_j}(u)\right).$$

An immediate consequence of expression (2.4) and Corollary 2.1 is the following.

Corollary 2.2. *The Gauss-Kronecker curvature of M at $p = \mathbf{x}(u)$ is*

$$K(u) = \frac{\det \mathcal{H}(h_{\mathbf{v}})(u)}{\det(g_{ij}(u))}.$$

Proposition 2.5. *With notation as above and with $\mathbf{v} = N(u)$ at $p = \mathbf{x}(u)$,*
(i) *the point p is a parabolic point if and only if* $\det(\mathcal{H}h_{\mathbf{v}})(u) = 0$;
(ii) *the point p is a flat umbilic point if and only if* $\operatorname{rank} \mathcal{H}(h_{\mathbf{v}})(u) = 0$.

Proof. Statement (i) follows from Corollary 2.2 as $\det \mathcal{H}(h_{\mathbf{v}})(u) = 0$ if and only if $K(u) = 0$. Statement (ii) follows from the expressions (2.4) and Proposition 2.3. □

We turn now to umbilic points and consider the family of distance squared functions $D : U \times \mathbb{R}^n \to \mathbb{R}$ on M, given by

$$D(u, a) = \langle \mathbf{x}(u) - a, \mathbf{x}(u) - a\rangle.$$

We label d_a the function defined on U by $d_a(u) = D(u, a)$.

Proposition 2.6. *Let M be a hypersurface parametrised by $\mathbf{x} : U \to \mathbb{R}^{n+1}$. Then, for a given u in U, $\partial d_a/\partial u_i(u) = 0$ for $i = 1, \ldots, n$ if and only if there exists a real number λ such that $a = \mathbf{x}(u) + \lambda N(u)$.*

Proof. The partial derivatives of the distance squared function D_a are

$$\frac{\partial d_a}{\partial u_i}(u) = 2\langle \mathbf{x}_{u_i}(u), \mathbf{x}(u) - a\rangle,\ i = 1, \ldots, n.$$

Thus, $\partial d_a/\partial u_i(u) = 0$ for $i = 1, \ldots, n$ if and only if $\mathbf{x}(u) - a$ is a normal vector to M at $p = \mathbf{x}(u)$. Equivalently, if and only if there exists a real number λ such that $a = \mathbf{x}(u) + \lambda N(u)$. □

We take now $a = \mathbf{x}(u) + \lambda N(u)$ at $p = \mathbf{x}(u)$. The second order partial derivatives of the distance squared function d_a are

$$\begin{aligned}\frac{\partial^2 d_a}{\partial u_i \partial u_j}(u) &= 2(\langle \mathbf{x}_{u_i u_j}(u), \mathbf{x}(u) - a\rangle + \langle \mathbf{x}_{u_j}(u), \mathbf{x}_{u_j}(u)\rangle)\\ &= 2(-\lambda h_{ij}(u) + g_{ij}(u)),\end{aligned} \quad (2.5)$$

where g_{ij} (resp. h_{ij}) are the coefficients of the first (resp. second) fundamental form as defined in Definition 2.2 (resp. 2.7). The hessian matrix of D_a at u is denoted by

$$\mathcal{H}(d_a)(u) = \left(\frac{\partial^2 d_a}{\partial u_i \partial u_j}(u)\right).$$

Proposition 2.7. *Let $p = \mathbf{x}(u)$ be a non-parabolic point on M and let $a = \mathbf{x}(u) + \lambda N(u)$ for some $\lambda \in \mathbb{R}$. Then p is an umbilic point if and only if* $\operatorname{rank} \mathcal{H}(d_a)(u) = 0$.

Proof. From the expressions (2.5), $\operatorname{rank} \mathcal{H}(d_a)(u) = 0$ if and only if $\lambda \neq 0$ and $h_{ij}(u) = g_{ij}(u)/\lambda$ for all $i = 1, \ldots, n$. Using matrix notation, $\operatorname{rank} \mathcal{H}(d_a)(u) = 0$ if and only if

$$(h_{ij}(u)) = \frac{1}{\lambda}(g_{ij}(u)).$$

This is equivalent to the matrix $(h_i^j(u))$ of the shape operator in the basis $B(\mathbf{x})$ being given by

$$(h_i^j(u)) = (h_{ij}(u))(g_{ij}(u))^{-1} = \frac{1}{\lambda} I_n,$$

where I_n is the identity matrix. This in turn is equivalent to p being an umbilic point with all it principal curvatures equal to $1/\lambda$. □

2.2 Higher codimension submanifolds of $\mathbb{R}^{n+r}$

Let now M be a submanifold of codimension r, with $r > 1$. At each point p on M, the tangent space T_pM is an n-dimensional vector subspace of $\mathbb{R}^{n+r}$. Here too we have the induced scalar product of $\mathbb{R}^{n+r}$ in T_pM and define the first fundamental form of M as for the hypersurfaces case.

Let $\mathbf{x} : U \to \mathbb{R}^{n+r}$ be an embedding of an open subset $U \subset \mathbb{R}^n$ in $\mathbb{R}^{n+r}$. We write, as in the case of hypersurfaces, $M = \mathbf{x}(U)$ and identify M and U via the embedding $\mathbf{x}$. The partial derivatives of $\mathbf{x}$ at u form a basis of the tangent space T_pM which we still denote by $B(\mathbf{x})$.

The first fundamental form of M, with respect to the basis $B(\mathbf{x})$, is given by $\mathrm{I}_p(\mathbf{w}) = \mathbf{w}^T(g_{ij}(u))\mathbf{w}$ where $g_{ij}(u) = \langle \mathbf{x}_{u_i}(u), \mathbf{x}_{u_j}(u)\rangle$.

The tangent space T_pM has an orthogonal complement N_pM of dimension r. The space N_pM is called the *normal space* of M at p and we have

$$N_pM = \{\mathbf{v} \in \mathbb{R}^{n+r} \, : \, \langle \mathbf{u}, \mathbf{v}\rangle = 0 \text{ for all } \mathbf{u} \in T_pM\}.$$

The unit normal space of M at p is defined as

$$(N_pM)_1 = \{\mathbf{v} \in N_pM \, : \, \langle \mathbf{v}, \mathbf{v}\rangle = 1\}.$$

The *normal bundle* and the *unit normal bundle* over M are defined, respectively, as the sets

$$NM = \bigcup_{p\in M} N_pM \quad \text{and} \quad N_1M = \bigcup_{p\in M} (N_pM)_1.$$

We have the Whitney sum decomposition of the restriction of the tangent bundle of $\mathbb{R}^{n+r}$ to M

$$T\mathbb{R}^{n+r}|M = TM \oplus NM.$$

If $\mathbf{v} \in T_p\mathbb{R}^{n+r}$, we can write $\mathbf{v} = \mathbf{v}^T + \mathbf{v}^N$, where $\mathbf{v}^T \in T_pM$ and $\mathbf{v}^N \in N_pM$.

In what follows we denote by $\mathcal{X}(U)$ (resp. $\mathcal{N}(U)$) the set of differentiable vector fields tangent to M (resp. normal to M).

Let $\overline{\nabla}$ be the Riemannian connection on $\mathbb{R}^{n+r}$. If ω_1 and ω_2 are local vector fields on M, and $\overline{\omega}_1$ and $\overline{\omega}_2$ are local extensions to $\mathbb{R}^{n+r}$, let

$$\nabla_{\omega_1}\omega_2 = \left(\overline{\nabla}_{\overline{\omega}_1}\overline{\omega}_2\right)^T$$

be the Riemannian connection relative to the metric induced on M.

Definition 2.14. The second fundamental form of the immersion $\mathbf{x} : U \to \mathbb{R}^{n+r}$ is the mapping

$$\mathrm{B} : \mathcal{X}(U) \times \mathcal{X}(U) \to \mathcal{N}(U),$$

given by $\mathrm{B}(\omega_1, \omega_2) = \overline{\nabla}_{\overline{\omega}_1}\overline{\omega}_2 - \nabla_{\omega_1}\omega_2$.

Proposition 2.8 (Proposition 2.1, [do Carmo (1992)]). *The mapping* B *is bilinear and symmetric.*

Definition 2.15. For each $p \in M$, the mapping B induces a quadratic mapping $\mathrm{II}_p : T_pM \to N_pM$, called the *second fundamental form* at p, defined by

$$\mathrm{II}_p(\mathrm{w}) = B(\omega, \omega)(p),$$

where $\mathrm{w} = \omega(p)$, $\omega \in \mathcal{X}(U)$.

We consider smooth sections of the unit normal bundle N_1M, i.e. smooth normal vector fields $\tilde{\nu} : M \to N_1M$ on M given by

$$\tilde{\nu}(p) = (p, \nu(p)).$$

Definition 2.16. The quadratic form $\mathrm{II}_\mathrm{p}^\mathbf{v} : \mathrm{T_pM} \to \mathbb{R}$ defined by

$$\mathrm{II}_p^{\mathbf{v}}(\omega) = \langle \mathrm{B}(\omega, \omega), \mathbf{v} \rangle, \quad \omega \in T_pM,$$

is called the *second fundamental form* of $\mathbf{x}$ at p *along the normal vector* $\mathbf{v} = \nu(p)$.

All the concepts and results in section 2.1 derived from the Gauss map and its derivative can be carried over to similar concepts along a given normal vector field $\tilde{\nu}$ on a higher codimension submanifold M of $\mathbb{R}^{n+r}$.

Definition 2.17. The map $G^\nu : M \to S^{n+r-1}$ given by $G^\nu(p) = \nu(p)$ is called the *Gauss map* of M *with respect to the normal vector field* ν.

The derivative of the Gauss map G^ν at a point p in M is a linear map

$$(dG^\nu)_p : T_pM \to T_p\mathbb{R}^{n+r} = T_pM \oplus N_pM.$$

Let $\pi^T : T_pM \oplus N_pM \to T_pM$ denote the projection to the first component given by $\pi^T(\mathbf{u}, \mathbf{v}) = \mathbf{u}$, and consider the linear map

$$W_p^\nu = -\pi^T \circ (dG^\nu)_p : T_pM \to T_pM.$$

Definition 2.18. The linear map W_p^ν is called the shape operator of M at p along the normal vector field ν, or simply, the ν-shape operator of M at p.

Theorem 2.4. *The ν-shape operator $W_p^\nu : T_pM \to T_pM$ is a self-adjoint operator. As a consequence, it has n real eigenvalues $\kappa_i^\nu(p), i = 1, \ldots, n$, and T_pM admits an orthonormal basis $\{\mathbf{v}_1(p), \ldots, \mathbf{v}_n(p)\}$ of eigenvectors of W_p^ν.*

Proof. The proof is similar to that of Theorem 2.1. □

Definition 2.19. The eigenvalues $\kappa_i^\nu(p)$, $i = 1, \dots n$, (resp. eigenvectors) of the ν-shape operator W_p^ν are called the *ν-principal curvatures* (resp. *ν-principal directions*) at p. The *Lipschitz-Killing curvature* of M at p with respect to ν is defined to be $K^\nu(p) = \det W_p^\nu = \prod_{i=1}^n \kappa_i^\nu(p)$.

In a similar way to the hypersurface case, the second fundamental form of M along a normal vector field ν is the quadratic form induced by the ν-shape operator W_p^ν.

Proposition 2.9. *The following holds*

$$\mathrm{II}_p^\nu(\mathbf{w}, \mathbf{w}) = \langle W_p^\nu(\mathbf{w}), \mathbf{w}\rangle_p.$$

Proof. The ν-second fundamental form can be written, with respect to the basis $B(\mathbf{x})$, in the form

$$\mathrm{II}_p^\nu(\mathbf{w}) = \mathbf{w}^T(h_{ij}^\nu(u))\mathbf{w}$$

where

$$h_{ij}^\nu(u) = \langle \nu(u), \mathbf{x}_{u_i u_j}(u)\rangle.$$

As $\langle \nu(u), \mathbf{x}_{u_i u_j}(u)\rangle = \langle -\nu_{u_i}(u), \mathbf{x}_{u_j}(u)\rangle_p$, the result follows. □

The functions h_{ij}^ν are called the coefficients of the ν-second fundamental form.

Theorem 2.5. *Let $\mathbf{x} : U \to \mathbb{R}^{n+r}$ be a parametrisation of a submanifold M in $\mathbb{R}^{n+r}$ of codimension r, and let $\{\nu_1(u), \dots, \nu_r(u)\}$ be an orthonormal frame of M, with $\nu_1(u) = \nu(u)$. Then,*

$$(G^\nu)_{u_i} = -\sum_{j=1}^{n} (h_i^j)^\nu \mathbf{x}_{u_j} + \sum_{k=2}^{r} \langle \nu_k, \nu_{u_i}\rangle \nu_k$$

with

$$(h_i^j)^\nu = (h_{ik}^\nu)(g^{kj}),$$

where all the functions are evaluated at u. As a consequence, we have the following Weingarten formula along ν

$$(\pi^T \circ G^\nu)_{u_i} = -\sum_{j=1}^{n} (h_i^j)^\nu \mathbf{x}_{u_j}.$$

Proof. Since $\{\mathbf{x}_{u_1}(u), \dots, \mathbf{x}_{u_n}(u), \nu_1(u), \dots, \nu_r(u)\}$ is a basis of $T_p\mathbb{R}^{n+r}$ at $p = \mathbf{x}(u)$, there exist smooth functions α_i^j and β_i^k on U such that

$$G^\nu_{u_i} = \sum_{j=1}^{n} \alpha_i^j \mathbf{x}_{u_j} + \sum_{k=1}^{r} \beta_i^k \nu_k.$$

We have $\langle G^\nu, G^\nu \rangle = 1$, so that $\langle G^\nu, G^\nu_{u_i} \rangle = 0$, and consequently

$$\langle G^\nu, G^\nu_{u_i} \rangle = \langle \nu_1, G^\nu_{u_i} \rangle = \beta_i^1 = 0$$

for all $i = 1, \dots, n$. Furthermore, we have

$$\beta_i^k = \sum_{\ell=1}^{r} \beta_i^\ell \delta_{k\ell} = \sum_{\ell=1}^{r} \beta_i^\ell \langle \nu_k, \nu_\ell \rangle = \langle \nu_k, G^\nu_{u_i} \rangle = \langle \nu_k, \nu_{u_i} \rangle.$$

We have, from the definition of the coefficients of the ν-second fundamental form,

$$-h^\nu_{i\ell} = \langle \nu_{u_i}, \mathbf{x}_{u_\ell} \rangle = \langle G^\nu_{u_i}, \mathbf{x}_{u_\ell} \rangle = \sum_{j=1}^{s} \alpha_i^j \langle \mathbf{x}_{u_j}, \mathbf{x}_{u_\ell} \rangle = \sum_{j=1}^{s} \alpha_i^j g_{j\ell}.$$

Thus,

$$(\alpha_i^j) = -(h^\nu_{i\ell})(g_{\ell j})^{-1} = -(h_i^j)^\nu.$$

The expression for $(\pi^T \circ G^\nu)_{u_i}$ follows from the fact that the vectors $\nu_i(u)$ are normal vectors. □

Corollary 2.3. *The Lipschitz-Killing curvature of M at $p = \mathbf{x}(u)$ along the unit normal vector field ν is given by*

$$K^\nu(u) = \frac{\det(h^\nu_{ij}(u))}{\det(g_{\alpha\beta}(u))}.$$

Remark 2.2. Any unit normal vector ν_0 at a fixed point $p = \mathbf{x}(u_0)$ can be extended locally to a unit normal vector field ν along M with $\nu(u_0) = \nu_0$. Differentiating the identity $\langle \nu(u), \mathbf{x}_{u_j}(u) \rangle = 0$ and evaluating at u_0 yields

$$h^\nu_{ij}(u_0) = \langle \nu_0, \mathbf{x}_{u_j u_j}(u_0) \rangle.$$

It follows that the coefficients of the ν-second fundamental at p are independent of the local extension ν of ν_0. As a consequence, the shape operator at p depends only on ν_0. From this, we deduce that the Lipschitz-Killing curvature at p is also independent of the local extension of ν_0, and write $K^{\nu_0}(u_0)$ for the Lipschitz-Killing curvature with respect to ν_0.

2.2.1 *Totally ν-umbilic submanifolds*

Following section 2.1.3, we make the following definition, where ν is a unit normal vector field along the submanifold M.

Definition 2.20. We say that a point p on M is an *ν-umbilic point* if all the principal curvatures of the ν-shape operator W_p^ν are equal. The submanifold M is totally ν-umbilic if all its points are ν-umbilic points.

We seek to characterise, as in section 2.1.3, the totally ν-umbilic submanifolds. For this we require the following concept.

Definition 2.21. A unit normal vector field ν is said to be *parallel* at p if the N_pM-component $(dG^\nu)_p + W_p^\nu$ of $(dG^\nu)_p$ is zero. The unit normal vector field ν is parallel if it is parallel at all points on M.

Theorem 2.6. *Suppose that M is totally ν-umbilic submanifold and that ν is parallel. Then $\kappa^\nu(p)$ is a constant function κ^ν. Moreover,*
(i) *if $\kappa^\nu \neq 0$, then M is contained in a hypersphere.*
(ii) *if $\kappa^\nu = 0$, then M is contained in a hyperplane.*

Proof. The condition on ν to be parallel gives $\nu_{u_i}(u) = \pi^T \circ \nu_{u_i}(u)$, and adding to it the condition on the submanifold to be totally ν-umbilic yields

$$\nu_{u_i}(u) = -\kappa^\nu(u)\mathbf{x}_{u_i}(u),$$

for $i = 1, \ldots, n$. The proof now follows the same steps as that of Theorem 2.3.

We deduce from the above expression of $\nu_{u_i}(u)$ that $-\nu_{u_iu_j}(u) = \kappa^\nu_{u_j}(u)\mathbf{x}_{u_i}(u) + \kappa^\nu(u)\mathbf{x}_{u_iu_j}(u)$. Since $\nu_{u_iu_j}(u) = \nu_{u_ju_i}(u)$ and $\mathbf{x}_{u_iu_j}(u) = \mathbf{x}_{u_ju_i}(u)$, we have

$$\kappa^\nu_{u_j}(u)\mathbf{x}_{u_i}(u) - \kappa^\nu_{u_i}\mathbf{x}_{u_j}(u) = 0.$$

The vectors $\mathbf{x}_{u_1}(u), \mathbf{x}_{u_2}(u), \ldots, \mathbf{x}_{u_n}(u)$ are lineally independent, so $\kappa^\nu_{u_i}(u) = 0$ for $i = 1, \ldots, n$. This means that $\kappa^\nu(u)$ is a constant, say κ^ν.

Suppose that $\kappa^\nu \neq 0$. Then the map $\mathbf{x}(u) + (1/\kappa^\nu)\nu(u)$ is a constant map a, which implies that M is contained in the hypersphere of centre a and radius $1/|\kappa^\nu|$.

If $\kappa^\nu = 0$, then ν is a constant vector. It follows that $\langle \mathbf{x}(u), \nu\rangle$ is a constant map c, so M is contained in the hyperplane $H(\nu, c)$. □

2.2.2 *ν-parabolic and ν-umbilic points*

Following the arguments in Remark 2.2, the property of a point p being ν-umbilic (Definition 2.20) depends only on the value of ν_0 at p_0 and not on the vector field ν. We call then p a ν_0-umbilic point. Likewise, the point p is a ν_0-parabolic point if $K^{\nu_0}(p_0) = 0$. A ν_0-flat umbilic point is ν_0-umbilic point with $K^{\nu_0}(p_0) = 0$.

We shall characterise these special points using the families of height functions and distances squared functions. Here, for M parametrised $\mathbf{x} : U \to \mathbb{R}^{n+r}$, the family of height functions $H : U \times S^{n+r-1} \to \mathbb{R}$ on M is given by

$$H(u, \mathbf{v}) = \langle \mathbf{x}(u), \mathbf{v} \rangle,$$

where S^{n+r-1} is the unit sphere in $\mathbb{R}^{n+r}$. We define the function $h_{\mathbf{v}}$ on U by $h_{\mathbf{v}}(u) = H(u, \mathbf{v})$ and denote its Hessian matrix by $\mathcal{H}(h_{\mathbf{v}})(u)$.

The following proposition is similar to Proposition 2.4 and Proposition 2.5 and the proof follows in the same line.

Proposition 2.10. *Let M be a submanifold of codimension r in $\mathbb{R}^{n+r}$ parametrised by $\mathbf{x} : U \to \mathbb{R}^{n+r}$. Then, for any $u \in U$, $\partial h_{\mathbf{v}}/\partial u_i(u) = 0$ for $i = 1, \ldots, n$ if and only if $\mathbf{v} \in (N_pM)_1$ at $p = \mathbf{x}(u)$.*

For $\mathbf{v} \in (N_pM)_1$,

(i)$K^{\mathbf{v}}(u) = \det \mathcal{H}(h_{\mathbf{v}})(u)/\det(g_{ij}(u))$;

(ii) p is a $\mathbf{v}$-parabolic point if and only if $\det \mathcal{H}_{\mathbf{v}})(u) = 0$;

(iii) p is a $\mathbf{v}$-flat umbilic point if and only if $\operatorname{rank} \mathcal{H}(h_{\mathbf{v}})(u) = 0$.

We have a similar result for the family of distance squared functions $D : U \times \mathbb{R}^{n+r} \to \mathbb{R}$, defined by $D(u, a) = \langle \mathbf{x}(u) - a, \mathbf{x}(u) - a \rangle$. We denote by d_a the function defined on U by $d_a(u) = D(u, a)$ and its Hessian matrix by $\mathcal{H}(d_a)(u)$.

Proposition 2.11. *Let M be a submanifold of codimension r in $\mathbb{R}^{n+r}$ parametrised by $\mathbf{x} : U \to \mathbb{R}^{n+r}$.*

(i) *For any $u \in U$, $\partial d_a/\partial u_i(u) = 0$ for $i = 1, \ldots, n$ if and only if there exist $\lambda \in \mathbb{R}$ and $v \in (N_pM)_1$ at $p = \mathbf{x}(u)$ such that $a = p + \lambda v$.*

(ii) *Suppose that p is not a v-parabolic point and that $a = p + \lambda v$. Then, p is a v-umbilic point if and only if $\operatorname{rank} \mathcal{H}(d_a)(u) = 0$.*

2.2.3 *The canal hypersurface*

For a given submanifold M of codimension r in $\mathbb{R}^{n+r}$, we can construct a hypersurface in $\mathbb{R}^{n+r}$ associated to M by taking the union of the $(r-1)$-

dimensional spheres in N_pM of radius $\varepsilon > 0$ and centre $p \in M$.

Definition 2.22. The *canal hypersurface* of M of radius ε is the set

$$CM(\varepsilon) = \{p + \varepsilon\nu \,:\, p \in M,\, \nu \in (N_pM)_1\}.$$

Theorem 2.7. *For ε small enough, the canal hypersurface of M is locally a smooth codimension 1 submanifold of $\mathbb{R}^{n+r}$.*

Proof. We parametrise the submanifold M (locally) by $\mathbf{x} : U \to \mathbb{R}^{n+r}$ and choose an orthonormal frame $\{\nu_1(u), \dots, \nu_r(u)\}$ of N_pM at $p = \mathbf{x}(u)$.

Let S^{r-1} be the unit sphere in $\mathbb{R}^r$ and denote the coordinate of its points by $\mu = (\mu_1, \dots, \mu_r)$, with $\mu_1^2 + \mu_2^2 + \cdots + \mu_r^2 = 1$.

At each point $p = \mathbf{x}(u)$, the map $N(u, -) : S^{r-1} \to (N_pM)_1$, given by

$$N(u, \mu) = \sum_{i=1}^{r} \mu_i \nu_i(u),$$

is a diffeomorphism.

Let $\mathbf{y} : U \times S^{r-1} \to \mathbb{R}^{n+r}$ be given by $\mathbf{y}(u, \mu) = \mathbf{x}(u) + \varepsilon N(u, \mu)$. We claim that $\mathbf{y}$ is an embedding for ε small enough. To show this, we first compute the partial derivatives of $\mathbf{y}$ with respect to the variables u_i. These are given by

$$\begin{aligned} \mathbf{y}_{u_i}(u, \mu) &= \mathbf{x}_{u_i}(u) + \varepsilon N_{u_i}(u, \mu) \\ &= \mathbf{x}_{u_i}(u) - \varepsilon \textstyle\sum_{i=1}^{r} \mu_i \sum_{j=1}^{n} (\lambda_{ij})^{\nu_i}(u) \mathbf{x}_{u_j}(u) \end{aligned}$$

where $(\lambda_{ij})^{\nu_i}$ are as in Theorem 2.5. As we are considering local properties of M, we can take U to be a bounded region in $\mathbb{R}^n$. Then, for sufficiently small $\varepsilon > 0$, the vectors $\mathbf{y}_{u_i}(u, \mu)$, $i = 1, \dots, n$, are linearly independent.

To compute the partial derivatives of $\mathbf{y}$ with respect to the variables μ_j, we take μ in one of the coordinate charts

$$\begin{aligned} U_k^+ &= \{\mu = (\mu_1, \dots, \mu_r) \in S^{r-1} \mid \mu_k > 0\}, \\ U_k^- &= \{\mu = (\mu_1, \dots, \mu_r) \in S^{r-1} \mid \mu_k < 0\}, \end{aligned}$$

with $k = 1, \dots, r$. Suppose that $\mu \in U_1^+$. Then $\mu_1 = \sqrt{1 - \sum_{j=2}^{r} \mu_j^2}$, and $N_{\mu_j}(u, \mu) = \nu_j(u) - (\mu_j/\mu_1)\nu_1(u)$ for $j = 2, \dots, r$. It follows that

$$\mathbf{y}_{\mu_j}(u, \mu) = \varepsilon(\nu_j(u) - \frac{\mu_j}{\mu_1}\nu_1(u)).$$

The vectors $\mathbf{y}_{u_i}(u, \mu)$, $i = 1, \dots, n$ and $\mathbf{y}_{\mu_j}(u, \mu)$, $j = 2 \dots, r$ are linearly independent. We obtain the same result if we take μ in the other coordinate charts of S^{r-1}. Therefore, the map $\mathbf{y}$ is an embedding for sufficiently small $\varepsilon > 0$ (and shrinking U if necessary). Its image is clearly the canal hypersurface $CM(\varepsilon)$. □

Proposition 2.12. *For ε small enough, the Gauss map $G : CM(\varepsilon) \to S^n$ of the canal hypersurface of $CM(\varepsilon)$ of M is given by*

$$G(p, \nu) = \nu.$$

Proof. Following the proof of Theorem 2.7, we take $\mathbf{y} : U \times S^{r-1} \to \mathbb{R}^{n+r}$, with $\mathbf{y}(u, \mu) = \mathbf{x}(u) + \varepsilon N(u, \mu)$, as a parametrisation of the canal hypersurface. Then the map G is given by $G(u, \mu) = N(u, \mu)$. (Observe that $N(u, \mu) \in (N_pM)_1$, with $p = \mathbf{x}(u)$, and $(N_pM)_1$ can be considered a subset of S^n.)

We have $\langle N(u, \mu), \mathbf{y}_{u_i}(u, \mu)\rangle = 0$ as $N(u, \mu) \in N_pM$ at $p = \mathbf{x}(u)$, and $\langle N(u, \mu), \mathbf{y}_{\mu_j}(u, \mu)\rangle = 0$ as the vectors $\nu_i(u)$, $i = 1, \ldots, r$ are orthonormal. It follows that $N(u, \mu)$ is a unit orthogonal vector to $CM(\varepsilon)$. □

Proposition 2.13. *Let $K(p, \nu)$ denote the Gauss-Kronecker curvature of the canal hypersurface $CM(\varepsilon)$ of M at (p, ν) and let $K^\nu(p)$ denote the Lipschitz-Killing curvature of M at p along the normal vector ν. Then,*

$$K(p, \nu) = K^\nu(p).$$

Proof. We take a parametrisation of M as before and choose an orthonormal frame along M in such a way that $\nu(u_0) = \nu_1(u_0)$ at the point $p = \mathbf{x}(u_0)$. Then, with the notation of the proof of Theorem 2.7, $N(u_0, \mu_0) = \nu(u_0)$, with $\mu_0 = (1, 0, \ldots, 0) \in U_1^+$.

By Theorem 2.5, we have, for $i = 1, \ldots, n$,

$$N_{u_i}(u_0, \nu(u_0)) = -\sum_{j=1}^{n} (\lambda_{ij})^\nu(u_0)\mathbf{x}_{u_j}(u_0)$$

and from the proof of Theorem 2.7, for $j = 2, \ldots, r$,

$$N_{\mu_j}(u_0, \mu_0) = \nu_j(u_0).$$

The matrix of the shape operator of $CM(\varepsilon)$ at (u_0, μ_0) with respect to the basis $\{\mathbf{x}_{u_1}(u_0), \ldots, \mathbf{x}_{u_n}(u_0), \nu_1(u_0), \ldots, \nu_r(u_0)\}$ has determinant $\det(-(\lambda_{ij})^\nu(u_0))$ which is precisely $K^\nu(u_0)$. □

Corollary 2.4. *Let p be a point on M and let ν_0 be a unit normal vector at p. Then p is a ν_0-parabolic point of M if and only if (p, ν_0) is a parabolic point of the canal hypersurface $CM(\varepsilon)$ of M, with ε small enough.*

Example 2.4 (The canal surface of a space curve). *Consider a smooth curve $\gamma : I \to \mathbb{R}^3$ parametrised by arc length. A parametrisation of the canal surface of γ of radius ε is given by $\mathbf{y} : I \times \mathbb{R} \to \mathbb{R}^3$, with*

$$\mathbf{y}(s, \theta) = \gamma(s) + \varepsilon(\cos\theta\mathbf{n}(s) + \sin\theta\mathbf{b}(s)),$$

where $\mathbf{n}(s)$ *and* $\mathbf{b}(s)$ *are, respectively, the principal normal and binormal vectors of* γ *at* s.

The first order partial derivatives of $\mathbf{y}$ *are*

$$\begin{aligned}\mathbf{y}_s &= (1-\varepsilon\kappa\cos\theta)\mathbf{t} + \varepsilon\tau\sin\theta\mathbf{n} - \varepsilon\tau\cos\theta\mathbf{b}\\ \mathbf{y}_\theta &= \varepsilon(-\sin\theta\mathbf{n} + \cos\theta\mathbf{b})\end{aligned}$$

where $\kappa, \tau, \mathbf{t}, \mathbf{n}, \mathbf{b}$ *are evaluated at* s. *The matrix of the first fundamental form is given by*

$$(g_{ij}) = \begin{pmatrix} (1-\varepsilon\kappa\cos\theta)^2 + \varepsilon^2\tau^2 & -\varepsilon^2\tau \\ -\varepsilon^2\tau & \varepsilon^2 \end{pmatrix}.$$

The second order partial derivatives of $\mathbf{y}$ *are given by*

$$\begin{aligned}\mathbf{y}_{ss} &= -\varepsilon(\kappa'\cos\theta + \tau^2\sin\theta)\mathbf{t} + (\varepsilon\tau'\sin\theta + \kappa(1-\varepsilon\kappa\cos\theta) - \varepsilon\tau^2\cos\theta)\mathbf{n}\\ &\quad -\varepsilon(\tau'\cos\theta + \tau^2\sin\theta)\mathbf{b},\\ \mathbf{y}_{s\theta} &= \varepsilon(\kappa\sin\theta\mathbf{t} + \tau\cos\theta\mathbf{n} + \tau\sin\theta\mathbf{t}),\\ \mathbf{y}_{\theta\theta} &= -\varepsilon(\cos\theta\mathbf{n} + \sin\theta\mathbf{b}).\end{aligned}$$

The Gauss map of the canal surface is given by $N(s,\theta) = \cos\theta\mathbf{n}(s) + \sin\theta\mathbf{b}(s)$, *so we can compute the matrix of the second fundamental form* (h_{ij}) *using the relations (2.2) and obtain*

$$(h_{ij}) = \begin{pmatrix} \kappa(1-\varepsilon\kappa\cos\theta)\cos\theta - \varepsilon\tau^2 & -\varepsilon\tau \\ -\varepsilon^2\tau & -\varepsilon \end{pmatrix}.$$

It follows by Corollary 2.1 that the Gaussian curvature of the canal surface is given by

$$K(s,\theta) = -\frac{\kappa(s)\cos\theta}{\varepsilon(1-\varepsilon\kappa(s)\cos\theta)}.$$

In particular, the parabolic set of the canal surface consists of the two curves $\gamma(s) \pm \mathbf{b}(s)$.

Chapter 3

Singularities of germs of smooth mappings

The birth of singularity theory can be traced back to the pioneering work of Whitney [Whitney (1955)] where he showed that maps from the plane to the plane have, in general, only folds and cusps singularities. Mather introduced in his seminal papers [Mather (1968, 1969a,b,c, 1970)] groups that act on the set of map-germs and set the foundations for the study of finite determinacy of map-germs, a concept linked to the existence of versal deformations and versal unfoldings. A versal deformation, is in some sense, one that contains all possible deformations of the initial map-germ. This has a wide range of applications in mathematics and other fields of science (see for example [Koenderink (1990); Poston and Stewart (1996); Thom (1983)]).

We give in this chapter some basic definitions and state the results that we need in other chapters. For beginners in singularity theory, we recommend the books [Arnol'd, Guseĭn-Zade and Varchenko (1985); Bröcker (1975); Gibson (1979); Martinet (1982)] and [Bruce and Giblin (1992)] for application to the geometry of curves. C. T. C. Wall's survey article [Wall (1981)] remains the first port of call for people embarking on the study of finite determinacy of map-germs.

The definitions and results are presented for maps $f : U \to \mathbb{R}^m$ of class C^∞ from an open subset U of $\mathbb{R}^n$ to $\mathbb{R}^m$, but they also hold for smooth map $f : M \to N$, where M and N are any smooth dimensional manifolds.

3.1 Germs of smooth mappings

Let X and Y be two subsets of $\mathbb{R}^n$ containing a point $p \in \mathbb{R}^n$. We say that X is equivalent to Y if there exists an open set $U \subset \mathbb{R}^n$ containing p such that $X \cap U = Y \cap U$. This defines an equivalence relation among subsets of $\mathbb{R}^n$ containing the point p. The equivalence class of a subset X is called the *germ* of X at p and is denoted by (X, p).

Let U and V be two open subsets of $\mathbb{R}^n$ containing a point $p \in \mathbb{R}^n$, and let $f : U \to \mathbb{R}^m$ and $g : V \to \mathbb{R}^m$ be two smooth maps. We say that $f \sim g$ if there exists an open set $W \subset U \cap V$ containing p such that $f = g$ on W, that is $f|_W = g|_W$.

The relation $\sim$ is an equivalence relation and a germ at p of a smooth map is by definition an equivalent class under this equivalence relation. A map-germ at p is denoted by

$$f : (\mathbb{R}^n, p) \to \mathbb{R}^m.$$

We shall denote by $\tilde{f} : U \to \mathbb{R}^m$ a representative of a germ f in a neighbourhood U of p.

Sometimes we require that all the elements of the equivalence classes have the same value at p, say q. Then we write

$$f : (\mathbb{R}^n, p) \to (\mathbb{R}^m, q).$$

Let $\mathcal{E}_n$ denote the set of germs, at the origin 0 in $\mathbb{R}^n$, of smooth functions $(\mathbb{R}^n, 0) \to \mathbb{R}$,

$$\mathcal{E}_n = \{f : (\mathbb{R}^n, 0) \to \mathbb{R} \mid f \text{ is the germ of a smooth function}\}.$$

With the addition and multiplication operations, $\mathcal{E}_n$ becomes a commutative $\mathbb{R}$-algebra with a unit. It has a maximal ideal $\mathcal{M}_n$ which is the subset of germs of functions that vanish at the origin. We have

$$\mathcal{M}_n = \{f \in \mathcal{E}_n \mid \tilde{f}(0) = 0\}.$$

Since $\mathcal{M}_n$ is the unique maximal ideal of $\mathcal{E}_n$, $\mathcal{E}_n$ is a local algebra. (It is, however, not a Nöether algebra [Gibson (1979)].) If $(x_1, \ldots, x_n)$ is a system of local coordinates of $(\mathbb{R}^n, 0)$, then $\mathcal{M}_n$ is generated by the germs of functions x_i, $i = 1, \ldots, n$, that is,

$$\mathcal{M}_n = \mathcal{E}_n \cdot \{x_1, \ldots, x_n\}.$$

For a given positive integer k, the kth-power of the maximal ideal $\mathcal{M}_n$ is denoted by $\mathcal{M}_n^k$. It is the set of germs of functions $f \in \mathcal{M}_n$ with zero partial derivatives of order less or equal to $k-1$ at the origin. We also have

$$\mathcal{M}_n^k = \mathcal{E}_n \cdot \{x_1^{i_1} \cdots x_n^{i_n} \, : \, i_1 + \cdots + i_n = k\}.$$

The set of all smooth map-germs $f : (\mathbb{R}^n, 0) \to \mathbb{R}^m$ is denoted by $\mathcal{E}(n, m)$. It is the direct product of m-copies of $\mathcal{E}_n$, that is,

$$\mathcal{E}(n, m) = \underbrace{\mathcal{E}_n \times \cdots \times \mathcal{E}_n}_{m} = (\mathcal{E}_n)^m.$$

We denote by $\mathcal{M}_n^{k+1} \cdot \mathcal{E}(n,m) = (\mathcal{M}_n^{k+1})^m$ the set of map-germs $f : (\mathbb{R}^n, 0) \to (\mathbb{R}^m, 0)$ with vanishing partial derivatives of order less or equal to k at the origin.

The k-jet space of smooth map-germs $(\mathbb{R}^n, 0) \to (\mathbb{R}^m, 0)$ is defined as

$$J^k(n,m) = \mathcal{M}_n \cdot \mathcal{E}(n,m)/\mathcal{M}_n^{k+1} \cdot \mathcal{E}(n,m).$$

The map $j^k : \mathcal{M}_n \cdot \mathcal{E}(n,m) \to J^k(n,m)$ assigns to each map-germ f its kth-jet. An element in $J^k(n,m)$ corresponding to a map-germ $f \in \mathcal{M}_n \cdot \mathcal{E}(n,m)$ is denoted by $j^k f$.

The set $J^k(n,m)$ can be identified with the set of polynomials of degree less than or equal to k, without the constant terms. Given $f \in \mathcal{M}_n \cdot \mathcal{E}(n,m)$, $j^k f$ is simply its Taylor polynomial of degree k at the origin.

3.2 Multi-germs of smooth mappings

Let S be a collection of s distinct points $p_1, \ldots, p_s$ in $\mathbb{R}^n$. The equivalence relation "$\sim$" in section 3.1 can be extended in a natural way to the collection S. Let U_i (resp. V_i), $i = 1, \ldots, s$, be pairwise disjoint open subsets of $\mathbb{R}^n$ with $p_i \in U_i \cap V_i$. Let $f_i : U_i \to \mathbb{R}^m$ and $g_i : V_i \to \mathbb{R}^m$, $i = 1, \ldots, s$ be a collection of smooth maps. Define $f = (f_i) : \cup_{i=1}^s U_i \to \mathbb{R}^m$ and $g = (g_i) : \cup_{i=1}^s V_i \to \mathbb{R}^m$ by $f|_{U_i} = f_i$ and $g|_{U_i} = g_i$.

We say that $f \sim_S g$ if and only if there exist open sets $W_i \subset U_i \cap V_i$ containing p_i, $i = 1, \ldots, s$, such that $f_i|_{W_i} = g_i|_{W_i}$. This is an equivalence relation and a multi-germ at S is by definition a representative of an equivalent class under this equivalence relation. We denote by $(\mathbb{R}^n, S) = \cup_{i=1}^s (\mathbb{R}^n, p_i)$ and write $f = (f_i) : (\mathbb{R}^n, S) \to \mathbb{R}^m$ for a multi-germ at S. When we require the multi-germ to fix a point q in the target, we write $f : (\mathbb{R}^n, S) \to (\mathbb{R}^m, q)$.

We take s independent local coordinate systems of $(\mathbb{R}^n, p_i)$, with all the p_i's set to be the origin. We denote a multi-germ in this case by $f : (\mathbb{R}^n, 0_s) \to \mathbb{R}^m$ and denote by $\mathcal{E}(n,m)_s$ the set of all these multi-germs. The multi-jets space $J^k(n,m)_s$ is defined as the set

$$J^k(n,m)_s = (\mathcal{M}_n \cdot \mathcal{E}(n,m))_s/(\mathcal{M}_n^{k+1} \cdot \mathcal{E}(n,m))_s.$$

An element $j^k f_s$ is the Taylor expansion of degree k of $f \in (\mathcal{M}_n \cdot \mathcal{E}(n,m))_s$.

3.3 Singularities of germs of smooth mappings

Let $f : U \subset \mathbb{R}^n \to \mathbb{R}^m$ be a smooth map and denote by $df : TU \to T\mathbb{R}^m$ its derivative map. The map f is *singular* at $p \in U$ if the rank of the linear map

$$(df)_p : \mathbb{R}^n \to \mathbb{R}^m$$

is not maximal, that is, if $\text{rank}(df)_p < \min(n, m)$. The point p is then said to be a *singular point* of f. Otherwise, we say that f is non-singular at p and p is a *regular point* of f. The *critical set* of f, denoted by $\Sigma(f)$, is the set of singular points of f, that is,

$$\Sigma(f) = \{p \in U | \text{ rank}(df)_p < \min(n, m)\}.$$

The *criminant* of f, denoted by $Cr(f)$, is

$$Cr(f) = \{p \in U \mid \text{rank}(df)_p < m \}.$$

When $n \geq m$, $Cr(f) = \Sigma(f)$, and when $n < m$, $Cr(f) = U$.

The *discriminant of* f, denoted by $\Delta(f)$, is the image of $Cr(f)$ by f:

$$\Delta(f) = f(Cr(f)).$$

Observe that when $n < m$, $\Delta(f) = f(U)$. The set $f(\Sigma(f))$ is called *the set of critical values* of f. The above definitions can be localised at a point $p \in \mathbb{R}^n$. A germ $f : (\mathbb{R}^n, p) \to \mathbb{R}^m$ is said to be *singular* if one of its representatives is singular at p. This definition does not depend on the choice of the representative of f at p as any two of these are identical in some neighbourhood of p.

The critical set (respectively, the criminant) of a map-germ f, still denoted by $\Sigma(f)$ (respectively, $Cr(f)$), is the set germ $(\Sigma(\tilde{f}), p)$ (respectively, $(Cr(\tilde{f}), p)$), where $\tilde{f}$ is a representative of f in some neighbourhood U of p. Again, this definition does not depend on the choice of the representative. Likewise, the discriminant of f is $\Delta(f) = (\Delta(\tilde{f}), f(p))$.

Example 3.1. (1) A germ of a function $f : (\mathbb{R}^n, 0) \to \mathbb{R}$ is singular if and only if all the partial derivatives of a representative of f vanish at the origin. Therefore, the set of all singular germs in $\mathcal{E}_n$ is the ideal $\mathcal{M}_n^2$.

(2) A germ of a curve $f : (\mathbb{R}, 0) \to \mathbb{R}^n$ is singular if and only if $f'(0)$ is the zero vector.

(3) Consider a map-germ $f : (\mathbb{R}^2, 0) \to (\mathbb{R}^2, 0)$ of rank 1 in the form $f(x, y) = (x, g(x, y))$. The graph of the germ g is the germ of the surface M in $\mathbb{R}^3$ parametrised by the map-germ $G_g : (\mathbb{R}^2, 0) \to (\mathbb{R}^3, 0)$ given by $G_g(x, y) = (x, y, g(x, y))$. Let (u, v, w) denote the coordinates in $\mathbb{R}^3$ and let $\pi(u, v, w) = (u, w)$ be the projection to the (u, w)-plane. Then $f = \pi \circ G_g$, so that f is singular if and only if π restricted to M is singular. (The map-germ G_g can also be viewed as a 1-parameter unfolding of the map-germ $(y, g(0, y))$, with x being the parameter; see section 3.8 for definition.)

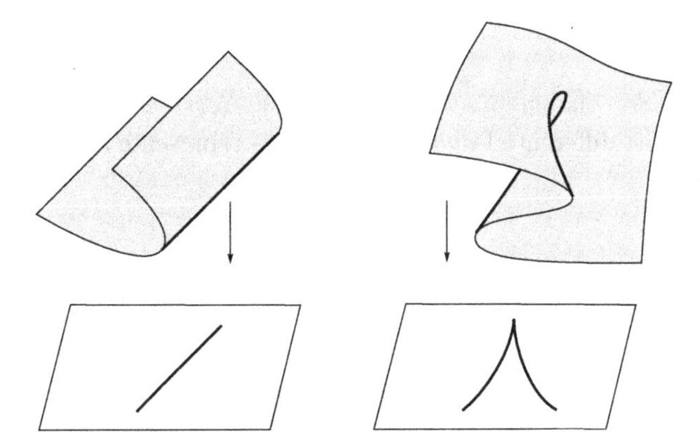

Fig. 3.1 The fold (left) and cusp (right) singularities realised by the projection of a surface to a plane. The singular sets and discriminants are the thick curves.

Consider the following special cases
(a) $f(x, y) = (x, y^2)$,
(b) $f(x, y) = (x, xy + y^3)$.

The map-germ $f(x, y) = (x, y^2)$ is called the *fold* and the map-germ $f(x, y) = (x, xy + y^3)$ is called the *cusp* or *pleat* ([Whitney (1955)]).

For the fold map-germ, the differential map df at (x, y) is represented by the matrix

$$\begin{pmatrix} 1 & 0 \\ 0 & 2y \end{pmatrix}.$$

This is singular if and only if $y = 0$, so the critical set $\Sigma(f)$ is the germ, at the origin in the source, of the curve $y = 0$. The discriminant of f is the germ of the curve $w = 0$ in the (u, w)-plane. This can be depicted by considering the projection of the surface M to the (u, w)-plane. The singular set is the critical set of the projection $\pi|_M$; see Figure 3.1, left.

For the cusp map-germ, the differential map df at (x, y) is represented by the matrix

$$\begin{pmatrix} 1 & 0 \\ y & x + 3y^2 \end{pmatrix}.$$

This is singular if and only if $x + 3y^2 = 0$, so the singular set $\Sigma(f)$ is the germ, at the origin in the source, of the parabola $x + 3y^2 = 0$. Then the discriminant of f is the germ, at the origin in the target, of the cusp curve $27w^2 + 4u^3 = 0$ (Figure 3.1, right). The cusp in the discriminant can be seen in Figure 3.1, right, as the image of the critical set of the projection $\pi|M$.

(4) Consider the map-germ $f : (\mathbb{R}^2, 0) \to (\mathbb{R}^3, 0)$ with $f(x, y) = (x^2, y, xy)$. The differential map df at (x, y) is represented by the matrix

$$\begin{pmatrix} 2x & 0 \\ 0 & 1 \\ y & x \end{pmatrix}.$$

This is singular if and only if $x = y = 0$, so the critical set $\Sigma(f)$ is the origin in the source. Its discriminant, which is its image, is as shown in Figure 3.2 and is called a cross-cap or a Whitney umbrella.

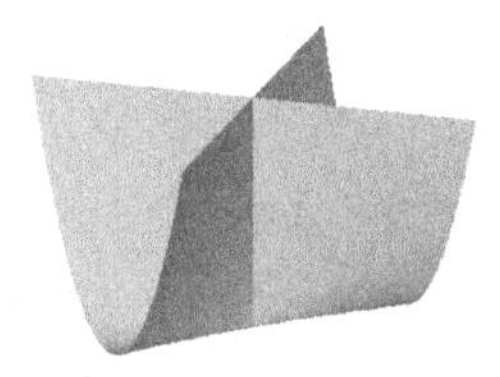

Fig. 3.2 A cross-cap.

Let $f = (f_i) : (\mathbb{R}^n, S) \to \mathbb{R}^m$ be a multi-germ, where $S = \{p_1, \ldots, p_s\}$. We say that f is *singular* if $f_i : (\mathbb{R}^n, p_i) \to \mathbb{R}^m$ is singular for some $p_i \in S$.

The *critical set* (respectively, the *criminant*) of f is the union of the critical sets (respectively, the criminants) of the map-germs f_i, $i = 1, \ldots, s$, that is

$$\Sigma(f) = \cup_{i=1}^{s} \Sigma(f_i), \quad \text{(respectively, } Cr(f) = \cup_{i=1}^{s} Cr(f_i)),$$

and its *discriminant* is $\Delta(f) = f(Cr(f))$.

Example 3.2. Let $f_1(x_1, y_1) = (x_1^2, y_1)$ and $f_2(x_2, y_2) = (x_2, x_2 y_2 + y_2^3)$ be two smooth map-germs, at the origin, from the plane to the plane and consider the bi-germ $f = (f_1, f_2)$. The critical set of f_1 is the germ, at the origin, of the curve $x_1 = 0$. The critical set of f_2 is the germ, at the origin, of the curve $x_2 + 3y_2^2 = 0$. Thus, the discriminant of f is the germ, at the origin in the target, of the curve $u(27w^2 - 4u^3) = 0$.

3.4 The Thom-Boardman symbols

We shall make use of the Thom-Boardman symbols of a smooth map $f : X \to Y$ between two smooth manifolds X and Y (see Chapter VI in [Golubitsky and Guillemin (1973)] for details).

The map f is said to have a singularity at $p \in X$ of type S_r if $(df)_p$ drops rank by r, that is, if $\operatorname{rank}(df)_p = \min(\dim X, \dim Y) - r$.

The subset of X of the singularities of f of type S_r is denoted by $S_r(f)$. For a generic map f, the subsets $S_r(f)$ are submanifolds of X of codimension $r^2 + er$, where $e = |\dim X - \dim Y|$ ([Golubitsky and Guillemin (1973)]) .

When $S_r(f)$ is a submanifold, we can consider the restriction $f|_{S_r(f)} : S_r(f) \to Y$. The subset of $S_r(f)$ where $f|_{S_r(f)}$ drops rank by s is denoted by $S_{r,s}(f)$. Under some genericity conditions, these subsets are submanifolds of $S_r(f)$. We can continue this process and define inductively a nested family of submanifolds $S_{i_1,\cdots,i_q}(f)$ of X. The points of $S_{i_1,\cdots,i_q}(f)$ are called singularities of f with *Thom-Boardman symbol* $S_{i_1,\cdots,i_q}$.

3.5 Mather's groups

Let $\mathcal{R}$ denote the group of germs of diffeomorphisms $(\mathbb{R}^n, 0) \to (\mathbb{R}^n, 0)$. This group is labelled the "right group" and acts smoothly on $\mathcal{E}(n,m)$ by

$$h \cdot f = f \circ h^{-1}$$

for any $h \in \mathcal{R}$ and $f \in \mathcal{E}(n,m)$.

The "left group" $\mathcal{L}$ of germs of diffeomorphisms $(\mathbb{R}^m, 0) \to (\mathbb{R}^m, 0)$ acts smoothly on $\mathcal{M}_n.\mathcal{E}(n,m)$ by

$$k \cdot f = k \circ f$$

for any $k \in \mathcal{L}$ and $f \in \mathcal{M}_n \cdot \mathcal{E}(n,m)$.

We denote by $\mathcal{A} = \mathcal{R} \times \mathcal{L}$ the direct product of $\mathcal{R}$ and $\mathcal{L}$. The group $\mathcal{A}$ is referred to as the *right-left* group. It acts smoothly on $\mathcal{M}_n \cdot \mathcal{E}(n,m)$ by

$$(h,k) \cdot f = k \circ f \circ h^{-1}$$

for any $(h,k) \in \mathcal{A}$ and $f \in \mathcal{M}_n \cdot \mathcal{E}(n,m)$.

We have another group of interest, namely the contact group $\mathcal{K}$. The group $\mathcal{K}$ is the set of germs of diffeomorphisms of $(\mathbb{R}^n \times \mathbb{R}^m, (0,0))$ which can be written in the form $H(x,y) = (h(x), H_1(x,y))$, with $h \in \mathcal{R}$ and $H_1(x,0) = 0$ for x near 0. We have $\pi \circ H = h \circ \pi$ where $\pi : \mathbb{R}^n \times$

$\mathbb{R}^m \to \mathbb{R}^n$ is the canonical projection. Thus H is a fibred mapping over the diffeomorphism h and preserves the 0-section $\mathbb{R}^n \times \{0\}$. The set of germs of diffeomorphisms of $(\mathbb{R}^n \times \mathbb{R}^m, (0,0))$ in the form (I, H), where I is the germ at 0 of the identity map of $\mathbb{R}^n$, is denoted by $\mathcal{C}$. The group $\mathcal{K}$ is the semi-direct product of $\mathcal{R}$ and $\mathcal{C}$, and we write $\mathcal{K} = \mathcal{R} \rtimes \mathcal{C}$.

The group $\mathcal{K}$ acts on $\mathcal{M}_n.\mathcal{E}(n,m)$ as follows. Given $f, g \in \mathcal{M}_n.\mathcal{E}(n,m)$ and $(h, H) \in \mathcal{K}$, $g = (h, H).f$ if and only if

$$(x, g(x)) = H(h^{-1}(x), f(h^{-1}(x))).$$

The diffeomorphism H sends the graph of the map-germ f to that of g.

Remark 3.1. The group $\mathcal{K}$ is a natural one to use when one seeks to understand the singularities of the zero level-sets of map-germs in $\mathcal{M}_n.\mathcal{E}(n,m)$. If two germs are $\mathcal{K}$-equivalent, then their zero level-sets are diffeomorphic The action of the group $\mathcal{A}$ is finer than that of $\mathcal{K}$. If two map-germs F and G are $\mathcal{A}$-equivalent, then $G = k \circ F \circ h^{-1}$ for some $(h, k) \in \mathcal{A}$, so the level sets $G^{-1}(c)$ and $F^{-1}(k^{-1}(c))$ are diffeomorphic for any c close to $0 \in \mathbb{R}^m$. Therefore, the group $\mathcal{A}$ preserves also the smooth structure of nearby level sets to the zero level set.

The groups $\mathcal{R}, \mathcal{L}, \mathcal{A}, \mathcal{C}, \mathcal{K}$ are known as the Mather's groups. The groups $\mathcal{R}$ and $\mathcal{L}$ can be considered in a natural way as subgroups of $\mathcal{A}$ and the groups $\mathcal{C}$ and $\mathcal{A}$ (and so $\mathcal{R}$ and $\mathcal{L}$) as subgroups of $\mathcal{K}$.

Let $\mathcal{G}$ be one of the Mather's groups. Two germs f, g are said to be *$\mathcal{G}$-equivalent* if they are in the same $\mathcal{G}$-orbit.

Let $\mathcal{G}_k$ be the subgroup of a Mather group $\mathcal{G}$ whose elements have k-jets the germ of the identity. The group $\mathcal{G}_k$ is a normal subgroup of $\mathcal{G}$. Define $J^k\mathcal{G} = \mathcal{G}/\mathcal{G}_k$. The elements of $J^k\mathcal{G}$ are the k-jets of the elements of $\mathcal{G}$.

The action of a Mather group $\mathcal{G}$ on $\mathcal{M}_n.\mathcal{E}(n,m)$ induces an action of $J^k\mathcal{G}$ on $J^k(n,m)$ as follows. For $j^kf \in J^k(n,m)$ and $j^kh \in J^k\mathcal{G}$,

$$j^kh.j^kf = j^k(h.f).$$

3.6 Tangent spaces to the $\mathcal{G}$-orbits

Let G be a Lie group acting smoothly on a smooth manifold M. Assume that the orbits of the action of G are smooth submanifolds of M. Then the tangent space $T_p(G.p)$ at $p \in M$ to the orbit $G.p$ of p is well defined. Also, there exists locally at p a smooth submanifold S of M containing p such that $T_p(G.p) \oplus T_pS = T_pM$ and a submanifold H of G containing the

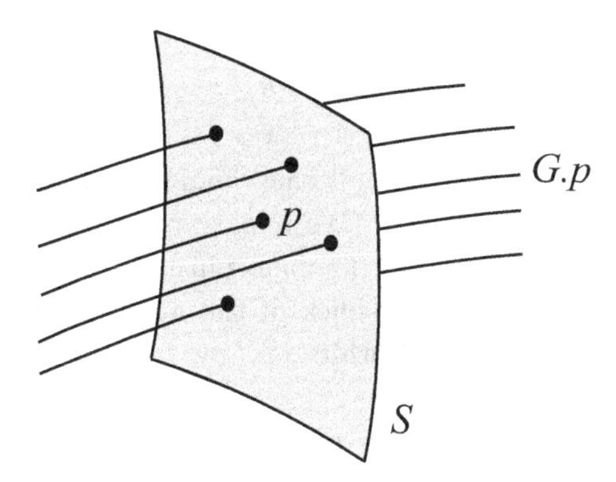

Fig. 3.3 Orbits of a Lie group action.

identity such that $H \times S$ is diffeomorphic to a neighbourhood of p in M. The submanifold S can then be used to parametrise the orbits of G near $G.p$ (see Figure 3.3 and [Gibson (1979)] for details).

The Mather groups are not Lie groups and $\mathcal{E}(n,m)$ is not a finite dimensional manifold (it is not even a Banach manifold). We proceeded as follows to define the tangent space to an orbit of one of the Mather's groups $\mathcal{G}$. Consider the group $\mathcal{R}$ acting on $\mathcal{E}_n$ and let $\mathcal{R}.f$ denote the $\mathcal{R}$ orbit $f \in \mathcal{E}_n$. Let $f \circ h_t$, $t \in (-\epsilon, \epsilon)$, be a path in $\mathcal{R}.f$ through f, where h_t is a path in $\mathcal{R}$ with h_0 being the identity map-germ. The germ of the vector field $\frac{\partial}{\partial t}(f \circ h_t)|_{t=0}$ is defined as a "tangent vector" to $\mathcal{R}.f$ at f. The union of these tangent vectors over all paths in $\mathcal{R}.f$ through f is defined to be the "tangent space" to the orbit $\mathcal{R}.f$ at f. One can proceed similarly for the other groups and define algebraically the tangent space to a $\mathcal{G}$-orbit as follows.

Let $\pi : T\mathbb{R}^m \to \mathbb{R}^m$ be the tangent bundle over $\mathbb{R}^m$. A map-germ $\xi : \mathbb{R}^n, 0 \to T\mathbb{R}^m$ is said to be a germ of *vector field along* $f \in \mathcal{E}(n,m)$ if $\pi \circ \xi = f$. The tangent space θ_f to $\mathcal{E}(n,m)$ at f is defined to be the $\mathcal{E}_n$-module of germs of vector fields along f.

Let $\theta_n = \theta_{id_{(\mathbb{R}^n,0)}}$ and $\theta_m = \theta_{id_{(\mathbb{R}^m,0)}}$, where $id_{(\mathbb{R}^n,0)}$ and $id_{(\mathbb{R}^m,0)}$ denote the germs of the identity maps on $(\mathbb{R}^n, 0)$ and $(\mathbb{R}^m, 0)$ respectively. We remark that θ_n is nothing but the set of germs of the vector field on $\mathbb{R}^n$ at the origin. The map

$$\begin{aligned} tf : \theta_n &\to \theta_f \\ \phi &\mapsto df \circ \phi \end{aligned}$$

is a $\mathcal{E}_n$-homomorphism, and the map

$$\begin{aligned} wf : \theta_m &\to \theta_f \\ \psi &\mapsto \psi \circ f \end{aligned}$$

is a $\mathcal{E}_m$-homomorphism via the pull-back homomorphism $f^* : \mathcal{E}_m \to \mathcal{E}_n$, where $f^*(\alpha) = \alpha \circ f$ for $\alpha \in \mathcal{E}_m$. (We remark that f^* is a ring homomorphism, but it is not an $\mathbb{R}$-algebra homomorphism.)

Let $f^*(\mathcal{M}_m)$ denote the pull-back of the maximal ideal in $\mathcal{E}_m$. The *tangent spaces* $L\mathcal{G} \cdot f$ to the $\mathcal{G}$-orbits of f at the germ f are defined as follows:

$$L\mathcal{R} \cdot f = tf(\mathcal{M}_n.\theta_n) \qquad L\mathcal{L} \cdot f = wf(\mathcal{M}_m.\theta_m) \qquad L\mathcal{C} \cdot f = f^*(\mathcal{M}_m).\theta_f$$
$$L\mathcal{A} \cdot f = L\mathcal{R} \cdot f + L\mathcal{L} \cdot f \qquad L\mathcal{K} \cdot f = L\mathcal{R} \cdot f + L\mathcal{C} \cdot f$$

The diffeomorphisms in a Mather group keep the origin fixed (in the source and or in the target). When studying deformations (see §3.8), the singularity can move away from the origin, so the vector fields involved in defining the tangent spaces are allowed not to fix the origin. For this reason, the *extended tangent spaces* are defined as follows:

$$L_e\mathcal{R} \cdot f = tf(\theta_n) \qquad L_e\mathcal{L} \cdot f = wf(\theta_m) \qquad L_e\mathcal{C} \cdot f = f^*(\mathcal{M}_m).\theta_f$$
$$L_e\mathcal{A} \cdot f = L_e\mathcal{R} \cdot f + L_e\mathcal{L} \cdot f \qquad L_e\mathcal{K} \cdot f = L_e\mathcal{R} \cdot f + L_e\mathcal{C} \cdot f$$

If we choose a system of coordinates $(y_1, \ldots, y_m)$ in $(\mathbb{R}^m, 0)$, the germs of the vector fields

$$\left(\frac{\partial}{\partial y_1}\right) \circ f, \ldots, \left(\frac{\partial}{\partial y_m}\right) \circ f$$

along f form a free basis of θ_f. Then θ_f can be identified canonically with $\mathcal{E}(n,m)$ (θ_f is a free $\mathcal{E}_n$-module of rank m) and we have

$$\begin{aligned}
L\mathcal{R} \cdot f &= \mathcal{M}_n.\left\{\tfrac{\partial f}{\partial x_1}, \ldots, \tfrac{\partial f}{\partial x_n}\right\}, & L_e\mathcal{R} \cdot f &= \mathcal{E}_n.\left\{\tfrac{\partial f}{\partial x_1}, \ldots, \tfrac{\partial f}{\partial x_n}\right\}, \\
L\mathcal{L} \cdot f &= f^*(\mathcal{M}_m).\{e_1, \ldots, e_m\}, & L_e\mathcal{L} \cdot f &= f^*(\mathcal{E}_m).\{e_1, \ldots, e_m\}, \\
L\mathcal{C} \cdot f &= f^*(\mathcal{M}_m).\mathcal{E}_n.\{e_1, \ldots, e_m\}, & L_e\mathcal{C} \cdot f &= f^*(\mathcal{M}_m).\mathcal{E}_n.\{e_1, \ldots, e_m\},
\end{aligned}$$

where $e_1, \ldots, e_m$ are the standard basis vectors of $\mathbb{R}^m$ considered as elements of $\mathcal{E}(n,m)$ and $(x_1, \ldots, x_n)$ is a coordinate system in $(\mathbb{R}^n, 0)$.

We remark that when $m = 1$, $L\mathcal{C} \cdot f = f^*(\mathcal{M}_1).\mathcal{E}_n = \mathcal{E}_n.\{f\}$, so the group $\mathcal{C}$ acts by multiplication by germs of functions in $\mathcal{E}_n$.

The *codimension of the orbit* of f is defined by

$$\mathrm{cod}(f, \mathcal{G}) = \dim_{\mathbb{R}}(\mathcal{M}_n.\mathcal{E}(n,m)/L\mathcal{G} \cdot f)$$

and the *codimension of the extended orbit* of f is defined by

$$\mathrm{cod}_e(f, \mathcal{G}) = \dim_{\mathbb{R}}(\mathcal{E}(n,m)/L_e\mathcal{G} \cdot f)\,.$$

See [Wall (1981)] for the relation between $\mathrm{cod}(f, \mathcal{G})$ and $\mathrm{cod}_e(f, \mathcal{G})$. For example, $\mathrm{cod}(f, \mathcal{A}) = \mathrm{cod}_e(f, \mathcal{A}) + m$.

3.7 Finite determinacy

A germ f is said to be $k - \mathcal{G}$*-determined* if any g with $j^k g = j^k f$ is $\mathcal{G}$-equivalent to f (notation: $g \sim_{\mathcal{G}} f$). The k-jet of f is then called a sufficient jet. The least integer k with this property is called the degree of determinacy of f. A $\mathcal{G}$-determined germ is a $k - \mathcal{G}$-determined germ for some integer k.

Some important properties of finitely determined map-germs are the following ([Wall (1981)], Theorem 1.2).

Theorem 3.1. *For each f and $\mathcal{G}$, the following are equivalent:*
(i) *f is $\mathcal{G}$-determined,*
(ii) *for some k, $\mathcal{M}_n^k.\mathcal{E}(n,m) \subset L\mathcal{G} \cdot f$,*
(iii) $\mathrm{cod}(f, \mathcal{A}) < \infty$,
(iv) $\mathrm{cod}_e(f, \mathcal{A}) < \infty$.

A great deal of work is carried out to find out whether a map-germ is $\mathcal{G}$-finitely determined and to find its degree of determinacy. There is a detailed account on determinacy of map-germs in C.T.C. Wall survey article [Wall (1981)], which we refer to for details and references. The algebraic structure of the tangent space $L\mathcal{G} \cdot f$ plays a key role. When $\mathcal{G}$ is $\mathcal{R}, \mathcal{C}$ or $\mathcal{K}$, the tangent space $L\mathcal{G} \cdot f$ is a $\mathcal{E}_n$-module and one can apply Nakayama's Lemma to prove the following.

Theorem 3.2. *Let $f \in \mathcal{M}_n.\mathcal{E}(n,m)$ and $\mathcal{G} = \mathcal{R}, \mathcal{C}$ or $\mathcal{K}$. Suppose that*

$$\mathcal{M}_n^k.\mathcal{E}(n,m) \subset L\mathcal{G} \cdot f + \mathcal{M}_n^{k+1}.\mathcal{E}(n,m).$$

Then f is k-$\mathcal{G}$-determined.

We observe that for the group $\mathcal{R}$, there are no $\mathcal{R}$-finitely determined map-germs in $\mathcal{M}_n.\mathcal{E}(n,m)$ if $m > 1$. Therefore, the group $\mathcal{R}$ is useful only when considering germs of functions.

It is shown in [Mather (1969c)] that two map-germs f and g are $\mathcal{C}$-equivalent if and only if $f^*(\mathcal{M}_m).\mathcal{E}_n = g^*(\mathcal{M}_m).\mathcal{E}_n$. That leads to the following result for the group $\mathcal{K}$.

Theorem 3.3. *The following assertions are equivalent.*
(i) *Two map-germs f and g are $\mathcal{K}$-equivalent.*
(ii) *There exists a germ of a diffeomorphism $\phi : (\mathbb{R}^n, 0) \to (\mathbb{R}^n, 0)$ such that $\phi^*(f^*(\mathcal{M}_m).\mathcal{E}_n) = g^*(\mathcal{M}_m).\mathcal{E}_n$.*

(iii) *There exists a germ of a diffeomorphism $\phi : (\mathbb{R}^n, 0) \to (\mathbb{R}^n, 0)$ such that ϕ induces an $\mathbb{R}$-algebra isomorphism*

$$\widetilde{\phi^*} : \mathcal{E}_n/f^*(\mathcal{M}_m).\mathcal{E}_n \to \mathcal{E}_n/g^*(\mathcal{M}_m).\mathcal{E}_n.$$

When $\mathcal{G} = \mathcal{L}$, the tangent space $L\mathcal{G} \cdot f$ is an $f^*(\mathcal{E}_m)$-module. In fact, if a map-germ is $\mathcal{L}$-finite, then $m \geq 2n$ ([Wall (1981)]).

Let $\mathcal{G}_s$ denote the subgroup of $\mathcal{G}$ whose elements have s-jet the identity.

Theorem 3.4. *A map-germ $f \in \mathcal{M}_n.\mathcal{E}(n,m)$ is k-$\mathcal{L}_1$-determined if and only if*

$$\mathcal{M}_n^{k+1}.\mathcal{E}(n,m) \subset L\mathcal{L}_1 \cdot f + \mathcal{M}_n^{k+1}.(f^*(\mathcal{M}_m).\mathcal{E}_n + \mathcal{M}_n^{k+1})\mathcal{E}(n,m).$$

When $\mathcal{G} = \mathcal{A}$, $L\mathcal{A} \cdot f$ has a mixed type module structure and this makes the estimation of the degree of determinacy much harder to deal with. The question of determining the exact degree of determinacy of a map-germ is solved in [Bruce, du Plessis and Wall (1987)] by considering unipotent actions of subgroups of $\mathcal{G}$.

The following corollary of the main determinacy result in [Bruce, du Plessis and Wall (1987)] can be used in practice to estimate the degree of $\mathcal{A}$-determinacy of map-germs.

Corollary 3.1 ([Bruce, du Plessis and Wall (1987)]). *If f satisfies*

$$\mathcal{M}_n^l.\mathcal{E}(n,m) \subset L\mathcal{K} \cdot f$$
$$\mathcal{M}_n^{r+1}.\mathcal{E}(n,m) \subset L\mathcal{A}_1 \cdot f + \mathcal{M}_n^{l+r+1}.\mathcal{E}(n,m)$$

then f is r-$\mathcal{A}_1$-determined.

Proof. See [Bruce, du Plessis and Wall (1987)], Corollary 2.5.2. □

3.8 Versal unfoldings

One of the most important concepts in the singularities of map-germs is that of versal unfoldings and versal deformations.

Definition 3.1. Let $f \in \mathcal{M}_n.\mathcal{E}(n,m)$. An a-parameter unfolding (a, F) of f is a map-germ

$$F : (\mathbb{R}^n \times \mathbb{R}^a, (0,0)) \to (\mathbb{R}^m \times \mathbb{R}^a, (0,0))$$

in the form $F(x,u) = (\bar{f}(x,u), u)$, with $\bar{f}(x,0) = f(x)$. The family

$$\bar{f} : (\mathbb{R}^n \times \mathbb{R}^a, (0,0)) \to (\mathbb{R}^m, 0)$$

is called an a-parameter deformation of f.

It is important to clarify the following about the deformation $\tilde{f}$ in Definition 3.1. Let $\bar{f} : U \times W \to V$ be a representative of the map-germ $\bar{f}$, where $U \times W$ is a neighbourhood of the origin $(0,0) \in \mathbb{R}^n \times \mathbb{R}^a$ and V is a neighbourhood of the origin in $\mathbb{R}^m$. Denote by $\bar{f}_u : U \to V$ the smooth map given $\bar{f}_u(x) = \bar{f}(x,u)$. Then $\bar{f}_0(0) = 0$ but $\bar{f}_u(0)$ is not necessarily the origin in $\mathbb{R}^m$ for $u \neq 0$. This means that the fibre $0 \times \mathbb{R}^a$ is not necessarily preserved by F. Also, the singularities of $\bar{f}_u$ may no longer be at the origin. This is why one needs to consider the extended groups $\mathcal{G}_e$.

Definition 3.2. Let $\mathcal{G}$ be a Mather group and I the identity in $\mathcal{G}$.

(i) A morphism between two unfoldings (a, F) and (b, G) is a pair $(\alpha, \psi) : (a, F) \to (b, G)$ with $\alpha : (\mathbb{R}^a, 0) \to (\mathcal{G}, I)$, $\psi : (\mathbb{R}^a, 0) \to (\mathbb{R}^b, 0)$, such that

$$\bar{f}_u = \alpha(u) \cdot \bar{g}_{\psi(u)}.$$

The unfolding (a, F) is then said to be induced from (b, G) by (α, ψ).

(ii) Two unfoldings (a, F) and (b, G) are $\mathcal{G}$-equivalent if there exists a morphism $(\alpha, \psi) : (a, F) \to (b, G)$ where ψ is invertible.

(iii) An unfolding (a, F) of a map-germ f is said to be $\mathcal{G}$-versal if any unfolding (b, G) of f can be induced from (a, F).

An analogous definition can be made for the extended group $\mathcal{G}_e$ by substituting $\mathcal{G}$ with $\mathcal{G}_e$ in Definition 3.2.

We need the following notion, which can also be defined for maps between manifolds.

Definition 3.3. A suspension of a map-germ $f : (\mathbb{R}^n, 0) \to (\mathbb{R}^m, 0)$ by a germ of a manifold $(\mathbb{R}^p, 0)$ is the map-germ $(f, \mathrm{id}) : (\mathbb{R}^n \times \mathbb{R}^p, (0,0)) \to (\mathbb{R}^m \times \mathbb{R}^p, (0,0))$ given by $(f, \mathrm{id})(x, u) = (f(x), u)$, where id is the identity map-germ.

Theorem 3.5. (i) *For each map-germ f and a Mather group $\mathcal{G}$, the following are equivalent:*

(a) *f is $\mathcal{G}$-finite;*

(b) *f has a $\mathcal{G}$-versal unfolding ;*

(c) *f has a $\mathcal{G}_e$-versal unfolding.*

(ii) *The least number a_0 of parameters for a $\mathcal{G}$-versal (resp. $\mathcal{G}_e$-versal) unfolding is* $\mathrm{cod}(f, \mathcal{G})$ *(resp.* $\mathrm{cod}(f, \mathcal{G}_e)$*). A versal unfolding (a_0, F) is called miniversal.*

(iii) *Miniversal unfoldings are unique up to equivalence. Any versal unfolding is equivalent to a suspension of a miniversal unfolding. Versal unfoldings (a, F) and (a, G) of f are equivalent.*

Proof. See [Wall (1981)], Theorem 3.4. □

The importance of a $\mathcal{G}$-miniversal unfolding (resp. $\mathcal{G}_e$miniversal unfolding) of a map-germ f (Theorem 3.5(iii)) is that it provides a $\mathcal{G}$ (resp. $\mathcal{G}_e$) model of all possible local deformations of the map-germ f.

Given an unfolding (a, F) of a map-germ f in $\mathcal{M}_n.\mathcal{E}(n, m)$, denote by $\dot{F}_i$, $i = 1, \dots, a$, the map-germs in $\mathcal{M}_n.\mathcal{E}(n, m)$ given by

$$\dot{F}_i(x) = \frac{\partial \bar{f}}{\partial u_i}(x, 0).$$

Theorem 3.6. *An unfolding (a, F) of a map-germ f in $\mathcal{M}_n.\mathcal{E}(n, m)$ is $\mathcal{G}$-versal if and only if*

$$L\mathcal{G} \cdot f + \mathbb{R}.\{\dot{F}_1, \dots, \dot{F}_a\} = \mathcal{M}_n.\mathcal{E}(n, m)$$

and $\mathcal{G}_e$-versal if and only if

$$L\mathcal{G}_e \cdot f + \mathbb{R}.\{\dot{F}_1, \dots, \dot{F}_a\} = \mathcal{E}(n, m).$$

Proof. See [Wall (1981)], Theorem 3.3. □

Theorem 3.5 and Theorem 3.6 offer a way of finding miniversal unfoldings. Let f be a $\mathcal{G}$-finite map-germ and let $h_1, \dots, h_c$ in $\mathcal{M}_n.\mathcal{E}(n, m)$ be an $\mathbb{R}$-basis of $\mathcal{M}_n.\mathcal{E}(n, m)/L\mathcal{G} \cdot f$ (so $c = \text{cod}(f, \mathcal{G})$). Define the unfolding

$$F(x, u) = (f(x) + \sum_{i=1}^{c} u_i h_i(x), u).$$

Similarly, let $g_1, \dots, g_d$ in $\mathcal{E}(n, m)$ be an $\mathbb{R}$-basis of $\mathcal{E}(n, m)/L\mathcal{G}_e \cdot f$ (so $d = \text{cod}(f, \mathcal{G}_e)$) and define the unfolding

$$G(x, u) = (f(x) + \sum_{i=1}^{d} u_i g_i(x), u).$$

Theorem 3.7. *The unfolding $F(x, u) = (f(x) + \sum_{i=1}^{c} u_i h_i(x), u)$ of the map-germ f is $\mathcal{G}$-miniversal and the unfolding $G(x, u) = (f(x) + \sum_{i=1}^{d} u_i g_i(x), u)$ is $\mathcal{G}_e$-miniversal.*

Proof. The proof follows by applying Theorem 3.6. □

An unfolding (a, F) of f is said to be $\mathcal{G}$-*trivial* (resp. $\mathcal{G}_e$-*trivial*) if it is $\mathcal{G}$-equivalent (resp. $\mathcal{G}_e$-equivalent) to the constant unfolding (a, f).

A map-germ f is $\mathcal{G}$-*stable* (resp. $\mathcal{G}_e$-*stable*) if all its unfoldings are trivial.

Theorem 3.8. *A map-germ f is $\mathcal{G}$-stable (resp. $\mathcal{G}_e$-stable) if and only if* $\text{cod}(f, \mathcal{G}) = 0$ *(resp.* $\text{cod}(f, \mathcal{G}_e) = 0$*).*

Proof. See [Mather (1970)]. □

Mather proved the following results which characterises the $\mathcal{A}$-stable map-germs.

Theorem 3.9. *Let $f : (\mathbb{R}^n, 0) \to (\mathbb{R}^m, 0)$ be a map-germ and F be an m-parameter unfolding of f given by $F(x,u) = (f(x) - u, u)$. Then,*

(i) *The map-germ f is $\mathcal{A}$-stable if and only if the unfolding (m, F) is $\mathcal{K}$-versal.*

(ii) *If f is $\mathcal{A}$-stable, then f is $(m+1)$-$\mathcal{A}$-determined.*

(iii) *Suppose that f and g are $\mathcal{A}$-stable map-germs. Then f and g are $\mathcal{A}$-equivalent if and only if they are $\mathcal{K}$-equivalent.*

Proof. See [Mather (1969b,c)]. □

It follows from the statements (ii) and (iii) of Theorem 3.9, that the $\mathbb{R}$-algebra isomorphism class of the local ring

$$Q_{m+1}(f) = \mathcal{E}_n/(f^*(\mathcal{M}_m)\mathcal{E}_n + \mathcal{M}_n^{m+2})$$

is a complete invariant for the $\mathcal{A}$-stable map-germs (i.e., it determines the $\mathcal{A}$-stable orbit).

3.9 Classification of singularities

Finding model objects is one of the major activities in mathematics. Given an equivalence relation in a set, one seeks representatives of the equivalent classes under this relation. The representatives are chosen in a simple form. In topology for example, the 2-sphere S^2 is a model for closed and simply connected surfaces under homeomorphisms. In matrix algebra, we know that the diagonal matrices are models for real symmetric matrices X under the equivalence relation AXA^T, where A is an invertible matrix. Here we seek models of map-germs under $\mathcal{G}$-equivalence, where $\mathcal{G}$ is one of Mather's group. This activity is called *classification* and consists of listing representatives of the orbits of the action of the group $\mathcal{G}$ on $\mathcal{M}_n.\mathcal{E}(n,m)$. The representatives are sometimes referred to as *normal forms.*

It is natural to start by finding models for non-singular map-germs. The following result is a consequence of the implicit functions theorem.

Theorem 3.10. *Suppose that $f : (\mathbb{R}^n, 0) \to (\mathbb{R}^m, 0)$ is non-singular.*

(1) *If $n \geq m$, then f is $\mathcal{R}$-equivalent to the germ of the projection $\pi(x_1, \ldots, x_n) = (x_1, \ldots, x_m)$.*

(2) *If $n < m$, then f is $\mathcal{A}$-equivalent to the germ of the immersion* $i(x_1, \dots, x_n) = (x_1, \dots, x_n, 0, \dots, 0)$.

The listing of representatives of all the $\mathcal{G}$-orbits is an impossible task, so some restrictions are needed. These are imposed, in general, by the problems under investigation. For problems arising in differential geometry, we seek $\mathcal{G}$-finite germs of lower $\mathcal{G}$ or $\mathcal{G}_e$-codimension. In other context, $\mathcal{G}$-finite *simple germs* are sought. The notion of simple germs is defined in [Arnol'd, Guseĭn-Zade and Varchenko (1985)] as follows. Let X be a manifold and G a Lie group acting on X. The modality of a point $x \in X$ under the action of G on X is the least number m such that a sufficiently small neighbourhood of x may be covered by a finite number of m-parameter families of orbits. The point x is said to be *simple* if its modality is 0, that is, a sufficiently small neighbourhood intersects only a finite number of orbits. The modality of a finitely determined map-germ is the modality of a sufficient jet in the jet-space under the action of the jet-group.

Each pair of dimensions (n, m) needs to be considered separately when carrying out the classification task. There are lists for most cases with $n + m \leq 6$. N.P. Kirk developed a Maple 5 computer programme ([Kirk (2000)]) called "Transversal" to carry out classifications of $\mathcal{G}$-singularities of map-germs. (The computer programme also deals with Damon's subgroups of $\mathcal{G}$, see section 3.10.) Tables that took months to produce by hand could be obtained in few days! The programme in [Kirk (2000)] is based on the complete transversal results established in ([Bruce, Kirk and du Plessis (1997)]). The classification is carried out inductively on the jet level and returns sufficient jets and other information such as their $\mathcal{G}$ or $\mathcal{G}_e$-codimensions and $\mathcal{G}$-versal unfoldings.

3.9.1 *Germs of functions*

We deal here in some details with the case of germs of functions (i.e., $m = 1$). The Hessian matrix of a germ of a function $f : (\mathbb{R}^n, 0) \to (\mathbb{R}, 0)$ is given by

$$\mathcal{H}(f)(0) = \left(\frac{\partial^2 f}{\partial x_i \partial x_j}(0) \right).$$

If f is singular at the origin, we say that this singularity is *non-degenerate* if $\operatorname{rank} \mathcal{H}(f)(0) = n$, equivalently, if and only if $\det \mathcal{H}(f)(0) \neq 0$.

Theorem 3.11 (The Morse Lemma). *Suppose that $f : (\mathbb{R}^n, 0) \to$*

$(\mathbb{R}, 0)$ *has a non-degenerate singularity. Then* f *is* $\mathcal{R}$*-equivalent to the non-degenerate quadratic form*

$$Q(x_1, \dots, x_n) = \pm x_1^2 \pm \cdots \pm x_n^2.$$

Proof. See for example [Martinet (1982)], pp 18-24. □

We consider next the case where the singularity of f is degenerate and define the corank of f by

$$\text{corank}(f)(0) = n - \text{rank}\mathcal{H}(f)(0),$$

In particular, the singularity of f is non-degenerate if and only if $\text{corank}(f)(0) = 0$.

Lemma 3.1 (Thom's splitting lemma). *Suppose that* $f : (\mathbb{R}^n, 0) \to (\mathbb{R}, 0)$ *has a singularity of* $\text{corank}(f)(0) = r$ *at the origin. Then* f *is* $\mathcal{R}$*-equivalent to a germ of the form*

$$g(x_1, \dots, x_r) + Q(x_{r+1}, \dots, x_n),$$

where $g \in \mathcal{M}_r^3$ *and* $Q(x_{r+1}, \dots, x_n) = \pm x_{r+1}^2 \pm \cdots \pm x_n^2$.

Proof. See for example [Bröcker (1975)], p 125. □

Given two germs g_1 and g_2 in $\mathcal{M}_r^3$ and Q as in Lemma 3.1, then $g_1 + Q$ and $g_2 + Q$ are $\mathcal{R}$-equivalent if and only if g_1 and g_2 are $\mathcal{R}$-equivalent. The Splitting Lemma reduces thus the dimension of the source where the classification is to be carried out.

An extensive list of $\mathcal{R}$-finite germs of functions is given in V.I. Arnold [Arnol'd, Guseĭn-Zade and Varchenko (1985)]. Table 3.1 shows the simple $\mathcal{R}$-finite germs of functions $(\mathbb{R}^n, 0) \to (\mathbb{R}, 0)$. We observe that $\text{cod}(g + Q, \mathcal{R}) = \text{cod}(g, \mathcal{R})$ for any $\mathcal{R}$-finite germ $g \in \mathcal{M}_r^3$.

The normal forms in Table 3.1 are also the normal forms for the simple $\mathcal{K}$-finite germs of functions.

We consider now deformations of germs of functions. We call the deformation $F : (\mathbb{R}^n \times \mathbb{R}^a, (0, 0)) \to (\mathbb{R}, 0)$ of the germ $f = F|_{\mathbb{R}^n \times \{0\}}$ a *family of germs of functions*. In some textbooks, F is referred to as an unfolding of f.

For germs of functions, it is important, as we shall see in Chapter 5, to consider also the direct product of the group $\mathcal{R}$ with translations, which we denote by $\mathcal{R}^+$. (This is denoted by $\mathcal{R}_e^{aug}$ in [Wall (1981)].)

Table 3.1 Simple germs of functions ([Arnol'd, Guseĭn-Zade and Varchenko (1985)]).

Name	Normal form	cod$(f,\mathcal{R})$
A_k, $k \geq 0$	$\pm x_1^{k+1} + Q(x_2,\ldots,x_n)$	k
D_k, $k \geq 4$	$x_1^2 x_2 \pm x_2^{k-1} + Q(x_3,\ldots,x_n)$	k
E_6	$x_1^3 + x_2^4 + Q(x_3,\ldots,x_n)$	6
E_7	$x_1^3 + x_1 x_2^4 + Q(x_3,\ldots,x_n)$	7
E_8	$x_1^3 + x_2^5 + Q(x_3,\ldots,x_n)$	8

$Q(x_r,\ldots,x_n) = \pm x_r^2 \pm \cdots \pm x_n^2$

Definition 3.4. Two families of germs of functions F and $G : (\mathbb{R}^n \times \mathbb{R}^a,(0,0)) \to (\mathbb{R},0)$ are P-$\mathcal{R}^+$-equivalent if there exist a germ of a diffeomorphism $\Phi : (\mathbb{R}^n \times \mathbb{R}^a,(0,0)) \to (\mathbb{R}^n \times \mathbb{R}^a,(0,0))$ of the form $\Phi(x,u) = (\alpha(x,u),\psi(u))$ and a germ of a function $c : (\mathbb{R}^a,0) \to \mathbb{R}$ such that

$$G(x,u) = F(\Phi(x,u)) + c(u).$$

In Definition 3.4, the letter "P" stands for parametrised (as we have a family of germs of diffeomorphisms $\alpha(-,u)$ of $\mathbb{R}^n$ parametrised by u) and "$+$" stands for the addition of the term $c(u)$.

Given a family of germs of functions F, we write

$$\dot{F}_i(x) = \frac{\partial F}{\partial u_i}(x,0).$$

Theorem 3.6 can be adapted as follows for families of functions.

Theorem 3.12. *A deformation $F : (\mathbb{R}^n \times \mathbb{R}^a,(0,0)) \to (\mathbb{R},0)$ of a germ of a function f in $\mathcal{M}_n$ is $\mathcal{R}^+$-versal if and only if*

$$L\mathcal{R}_e \cdot f + \mathbb{R}.\left\{1,\dot{F}_1,\ldots,\dot{F}_a\right\} = \mathcal{E}_n.$$

Definition 3.4 of P-$\mathcal{R}^+$-equivalence can be extended to families $F : (\mathbb{R}^n \times \mathbb{R}^a,(0,0)) \to (\mathbb{R},0)$ and $G : (\mathbb{R}^p \times \mathbb{R}^a,(0,0)) \to (\mathbb{R},0)$ with $n \neq p$. We add a non-degenerate quadratic form $Q(y_{n+1},\ldots,y_{n+p})$ to F and a non-degenerate quadratic form $Q'(z_{p+1},\ldots z_{n+p})$ to G and consider the two families of germs $F+Q$ and $G+Q'$ from $(\mathbb{R}^{n+p} \times \mathbb{R}^a,(0,0)) \to (\mathbb{R},0)$.

Definition 3.5. We say that the two families F and G are *stably P-$\mathcal{R}^+$-equivalent* if $F+Q$ and $G+Q'$ are P-$\mathcal{R}^+$-equivalent.

Theorem 3.13. *Let $f \in \mathcal{M}_n^2$ be an $\mathcal{R}$-finitely determined germ with $1 \leq \operatorname{cod}_e(f,\mathcal{R}) \leq 4$. Then any $\mathcal{R}^+$-miniversal unfolding of f is P-$\mathcal{R}^+$-equivalent to one of the following families of germs, where $Q(x_r,\ldots,x_n) = \pm x_r^2 \pm \cdots \pm x_n^2$:*

(1) A_2: $Q(x_2, \ldots, x_n) + x_1^3 + u_1 x_1$,
(2) A_3: $Q(x_2, \ldots, x_n) \pm x_1^4 + u_2 x_1^2 + u_1 x_1$,
(3) A_4: $Q(x_2, \ldots, x_n) + x_1^5 + u_3 x_1^3 + u_2 x_1^2 + u_1 x_1$,
(4) A_5: $Q(x_2, \ldots, x_n) \pm x_1^6 + u_4 x_1^4 + u_3 x_1^3 + u_2 x_1^2 + u_1 x_1$,
(5) D_4^-: $Q(x_3, \ldots, x_n) + x_1^3 - x_1 x_2^2 + u_3(x_1^2 + x_2^2) + u_2 x_2 + u_1 x_1$,
(6) D_4^+: $Q(x_3, \ldots, x_n) + x_1^3 + x_2^3 + u_3 x_1 x_2 + u_2 x_2 + u_1 x_1$,
(7) D_5: $Q(x_3, \ldots, x_n) + x_1^2 x_2 + x_2^4 + u_4 x_2^2 + u_3 x_1^2 + u_2 x_2 + u_1 x_1$.

Proof. All germs of functions f of $\text{cod}_e(f, \mathcal{R}) \leq 4$ are simple [Arnol'd, Guseĭn-Zade and Varchenko (1985)]. The result follows by applying Theorem 3.12 and Theorem 3.7 to the normal forms in Table 3.1. □

The singularities in Theorem 3.13 are referred to as the *Thom's seven elementary catastrophes.*

We turn now to the group $\mathcal{K}$.

Definition 3.6. Two families of germs functions F and $G : (\mathbb{R}^n \times \mathbb{R}^a, (0,0)) \to (\mathbb{R}, 0)$ are P-$\mathcal{K}$-equivalent if there exist a germ of a diffeomorphism $\Phi : (\mathbb{R}^n \times \mathbb{R}^a, (0,0)) \to (\mathbb{R}^n \times \mathbb{R}^a, (0,0))$ of the form $\Phi(x,u) = (\alpha(x,u), \psi(u))$ and a germ of a function $\lambda : (\mathbb{R}^n \times \mathbb{R}^a, (0,0)) \to \mathbb{R}$, with $\lambda(0,0) \neq 0$, such that

$$\lambda(x,u) \cdot G(x,u) = F(\Phi(x,u)).$$

Theorem 3.6 can be written as follows for families of functions.

Theorem 3.14. *A deformation $F : (\mathbb{R}^n \times \mathbb{R}^a, (0,0)) \to (\mathbb{R}, 0)$ of a germ of a function f in $\mathcal{M}_n$ is $\mathcal{K}$-versal if and only if*

$$\mathcal{E}_n.\left\{\frac{\partial f}{\partial x_1}, \ldots, \frac{\partial f}{\partial x_n}, f\right\} + \mathbb{R}.\left\{\dot{F}_1, \ldots, \dot{F}_a\right\} = \mathcal{E}_n.$$

Here too Definition 3.6 can also be extended to families $F : (\mathbb{R}^n \times \mathbb{R}^a, 0) \to (\mathbb{R}, 0)$ and $G : (\mathbb{R}^p \times \mathbb{R}^a, (0,0)) \to (\mathbb{R}, 0)$ with $n \neq p$. We add a non-degenerate quadratic form $Q(y_{n+1}, \ldots, y_{n+p})$ to F and a non-degenerate quadratic form $Q'(z_{p+1}, \ldots z_{n+p})$ to G and consider the two families of germs $F + Q$ and $G + Q'$ from $(\mathbb{R}^{n+p} \times \mathbb{R}^a, (0,0)) \to (\mathbb{R}, 0)$.

Definition 3.7. We say that the two families F and G are *stably P-$\mathcal{K}$-equivalent* if $F + Q$ and $G + Q'$ are P-$\mathcal{K}$-equivalent.

Theorem 3.15. *Let $f \in \mathcal{M}_n^2$ be a $\mathcal{K}$-finitely determined germ with $1 \leq \text{cod}_e(f, \mathcal{K}) \leq 4$. Then any $\mathcal{K}$-miniversal unfolding of f is P-$\mathcal{K}$-equivalent* *to one of the following families of germs, where* $Q(x_r,\dots,x_n) = \pm x_r^2 \pm \cdots \pm x_n^2$:

(1) A_1: $Q(x_1,\dots,x_n) + u_1$,

(2) A_2: $Q(x_2,\dots,x_n) + x_1^3 + u_2x_1 + u_1$,

(3) A_3: $Q(x_2,\dots,x_n) + x_1^4 + u_3x_1^2 + u_2x_1 + u_1$,

(4) A_4: $Q(x_2,\dots,x_n) + x_1^5 + u_4x_1^3 + u_3x_1^2 + u_2x_1 + u_1$,

(5) D_4^- : $Q(x_3,\dots,x_n) + x_1^3 - x_1x_2^2 + u_4(x_1^2 + x_2^2) + u_3x_2 + u_2x_1 + u_1$,

(6) D_4^+: $Q(x_3,\dots,x_n) + x_1^3 + x_2^3 + u_4x_1x_2 + u_3x_2 + u_2x_1 + u_1$.

Proof. The result follows by applying Theorem 3.14 and Theorem 3.7 to the normal forms in Table 3.1. □

3.9.2 *Discriminants and bifurcation sets*

We associate to a family of germs of functions F some germs of sets. For the families of height functions and distance squared functions on a hypersurface, these sets captures geometric information about the hypersurface.

The *catastrophe set* C_F of a family $F : (\mathbb{R}^n \times \mathbb{R}^a,(0,0)) \to (\mathbb{R},0)$ is defined by

$$C_F = \left\{(x,u) \in (\mathbb{R}^n \times \mathbb{R}^a,(0,0)) \mid \frac{\partial F}{\partial x_1}(x,u) = \cdots \frac{\partial F}{\partial x_n}(x,u) = 0\right\}.$$

The *bifurcation set* of F is defined by

$$B_F = \left\{u \in (\mathbb{R}^a,0) \middle| \ \exists (x,u) \in C_F \ \text{ and } \ \operatorname{rank}\left(\frac{\partial^2 F}{\partial x_i \partial x_j}(x,u)\right) < n\right\}.$$

The *discriminant* of F is defined as

$$D_F = \left\{u \in (\mathbb{R}^a,0) \middle| \ \exists x \in (\mathbb{R}^n,0) \text{ and } F = \frac{\partial F}{\partial x_1} = \cdots \frac{\partial F}{\partial x_n} = 0 \text{ at } (x,u)\right\}.$$

Let $\pi_{C_F} = \pi_2|_{C_F} : C_F \to (\mathbb{R}^a,0)$, where $\pi_2 : (\mathbb{R}^n \times \mathbb{R}^a,(0,0)) \to (\mathbb{R}^a,0)$ is the projection to the second component. We call π_{C_F} the *catastrophe map-germ* of F.

Proposition 3.1. *Let F and G be two families of germs of functions $(\mathbb{R}^n \times \mathbb{R}^a,0) \to (\mathbb{R},0)$ such that their catastrophe sets C_F and C_G are smooth submanifolds. Suppose that F and G are P-$\mathcal{R}^+$-equivalent. Then the catastrophe map-germs π_{C_F} and π_{C_G} are $\mathcal{A}$-equivalent. Moreover, there exists a germ of a diffeomorphism $\phi : (\mathbb{R}^n,0) \to (\mathbb{R}^n,0)$ such that $\phi(B_F) = B_G$.*

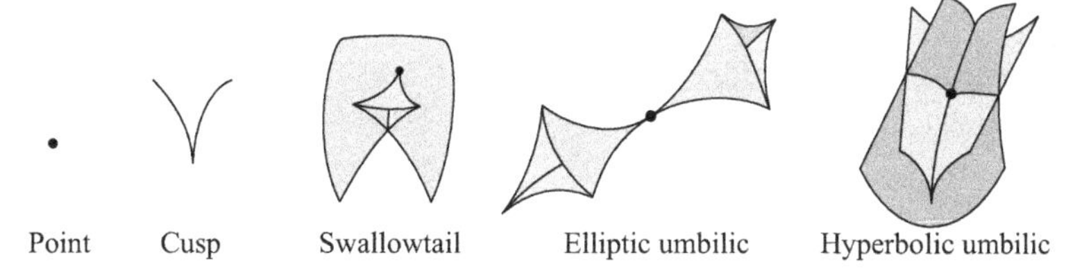

Fig. 3.4 Bifurcation sets of miniversal unfoldings of germs of $\mathcal{R}_e$-codimension ≤ 3.

Proof. By Definition 3.4, there exist a germ of a diffeomorphism $\Phi : (\mathbb{R}^n \times \mathbb{R}^a, 0) \to (\mathbb{R}^n \times \mathbb{R}^a, 0)$ of the form $\Phi(x,u) = (\alpha(x,u), \psi(u))$ and a germ of a function $c : (\mathbb{R}^a, 0) \to (\mathbb{R}, 0)$ such that $G(x,u) = F(\alpha(x,u), \psi(u)) + c(u)$. We have

$$\frac{\partial G}{\partial x_i}(x,u) = \sum_{\ell=1}^{n} \frac{\partial F}{\partial x_\ell}(\Phi(x,u)) \frac{\partial \alpha_\ell}{\partial x_i}(x,u).$$

The matrix $(\partial \alpha_\ell / \partial x_i)(0,0)$ is regular, so $\frac{\partial G}{\partial x_i}(x,u) = 0$, for $i = 1, \dots, n$, if and only if $\frac{\partial F}{\partial x_i}(\Phi(x,u)) = 0$, for $i = 1, \dots, n$. Therefore, $\Phi(C_G) = C_F$ and $\psi \circ \pi_{C_G} = \pi_{C_F} \circ (\Phi|_{C_G})$. This means that π_{C_G} and π_{C_F} are $\mathcal{A}$-equivalent.

At $(x,u) \in C_G$, we have

$$\frac{\partial^2 G}{\partial x_i \partial x_j}(x,u) = \sum_{\ell=1}^{n} \left(\sum_{h=1}^{n} \frac{\partial^2 F}{\partial x_h \partial x_\ell}(\Phi(x,u)) \frac{\partial \alpha_h}{\partial x_j}(x,u) \right) \frac{\partial \alpha_\ell}{\partial x_i}(x,u).$$

Again, since the matrix $(\partial \alpha_\ell / \partial x_i)(0,0)$ is regular,

$$\text{rank} \left(\frac{\partial^2 G}{\partial x_i \partial x_j}(x,u) \right) = \text{rank} \left(\frac{\partial^2 F}{\partial x_i \partial x_j}(\Phi(x,u)) \right).$$

Therefore, $u \in B_G$ if and only if $\psi(u) \in B_F$, that is, $B_F = \psi(B_G)$. □

The catastrophe sets and bifurcation sets of the Thom's seven elementary catastrophes in Theorem 3.13 are computed in [Bröcker (1975)]. We re-do the calculations here. We have $\frac{\partial F}{\partial x_i} = \pm 2x_i$, $i = r, \dots, n$ for all the models in Theorem 3.13, so we set the redundant variable $x_i = 0$, $i = r, \dots, n$. We draw in Figure 3.4 the bifurcation sets of the miniversal unfoldings in Theorem 3.13 of germs of $\mathcal{R}_e$-codimension ≤ 3.

(1) *The A_2-singularity*
For $F(x,u) = x_1^3 + u_1 x_1$, $\frac{\partial F}{\partial x_1} = 3x_1^2 + u_1$ and $\frac{\partial^2 F}{\partial x_1^2} = 6x_1$, so

$$C_F = \{(x_1, u_1) \,|\, u_1 = -3x_1^2\},$$
$$B_F = \{0\}.$$

The bifurcation set is a point in this case (Figure 3.4).

(2) *The A_3-singularity*

For $F(x,u) = \pm x_1^4 + u_2x_1^2 + u_1x_1$, $\frac{\partial F}{\partial x_1} = \pm 4x_1^3 + 2u_2x_1 + u_1$ and $\frac{\partial^2 F}{\partial x_1^2} = \pm 12x_1^2 + 2u_2$, so

$$\begin{aligned} C_F &= \{(x_1,u_1,u_2) \mid u_1 = \mp 4x_1^3 - 2u_2x_1\}, \\ B_F &= \{(u_1,u_2) \mid 27u_1^2 + 8u_2^3 = 0\}. \end{aligned}$$

It follows that the bifurcation set is a cusp (Figure 3.4).

(3) *The A_4-singularity*

For $F(x,u) = x_1^5 + u_3x_1^3 + u_2x_1^2 + u_1x_1$, $\frac{\partial F}{\partial x_1} = 5x_1^4 + 3u_3x_1^2 + 2u_2x_1 + u_1$ and $\frac{\partial^2 F}{\partial x_1^2} = 20x_1^3 + 6u_3x_1 + 2u_2$, so

$$C_F = \{(x_1,u_1,u_2,u_3) \mid u_1 = -5x_1^4 - 3u_3x_1^2 - 2u_2x_1\},$$

and B_F is the germ of the surface parametrised by

$$(x_1,u_3) \mapsto (u_1,u_2,u_3) = (15x_1^4 + 3u_3x_1^2, -10x_1^3 - 3u_3x_1, u_3),$$

and called a swallowtail surface (Figure 3.4).

(4) *The A_5-singularity*

For $F(x,u) = \pm x_1^6 + u_4x_1^4 + u_3x_1^3 + u_2x_1^2 + u_1x_1$, we have

$$\begin{aligned} \frac{\partial F}{\partial x_1} &= \pm 6x_1^5 + 4u_4x_1^3 + 3u_3x_1^2 + 2u_2x_1 + u_1, \\ \frac{\partial^2 F}{\partial x_1^2} &= \pm 30x_1^4 + 12u_4x_1^2 + 6u_3x_1 + 2u_2, \end{aligned}$$

so

$$C_F = \{(x_1,u_1,u_2,u_3) \mid u_1 = \mp 6x_1^5 - 4u_4x_1^3 - 3u_3x_1^2 - 2u_2x_1\},$$

and B_F is the germ of the 3-dimensional variety parametrised by (x_1,u_3,u_4) and given by

$$(u_1,u_2,u_3,u_4) = (\pm 24x_1^4 + 8u_4x_1^3 + 3u_3x_1^2, \mp 15x_1^3 - 6u_4x_1^2 - 3u_3x_1, u_3, u_4).$$

(5) *The D_4^--singularity*

To simplify calculations, we take a slightly different normal form here and consider $F(x,u) = \frac{1}{3}x_1^3 - x_1x_2^2 + u_3(x_1^2 + x_2^2) + u_2x_2 + u_1x_1$. Then we have

$$\begin{aligned} \frac{\partial F}{\partial x_1} &= x_1^2 - x_2^2 + 2u_3x_1 + u_1, \\ \frac{\partial F}{\partial x_2} &= -2x_1x_2 + 2u_3x_2 + u_2, \end{aligned}$$

and

$$\begin{pmatrix} \frac{\partial^2 F}{\partial x_1^2} & \frac{\partial^2 F}{\partial x_1 \partial x_2} \\ \frac{\partial^2 F}{\partial x_2 \partial x_1} & \frac{\partial^2 F}{\partial x_2^2} \end{pmatrix} = \begin{pmatrix} 2x_1 + 2u_3 & -2x_2 \\ -2x_2 & -2x_1 + 2u_3 \end{pmatrix}.$$

It follows that

$$C_F = \{(x_1, x_2, u_1, u_2, u_3) \mid u_1 = -x_1^2 + x_2^2 - 2u_3x_1, \\ u_2 = 2x_1x_2 - 2u_3x_2\}$$

and

$$\begin{aligned} B_F = \{(u_1, u_2, u_3) \mid u_1 &= -x_1^2 + x_2^2 - 2u_3x_1, \\ u_2 &= 2x_1x_2 - 2u_3x_2, \\ u_3^2 &= x_1^2 + x_2^2, \text{ with } (x_1, x_2) \in (\mathbb{R}^2, 0) \ \}. \end{aligned}$$

The singularity of $F(x, 0)$ is called *elliptic umbilic* and the bifurcation set of F is as in Figure 3.4 (called pyramid).

(6) *The D_4^+-singularity*

For $F(x, u) = x_1^3 + x_3^2 + u_3x_1x_2 + u_2x_2 + u_1x_1$, we have

$$\begin{aligned} \frac{\partial F}{\partial x_1} &= 3x_1^2 + u_3x_2 + u_1, \\ \frac{\partial F}{\partial x_1} &= 3x_2^2 + u_3x_1 + u_2, \end{aligned}$$

and

$$\begin{pmatrix} \frac{\partial^2 F}{\partial x_1^2} & \frac{\partial^2 F}{\partial x_1 \partial x_2} \\ \frac{\partial^2 F}{\partial x_2 \partial x_2} & \frac{\partial^2 F}{\partial x_2^2} \end{pmatrix} = \begin{pmatrix} 6x_1 & u_3 \\ u_3 & 6x_2 \end{pmatrix}.$$

Then,

$$C_F = \{(x_1, x_2, u_1, u_2, u_3) \mid u_1 = -3x_1^2 - u_3x_2, \\ u_2 = -3x_2^2 - u_3x_1\}$$

and

$$\begin{aligned} B_F = \{(u_1, u_2, u_3) \mid u_1 &= -3x_1^2 - u_3x_2, \\ u_2 &= -3x_2^2 - u_3x_1 \\ u_3^2 &= 36x_1x_2, \text{ with } (x_1, x_2) \in (\mathbb{R}^2, 0) \ \}. \end{aligned}$$

The singularity of $F(x, 0)$ is called *hyperbolic umbilic* and the bifurcation set of F is as in Figure 3.4 (called purse).

(7) *The D_5-singularity*

Here too we take a different normal form $F(x, u) = x_1^2x_2 + \frac{1}{6}x_2^4 + u_4x_2^2 + u_3x_1^2 + u_2x_2 + u_1x_1$. We have

$$\begin{aligned} \frac{\partial F}{\partial x_1} &= 2x_1x_2 + 2u_3x_1 + u_1, \\ \frac{\partial F}{\partial x_2} &= x_1^2 + \tfrac{2}{3}x_2^3 + 2u_4x_2 + u_2, \end{aligned}$$

and

$$\begin{pmatrix} \frac{\partial^2 F}{\partial x_1^2} & \frac{\partial^2 F}{\partial x_1 \partial x_2} \\ \frac{\partial^2 F}{\partial x_2 \partial x_1} & \frac{\partial^2 F}{\partial x_2^2} \end{pmatrix} = \begin{pmatrix} 2x_2 + 2u_3 & 2x_1 \\ 2x_1 & 2x_2^2 + 2u_4 \end{pmatrix}.$$

We deduce that

$$C_F = \{(x_1, x_2, u_1, u_2, u_3, u_4) \,|\, u_1 = -2x_1x_2 - 2u_3x_1, \\ u_2 = -x_1^2 - \tfrac{2}{3}x_2^3 - 2u_4x_2\}$$

and

$$B_F = \{(u_1, u_2, u_3, u_4) \,|\, u_1 = -2x_1x_2 - 2u_3x_1, \\ u_2 = -x_1^2 - \tfrac{2}{3}x_2^3 - 2u_4x_2, \\ (x_2 + u_3)(x_2^2 + u_4) - x_1^2 = 0, \text{ with } (x_1, x_2) \in (\mathbb{R}^2, 0)\ \}.$$

Theorem 3.16. *Let F and G be two P-$\mathcal{K}$-equivalent families of germs of functions. Then the discriminants D_F and D_G are diffeomorphic.*

Proof. By Definition 3.6, there exist a germ of a diffeomorphism $\Phi : (\mathbb{R}^n \times \mathbb{R}^a, 0) \to (\mathbb{R}^n \times \mathbb{R}^a, 0)$ of the form $\Phi(x, u) = (\alpha(x, u), \psi(u))$ and a germ of a function $\lambda : (\mathbb{R}^n \times \mathbb{R}^a, (0, 0)) \to (\mathbb{R}, 0)$, with $\lambda(0, 0) \neq 0$, such that $\lambda(x, u) \cdot G(x, u) = F(\Phi(x, u))$. Then, $G(x, u) = 0$ if and only if $F(\Phi(x, u)) = 0$.

We have

$$\frac{\partial \lambda}{\partial x_i}(x, u) \cdot G(x, u) + \lambda(x, u) \cdot \frac{\partial G}{\partial x_i}(x, u) = \sum_{j=1}^{n} \frac{\partial F}{\partial x_j}(\Phi(x, u)) \frac{\partial \alpha_j}{\partial x_i}(x, u).$$

The matrix $(\partial \alpha_\ell / \partial x_i)(0, 0)$ is regular, so $G(x, u) = 0$ and $\partial G / \partial x_i(x, u) = 0$ for $i = 1, \dots, n$, if and only if $F(\Phi(x, u)) = 0$ and $\partial F / \partial x_i(\Phi(x, u)) = 0$ for $i = 1, \dots, n$. This means that $D_F = \psi(D_G)$. □

We make use of the calculations for the bifurcation sets of the families in Theorem 3.13 to compute the discriminants of the families in Theorem 3.15.

(1) *A_1-singularity*
For $F(x, u) = u_1$, $D_F = \{0\}$.

(2) *A_2-singularity*
For $F(x, u) = x_1^3 + u_2x_1 + u_1$,

$$D_F = \{(u_1, u_2) \,|\, 27u_1^2 + 8u_2^3 = 0\},$$

which is a cusp.

(3) *A_3-singularity*
For $F(x, u) = x_1^4 + u_3x_1^2 + u_2x_1 + u_1$,

$$D_F = \{(u_1, u_2, u_3) \,|\, u_1 = 3x_1^4 + u_3x_1, \\ u_2 = -4x_1^3 - 2u_3x_1^2, \text{ with } (x_1, u_3) \in (\mathbb{R}^2, 0)\ \},$$

which is a swallowtail surface.

(4) A_4-*singularity*
For $F(x,u) = x_1^5 + u_4x_1^3 + u_3x_1^2 + u_2x_1 + u_1$,

$$D_F = \{(u_1,u_2,u_3,u_4) \mid u_1 = 4x_1^5 + 2u_4x_1^3 + u_3x_1^2,\ u_2 = -5x_1^4 - 3u_4x_1^2 - 2u_3x_1,\ \text{with } (x_1,u_3,u_4) \in (\mathbb{R}^3,0)\ \}.$$

(5) D_4^--*singularity*
For $F(x,u) = \frac{1}{3}x_1^3 - x_1x_2^2 + u_4(x_1^2 + x_2^2) + u_3x_2 + u_2x_1 + u_1$,

$$D_F = \{(u_1,u_2,u_3,u_4) \mid u_1 = \tfrac{2}{3}x_1^3 - 2x_1x_2^2 + u_4(x_1^2 + x_2^2),\ u_2 = -x_1^2 + x_2^2 - 2u_4x_1,\ u_3 = 2x_1x_2 - 2u_4x_2,\ \text{with } (x_1,x_2,u_4) \in (\mathbb{R}^3,0)\ \}.$$

(6) D_4^+-*singularity* For $F(x,u) = x_1^3 + x_3^2 + u_4x_1x_2 + u_3x_2 + u_2x_1 + u_1$,

$$D_F = \{(u_1,u_2,u_3,u_4) \mid u_1 = 2x_1^3 + 2x_3^2 + u_4x_1x_2,\ u_2 = -3x_1^2 - u_4x_2,\ u_3 = -3x_2^2 - u_4x_1,\ \text{with } (x_1,x_2,u_4) \in (\mathbb{R}^3,0)\ \}.$$

Remark 3.2. It is not difficult to check that the Boardman symbol of the catastrophe map-germ associated to an A_k-singularity is $S_{1_{k-1}}$, where the subindex $k-1$ refers to the number of entries of the subindex 1 in the symbol, so $S_{1_1} = S_{1,0}$, $S_{1_2} = S_{1,1,0}$, $S_{1_3} = S_{1,1,1,0}$ and so on. The Boardman symbol of the catastrophe map-germ associated to a D_k-singularity is S_2.

3.10 Damon's geometric subgroups

The results on finite determinacy, complete transversal and versal unfoldings are stated here for the Mather groups $\mathcal{R}$, $\mathcal{L}$, $\mathcal{C}$, $\mathcal{K}$ and $\mathcal{A}$. However, in various situations one has to deal with subgroups of one of these groups. For example, consider germs of a function on a variety $(X,0)$ in $\mathbb{R}^n$. Any relevant diffeomorphism in $\mathcal{R}$ should preserve the germ of the variety $(X,0)$. J. Damon showed that the results we presented in this chapter are valid for a large class of subgroups of $\mathcal{K}$ and $\mathcal{A}$, which he called geometric subgroups of $\mathcal{K}$ and $\mathcal{A}$ ([Damon (1984)]). For instance, the subgroup of $\mathcal{R}$ of diffeomorphisms preserving a variety $(X,0)$ in $\mathbb{R}^n$ is a geometric subgroup. Another example of a geometric subgroup is the group of equivariant diffeomorphisms with respect to a compact Lie group action.

3.11 Notes

Our aim in this chapter is to set the singularity theory notation and state the results we need in this book. It is far short from being a survey of results in the area of singularities of map-germs. We give below some of the research directions in this area. The few mentioned references (mainly books) are meant as appetisers.

We stated the results for C^∞ (real) map-germs as these arise in applications to differential geometry. The study of C^r map-germs and of complex holomorphic map-germs is very rich. The survey article of Wall ([Wall (1981)]) touches on this.

J. Milnor ([Milnor (1968)]) considered the topology of the fibre of a holomorphic function. If the germ of a map f at its singular point is $\mathcal{R}$-finite, then the intersection of the singular fibre with a small sphere has the homotopy type of the wedge of $\mu(f)$-spheres, where

$$\mu(f) = \operatorname{cod}(f, \mathcal{R})$$

($\mu(f)$ is called the Milnor number). Furthermore, given an $\mathcal{R}$-finite holomorphic map-germ $f\colon (\mathbb{C}^n, 0) \to (\mathbb{C}, 0)$, one has a locally trivial fibre bundle

$$f/\|f\|\colon S_\varepsilon^{n-1} \setminus K \to S^1$$

on any sufficiently small sphere S_ε^{n-1}, outside some small tubular neighbourhood N_K of the link $K = f^{-1}(0) \cap S_\varepsilon^{n-1}$. This is known as Milnor Fibration Theorem. (See [Brieskorn and Knörrer (1986); Wall (2004)] for the case $n = 2$, and [Seade (2007)] for the real case.)

A classical example of a family of non-equivalent $\mathcal{R}$-finite germs is $f_\lambda(x, y) = xy(x - y)(x + \lambda y)$, for $\lambda \neq 0, 1$. The parameter λ is a modulus and represents the cross-ratio of the four lines $f_\lambda^{-1}(0)$. If one relaxes the equivalence relation and considers germs of homeomorphisms in the source instead of diffeomorphisms, then all the members of the family f_λ become topologically equivalent. We say that the family f_λ is topologically trivial. The problem of determining whether a family of map-germs is topologically trivial was and still remains a major problem in singularity theory ([du Plessis and Wall (1995)]; see also [Damon (1988, 1992)] for topological triviality for Damon's geometric subgroups).

A $\mathcal{G}$-*invariant* of a map-germ f is a number that depends only on the $\mathcal{G}$-orbit of f. For example, the Milnor number μ is an $\mathcal{R}$-invariant of germs of functions (it is also a topological invariant). It turns out that the constancy of μ in a family of germs of functions implies that the family is topologically trivial (in fact it implies the equisingularity of the family) when $n \neq 3$ ([Lê and Ramanujam (1976)]). One approach to tackle the problem of topological triviality or equisingularity is to determine a collection of invariants whose constancy in the family implies the triviality of the family (see [Gaffney and Massey (1999)]).

Metric properties of singular sets (semi-algebraic, sub-analytic etc) is also well developed. Here, bi-Lipschitz maps play a key role, and one has the notion of bi-Lipschitz equivalence and bi-Lipschitz equisingularity ([Birbrair (2007); Mostowski (1985)]).

There is also a great interest in map-germs with non-isolated singularities, see for example [Massey (1995)].

H. Whitney showed that maps from the plane to the plane have only singularities of type fold, cusps or double folds ([Whitney (1955)]). This is the first result on the singularities of maps (rather than map-germs). See [Saeki (2004)] for more recent results.

There is also a branch of research of global topological invariants which measure how the global topology of spaces forces the singularities of mappings. Each singularity type determines a universal characteristic class, which is a polynomial in the generators of the cohomology ring of a classifying space. The resulting polynomial is the Thom polynomial of the singularity. See for example, [Kazarian (2006, 2001)].

There are applications of singularity theory to bifurcations of stationary solutions of ordinary differential equations in [Golubitsky and Schaeffer (1985)] and to solutions of quasi-linear first order partial differential equations in [Izumiya and Kossioris (1997b)].

Chapter 4

Contact between submanifolds of $\mathbb{R}^n$

Following Felix Klein's approach, a geometry of submanifolds in an ambient space X depends on the description of their properties which are invariant under the action of a transformation group of X ([Klein (1974)]). Each geometry has its class of model submanifolds which is invariant under the action of the transformation group of X. These submanifolds are homogeneous spaces (they look locally the same at all points). We can then associate geometrical invariants to a submanifold M by comparing it with the models in such a way that an invariant at a given point p in M is defined as that of the model that better approximates M at p. For instance, the curvature of a plane curve at a point p is the curvature of the osculating circle at p. The osculating circle has the highest possible contact with the curve at p among all circles passing through the point p. The model submanifolds do not need to be unique at each point. This is illustrated by the focal hyperspheres at a point p on a hypersurface in $\mathbb{R}^n$, $n \geq 3$. There are n focal hyperspheres at p and this leads to the hypersurface having n-principal curvatures, where each principal curvature is the curvature of an osculating hypersphere. The osculating hyperspheres at a point p are the hyperspheres with highest contact with a hypersurface at p. We investigate in this chapter the concept of contact between submanifolds as a singularity theory tool for the study of differential geometry of submanifold of $\mathbb{R}^n$ (or of any manifold).

The relation between the contact of equidimensional submanifolds of a given manifold and $\mathcal{K}$-singularities of maps was introduced by Mather in [Mather (1969b)] in his investigation of properties of the contact group $\mathcal{K}$. The general theory of contact between submanifolds of any dimensions of a given manifold was developed by Montaldi in his PhD thesis [Montaldi (1983)]. The results that we need in this book can also be found in [Montaldi

(1986a)] (see also [Montaldi (1991)]) and some of them are reproduced in section 4.1.

The maps defining the contact between a submanifold of $\mathbb{R}^n$ and model submanifolds come naturally in a family of map-germs. One can ask if such a family is versal and what possible singularities one can expect in the members of the family. This leads to the notion of transversality and genericity which are dealt with in section 4.2. The remaining sections give applications of Montaldi's genericity theorems to the study of the families of height functions, distance squared functions and projections of hypersurfaces to $\mathbb{R}^n$.

4.1 Contact between submanifolds

We start with the example of plane curves. Let $\alpha(t) = (x(t), y(t))$ be a regular plane curve and let β be another plane curve given as the zero set of a smooth function $F : \mathbb{R}^2 \to \mathbb{R}$. We say that the curve α has $(k+1)$-*point contact* at t_0 with the curve β if t_0 is a zero of order k of the function $g(t) = F(\alpha(t)) = F(x(t), y(t))$, that is,

$$g(t_0) = g'(t_0) = \ldots = g^{(k)}(t_0) = 0 \text{ and } g^{(k+1)}(t_0) \neq 0,$$

where $g^{(i)}$ denotes the ith-derivative of the function g. Using the singularity theory terminology in Chapter 3, the curve α has $(k+1)$-point contact at t_0 with β if and only if the function g has an A_k-singularity at t_0.

A particular case of interest is when the curve β is a circle or a line. If for instance β is a circle, then the singularities of g reveal geometric information about the curve α (see Proposition 1.3 in Chapter 1).

We extend in this chapter the notion of contact between plane curves and define the contact between two submanifolds of an n-dimensional manifold.

We start with the following definition.

Definition 4.1 ([Montaldi (1983, 1986a)]). *Let M_i, N_i, $i = 1, 2$, be submanifolds of $\mathbb{R}^n$ with* $\dim(M_1) = \dim(M_2) = m$ *and* $\dim(N_1) = \dim(N_2) = d$. *We say that the contact of M_1 and N_1 at y_1 is of the same type as the contact of M_2 and N_2 at y_2 if there is a germ of a diffeomorphism $\Phi : (\mathbb{R}^n, y_1) \to (\mathbb{R}^n, y_2)$ such that $\Phi(M_1) = M_2$ and $\Phi(N_1) = N_2$. In this case we write $K(M_1, N_1; y_1) = K(M_2, N_2; y_2)$.*

The definition of contact between submanifolds in Definition 4.1 is local in nature, so the ambient space $\mathbb{R}^n$ can be replaced by any manifold Z.

Following the setting for plane curves, suppose that a submanifold of $\mathbb{R}^n$ is given locally as the image of an immersion-germ $g : (M, x) \to (\mathbb{R}^n, 0)$ of some manifold M and that another submanifold N is given locally as the zero set of a submersion-germ $f : (\mathbb{R}^n, 0) \to (\mathbb{R}^k, 0)$, that is, $N = f^{-1}(0)$. We say that f *cuts out* N. We consider, as in the case of plane curves, the germ of the composite map $f \circ g$ at x with $g(x) = y \in f^{-1}(0)$ and analyse its $\mathcal{K}$-singularities, where $\mathcal{K}$ is the contact group (see Chapter 3). We call the composite map-germ $f \circ g$ the *contact map-germ.*

One can ask if different choices of immersions and/or submersions of the submanifolds lead to $\mathcal{K}$-equivalent contact map-germs. Also, one can choose to immerse N and cut out M. Following [Montaldi (1986a)], we get a positive answer to these questions from the following lemma (Symmetry Lemma, in [Montaldi (1986a)]).

Lemma 4.1 (Symmetry Lemma in [Montaldi (1986a)]). *Let M and N be submanifold-germs of $\mathbb{R}^n$ at 0. Let*

$$g : (M, x_0) \to (\mathbb{R}^n, 0) \quad \textit{and} \quad \bar{g} : (N, 0) \to (\mathbb{R}^n, 0)$$

be immersion-germs, and let

$$f : (\mathbb{R}^n, 0) \to (\mathbb{R}^{k_1}, 0) \quad \textit{and} \quad \bar{f} : (\mathbb{R}^n, 0) \to (\mathbb{R}^{k_2}, 0)$$

be submersion-germs such that $M = \bar{f}^{-1}(0)$ and $N = f^{-1}(0)$. If $k_2 \geq k_1$, then $\bar{f} \circ \bar{g}$ is $\mathcal{K}$-equivalent to a suspension of $f \circ g$ by $\mathbb{R}^{k_2 - k_1}$.

Proof. Consider the following commutative diagram

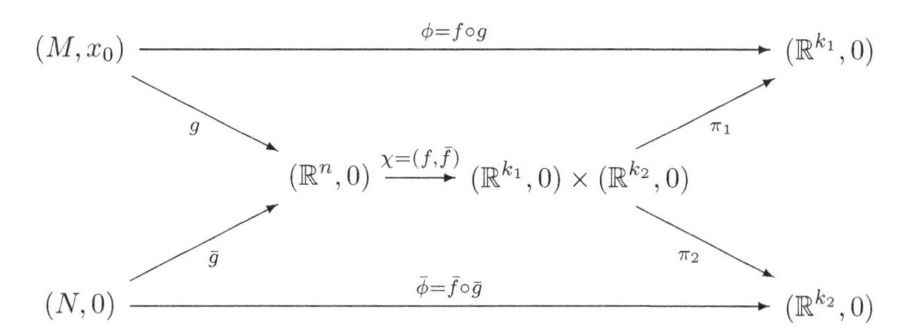

where π_1 and π_2 are the canonical projections to the first and second components respectively.

With convenient choices of coordinates in M and $(\mathbb{R}^n, 0)$ the immersion $g : (M, x_0) \to (\mathbb{R}^n, 0)$ can be written in the form $g(x) = (x, 0)$. Since $\bar{f} \circ g = 0$, we also get $\bar{f}(x, z) = z$. Then, $\chi(x, z) = (f(x, z), z)$ and $\chi(x, 0) = (f \circ g(x), 0)$. It follows that $\chi(x, z)$ can be considered as an unfolding of $\phi = f \circ g$. Since an unfolding of any map is $\mathcal{K}$-equivalent to a suspension of that map, χ is $\mathcal{K}$-equivalent to a suspension of ϕ. Similarly, χ can also be considered as an unfolding of $\bar{\phi} = \bar{f} \circ \bar{g} : (M, x_0) \to (\mathbb{R}^{k_2}, 0)$, and we get χ is $\mathcal{K}$-equivalent to a suspension of $\bar{\phi}$. Since $k_2 \geq k_1$, it follows that $\bar{\phi}$ is $\mathcal{K}$-equivalent to a suspension of ϕ as required. □

Lemma 4.2 follows from Lemma 4.1.

Lemma 4.2 ([Montaldi (1986a)], p. 196). *For any pair of germs of submanifold in $\mathbb{R}^n$, the $\mathcal{K}$-class of the contact map depends only on the submanifold-germs themselves and not on the contact map.*

Montaldi proved the following result which shows that the $\mathcal{K}$-equivalence class of the contact map-germ $f \circ g$ determines the contact class of the pair of submanifolds M and N.

Theorem 4.1 ([Montaldi (1986a)], p. 195). *Let $g_i : (M_i, x_i) \to (\mathbb{R}^n, 0)$ be immersion-germs and $f_i : (\mathbb{R}^n, 0) \to (\mathbb{R}^k, 0)$ submersion-germs, with $N_i = f_i^{-1}(0)$, $i = 1, 2$. Then the pairs (M_1, N_1) and (M_2, N_2) have the same contact type if and only if $f_1 \circ g_1$ and $f_2 \circ g_2$ are $\mathcal{K}$-equivalent.*

The proof of Theorem 4.1 is done in two steps. The first step deals with the equidimensional case, i.e., when $\dim M_i = \dim N_i = m$, and the second deals with the general case.

Proof of Theorem 4.1 for the equidimensional case. Suppose that the pairs (M_1, N_1) and (M_2, N_2) have the same contact type. Let H be the diffeomorphism of $(\mathbb{R}^n, 0)$ taking $g_1(M_1)$ to $g_2(M_2)$ and N_1 to N_2. As $H_{|g_1(M_1)} : g_1(M_1) \to g_2(M_2)$ is a diffeomorphism, there exists a diffeomorphism $h : M_1 \to M_2$, such that $H \circ g_1 = g_2 \circ h$. We also have $(f_2 \circ H)^{-1}(0) = f_1^{-1}(0)$. Then we can write each coordinate function $(f_2 \circ H)_j$ of $f_2 \circ H$ as

$$(f_2 \circ H)_j(y) = \Sigma_{i=1}^k f_{1i}(y) a_{ij}(y), \tag{4.1}$$

where $f_1 = (f_{11}, \dots, f_{1k})$ and $a_{ij} : \mathbb{R}^k \to \mathbb{R}$, $i = 1, \dots, k$, are germs of smooth functions. To show this, we use the hypothesis that $f_1 : (\mathbb{R}^n, 0) \to (\mathbb{R}^k, 0)$ is a submersion-germ, and choose a coordinate system in $(\mathbb{R}^n, 0)$

such that $f^{-1}(0)$ is the linear subspace $\mathbb{R}^{n-k} \times \{0\}$. Then, equation (4.1) follows from Hadamard's Lemma. Now, the map $f_2 \circ H$ is also a submersion, and as $f_1(0) = 0$, it follows that the $k \times k$ matrix $[a_1(y), \dots, a_k(y)]$ is invertible in a neighbourhood of 0, where the vectors $a_i(y)$ have coordinates $(a_{ij}(y))$.

We now define $\theta : \mathbb{R}^n \times \mathbb{R}^k \to \mathbb{R}^k$ by

$$\theta(y, z) = \Sigma_{i=1}^{k} z_i a_i(y),$$

and $\theta' : M_1 \times \mathbb{R}^k \to \mathbb{R}^k$ by

$$\theta'(x, z) = \theta(g_1(x), z).$$

Then, $\theta'(x, f_1 \circ g_1(x)) = f_2 \circ g_2 \circ h(x)$. It follows that $f_1 \circ g_1 \sim_{\mathcal{K}} f_2 \circ g_2$ with (h, θ') the diffeomorphism in $\mathcal{K}$ taking $f_1 \circ g_1$ to $f_2 \circ g_2$.

Suppose now that $f_1 \circ g_1$ and $f_2 \circ g_2$ are $\mathcal{K}$-equivalent and assume that $\dim M_i = \dim N_i = m$. We shall write each $g_i(M_i)$ as the graph of some map $\phi_i : \mathbb{R}^m \to \mathbb{R}^k$, $k = n - m$, and the N_i as $f_i^{-1}(0)$, where $f_i : \mathbb{R}^n \to \mathbb{R}^k$ is the projection $f_i(x_1, \dots, x_m, x_{m+1}, \dots, x_n) = (x_{m+1}, \dots, x_n)$, $i = 1, 2$. To do this we first choose a coordinate system in $\mathbb{R}^n$ so that $N_1 = \mathbb{R}^m \times \{0\}$. We then choose a k-dimensional vector subspace $V_1 \subset \mathbb{R}^n$ transverse to both $g_1(M_1)$ and N_1 and write $\mathbb{R}^n = N_1 \times V_1$. Let $\pi : \mathbb{R}^n \to N_1$ be the canonical projection to the first factor. Then $\pi_{|M_1} : g_1(M_1) \to N_1$ is a diffeomorphism which induces a coordinate system on $g_1(M_1)$, hence on M_1. With respect to these coordinate system $N_1 = \mathbb{R}^m \times \{0\}$, while M_1 is the graph of the map $f_1 \circ g_1$ (by considering f_1 as the projection: $N_1 \times V_1 \to V_1$, which in the chosen coordinate systems are respectively $\mathbb{R}^m \times \mathbb{R}^k$ and $\mathbb{R}^k$). A similar construction can be done for M_2 and N_2. Then any diffeomorphism $H : \mathbb{R}^m \times \mathbb{R}^k \to \mathbb{R}^m \times \mathbb{R}^k$ preserving $\mathbb{R}^m \times \{0\}$ and taking the graph of $f_1 \circ g_1$ to the graph of $f_2 \circ g_2$ is a diffeomorphism taking M_1 to M_2 and N_1 to N_2, and this concludes the proof for the equidimensional case. □

The proof of the general case of Theorem 4.1 requires the following two lemmas which relate the contact between two submanifolds in $\mathbb{R}^n$ with that of their suspensions.

Lemma 4.3 (Lemma A in [Montaldi (1986a)]). *For any positive integer a, let $M_i' = M_i \times \mathbb{R}^a$, $N_i' = N_i \times \{0\}$, with $y_i' = (y_i, 0)$ all in $\mathbb{R}^n \times \mathbb{R}^a$ for $i = 1, 2$. Then $K(M_1, N_1; y_1) = K(M_2, N_2; y_2)$ if and only if $K(M_1', N_1'; y_1') = K(M_2', N_2'; y_2')$.*

Proof. The sufficient part of the statement follows from the fact that the suspension of the diffeomorphism taking M_1 to M_2 and N_1 to N_2 takes M_1' to M_2' and N_1' to N_2'.

For the converse, let $H' : (\mathbb{R}^n \times \mathbb{R}^a, y_1') \to (\mathbb{R}^n \times \mathbb{R}^a, y_2')$ be the diffeomorphism-germ such that $H'(M_1', y_1') = (M_2', y_2')$, $H'(N_1', y_1') = (N_2', y_2')$. We write $H' = (H_1, H_2)$, where $H_1 : (\mathbb{R}^{n+a}, y_1') \to (\mathbb{R}^n, y_1)$ and $H_2 : (\mathbb{R}^{n+a}, y_1') \to (\mathbb{R}^a, 0)$. As H' is a diffeomorphism it follows that H_1 and H_2 are submersions.

Suppose that there exists a map-germ $\eta : (\mathbb{R}^n, y_1) \to (\mathbb{R}^a, 0)$ such that

(a) $\eta_{|N_1} = 0$ and
(b) the map $H : (\mathbb{R}^n, y_1) \to (\mathbb{R}^n, y_2)$, given by $H(x) = H_1(x, \eta(x))$ is a diffeomorphism-germ.

Then the map H will be the required map. Indeed, for all $x \in (M_1, y_1)$ we get $(x, \eta(x)) \in M_1'$, then $H'(x, \eta(x)) \in M_2'$ and hence $H(x) = H_1(x, \eta(x)) \in M_2$. Similarly, $y \in N_1$ gives $(y, \eta(y)) = (y, 0) \in N_1'$, hence $H'(y, 0) \in N_2'$ and then $H(y) \in N_2$.

We show now that the map η does exist. For (b) it is enough to show that the derivative map dH is injective. We write $U = d\eta$ and $dH_1 = (A, B)$, where $A : \mathbb{R}^n \to \mathbb{R}^n$ and $B : \mathbb{R}^a \to \mathbb{R}^n$ are linear mappings. Since (A, B) has rank n, we require U so that $A + BU$ has rank n.

For condition (a) we require U to be zero on TN_1. Since A restricted to TN_1 is injective, we can show that such U exists. □

Lemma 4.4 (Lemma B in [Montaldi (1986a)]). *Let M_i, N_i, f_i and g_i as above, with $i = 1, 2$. Let a and b be non-negative integers. Consider the submanifold-germs of $\mathbb{R}^n \times \mathbb{R}^a \times \mathbb{R}^b$ at the origin $M_i' = M_i \times \mathbb{R}^a \times \{0\}$ and $N_i' = N_i \times \{0\} \times \mathbb{R}^b$.*

Let g_i' be immersion-germs with image M_i' and f_i' submersion-germs with zero-set N_i'. Then $f_1' \circ g_1'$ and $f_2' \circ g_2'$ are $\mathcal{K}$-equivalent if and only if $f_1 \circ g_1$ and $f_2 \circ g_2$ are $\mathcal{K}$-equivalent.

Proof. By Lemma 4.2 we can choose

$$\begin{aligned} g_i'(x, u) &\mapsto (g_i(x), u, 0), \\ f_i'(y, u, v) &\mapsto (f_i(y), u), \end{aligned}$$

without changing the $\mathcal{K}$ class of the contact maps. Then $f_i' \circ g_i'(x, u) = (f_i \circ g_i(x), u)$, so $f_i' \circ g_i'$ is a suspension of $f_i \circ g_i$, and the result follows by Theorem 3.3 in Chapter 3. □

Proof of Theorem 4.1 of the general case. The proof necessary part is the same as that of the equidimensional case. For the sufficient part, if $\dim(M_i) \neq \dim(N_i)$, we can suspend the submanifold of lower dimension, say M_i, with $\mathbb{R}^a$, where $a = \dim(N_i) - \dim(M_i)$. This gives $M_i' = M_i \times \mathbb{R}^a$ and $N_i' = N_i \times \{0\}$ in $\mathbb{R}^n \times \mathbb{R}^a$ and the map-germs $g_i'(x, a) = (g_i(x), a)$ and $f_i'(y, a) = (f_i(y), a)$.

We then have the following implications, which give the result:

$$\begin{array}{ccc} f_1' \circ g_1' \sim_{\mathcal{K}} f_2' \circ g_2' & \stackrel{(b)}{\Longrightarrow} & K(M_1', N_1') = K(M_2', N_2') \\ \Uparrow(a) & & \Downarrow(c) \\ f_1 \circ g_1 \sim_{\mathcal{K}} f_2 \circ g_2 & & K(M_1, N_1) = K(M_2, N_2) \end{array}$$

where (a) follows by Lemma 4.4, (b) by the proof of the result in the equidimensional case and (c) by Lemma 4.3. □

Example 4.1. Let C be a curve in $\mathbb{R}^n$ given as the image of an immersion $g : \mathbb{R} \to \mathbb{R}^n$ and let N be a submanifold given by the zero set of a submersion $f : \mathbb{R}^n \to \mathbb{R}^k$. The resulting contact map is the map $h = f \circ g : \mathbb{R} \to \mathbb{R}^k$. Any $\mathcal{K}$-finitely determined map-germ $(\mathbb{R}, 0) \to \mathbb{R}^k$ is $\mathcal{K}$-equivalent to a map-germ $h(t) = (t^{s+1}, 0, \ldots, 0)$, for some s. The local ring $Q(h)$ is isomorphic to $\mathbb{R}\{1, t, \ldots, t^s\}$, and we say that h has an A_s-singularity. As in the case of plane curves, we say that the curve C has an $(s+1)$-point contact with N. This coincides with the notion of $(s+1)$-point contact in the case of plane curves.

We consider now the contact of a submanifold in $\mathbb{R}^n$ with foliations. This is motivated by problems involving contact of the submanifold and a family of model submanifolds. In many situations, it is necessary to replace the group $\mathcal{K}$ by a group that gives information not only about the contact of a submanifold with the zero fibre of a submersion, but also on its contact with nearby fibres.

Here we consider the relationship between the contact of a submanifold with foliations and the $\mathcal{R}^+$-class of functions. Let M_i $(i = 1, 2)$ be manifolds with $\dim M_1 = \dim M_2$, and let $g_i : (M_i, x_i) \to (\mathbb{R}^n, y_i)$ be germs of immersions. Let $f_i : (\mathbb{R}^n, y_i) \to (\mathbb{R}, 0)$ be germs of submersions.

For a germ of a submersion $f : (\mathbb{R}^n, 0) \to (\mathbb{R}, 0)$, we denote by $\mathcal{F}_f$ the regular foliation $\mathcal{F}_f = \{f^{-1}(c) \,|\, c \in (\mathbb{R}, 0)\}$. We say that *the contact of $g_1(M_1)$ with the foliation $\mathcal{F}_{f_1}$* at y_1 is of the *same type* as *the contact of $g_2(M_2)$ with the foliation $\mathcal{F}_{f_2}$ at y_2* if there exists a germ of a diffeomorphism $\Phi : (\mathbb{R}^n, y_1) \to (\mathbb{R}^n, y_2)$ such that $\Phi(g_1(M_1)) = g_2(M_2)$ and

$\Phi(Y_1(c)) = Y_2(c)$, where $Y_i(c) = f_i^{-1}(c)$ for all $c \in (\mathbb{R}, 0)$. In this case we write $K(g_1(M_1), \mathcal{F}_{f_1}; y_1) = K(g_2(M_2), \mathcal{F}_{f_2}; y_2)$.

Here too we can replace the ambient space $\mathbb{R}^n$ by any manifold. The following result characterises the contact of a hypersurface with a foliation in terms of the $\mathcal{R}^+$-singularities of functions.

Proposition 4.1 ([Goryunov (1990)], Appendix). *Let M_i, $i = 1, 2$, be two manifolds with* $\dim(M_1) = \dim(M_2) = n - 1$, *and $g_i : (M_i, x_i) \to (\mathbb{R}^n, y_i)$ be germs of immersion. Let $f_i : (\mathbb{R}^n, y_i) \to (\mathbb{R}, 0)$ be germs of submersions. Then $K(g_1(M_1), \mathcal{F}_{f_1}; y_1) = K(g_2(M_2), \mathcal{F}_{f_2}; y_2)$ if and only if $f_1 \circ g_1$ and $f_2 \circ g_2$ are $\mathcal{R}^+$-equivalent.*

Proof. See [Goryunov (1990)]. □

4.2 Genericity

The space $C^\infty(X, Y)$ of smooth maps between two manifolds X and Y is endowed with the so-called Whitney C^∞-topology which is defined as follows (more details can be found in [Golubitsky and Guillemin (1973)]).

For $p \in X$ and $q \in Y$, and for a non-negative integer k, denote by $J^k(X, Y)_{p,q}$ the set of k-jets of map-germs $(X, p) \to (Y, q)$. The k-jet space of mappings from X to Y is defined as $J^k(X, Y) = \cup_{p \in X, q \in Y} J^k(X, Y)_{p,q}$. The set $J^k(X, Y)$ is a smooth manifold (Theorem 2.7 in [Golubitsky and Guillemin (1973)]). The topology of $J^k(X, Y)$ is used to define a topology on $C^\infty(X, Y)$ as follows. Let U be an open set in $J^k(X, Y)$ and denote by

$$M(U) = \{f \in C^\infty(X, Y) \,|\, j^k f(X) \subset U\}.$$

The family of sets $\{M(U)\}$ where U is an open set of $J^k(X, Y)$ forms a basis for a topology on $C^\infty(X, Y)$ (note that $M(U) \cap M(V) = M(U \cap V)$). This topology is called the Whitney C^k-topology. Denote by W_k the set of open subsets of $C^\infty(X, Y)$ in the Whitney C^k-topology. The Whitney C^∞-topology on $C^\infty(X, Y)$ is the topology whose basis is $W = \bigcup_{k=0}^{\infty} W_k$.

Let $\mathrm{Imm}(X, Y)$ denote the subset of $C^\infty(X, Y)$ whose elements are proper C^∞-immersions from X to Y, and $\mathrm{Emb}(X, Y)$ the space of proper C^∞-embeddings of X into Y. The sets $\mathrm{Imm}(X, Y)$ and $\mathrm{Emb}(X, Y)$ are given the induced Whitney C^∞-topology.

With this topology, the set $\mathrm{Imm}(X, Y)$ is an open subset of $C^\infty(X, Y)$ and when $\dim(Y) \geq 2\dim(X)$, $\mathrm{Imm}(X, Y)$ is also dense in $C^\infty(X, Y)$ (Whitney Immersion Theorem [Golubitsky and Guillemin (1973)]). When

$\dim(Y) \geq 2\dim(X) + 1$, the set $\mathrm{Emb}(X, Y)$ is dense in $C^\infty(X, Y)$, and when X is compact, it is also open. See [Golubitsky and Guillemin (1973)] for proofs.

A property P in the topological space $\mathrm{Imm}(X, Y)$ (resp. $\mathrm{Emb}(X, Y)$) is said to be *generic* if it is satisfied by a residual subset of $\mathrm{Imm}(X, Y)$ (resp. $\mathrm{Emb}(X, Y)$). A *residual* subset of a topological space is a countable intersection of open dense subsets. Since $\mathrm{Imm}(X, Y)$ is a Baire space, a residual subset of $\mathrm{Imm}(X, Y)$ is dense set. Similarly, a residual subset of $\mathrm{Emb}(X, Y)$ is also dense ([Golubitsky and Guillemin (1973)] p. 44).

Given a generic property P, we call immersions (resp. embeddings) that satisfy P *generic immersions* (resp. *generic embeddings*).

We shall apply the results in section 4.1 in the following way. We consider the contact of a manifold M immersed in $\mathbb{R}^n$ with families of submanifolds (more specifically, k-spheres and k-planes) and show that this contact is generic for a dense subset of immersions of M in $\mathbb{R}^n$. We also list the generic contacts.

First we recall the notion of transversality and Thom's transversality theorem ([Golubitsky and Guillemin (1973)]).

Definition 4.2. Let $f : X \to Y$ be a C^∞-map and let $Z \subset Y$ be a submanifold. The map f is transverse to Z at $x \in X$ if one of the following conditions holds:

(i) $f(x)$ is not in Z.
(ii) $f(x) \in Z$ and $df_x(T_xX) + T_{f(x)}Z = T_{f(x)}Y$.

If f is transverse to Z for every $x \in X$, we say that f is transverse to Z and write $f \pitchfork Z$.

Theorem 4.2. *Let* $g : X \to Y$ *be a smooth map between smooth manifolds. Let* $Z \subset J^r(X, Y)$ *be a submanifold. Then, the set*

$$\mathcal{T}_Z = \{g \in C^\infty(X, Y) \,|\, j^r g \pitchfork Z\}$$

is a dense subset of $C^\infty(X, Y)$. *Moreover, if* Z *is a closed subset of* $J^r(X, Y)$, *then* $\mathcal{T}_Z$ *is open in* $C^\infty(X, Y)$.

The key to the results in this section is the next proposition which is a variant of Thom's transversality theorem ([Thom (1956)], [Golubitsky and Guillemin (1973)]).

Proposition 4.2. *Let $\mathcal{B}, X$ and Y be any smooth manifolds, and let Z be a submanifold of $J^r(X,Y)\times\mathcal{B}$. Then the set*

$$T_Z = \{g \in C^\infty(X,Y) \,|\, (j^r g, \mathrm{id}) \pitchfork Z\}$$

is residual in $C^\infty(X,Y)$, where $(j^r g, \mathrm{id}) : X \times \mathcal{B} \to J^r(X,Y)\times\mathcal{B}$ is the product map and $\mathrm{id} : \mathcal{B} \to \mathcal{B}$ is the identity map. Moreover, if Z is closed and $\mathcal{B}$ is compact, then T_Z is open and dense.

Let $(N_b)_{b\in\mathcal{B}}$ denote our chosen model submanifolds of $\mathbb{R}^n$, where $\mathcal{B}$ is a smooth manifold and let $g \in \mathrm{Imm}(M, \mathbb{R}^n)$.

To study the contact between $g(M)$ and N_b, $b \in \mathcal{B}$, for each $b \in \mathcal{B}$, we suppose that there is a submersion $f_b : \mathbb{R}^n \to \mathbb{R}^k$, with $N_b = f_b^{-1}(0)$. We also suppose that the family of maps $F : \mathbb{R}^n \times \mathcal{B} \to \mathbb{R}^k$, given by

$$F(y,b) = f_b(y) \tag{4.2}$$

is smooth. By Theorem 4.1, the contact between $g(M)$ and a model submanifold N_b at the point $g(x)$ is given by the $\mathcal{K}$-singularities of the composite map $f_b \circ g$ at x.

We denote by $\phi_{g,b} : M \to \mathbb{R}^k$ the map $\phi_{g,b}(x) = f_b \circ g(x)$. We also denote by $\phi_g : M \times \mathcal{B} \to \mathbb{R}^k$ the family of maps given by

$$\phi_g(x,b) = \phi_{g,b}(x) = f_b \circ g(x).$$

Denote by $J^r_y(M,\mathbb{R}^k)$ the subset of the jet space $J^r(M,\mathbb{R}^k)$ of jets with target y. In our application, $0 \in \mathbb{R}^k$ is a preferred target as it is the target of $f_b \circ g$. We consider all maps with non-zero target as being $\mathcal{K}$-equivalent, indeed if two map-germs have non-zero target their local algebras are isomorphic and equal to $\mathcal{E}(n)$. Moreover if one map has target 0, and another does not, then they are not $\mathcal{K}$-equivalent. Thus, any $\mathcal{K}$-invariant submanifold of $J^r(M,\mathbb{R}^k)$ is either all of the complement of $J^r_0(M,\mathbb{R}^k)$ or is a submanifold of $J^r_0(M,\mathbb{R}^k)$.

Theorem 4.3 (Theorem 2.2 in [Montaldi (1983)]). *Let $M, \mathcal{B}, f_b, g$ and ϕ_g be as above. Let W be either a $\mathcal{K}$-invariant submanifold of $J^r_0(M,\mathbb{R}^k)$ or all of its complement. Then the set*

$$R_W = \{g \in \mathrm{Imm}(M,\mathbb{R}^n) : j^r_1\phi_g \pitchfork W\}$$

is residual in $\mathrm{Imm}\,(M,\mathbb{R}^n)$, where j^r_1 is the r-jet with respect to the first variable, so $j^r_1\phi_g$ maps $M\times\mathcal{B}$ to $J^r(M,\mathbb{R}^k)$. Moreover, if $\mathcal{B}$ is compact and W is closed then R_W is open and dense.

Proof. Define the map $\Gamma : M \times C^\infty(M, \mathbb{R}^n) \times \mathcal{B} \to M \times C^\infty(M, \mathbb{R}^k)$ by

$$\Gamma(x, g, b) = (x, f_b \circ g)$$

and let $\Gamma^r : J^r(M, \mathbb{R}^n) \times \mathcal{B} \to J^r(M, \mathbb{R}^k)$ be the induced map in jet space given by

$$\Gamma^r(j^r g(x), b) = j^r(f_b \circ g)(x) = j^r \phi_{g,b}(x).$$

The main idea of the proof is as follows. If $\Gamma^r \pitchfork W$, then $Z = (\Gamma^r)^{-1}(W) \subset J^r(M, \mathbb{R}^n) \times \mathcal{B}$ is a smooth submanifold ([Golubitsky and Guillemin (1973)]) and the theorem follows from Proposition 4.2 as $R_W = T_Z \cap \mathrm{Imm}(M, \mathbb{R}^n)$. Note that since Imm $(M, \mathbb{R}^n)$ is open in $C^\infty(M, \mathbb{R}^n)$, the set $T_Z \cap \mathrm{Imm}(M, \mathbb{R}^n)$ is a residual set in $\mathrm{Imm}(M, \mathbb{R}^n)$.

We now prove that the condition that $f_b : \mathbb{R}^n \to \mathbb{R}^k$ is a submersion imply that $\Gamma^r \pitchfork W$. If W is the complement of $J_0^r(M, \mathbb{R}^k)$, then it is open and the transversality condition $\Gamma^r \pitchfork W$ holds trivially. One therefore needs to show that $\Gamma^r \pitchfork W$ whenever W is a submanifold of $J_0^r(M, \mathbb{R}^k)$. For this, it is enough to show that for any $b \in \mathcal{B}$, Γ_b^r is a submersion at x if $\Gamma_b^r(x) \in J_0^r(M, \mathbb{R}^k)$.

This is a local condition, so we can suppose x to be the origin and the tangent space to the fibre over x in $J^r(M, \mathbb{R}^k)$ at $j^r g(x)$ can be identified with $\theta_g / \mathcal{M}^{r+1}.\theta_g$, where θ_g is the set of germs at x of vector fields along g. (Observe that θ_g is isomorphic to $\mathcal{E}(m, n)$, hence $\theta_g / \mathcal{M}_m^{r+1}.\theta_g$ is isomorphic to $\mathcal{E}(m, n)/\mathcal{M}_m^{r+1}.\mathcal{E}(m, n)$).

Now for each $b \in \mathcal{B}$, the map $\Gamma_b = \Gamma(.,., b)$ induces a map

$$\begin{aligned} \Gamma_{b*} : \theta_g &\longrightarrow \theta_\phi \\ \xi &\longmapsto df_b(\xi), \end{aligned}$$

where $\phi = \phi_{g,b}$ and the derivative map is taken at $g(x) = g(0) = 0$. The map

$$\Gamma_{b*}^r : \theta_g / \mathcal{M}^{r+1}.\theta_g \to \theta_\phi / \mathcal{M}^{r+1}.\theta_\phi$$

is the derivative of Γ_b^r restricted to the fibre over x.

Then Γ_{b*} is surjective if f_b is a submersion at $g(x)$. Indeed, suppose that f_b is a submersion at $y = g(x)$, and choose coordinates in $\mathbb{R}^n$ and $\mathbb{R}^k$ so that f_b takes the form

$$f_b(y_1, \ldots, y_n) = (y_1, \ldots, y_k).$$

Let $\xi = (h_1(x), \ldots, h_n(x))$, then $\Gamma_{b*}(\xi) = (h_1(x), \ldots, h_k(x))$, so Γ_{b*} is clearly surjective. It follows that Γ_b is a submersion, and consequently so is Γ_b^r as required.

To complete the proof, we consider the following diagram, where $I^r(M,\mathbb{R}^n)$ is the set of r-jets of immersions $g: M \to \mathbb{R}^n$:

$$\begin{array}{ccccc} M \times \mathcal{B} & \xrightarrow{j^r g \times id} & I^r(M \times \mathbb{R}^n) \times \mathcal{B} & \xrightarrow{\Gamma^r} & J^r(M, \mathbb{R}^k) \\ \downarrow & & \downarrow \pi & & \\ M & \xrightarrow{j^r g} & I^r(M, \mathbb{R}^n) & & \end{array}$$

We now suppose that W is closed and $\mathcal{B}$ is compact. Then the restriction of the projection $\pi : I^r(M,\mathbb{R}^n) \times \mathcal{B} \to I^r(M,\mathbb{R}^n)$ to the closed set $W' := {\Gamma^r}^{-1}(W)$ is a proper map. Hence R_W is open and dense. □

The results in Theorem 4.3 are extended in [Montaldi (1986a)] to cover the Mather's subgroups of $\mathcal{K}$, $\mathcal{G} = \mathcal{R}, \mathcal{L}, \mathcal{A}, \mathcal{C}, \mathcal{K}$. (The case $\mathcal{G} = \mathcal{R}^+$ is dealt with in [Looijenga (1974)].)

We need the following definition before stating Montaldi's Theorem for the Mather's groups.

Definition 4.3. Let $\mathcal{G}$ be one of Mather's subgroups of $\mathcal{K}$. A family of maps $F : \mathbb{R}^n \times \mathcal{B} \to \mathbb{R}^k$, given by $F(y,b) = f_b(y)$, is said to be locally $\mathcal{G}$-versal if for every $(y,b) \in \mathbb{R}^n \times \mathcal{B}$, the germ of F at (y,b) is a $\mathcal{G}$-versal unfolding of f_b at y. The family F is said to be $\mathcal{G}$-versal if for every $b \in \mathcal{B}$ and every finite subset $S \subset \mathbb{R}^n$, the multi-germ of F at $S \times \{b\}$ is a $\mathcal{G}$-versal unfolding of the multigerm of f_b at S.

Let $g : M \to \mathbb{R}^n$ be an immersion and denote, as before, by $\phi_g : M \times \mathcal{B} \to \mathbb{R}^k$ the map given by

$$\phi_g(x,b) = F(g(x),b),$$

with F as in Definition 4.3.

Theorem 4.4 ([Montaldi (1986a)]). *Let $\mathcal{G} = \mathcal{R}, \mathcal{R}^+, \mathcal{L}, \mathcal{A}, \mathcal{C}, \mathcal{K}$ and let M, F and ϕ be as above.*
(i) *Suppose that F is locally $\mathcal{G}$-versal and let $W \subset J^r(M,\mathbb{R}^k)$ be a $\mathcal{G}$-invariant submanifold. Then the set*

$$R_W = \{g \in \operatorname{Imm}(M,\mathbb{R}^n) \,|\, j_1^r \phi_g \pitchfork W\}$$

is a residual set in $\operatorname{Imm}(M,\mathbb{R}^n)$.
(ii) *Suppose that F is $\mathcal{G}$-versal and let $W \subset J^r(M,\mathbb{R}^k)_s$ be a $\mathcal{G}$-invariant submanifold, with $s \geq 1$. Then the set*

$$R_W = \{g \in \operatorname{Emb}(M,\mathbb{R}^n) \,|\, j_1^r \phi_g \pitchfork W\}$$

is residual in $\mathrm{Emb}(M, \mathbb{R}^n)$.

If W *is closed and* M *is compact, then in* (i) *and* (ii) R_W *is open and dense.*

Proof. See [Montaldi (1986a)]. The arguments of the proof of (ii) hold only for embeddings $g : M \to \mathbb{R}^n$. □

Remark 4.1. Theorem 4.4 still holds if one replaces the ambient space $\mathbb{R}^n$ by a manifold Z and the target space by another manifold Q.

4.3 The meaning of generic immersions

We clarify here what is meant by a *generic immersion* of a given m-dimensional manifold M in $\mathbb{R}^n$ with respect to a given family of model submanifolds of $\mathbb{R}^n$ of codimension k parametrized by a manifold $\mathcal{B}$, with $\dim \mathcal{B} = d$. Let $F : \mathbb{R}^n \times \mathcal{B} \to \mathbb{R}^k$ be the family of submersions defining the model manifolds, i.e. $N_b = F_b^{-1}(0)$, and $\Phi_g : M \times \mathcal{B} \to \mathbb{R}^k$ the composite $\Phi_g = F \circ g$, where $g : M \to \mathbb{R}^n$ is an immersion of M in $\mathbb{R}^n$. For each fixed $b \in \mathcal{B}$, Φ_g measures the contact of M and N_b at the given point. The desirable properties of a generic immersion are described in terms of the $\mathcal{K}_e$-versality of the family Φ_g. More precisely, we say that g is generic (with respect to its contacts with the family of model submanifolds) if Φ_g is $\mathcal{K}_e$-versal. A versal unfolding of a map-germ provides a model of all possible singularities appearing under small perturbations of the map-germ. In particular, a $\mathcal{K}_e$-versal family gives information on all local contacts of M and the elements of the family $(N_b)_{b \in \mathcal{B}}$.

The applications of this setting to differential geometry of M may require finer equivalence relations than contact equivalence. This is the case, for instance, when we are concerned with geometric information derived from the contact of M not only with $F_b^{-1}(0)$, but also on nearby level sets of F_b. In this case, we use $\mathcal{A}$ (or $\mathcal{A}_e$) equivalence of the composite map $\phi_{g,b}$, when the dimension of the target is bigger than 1, and $\mathcal{R}^+$-equivalence otherwise. If two map-germs are $\mathcal{A}$ (respectively $\mathcal{R}^+$) equivalent, the foliations defined by their level sets are isomorphic. In this book we are mainly concerned with the action of the three groups $\mathcal{G} = \mathcal{K}, \mathcal{A}, \mathcal{R}^+$. A generic immersion with respect to $\mathcal{G}$-equivalence is an immersion $g : M \to \mathbb{R}^n$ for which the corresponding composite map $\Phi_{g,b}$ is $\mathcal{G}_e$-versal.

An unfolding is $\mathcal{G}$-versal if and only if it is transversal to the $\mathcal{G}$-orbits, (see Theorem 3.6), this transversality condition being verified at the level of jet spaces. Mather in a series of papers found the tools to build the bridge between versality and transversality. His results give a method to describe the generic singularities in terms of transversality of the r jet of the mapping to a $J^r\mathcal{G}$ invariant stratification of $J^r(M,\mathbb{R}^k)$. The idea is to determine the pair of dimensions (m,k) for which the relevant strata of this stratification are the r-jets of simple $\mathcal{G}$-orbits. When $\mathcal{G}=\mathcal{K}$, Mather computed in [Mather (1971)] the codimension $\sigma(m,k)$ of the set of $\mathcal{K}$ non-simple singularities in $J_0^r(M,\mathbb{R}^k)$ - the fibre of $J^r(M,\mathbb{R}^k)$ over a point $(x,0)$ in $M\times\mathbb{R}^k$ - for sufficiently large r. As we will see, when $\sigma(m,k)$ is sufficiently large, transversality to the $\mathcal{K}$-invariant stratification in jet space means avoiding non-simple singularities. The pairs (m,k) satisfying this condition form the *nice dimensions*.

The action of the group $\mathcal{K}$ does not move the origin in $\mathbb{R}^k$, so the $\mathcal{K}_e$-tangent space to any singular jet is contained in $J_0^r(M,\mathbb{R}^k)$. Hence, the $\mathcal{K}_e$-codimension of the set of all non-simple singularities in the jet space $J^r(M,\mathbb{R}^k)$ is $\sigma(m,k)+k$.

For a given immersion $g:M\to\mathbb{R}^n$, it also follows that the dimension of the image of the associated jet-extension map $j_1^r\Phi_g:M\times\mathcal{B}\to J^r(M,\mathbb{R}^k)$ is $m+d$, where $d=\dim\mathcal{B}$, and Φ_g is the composite $F\circ g$.

Suppose that $m+d<\sigma(m,k)+k$ (this will be the case for all the applications we consider in this book). Let $\{W_1,\dots,W_s\}$ be the finite set of all the $\mathcal{K}$-orbits in $J^r(M,\mathbb{R}^k)$ of $\mathcal{K}_e$-codimension less than $m+d$ and let $\{W_{s+1},\dots,W_t\}$ be a finite stratification of the complement of $W_1\cup\ldots\cup W_s$. Let R be the residual set of immersions given by the intersection of the sets R_{W_i}, $i=1,\dots,t$ in Theorem 4.3. For $g\in R$, it follows from the condition $m+d<\sigma(m,k)+k$ that the associated jet-extension map $j_1^r\Phi_g$ misses the strata W_i for $i>s$, and is transverse to the strata W_i for $i\le s$. We call the immersions $g\in R$ *generic immersions*.

When $m+d\ge\sigma(m,k)+k$, a generic embedding does not avoid in general the non-simple singularities. In this case, a stratification of $J_0^r(M,\mathbb{R}^k)$ is given by the strata which are the simple $\mathcal{K}$-orbits together with strata which are the union of the non-simple $\mathcal{K}$-orbits parametrised by the moduli (usually excluding some exceptional values of the moduli).

When the dimensions m, d, k are such that the non-simple singularities are not encountered for a generic immersion, the strata of codimension smaller than or equal to $m+d$ are $\mathcal{K}$-orbits, and the transversality of $j_1^r\Phi_g$ to the stratification in jet space, for sufficiently high values of r gives that Φ_g is $\mathcal{K}_e$-versal. If, on the other hand, the non-simple singularities are present, then all the singularities that arise for a generic immersion cannot be presented transversely.

The formulae for the codimension $\sigma_r(m,k)$ of the algebraic variety consisting of all non-simple singularities in $J_0^r(M,\mathbb{R}^k)$, are given by Mather in [Mather (1971)]. The number $\sigma_r(m,k)$ is a decreasing function of r, and $\sigma(m,k)$ is defined to be $\inf_r \sigma_r(m,k)$. The formulae for $\sigma(m,k)$ are as follows:

Case I: $m \leq k$

$$\begin{aligned}\sigma(m,k) &= 6(k-m)+8 \quad \text{if } k-m \geq 4 \text{ and } m \geq 4\\ &= 6(k-m)+9 \quad \text{if } 0 < k-m < 3 \text{ and } m > 4, \text{ or}\\ &\qquad\qquad\qquad\quad\ \text{if } m=3\\ &= 7(k-m)+10 \ \text{if } m=2\\ &= \infty \qquad\qquad\qquad \text{if } m=1\end{aligned}$$

Case II: $m > k$

$$\begin{aligned}\sigma(m,k) &= 9 \qquad\qquad \text{if } m-k=1\\ &= 8 \qquad\qquad \text{if } m-k=2\\ &= m-k+7 \quad \text{if } m-k \geq 3\end{aligned}$$

Remark 4.2. In some applications in this book, for instance when considering the family of projections into linear spaces (section 4.6), we take $\mathcal{G}=\mathcal{A}$ or $\mathcal{R}^+$ if $k=1$. The orbits of the extended actions of these groups are invariant by translations $y \to y+c$, y and c in $\mathbb{R}^k$. Then, for $\mathcal{G}=\mathcal{A},\mathcal{R}^+$, and $f:(\mathbb{R}^m,0)\to(\mathbb{R}^k,0)$, the $\mathcal{G}_e$-tangent space of f contains the vector subspace of θ_f generated by $\{\frac{\partial}{\partial y_i}\circ f\}$, $i=1,\ldots,k$ where $\{\frac{\partial}{\partial y_i},\ i=1,\ldots,k\}$ are the generators of θ_k.

When $k=1$, the $\mathcal{R}^+$-simple germs coincide with the $\mathcal{K}$-simple germs coincide. When $k>1$ and $\mathcal{G}=\mathcal{A}$, a formula for the codimension $\sigma_{\mathcal{A}}(m,k)$ of the set of the $\mathcal{A}_e$-non-simple singularities is not known in general (see [Rieger and Ruas (2005)], [Oset Sinha, Ruas and Atique (2015)] for some partial results). In this case, we proceed as follows. With similar arguments as above, for each pair (m,k), if the $\mathcal{A}$-classification of map-germs $(\mathbb{R}^m,0)\to(\mathbb{R}^k,0)$ of $\mathcal{A}_e$-codimension $\leq d=\dim(\mathcal{B})$ is finite, that is, all orbits in this classification are $\mathcal{A}$-simple, then we let $\{W_1,\ldots,W_s\}$ to be the finite set of $\mathcal{A}$-orbits in $J^r(M,\mathbb{R}^k)$ of $\mathcal{A}_e$-codimension $\leq d$, and let $\{W_{s+1},\ldots,W_t\}$ to be a finite stratification of the complement of $W_1\cup\ldots\cup W_s$. In this case, we set $\mathcal{W}$ to be this stratification. Again, g is a *generic immersion* if $j^r\Phi_g \pitchfork \mathcal{W}$.

4.4 Contact with hyperplanes

We take the model submanifolds to be the hyperplanes in $\mathbb{R}^n$. A hyperplane in $\mathbb{R}^n$ is determined, in a unique way, by a unit vector $\mathbf{v}$ in $\mathbb{R}^n$ and a scalar r. If $H(\mathbf{v}, r)$ denotes such a hyperplane, then

$$H(\mathbf{v}, r) = \{y \in \mathbb{R}^n \mid \langle y, \mathbf{v}\rangle - r = 0\}.$$

The family (4.2) can be written as $\widetilde{\mathcal{H}} : \mathbb{R}^n \times S^{n-1} \times \mathbb{R} \to \mathbb{R}$ with

$$\widetilde{\mathcal{H}}(y, \mathbf{v}, r) = \langle y, \mathbf{v}\rangle - r. \tag{4.3}$$

The family $\widetilde{\mathcal{H}}$ is called the (universal) *extended family of height functions*. Here the manifold $\mathcal{B}$ in section 4.1 is $S^{n-1} \times \mathbb{R}$ and the function $\widetilde{\mathfrak{h}}_{(\mathbf{v},r)} : \mathbb{R}^n \to \mathbb{R}$, given by $\widetilde{\mathfrak{h}}_{(\mathbf{v},r)}(y) = \widetilde{\mathcal{H}}(y, \mathbf{v}, r)$, is clearly a submersion for any $(\mathbf{v}, r) \in S^{n-1} \times \mathbb{R}$. The zero fibre of this function is precisely the hyperplane $H(\mathbf{v}, r)$.

We are also interested in the contact of submanifolds with families of parallel hyperplanes. This is why we also consider the (universal) *family of height functions* $\mathcal{H} : \mathbb{R}^n \times S^{n-1} \to \mathbb{R}$ given by

$$\mathcal{H}(y, \mathbf{v}) = \langle y, \mathbf{v}\rangle.$$

Given an immersion $g : M \to \mathbb{R}^n$ of a submanifold M into $\mathbb{R}^n$, we consider the *family of height functions* $H : M \times S^{n-1} \to \mathbb{R}$ on M defined by

$$H(p, \mathbf{v}) = \mathcal{H}(g(p), \mathbf{v}) = \langle g(p), \mathbf{v}\rangle. \tag{4.4}$$

The *extended family of height functions* $\widetilde{H} : M \times S^{n-1} \times \mathbb{R} \to \mathbb{R}$ on M is defined by

$$\widetilde{H}(p, \mathbf{v}, r) = \widetilde{\mathcal{H}}(g(p), \mathbf{v}, r) = \langle g(p), \mathbf{v}\rangle - r.$$

For $\mathbf{v}$ fixed, we denote by $h_{\mathbf{v}} : M \to \mathbb{R}$ the function given by $h_{\mathbf{v}}(p) = H(p, \mathbf{v})$. Following Theorem 4.1, the contact of $g(M)$ with the family of parallel hyperplanes determined by $\mathbf{v} \in S^{n-1}$ is measured by the $\mathcal{K}$-singularities of the function $h_{\mathbf{v}}$.

For low values of m and n, the singularities of $h_{\mathbf{v}}$ of a generic immersed m-dimensional submanifold in $\mathbb{R}^n$ can be obtained using Theorem 4.3. We give below the singularities of the height function in the case of generic immersions of curves and surfaces (i.e., $m = 1$ or $m = 2$).

Theorem 4.5. (i) *For an open and dense set of immersions of a smooth curve C in $\mathbb{R}^n$, $n \geq 2$, the family $\widetilde{H}$ (resp. H) is locally $\mathcal{K}_e$-versal (resp. P-$\mathcal{R}^+$-versal).*

(ii) *For an open and dense set of immersions of a 2-dimensional surface M in $\mathbb{R}^n$, with $3 \leq n \leq 7$, the family $\widetilde{H}$ (resp. H) is locally $\mathcal{K}_e$-versal (resp. P-$\mathcal{R}^+$-versal).*

Proof. Let $\mathcal{G} = \mathcal{K}$. The extended height function family $\widetilde{H}$ is a n parametric family. According to the discussion in section 4.3, we should find n such that the condition $m + d < \sigma(m, k) + k$ holds, where $m = 1$ in (i) and $m = 2$ in (ii), $d = n$ and $k = 1$. Now, $\sigma(1,1) = \infty$ and $\sigma(2,1) = 9$, and the result follows from Theorem 4.3.

The arguments for $\mathcal{G} = \mathcal{R}^+$ are similar. Notice that in this case $d = n-1$ and the condition $m + d < \sigma(m, k)$ gives the same inequality as above. □

Theorem 4.6. *For a generic immersed curve C in $\mathbb{R}^n$, the local $\mathcal{K}$-singularities of $h_{\mathbf{v}}$ are of type A_k, $k = 1, \ldots, n$.*

Proof. The simple singularities in this case are the A_k-singularities. These describe the simple orbits of $\mathcal{G} = \mathcal{K}$ and $\mathcal{R}^+$. The strata of the stratification of the jet space $J^r(1,1)$ are the A_k-orbits. Hence, for a generic embedding of a curve in $\mathbb{R}^n$, the r-jet of the extended family of the height functions, $J_1^r\widetilde{H} : C \times S^{n-1} \times \mathbb{R} \to J^r(C, \mathbb{R})$ avoids all strata of codimension bigger than n. □

Theorem 4.7. (i) *For a generic immersed surface M in $\mathbb{R}^3$, the local $\mathcal{K}$-singularities of $h_{\mathbf{v}}$ are of type A_k, $k = 1, 2, 3$.*

(ii) *For a generic immersed surface M in $\mathbb{R}^4$, the local $\mathcal{K}$-singularities of $h_{\mathbf{v}}$ are of type A_k, $k = 1, 2, 3, 4$ or D_4.*

(iii) *For a generic immersed surface M in $\mathbb{R}^5$, the local $\mathcal{K}$-singularities of $h_{\mathbf{v}}$ are A_k, $k = 1, 2, 3, 4, 5$, D_4 or D_5.*

Proof. The argument is similar to that in the proof of Theorem 4.6. Here $m = 2$ and $k = 1$ and one needs to stratify $J_0^r(M, \mathbb{R})$. From Theorem 4.5 the relevant strata of the stratification in jet space are the orbits of the simple singularities A_k, D_k, E_6, E_7, E_8.

Now, the generic singularities of the height function must have codimension less than or equal to n. Hence, when $n = 3, 4, 5$ we get respectively those in (i), (ii) and (iii). □

Remark 4.3. Theorem 4.7 gives the possible singularities of the height function on an immersed surface M in $\mathbb{R}^n$, $n = 3, 4, 5$. The singularities describe the contact of $g(M)$ with hyperplanes, where g is a generic immersion $M \to \mathbb{R}^n$. We extract in Chapters 6, 7, 8 extrinsic geometric properties of $g(M)$ from each of the generic singularities of the height function. We also do the same for the families of functions and maps in the subsequent sections.

4.5 The family of distance squared functions

In this section, the model submanifolds in $\mathbb{R}^n$ are taken to be hyperspheres. A hypersphere in $\mathbb{R}^n$ is determined, in a unique way, by its centre $a \in \mathbb{R}^n$ and its radius r. Let

$$S(a,r) = \{y \in \mathbb{R}^n \mid \langle y-a, y-a\rangle - r^2 = 0\}$$

denote such a hypersphere. Then the family (4.2) can be written as $\widetilde{\mathcal{D}} : \mathbb{R}^n \times \mathbb{R}^n \times \mathbb{R} \to \mathbb{R}$, with

$$\widetilde{\mathcal{D}}(y,a,r) = \langle y-a, y-a\rangle - r^2. \tag{4.5}$$

The family $\widetilde{\mathcal{D}}$ is called the (universal) *extended family of distance squared functions*. Here, the manifold $\mathcal{B}$ in section 4.1 is $\mathbb{R}^n \times \mathbb{R}$ and the function $\widetilde{\mathfrak{d}}_{a,r} : \mathbb{R}^n \to \mathbb{R}$, given by $\widetilde{\mathfrak{d}}_{a,r}(y) = \widetilde{\mathcal{D}}(y,a,r)$, is a submersion if and only if $r \neq 0$. The zero fibre of this function is the hypersphere $S(a,r)$, including the degenerate hypersphere when $r = 0$ (which is just the point a). This is the reason why we chose in section 4.2 to cut out N_b as we can in this way include singular fibres.

We are interested in the contact of a submanifold with a family of hyperspheres with the same centre and consider the (universal) *family of distance squared functions* $\mathcal{D} : \mathbb{R}^n \times \mathbb{R}^n \to \mathbb{R}$, given by

$$\mathcal{D}(y,a) = \langle y-a, y-a\rangle.$$

Given an immersion $g : M \to \mathbb{R}^n$ of a submanifold M into $\mathbb{R}^n$, we consider the *family of distance squared functions* $D : M \times \mathbb{R}^n \to \mathbb{R}$ on M defined by

$$D(p,a) = \mathcal{D}(g(p),a) = \langle g(p)-a, g(p)-a\rangle. \tag{4.6}$$

The *extended family of distance squared functions* $\widetilde{D} : M \times \mathbb{R}^n \times \mathbb{R} \to \mathbb{R}$ on M is defined by

$$\widetilde{D}(p,a,r) = \widetilde{\mathcal{D}}(g(p),a,r) = \langle g(p)-a, g(p)-a\rangle - r^2.$$

For $\mathbf{v}$ fixed, we denote by $d_a : M \to \mathbb{R}$ the function given by $d_a(p) = D(p,a)$. Following Proposition 4.1, the contact of $g(M)$ with the family of hyperspheres with the same centre $a \in \mathbb{R}^n$ is measured by the

$\mathcal{R}^+$-singularities of the function d_a. Here too, the singularities of the distance squared function d_a of a generic immersed m-dimensional submanifold in $\mathbb{R}^n$ can be obtained using Theorem 4.3. We give below the possible singularities of the distance squared function in the case of curves and surfaces.

The proofs of Theorems 4.8, 4.9, 4.10 are analogous to those of Theorems 4.6, 4.7 and are omitted.

Theorem 4.8. (i) *For an open and dense set of immersions of a curve C in $\mathbb{R}^n$, $n \geq 2$ the family $\tilde{D}$ (resp. D). is locally $\mathcal{K}_e$-versal (resp. P-$\mathcal{R}^+$-versal).*

(ii) *For an open and dense set of immersions of a 2-dimensional surface M in $\mathbb{R}^n$, with $3 \leq n \leq 6$, the family $\tilde{D}$ (resp. D) is locally $\mathcal{K}_e$-versal (resp. P-$\mathcal{R}^+$-versal).*

Theorem 4.9. *For a generic immersed curve C in $\mathbb{R}^n$, the $\mathcal{K}$-singularities of D_a are of type A_k, $k = 1, \ldots, n+1$.*

Theorem 4.10. *For a generic immersed surface M in $\mathbb{R}^3$, the $\mathcal{K}$-singularities of D_a are of type A_k, $k = 1, 2, 3, 4$ or D_4.*

One can relate the contact of a submanifold in $\mathbb{R}^n$ with hyperplanes and the contact between submanifolds in S^n with hyperspheres using the stereographic projection.

Denote by $Np = (0, \cdots, 0, 1)$ the north pole on the unit n-sphere S^n of $\mathbb{R}^{n+1}$. The stereographic projection $\psi : S^n \setminus \{Np\} \to \mathbb{R}^n \times \{0\}$ assigns to each point $q \in S^n \setminus \{Np\}$ the point $x \in \mathbb{R}^n \times \{0\} \equiv \mathbb{R}^n$ determined by the intersection of the hyperplane $\mathbb{R}^n$ with the line in $\mathbb{R}^{n+1}$ that passes through the points Np and q. The map ψ is a diffeomorphism and we denote its inverse by $\varphi : \mathbb{R}^n \to S^n \setminus \{Np\}$. The map φ maps a hypersphere in $\mathbb{R}^n$ to a hypersphere in S^n, and maps a hyperplane in $\mathbb{R}^n$ to a hypersphere in S^n passing through the north pole Np.

The image $\varphi(S(a, r))$ of the hypersphere of radius r and centre $a \in \mathbb{R}^n$ is a hypersphere in S^n and there is a unique hyperplane $H(\mathbf{a}, r)$ of $\mathbb{R}^{n+1}$ whose intersection with S^n is $\varphi(S(a, r))$.

Given an immersed manifold M in $\mathbb{R}^n$, its image $\varphi(M)$ is a submanifold of $\mathbb{R}^{n+1}$ contained in the hypersphere S^n. Since φ is a diffeomorphism, it follows from Definition 4.1 that it preserves the contacts between submanifolds. Hence the contact of M with any hypersphere $S(a, r)$ at a point p in $\mathbb{R}^n$ is the same as the contact of $\varphi(M)$ with the hypersphere $\varphi(S(a, r))$ at $\varphi(p)$ in S^n.

Theorem 4.11. *Given the hypersphere $S(a,r)$ in $\mathbb{R}^n$, suppose that $n(a,r)$ and $\rho(a,r)$ are respectively a normal unit vector and the distance to the origin of the hyperplane $H(\mathbf{a},r)$ determining $\varphi(S(a,r))$ in $\mathbb{R}^{n+1}$. Suppose that $g: M \to \mathbb{R}^n$ is an immersion of a manifold M in $\mathbb{R}^n$ and that $S(a,r)$ is tangent to $g(M)$ at the point $p_0 = g(q_0)$. Then the germs at q_0 of the extended distance squared function $\tilde{D}_{(a,r)}$ on M and of the extended height function $\tilde{H}_{(n(a,r),\rho(a,r))}$ on $\varphi(M)$ are $\mathcal{K}$-equivalent.*

Proof. Consider the function $\lambda_{a,r}: \mathbb{R}^{n+1} \to \mathbb{R}$ given by

$$\lambda_{a,r} = \langle u, n(a,r)\rangle - \rho(a,r).$$

We have $(\lambda_{a,r})^{-1}(0) = H(\mathbf{a},r)$, so $\lambda_{a,r} \circ i \circ \varphi \circ \boldsymbol{x}$ is a contact map of the pair $(\varphi(M), H(\mathbf{a},r))$ at p in $\mathbb{R}^{n+1}$, where i is the canonical inclusion of the hypersphere S^n in $\mathbb{R}^{n+1}$.

If we denote $\bar{h}_{a,r} = \lambda_{a,r}|_{S^n}$, we get $(\bar{h}_{a,r})^{-1}(0) = \varphi(S(a,r)) = H_{a,r} \cap S^n$. Hence $\bar{h}_{a,r} \circ \varphi \circ \boldsymbol{x}$ is the contact map of the pair $(\varphi(M), \varphi(S(a,r)))$ in S^n. As $\bar{h}_{a,r} \circ \varphi \circ \boldsymbol{x} = \lambda_{a,r} \circ i \circ \varphi \circ \boldsymbol{x}$, it follows that

$$h_{a,r} \circ \varphi \circ \boldsymbol{x} \sim_{\mathcal{K}} \lambda_{a,r} \circ i \circ \varphi \circ \boldsymbol{x}.$$

Since $\tilde{d}_{a,r}$ is also the contact map of M and $S(a,r)$ too and $\tilde{h}_{a,r} = \lambda_{a,r} \circ i \circ \varphi \circ \boldsymbol{x}$, it follows that

$$\tilde{d}_{a,r} \sim_{\mathcal{K}} \lambda_{a,r} \circ i \circ \varphi \circ \boldsymbol{x} \sim_{\mathcal{K}} \tilde{h}_{a,r}.$$ □

Remark 4.4. It is shown in [Romero Fuster (1997)] that the extended family of distance squared functions $\tilde{D}$ on M and the extended family height functions $\tilde{H}$ on $\varphi(M)$ are $\mathcal{K}$-equivalent as unfoldings.

4.6 The family of projections into hyperplanes

We consider here the contact of a submanifold in $\mathbb{R}^n$ with lines. We shall bundle together all parallel lines and represent them by their unit direction vectors $\mathbf{v} \in S^{n-1}$. A vector $\mathbf{v} \in S^{n-1}$ is the kernel of a linear map from $\mathbb{R}^n$ to an $(n-1)$-dimensional linear space V. We choose V to be the hyperplane orthogonal to $\mathbf{v}$. This is precisely $T_{\mathbf{v}}S^{n-1}$, the tangent space to the unit sphere S^{n-1} at $\mathbf{v}$. We take the linear map to be the orthogonal projection in $\mathbb{R}^n$ to $T_{\mathbf{v}}S^{n-1}$ parallel to $\mathbf{v}$. Thus, a point $y \in \mathbb{R}^n$ is projected to a point $q = y + \lambda\mathbf{v}$. As $q \in T_{\mathbf{v}}S^{n-1}$, $\langle q, \mathbf{v}\rangle = 0$ and this gives $\lambda = -\langle y, \mathbf{v}\rangle$. By varying $\mathbf{v}$ in S^{n-1}, we obtain the (universal) *family of orthogonal projections* $\mathcal{P}: \mathbb{R}^n \times S^{n-1} \to TS^{n-1}$ given by

$$\mathcal{P}(y,\mathbf{v}) = (\mathbf{v}, y - \langle y, \mathbf{v}\rangle\mathbf{v}).$$

Here the manifold $\mathcal{B}$ in section 4.1 is S^{n-1} and the map $\mathcal{P}_{\mathbf{v}} : \mathbb{R}^n \to T_{\mathbf{v}}S^{n-1}$, given by $\mathcal{P}_{\mathbf{v}}(y) = y - \langle y, \mathbf{v}\rangle \mathbf{v}$, is clearly a submersion for any $\mathbf{v} \in S^{n-1}$. The zero fibre of this map is the line through the origin parallel to $\mathbf{v}$.

Given an immersion $g : M \to \mathbb{R}^n$ of a submanifold M into $\mathbb{R}^n$, we consider the *family of orthogonal projections* $P : M \times S^{n-1} \to TS^{n-1}$ on M defined by

$$P(p, \mathbf{v}) = \mathcal{P}(g(p), \mathbf{v}) = (\mathbf{v}, g(p) - \langle g(p), \mathbf{v}\rangle). \tag{4.7}$$

Following Theorem 4.1, the contact of $g(M)$ with the family of parallel lines to $\mathbf{v} \in S^{n-1}$ is measured by the $\mathcal{K}$-singularities the mapping $\mathcal{P}_{\mathbf{v}}$ given by

$$P_{\mathbf{v}}(p) = g(p) - \langle g(p), \mathbf{v}\rangle.$$

Given a point $p_0 \in M$, we choose a local parametrisation $\boldsymbol{x} : U \subset \mathbb{R}^m \to M$ of M at p_0 with $\boldsymbol{x}(0) = p_0$ (we assume of course that $0 \in U$). We also identify $T_{\mathbf{v}}S^{n-1}$ with $\mathbb{R}^{n-1}$ and suppose that $P_{\mathbf{v}}(p_0) = 0$. Then the composite map $P_{\mathbf{v}} \circ \boldsymbol{x}$ is locally a map-germ $(\mathbb{R}^m, 0) \to (\mathbb{R}^{n-1}, 0)$.

There is an advantage in considering the $\mathcal{A}$-singularities of the map-germ $P_{\mathbf{v}} \circ \boldsymbol{x}$ instead of its $\mathcal{K}$-singularities when $n \geq 3$ as the dimension of the target is greater than 1. In fact, not only the zero sets, but also the singular sets and the discriminants of two $\mathcal{A}$-equivalent contact maps $P_{\mathbf{v}} \circ \boldsymbol{x}$ and $P_{\boldsymbol{v}'} \circ \boldsymbol{x}'$ are diffeomorphic.

We consider in some detail the case of surfaces ($\dim(M) = 2$) immersed in $\mathbb{R}^n$ with $n = 3, 4, 5$.

The family of projections of surfaces in $\mathbb{R}^3$ to planes was investigated by Koenderink and van Doorn [Koenderink and van Doorn (1976)], Gaffney and Ruas [Gaffney and Ruas (1979)] and Arnol'd [Arnol'd (1979)]. The family of projections of surfaces in $\mathbb{R}^4$ into 3-dimensional spaces was studied by D. Mond [Mond (1985)]. These results can be also be proved using Theorem 4.4.

Theorem 4.12. *For an open and dense set of immersions of a surface M in $\mathbb{R}^n$, $n = 3, 4, 5$, the family of orthogonal projections is locally $\mathcal{A}_e$-versal. The local $\mathcal{A}$-singularities of $P_{\mathbf{v}}$ are those in* Table 4.1 *for $n = 3$,* Table 4.2 *for $n = 4$ and* Table 4.3 *for $n = 5$.*

Proof. Following Remark 4.2, we proceed as follows. First, we verify if the $\mathcal{A}$-classification of map-germs $(\mathbb{R}^2, 0) \to (R^{n-1}, 0)$, $n = 3, 4, 5$ of $\mathcal{A}_e$-codimension less than or equal to $d = n - 1$ is finite, that is, if all

Table 4.1 Local singularities of projections of surfaces in $\mathbb{R}^3$ to planes ([Rieger (1987)]).

Name	Normal form	$\mathcal{A}_e$-codimension
Immersion	(x, y)	0
Fold	(x, y^2)	0
Cusp	$(x, xy + y^3)$	0
4_2 (Lips/Beaks)	$(x, y^3 \pm x^2 y)$	1
4_3 (Goose)	$(x, y^3 \pm x^3 y)$	2
5 (Swallowtail)	$(x, xy + y^4)$	1
6 (Butterfly)	$(x, xy + y^5 \pm y^7)$	2
11_5 (Gulls)	$(x, xy^2 + y^4 + y^5)$	2

Table 4.2 Local singularities of projections ot surfaces in $\mathbb{R}^1$ to 3-spaces ([Mond (1985)]).

Name	Normal form	$\mathcal{A}_e$-codimension
Immersion	$(x, y, 0)$	0
Cross-cap	(x, xy, y^2),	0
S_k	$(x, y^2, y^3 + x^{k+1} y)$, $k > 0$	k
B_k	$(x, y^2, x^2 y + y^{2k+1})$, $k > 2$	k
C_k	$(x, y^2, xy^3 \pm x^k y)$, $k \geq 2$	k
H_k	$(x, xy + \pm y^{3k-1}, y^3)$, $k \geq 2$	k

Table 4.3 Local singularities of projections of surfaces in $\mathbb{R}^5$ to 4-spaces ([Kirk (2000)] and [Rieger (2007)]).

Name	Normal form	$\mathcal{A}_e$-codimension
Immersion	$(x, y, 0, 0)$	0
I_k	(x, xy, y^2, y^{2k+1}), $k = 1, 2, 3, 4$	k
II_2	$(x, y^2, y^3, x^k y)$, $k = 2$	4
$III_{2,3}$	$(x, y^2, y^3 \pm x^k y, x^l y)$, $k = 2$, $l = 3$	4
VII_1	$(x, xy, xy^2 \pm y^{3k+1}, xy^3)$, $k = 1$	4

orbits in this classification are $\mathcal{A}$-simple. Then we let $\{W_1, \ldots, W_s\}$ to be the finite set of $\mathcal{A}$-orbits in $J^r(M^2, \mathbb{R}^{n-1})$ of $\mathcal{A}_e$-codimension $\leq n - 1$, and let $\{W_{s+1}, \ldots, W_t\}$ to be a finite stratification of the complement of $W_1 \cup \ldots \cup W_s$. The result then follows applying Theorem 4.4.

In fact, for $n = 3$, the relevant $\mathcal{A}$-orbits are given by the corank 1 singularities from the plane to the plane of $\mathcal{A}_e$-codimension less than or equal to 2, see Table 4.1. Similarly, for $n = 4$, the relevant singularities are those in Table 4.2. The classification of singularities $\mathbb{R}^2 \to \mathbb{R}^5$ of $\mathcal{A}_e$-codimension≤ 4 can be found in Table 4.3. □

4.7 Notes

We considered in this chapter the notion of contact and Thom's transversality theorem as main tools for investigating generic properties of submanifolds of $\mathbb{R}^n$.

The application of singularity theory to the study of the extrinsic geometry of submanifolds in Euclidean spaces started in the late 1960's. René Thom, in his investigation of parabolic points on surfaces in $\mathbb{R}^3$, pointed at the importance of the family of height functions as a tool for studying the differential geometry of submanifolds of $\mathbb{R}^n$. The idea of René Thom was taken up by Porteous in [Porteous (1983a,b)] where he considered the generic singularities of the distance squared function.

The subject has had a great development since then, and problems on extrinsic geometry of submanifolds in Euclidean spaces motivated several extensions of Thom's transversality theorems (see for example [Abraham (1963)], [Buchner (1974)], [Looijenga (1974)], [Mather (1970)], [Wall (1977)] for an account of the earlier works on this subject).

Bruce in [Bruce (1986)] extended the notion of contact between manifolds to the notion of contact of a submanifold with an eventually singular space, and defined *weak transversality*. With weak transversality, one can consider $\mathcal{K}$-finitely determined germs and not merely submersion-germs.

A more general result is given in [Montaldi (1986a)], in which it is allowed contact of submanifolds of $\mathbb{R}^n$ with model submanifolds given by an $\mathcal{G}$-versal family of mappings F. Montaldi's Theorem 4.4 holds for any Damon's geometric subgroup, allowing a wider range of applications to differential geometry.

Chapter 5

Lagrangian and Legendrian Singularities

A caustic in Euclidean space is the envelope of light rays reflected on a concave surface. It is the set of points in the space which are mostly illuminated by these rays. Caustics can be visible (the reader may try to catch one on a surface of a cup of coffee or tea).

A phenomenon closely related to caustics, but which is not visible, is wavefronts. If we think of light as particles propagating at unit speed along the normal lines to a surface, then these particles trace another surface at any given time t. This surface is called the wavefront of the original surface (which is called the initial front). Wavefronts may acquire singularities, and their singularities form the caustic.

The generic singularities that can occur in caustics and wavefronts and the way they deform as the original front is deformed are described by Arnol'd and Zakalyukin using the theory of Lagrangian and Legendrian singularities [Arnol'd, Guseĭn-Zade and Varchenko (1985); Zakalyukin (1976, 1984)]. This theory was initiated by L. Hörmander [Hörmander (1971)] and has wide applications [Arnol'd (1983); Izumiya (1993, 1995); Izumiya and Kossioris (1995, 1997a,b); Izumiya, Kossioris and Makrakis (2001); Izumiya and Janeczko (2003); Izumiya, Pei and Sano (2003); Zakalyukin (1995)]. We apply it here to study some aspects of the extrinsic geometry of a submanifold of Euclidean spaces.

We start by recalling some basic concepts in symplectic and contact geometries. We state a fundamental result that says that any Lagrangian or Legendrian manifold is defined locally by a generating family of functions. The generating families of interest here are the families of distance squared functions, the family of height functions and the extended family of height functions on a submanifold in an Euclidean space. We showed in Chapter 4 how these families measure the contact of a submanifold with model hypersurfaces (totally umbilic hypersurfaces). We consider here the theory of contact from the view point of the Lagrangian and Legendrian singularities.

5.1 Symplectic manifolds

A *symplectic form* ω on a smooth manifold M is a closed, non-degenerate and skew-symmetric 2-form.

It follows from the above definition of a symplectic form ω that $d\omega = 0$, ω^n is a volume form and that $\dim M = 2n$, for some positive integer n.

A manifold M equipped with a symplectic form ω is called a *symplectic manifold.* We also call ω a *symplectic structure* on M.

Example 5.1. Let $\mathbb{R}^{2n} = \{(x_1, \dots, x_n, p_1, \dots, p_n) \mid x_i, p_i \in \mathbb{R}\}$ and consider the 1-form

$$\lambda = \sum_{i=1}^{n} p_i dx_i.$$

This yields the symplectic form

$$\omega = -d\lambda = \sum_{i=1}^{n} dx_i \wedge dp_i$$

on $\mathbb{R}^{2n}$. The vector space $\mathbb{R}^{2n}$ equipped with this symplectic form is the standard model of a linear symplectic manifold.

Example 5.2 (The cotangent bundle). Let N be a smooth manifold and T^*N its cotangent bundle. There is a canonical symplectic structure on T^*N which is defined as follows (see for example [McDuff and Salamon (1995)] for details). Denote by $\rho : T^*N \to N$ the canonical projection defined by $\rho(q, v) = q$ for any $(q, v) \in T^*N$ (so $q \in N$ and $v : T_qN \to \mathbb{R}$ is an element of T_q^*N).

The differential map $d\rho : T(T^*N) \to TN$ determines at each point $(q, v) \in T^*N$ a linear map

$$d\rho_{(q,v)} : T_{(q,v)}(T^*N) \to T_qN.$$

The *canonical 1-form* (or, the *Liouville form*) λ on T^*N is defined at each point $(q, v) \in T^*N$ by the linear map $\lambda_{(q,v)} = v \circ d\rho_{(q,v)} : T_{(q,v)}(T^*N) \to \mathbb{R}$ given by

$$\lambda_{(q,v)}(w) = v(d\rho_{(q,v)}(w)).$$

The *canonical symplectic structure* on T^*N is given by the 2-form

$$\omega = -d\lambda.$$

Let $x : U \to \mathbb{R}^n$ be a local system of coordinates of N and write $x = (x_1, \ldots, x_n)$, where $x_i : U \to \mathbb{R}$, $i = 1, \ldots, n$, are smooth functions. The 1-forms $dx_i(q) : T_qN \to \mathbb{R}$, $i = 1, \ldots, n$, form a basis of T_q^*N, so any v in T_q^*N can be written in a unique way in the form

$$v = \sum_{i=1}^{n} p_i(q, v) dx_i(q).$$

The *dual coordinates* p_i are completely determined by q and v and yield a smooth map $p : T^*U \to \mathbb{R}^n$ given by

$$p(q, v) = (p_1(q, v), \ldots, p_n(q, v)).$$

We obtain a local system of coordinates $\phi : T^*U \to \mathbb{R}^n \times \mathbb{R}^n$ of T^*N given by

$$\phi(q, v)) = (x(q), p(q, v)).$$

In this system of coordinates, the 1-form λ is given by

$$\lambda = \sum_{i=1}^{n} p_i dx_i$$

and the symplectic form ω by

$$\omega = \sum_{i=1}^{n} dx_i \wedge dp_i.$$

Definition 5.1. Let (M_1, ω_1) and (M_2, ω_2) be two symplectic manifolds. A *symplectomorphism* between (M_1, ω_1) and (M_2, ω_2) is a diffeomorphism $\phi : M_1 \to M_2$ which sends the symplectic structure ω_2 on M_2 to the symplectic structure ω_1 on M_1, that is, $\varphi^*\omega_2 = \omega_1$. More specifically, $\omega_1(q, v_1, v_2) = \omega_2(\phi(q), d\phi_q(v_1), d\phi_q(v_2))$ for any $q \in M_1$ and v_1, v_2 in T_qM_1.

Theorem 5.1 (The Darboux Theorem). *Any two symplectic manifolds of the same dimension are locally symplectomorphic.*

Proof. See [Arnol'd, Guseĭn-Zade and Varchenko (1985)]. □

5.1.1 *Lagrangian submanifolds and Langrangian maps*

Definition 5.2. Let M be a $2n$-dimensional smooth manifold and let ω be a symplectic form on M. We say that a smooth submanifold L of M is a Lagrangian submanifold if $\dim L = n$ and $\omega_{|L} = 0$.

Example 5.3. Consider the cotangent bundle T^*N of a smooth manifold N endowed with the canonical symplectic structure (Example 5.2). Then the fibres of the bundle $\rho : T^*N \to N$ and the zero section of this bundle are *Lagrangian submanifolds.* We can construct other Lagrangian submanifolds of T^*N as follows.

Let $f : N \to \mathbb{R}$ be a smooth function. Then the differential of f can be considered as an embedding $Df : N \to T^*N$, given by $Df(x) = (x, df_x)$. In the local system of coordinates $x = (x_1, \ldots, x_n)$ and their dual coordinates $p = (p_1, \ldots, p_n)$, we have

$$Df(x) = \left(x, \frac{\partial f}{\partial x_1}(x), \ldots, \frac{\partial f}{\partial x_n}(x)\right),$$

so that the pull-back of the canonical 1-form λ by Df is given by

$$Df^*\lambda(x) = \sum_{i=1}^{n} \frac{\partial f}{\partial x_i}(x)dx_i = df_x.$$

Then $Df^*\omega = ddf = 0$, so $Df(N)$ is a Lagrangian submanifold of T^*N.

Definition 5.3. Let $\pi : E \to N$ be a fibre bundle such that E is a symplectic manifold. We say that $\pi : E \to N$ is a *Lagrangian fibration* if its fibres are Lagrangian submanifolds of E.

Example 5.4. Consider the linear symplectic space $\mathbb{R}^{2n}$ and let $\pi : \mathbb{R}^{2n} \to \mathbb{R}^n$ be defined by $\pi(x_1, \ldots, x_n, p_1, \ldots, p_n) = (x_1, \ldots, x_n)$. Then π is a Lagrangian fibration.

Example 5.5. The cotangent bundle $\rho : T^*N \to N$ is a Lagrangian fibration. Indeed, the canonical 1-form λ is zero along its fibres, and hence $\omega = d\lambda$ is also zero along the fibres.

Let $\pi : E \to N$ and $\pi' : E' \to N'$ be Lagrangian fibrations. A symplectomorphism $\Phi : E \to E'$ is said to be a *Lagrangian diffeomorphism* if there exists a diffeomorphism $\phi : N \to N'$ such that $\pi' \circ \Phi = \phi \circ \pi$.

Theorem 5.2. *All Lagrangian fibrations of a fixed dimension are locally Lagrangian diffeomorphic.*

Proof. See [Arnol'd, Guseĭn-Zade and Varchenko (1985)]. □

Let $\pi : E \to N$ be a Lagrangian fibration and consider a Lagrangian immersion $i : L \to E$ (i.e., $i^*\omega = 0$). The restriction of the projection π to $i(L)$, that is $\pi \circ i : L \to N$, is called a *Lagrangian map*.

A *caustic* is the set of critical values of a Lagrangian map. We denote the caustic of the Lagrangian map $\pi \circ i : L \to N$ by $C(i(L))$.

Two Lagrangian maps $\pi \circ i : L \to N$ and $\pi' \circ i' : L' \to N'$ are said to be *Lagrangian equivalent* if there is a Lagrangian diffeomorphism $\Phi : E \to E'$ such that $\Phi(i(L)) = i'(L')$. If $\pi \circ i : L \to N$ and $\pi' \circ i' : L' \to N'$ are Lagrangian equivalent, then the caustics $C(i(L))$ and $C(i'(L'))$ are diffeomorphic.

5.1.2 *Lagrangian singularities*

We consider in this section the local singularities of Lagragian maps. All the concepts in §5.1.1 can be defined for germs. We start by stating a key result in the theory of Lagrangian maps which describes the germ of a Lagrangian immersion $i : L \to E$ using germs of families of functions.

From Theorem 5.2, all Lagrangian fibrations are locally Lagrangian diffeomorphic. Therefore, we can work on the cotangent bundle $\pi : T^*\mathbb{R}^n \to \mathbb{R}^n$ and all the results there will be valid on any Lagrangian fibration.

Let $(x, p) = (x_1, \dots, x_n, p_1, \dots, p_n)$ denote the canonical coordinates on $T^*\mathbb{R}^n$, λ the canonical 1-form and ω the canonical symplectic form on $T^*\mathbb{R}^n$ (Example 5.1).

Let $F : (\mathbb{R}^k \times \mathbb{R}^n, 0) \to (\mathbb{R}, 0)$ be an n-parameter family of germs of functions from $(\mathbb{R}^k, 0)$ to $(\mathbb{R}, 0)$, and denote by $(q, x) = (q_1, \dots, q_k, x_1, \dots, x_n)$ the coordinates in $\mathbb{R}^k \times \mathbb{R}^n$. In Chapter 3, we associated some set germs the family of functions F. The germ of the catastrophe set of F is the set germ

$$(C_F, 0) = \left\{ (q, x) \in (\mathbb{R}^k \times \mathbb{R}^n, 0) \middle| \frac{\partial F}{\partial q_1}(q, x) = \cdots = \frac{\partial F}{\partial q_k}(q, x) = 0 \right\}$$

and its bifurcation set is the set germ

$$(B_F, 0) = \left\{ x \in (\mathbb{R}^n, 0) \middle| \ \exists (q, x) \in C_F \text{ such that rank} \left(\frac{\partial^2 F}{\partial q_i \partial q_j}(q, x) \right) < k \right\}.$$

Let $\pi_2 : (\mathbb{R}^k \times \mathbb{R}^n, 0) \to (\mathbb{R}^n, 0)$ denote the canonical projection and consider the map-germ π_{C_F} which is given by the restriction of the projection π_2 to $(C_F, 0)$. Thus,

$$\pi_{C_F} : (C_F, 0) \to (\mathbb{R}^n, 0)$$

with $\pi_{C_F}(q,x) = x$ for any $(q,x) \in (C_F, 0)$. The map-germ π_{C_F} is the catastrophe map of F.

We say that F is a *Morse family of functions* if the map-germ $\Delta F : (\mathbb{R}^k \times \mathbb{R}^n, 0) \to (\mathbb{R}^k, 0)$ given by

$$\Delta F(q,x) = \left(\frac{\partial F}{\partial q_1}, \ldots, \frac{\partial F}{\partial q_k}\right)(q,x)$$

is not singular. When F is a Morse family of functions, $(C_F, 0)$ is a germ of a smooth submanifold of $(\mathbb{R}^k \times \mathbb{R}^n, 0)$ of dimension n. We immerse $(C_F, 0)$ in the cotangent bundle $T^*\mathbb{R}^n$ by the map-germ $L(F) : (C_F, 0) \to T^*\mathbb{R}^n$ defined by

$$L(F)(q,x) = \left(x, \frac{\partial F}{\partial x_1}(q,x), \ldots, \frac{\partial F}{\partial x_n}(q,x)\right).$$

(One can check that this is indeed an immersion.) We have

$$L(F)^*\lambda = \sum_{i=1}^{n} \frac{\partial F}{\partial x_i} dx_i|_{C_F} = dF|_{C_F}$$

so that

$$L(F)^*\omega = L(F)^*d\lambda = dL(F)^*\lambda = d(dF|_{C_F}) = (ddF)|_{C_F} = 0.$$

This proves that $L(F)$ is a germ of a Lagrangian immersion.

The family of map-germs F is called the *generating family* of the germ of the Lagrangian submanifold $L(F)(C_F)$ of $T^*\mathbb{R}^n$.

We have the following fundamental theorem.

Theorem 5.3. *Let L be a germ of a Lagrangian submanifold of $T^*\mathbb{R}^n$. Then there exists a germ of a Morse family of functions F such that $L(F)(C_F) = L$.*

Proof. See [Arnol'd, Guseĭn-Zade and Varchenko (1985)]. □

Remark 5.1. The bifurcation set of F is the set of critical values of the catastrophe map-germ π_{C_F}. Since $\pi_{C_F}(q,x) = \pi \circ L(F)(q,x)$, it coincides with the caustic $C(L(F))$.

A germ of a Lagrangian immersion $i : (L,x) \to (T^*\mathbb{R}^n, p)$ (or a germ of Lagrangian map $\pi \circ i : (L,x) \to (\mathbb{R}^n, \pi(p)))$ is said to be *Lagrangian stable* if for any representative $\bar{i} : V \to T^*\mathbb{R}^n$ of i, there exist a neighbourhood W of $\bar{i}$ (in the Whitney C^∞-topology on the subset of Lagrangian immersions considered as a subspace of $C^\infty(\mathbb{R}^n, T^*\mathbb{R}^n)$) and a neighbourhood V of x

with the following property: For any Lagrangian immersion $\bar{j}$ in W, there exists $x' \in V$ such that $\pi \circ i$ and $\pi \circ j$ are Lagrangian equivalent, where $j : (L, x') \to (T^*\mathbb{R}^n, p')$ is the germ of $\bar{j}$ at x'.

Theorem 5.3 states that any germ of a Lagrangian submanifold of $T^*\mathbb{R}^n$ can be constructed from a generating family of functions. The notion of Lagrangian equivalence and of Lagrangian stability can be formulated in terms of generating families.

Theorem 5.4. (1) *Let $F : (\mathbb{R}^k \times \mathbb{R}^n, 0) \to (\mathbb{R}, 0)$ and $G : (\mathbb{R}^l \times \mathbb{R}^n, 0) \to (\mathbb{R}, 0)$ be two Morse families of functions. Then the Lagrangian map-germs $\pi \circ L(F)$ and $\pi \circ L(G)$ are Lagrangian equivalent if and only if F and G are stably P-$\mathcal{R}^+$-equivalent.*

(2) *The Lagrangian map-germ $\pi \circ L(F)$ is Lagrangian stable if and only if F is an $\mathcal{R}^+$- versal unfolding of $f(q) = F(q, 0)$.*

Proof. See [Arnol'd, Guseĭn-Zade and Varchenko (1985)]. □

If $k = l$ in Theorem 5.4(1), we replace stably P-$\mathcal{R}^+$-equivalent by P-$\mathcal{R}^+$-equivalent.

The classification of singularities of germs of functions and of their $\mathcal{R}^+$-versal unfoldings can now be used to obtain a list of normal forms of the singularities of Lagrangian map-germs in low dimensions.

Theorem 5.5. *Let $F : (\mathbb{R}^k \times \mathbb{R}^n, 0) \to (\mathbb{R}, 0)$ be a Morse family of functions. Suppose that $L(F) : (C(F), 0) \to T^*\mathbb{R}^n$ is Lagrangian stable and $n \leq 4$. Then $L(F)$ is Lagrangian equivalent to a germ of a Lagrangian submanifold whose generating family $G(q_1, \ldots, q_k, x_1, \ldots, x_n)$ is one of the following germs, where $Q(q_r, \ldots, q_k) = \pm q_r^2 \pm \cdots \pm q_k^2$,*

(1) $Q(q_2, \ldots, q_k) + q_1^3 + x_1 q_1$.
(2) $Q(q_2, \ldots, q_k) \pm q_1^4 + x_1 q_1 + x_2 q_1^2$.
(3) $Q(q_2, \ldots, q_k) + q_1^5 + x_1 q_1 + x_2 q_1^2 + x_3 q_1^3$.
(4) $Q(q_2, \ldots, q_k) \pm q_1^6 + x_1 q_1 + x_2 q_1^2 + x_3 q_1^3 + x_4 q_1^4$.
(5) $Q(q_3, \ldots, q_k) + q_1^3 - q_1 q_2^2 + x_1 q_1 + x_2 q_2 + x_3(q_1^2 + q_2^2)$.
(6) $Q(q_3, \ldots, q_k) + q_1^3 + q_2^3 + x_1 q_1 + x_2 q_2 + x_3 q_1 q_2$.
(7) $Q(q_3, \ldots, q_k) + q_1^2 q_2 + q_2^4 + x_1 q_1 + x_2 q_2 + x_3 q_1^2 + x_4 q_2^2$.

Proof. It follows from Theorem 5.4 that F is an $\mathcal{R}^+$-versal unfolding of $f = F|_{\mathbb{R}^k \times \{0\}}$. Since $n \leq 4$, $\text{cod}_e(f, \mathcal{R}) \leq 4$ and by Theorem 3.13 we have that F is P-$\mathcal{R}^+$-equivalent to one of the germs in the above list. □

The normal forms of the Lagrangian stable map-germs $\pi \circ L(G)$ when $n \leq 4$ are given in Table 5.1.

Table 5.1 Lagrangian stable singularities in $\mathbb{R}^n$, $n \leq 4$.

G singularity type	$\pi \circ L(G)$ singularity type	Normal form
A_2	Fold	q_1^2
A_3	Cusp	$(q_1^3 + x_2 q_1, x_2)$
A_4	Swallowtail	$(q_1^4 + x_2 q_1 + x_3 q_1^2, x_2, x_3)$
A_5	Butterfly	$(q_1^5 + x_2 q_1 + x_3 q_1^2 + x_4 q_1^3, x_2, x_3, x_4)$
D_4^+	Elliptic umbilic	$(q_1^2 - q_2^2 + x_3 q_1, q_1^2 q_2^2 + x_3 q_2)$
D_4^-	Hyperbolic Umbilic	$(q_1^2 + x_3 q_2, q_2^2 + x_3 q_1)$
D_5	Parabolic Umbilic	$(q_1 q_2 + x_3 q_1, q_1^2 + q_2^3 + x_4 q_2)$

On the other hand, Golubitsky and Guillemin [Golubitsky and Guillemin (1975)] gave an algebraic characterisation for the $\mathcal{R}^+$-equivalence among function germs. For any $f \in \mathcal{E}_n$, we denote by

$$J_f = L\mathcal{R}_e \cdot f = \mathcal{E}_n \left\{ \frac{\partial f}{\partial x_1}, \dots, \frac{\partial f}{\partial x_n} \right\},$$

which is called the *Jacobian ideal* of f in $\mathcal{E}_n$. We define a local ring by $\mathcal{R}^{(k)}(f) = \mathcal{E}_n / J_f^k$. Let $[f]$ be the image of f in this local ring. We say that f satisfies the *Milnor Condition* if $\dim_{\mathbb{R}} \mathcal{R}^{(1)}(f) < \infty$.

Proposition 5.1 ([Golubitsky and Guillemin (1975)]). *Let f and g be germs of functions in $\mathcal{E}_n$ satisfying the Milnor condition with $df(0) = dg(0) = 0$. Then f and g are $\mathcal{R}^+$-equivalent if and only if the ranks and the signatures of the Hessians $\mathcal{H}(f)(0)$ and $\mathcal{H}(g)(0)$ are equal, and there is an isomorphism $\gamma : \mathcal{R}^{(2)}(f) \to \mathcal{R}^{(2)}(g)$ such that $\gamma([f]) = [g]$.*

For Lagrangian stable Lagrangian map-germs, we have the following classification theorem.

Theorem 5.6. *Let $F : (\mathbb{R}^k \times \mathbb{R}^n, 0) \to (\mathbb{R}, 0)$ and $G : (\mathbb{R}^k \times \mathbb{R}^n, 0) \to (\mathbb{R}, 0)$ be two Morse families of functions. Suppose that the Lagrangian map-germs $\pi \circ L(F)$ and $\pi \circ L(G)$ are Lagrangian stable. Then the following statements are equivalent:*

(1) *$\pi \circ L(F)$ and $\pi \circ L(G)$ are Lagrangian equivalent,*
(2) *F and G are P-$\mathcal{R}^+$-equivalent,*
(3) *$f = F|\mathbb{R}^k \times \{0\}$ and $g = G|\mathbb{R}^k \times \{0\}$ are $\mathcal{R}$-equivalent,*
(4) (a) *the ranks and the signatures of the Hessians $\mathcal{H}(f)(0)$ and $\mathcal{H}(g)(0)$ are equal, and*
   (b) *there is an isomorphism $\gamma : \mathcal{R}_2(f) \to \mathcal{R}_2(g)$ such that $\gamma([f]) = [g]$.*

Proof. Statements (1) and (2) are equivalent by Theorem 5.4. By definition, statement (2) implies statement (3). Since $\pi \circ L(F)$ and $\pi \circ L(G)$

are Lagrangian stable, F and G are $\mathcal{R}^+$-versal unfoldings of f and g, respectively. The uniqueness of the versal deformation shows that statement (3) implies statement (2). Since F and G are $\mathcal{R}^+$-versal unfoldings of f and g, f and g satisfy the Milnor condition, so Proposition 5.1 implies that statements (3) and (4) are equivalent. $\square$

5.2 Contact manifolds

A *contact structure* on a manifold M is a maximally non-integrable field of hyperplanes K in the tangent spaces of M. If α is a 1-form that defines K locally (i.e., $\alpha^{-1}(0) = K_{|U}$ in an open set U of M), then the maximal non-integrability condition is equivalent to $\alpha \wedge d\alpha^n \neq 0$ in U. It is also equivalent to the 2-form $d\alpha$ being non-degenerate on each plane $\alpha = 0$. A consequence of this is that M must be an odd dimensional manifold.

The local 1-form α is called a *contact form.* A manifold M with a contact structure K is denoted by (M, K) and is called a *contact manifold.*

Example 5.6. Let $\mathbb{R}^{2n+1} = \{(x_1, \ldots, x_n, y, p_n, \ldots, p_n) \mid x_i, y, p_i \in \mathbb{R}\}$ and consider the 1-form

$$\alpha = dy - \sum_{i=1}^{n} p_i dx_i.$$

Then

$$\alpha \wedge d\alpha^n = (-1)^{n(n+1)/2} dx_1 \wedge \ldots \wedge dx_n \wedge dy \wedge dp_1 \wedge \ldots \wedge dp_n,$$

which is the standard volume form in $\mathbb{R}^{2n+1}$. Thus α is a contact form. It defines the *standard contact structure* on $M = \mathbb{R}^{2n+1}$.

Example 5.7. Let N be a C^∞-manifold of dimension n and let $J^1(N, \mathbb{R})$ be the manifold of 1-jets of functions on N. This is a $(2n+1)$-dimensional manifold and has a natural contact structure constructed as follows. We can identify canonically $J^1(N, \mathbb{R})$ with $T^*N \times \mathbb{R}$, so that we have a 1-form $\alpha = dy - \lambda$, where y is a coordinate of $\mathbb{R}$ and λ is the Liouville form on T^*N. Let $x = (x_1, \ldots, x_n)$ be a local system of coordinates in an open set U of N. We obtain local coordinates $(x_1, \ldots, x_n, p_1 \ldots, p_n, y)$ in $J^1(U, \mathbb{R})$ where $y = f(x)$ and $p_i = \partial f / \partial x_i(x)$, $i = 1, \ldots, n$, for any $f : N \to \mathbb{R}$ and $x \in U$. The local contact form $\alpha = dy - \sum_{i=1}^{n} p_i dx_i$ defines the natural contact structure on $J^1(N, \mathbb{R})$. This contact structure is independent of the local system of coordinates.

Example 5.8. Let $\pi : PT^*N \to N$ be the projective cotangent bundle over an n-dimensional manifold N. Consider the tangent bundle $\tau : T(PT^*N) \to PT^*N$ of PT^*N and the differential map $d\pi : T(PT^*N) \to TN$ of π.

For any X in $T(PT^*N)$, there exists an element $\xi \in T^*N$ such that $\tau(X) = [\xi]$, where $[\xi]$ denotes the equivalence class in the projective space represented by ξ.

For an element $v \in TN$, the property $\xi(v) = 0$ does not depend on the choice of a representative of the class $[\xi]$. Thus, we can define the canonical contact structure on PT^*N by the hyperplanes

$$K = \{X \in T(PT^*N) \mid \tau(X)(d\pi(X)) = 0\}.$$

Let $(x_1, \ldots, x_n)$ be a local system of coordinates in an open set U of N. Then we have a local trivialization $PT^*U \cong U \times P(\mathbb{R}^n)^*$ If $[\xi_1 : \ldots : \xi_n]$ denotes the homogeneous coordinates of the dual projective space $P(\mathbb{R}^n)^*$, then $((x_1, \ldots, x_n), [\xi_1 : \ldots : \xi_n])$ is a local system of coordinates in PT^*U.

In the above system of coordinates, we have $\tau(X) = [\xi] = [\xi_1 : \cdots : \xi_n]$, where $\xi = \sum_{i=1}^n \xi_i dx_i$. Therefore, $X \in K_{(x,[\xi])}$ if and only if $\sum_{i=1}^n \mu_i \xi_i = 0$, where $d\pi(X) = \sum_{i=1}^n \mu_i \frac{\partial}{\partial x_i}$.

Consider the open subset $V_1 = \{[\xi_1 : \cdots : \xi_n] \mid \xi_1 \neq 0\}$ of $P(\mathbb{R}^n)^*$. Then we have the affine local system of coordinates $\psi_1 : U \times V_1 \to U \times \mathbb{R}^{n-1}$ of PT^*U defined by

$$\psi_1((x_1, \ldots, x_n), [\xi_1 : \cdots : \xi_n]) = \left((x_1, \ldots, x_n), \left(\frac{\xi_2}{\xi_1}, \ldots, \frac{\xi_n}{\xi_1}\right)\right).$$

In this affine local system of coordinates, a contact form which defines locally the contact structure K is given by

$$\alpha = dx_1 + \sum_{i=2}^{n} \frac{\xi_i}{\xi_1} dx_i.$$

We obtain similar expressions for the contact forms in the coordinate neighbourhoods $U \times V_i$, where $V_i = \{[\xi_1 : \cdots : \xi_n] \mid \xi_i \neq 0\}$, $i = 2, \ldots n$.

For the projective cotangent bundle $\pi : PT^*(\mathbb{R}^n \times \mathbb{R}) \to \mathbb{R}^n \times \mathbb{R}$, we have the special open set $V_\eta = \{[\xi_1 : \cdots : \xi_n : \eta] \in P(\mathbb{R}^n \times \mathbb{R})^* | \eta \neq 0\}$. We then choose the system of coordinates $((x_1, \ldots, x_n, y), (p_1, \ldots p_n))$ of $\mathbb{R}^n \times \mathbb{R} \times V_\eta$, where $p_i = -\xi_i/\eta$, $i = 1, \ldots, n$. In this local system of coordinates, the contact form in $PT^*(\mathbb{R}^n \times \mathbb{R})$ is given by

$$\alpha = dy - \sum_{i=1}^{n} p_i dx_i.$$

The contact manifold $\mathbb{R}^n \times \mathbb{R} \times V_\eta$ can thus be identified with the 1-jet space $J^1(\mathbb{R}^n, \mathbb{R})$. Consequently, the 1-jet space $J^1(\mathbb{R}^n, \mathbb{R})$ can be considered as a special affine coordinate neighbourhood of $PT^*(\mathbb{R}^n \times \mathbb{R})$.

Definition 5.4. Let (M_1, K_1) and (M_2, K_2) be two contact manifolds. A diffeomorphism $\phi : M_1 \to M_2$ which sends the contact structure K_1 to the contact structure K_2, that is, $d\phi(K_1) = K_2$, is called a *contactomorphism* between (M_1, K_1) and (M_2, K_2). The two contact manifolds are then said to be *contactomorphic*.

We have the contact version of the Darboux theorem.

Theorem 5.7 (The contact Darboux Theorem). *Any two contact manifolds of the same dimension are locally contactomorphic.*

Proof. See [Arnol'd, Guseĭn-Zade and Varchenko (1985)]. □

5.2.1 *Legendrian submanifolds and Legendrian maps*

Definition 5.5. Let (M, K) be a $(2n+1)$-dimensional contact manifold. We say that a smooth submanifold $\mathscr{L}$ of M is a *Legendrian submanifold* if $\dim \mathscr{L} = n$ and $T_p\mathscr{L} \subset K_p$ for any $p \in \mathscr{L}$.

Example 5.9. Let N be a smooth manifold and $f : N \to \mathbb{R}$ a smooth function. Then the 1-jet map $j^1 f : N \to J^1(N, \mathbb{R})$ is an embedding of N in $J^1(N, \mathbb{R})$. This map is given in a local system of coordinates $(x_1, \dots, x_n, p_1, \dots, p_n, y)$ of $J^1(N, \mathbb{R})$ by

$$j^1 f(x) = \left(x, \frac{\partial f}{\partial x_1}(x), \dots \frac{\partial f}{\partial x_n}(x), f(x) \right).$$

Then the pull-back of the natural contact structure $\alpha = dy - \sum_{i=1}^n p_i dx_i$ in $J^1(N, \mathbb{R})$ by $j^1 f$ is

$$j^1 f^* \alpha = df(x) - \sum_{i=1}^n \frac{\partial f}{\partial x_i}(x) dx_i = df(x) - df(x) = 0.$$

Therefore $j^1 f(N)$ is a Legendrian submanifold of $J^1(N, \mathbb{R})$.

Definition 5.6. Let $\pi : E \to N$ be a fibre bundle such that E is a contact manifold. We say that $\pi : E \to N$ is a *Legendrian fibration* if its fibres are Legendrian submanifolds of E.

Example 5.10. Consider the standard contact manifold $\mathbb{R}^{2n+1}$ (Example 5.6) and let $\pi : \mathbb{R}^{2n+1} \to \mathbb{R}^{n+1}$ be defined by $\pi(x_1, \ldots, x_n, p_1, \ldots, p_n, y) = (x_1, \ldots, x_n, y)$. Then π is a Legendrian fibration.

Example 5.11. The canonical projection $\pi : PT^*N \to N$ is a Legendrian fibration with respect to the canonical contact structure of PT^*N (Example 5.8).

Let $\pi : E \to N$ and $\pi' : E' \to N'$ be two Legendrian fibrations. A $\Psi : E \to E'$ is said to be a *Legendrian diffeomorphism* if there exists a diffeomorphism $\psi : N \to N'$ such that $\pi' \circ \Psi = \psi \circ \pi$.

Theorem 5.8. *All Legendrian fibrations of a fixed dimension are locally Legendrian diffeomorphic.*

Proof. See [Arnol'd, Guseĭn-Zade and Varchenko (1985)]. □

Let $\pi : E \to N$ be a Legendrian fibration and consider a Legendrian immersion $i : \mathscr{L} \to E$. The composition of the projection with $\pi \circ i : \mathscr{L} \to N$ is called a *Legendrian map*.

A *wavefront* is the image of a Legendrian map. The wavefront of the Legendrian map $\pi \circ i : \mathscr{L} \to N$ is denoted by $W(i(\mathscr{L}))$. The map $i : \mathscr{L} \to E$ is called the *Legendrian lift* of $W_{\mathscr{L}}$. The Legendrian submanifold $i(\mathscr{L})$ is also referred to as the Legendrian lift of $W(i(\mathscr{L}))$.

Two Legendrian immersions $i : \mathscr{L} \to E$ and $i' : \mathscr{L}' \to E'$ (or, Legendrian maps) are said to be *Legendrian equivalent* if there is a Legendrian diffeomorphism $\Psi : E \to E'$ such that $\Psi(i(\mathscr{L})) = i'(\mathscr{L}')$. If $i : \mathscr{L} \to E$ and $i' : \mathscr{L}' \to E'$ are Legendrian equivalent, then their wavefronts $W(i(\mathscr{L}))$ and $W(i'(\mathscr{L}'))$ are diffeomorphic.

5.2.2 *Legendrian singularities*

We consider the local singularities of Legendrian maps and work with $E = PT^*\mathbb{R}^n$ (see Theorem 5.8). Here too we have a result which describes a germ of Legendrian immersion $i : \mathscr{L} \to E$ in terms of a germ of families of functions.

Let $F : (\mathbb{R}^k \times \mathbb{R}^n, 0) \to (\mathbb{R}, 0)$ be an n-parameter family of germs of smooth function from $(\mathbb{R}^k, 0)$ to $(\mathbb{R}, 0)$. We say that F is a *Morse family of hypersurfaces* if the map-germ $\Delta_* F : (\mathbb{R}^k \times \mathbb{R}^n, 0) \to (\mathbb{R} \times \mathbb{R}^k, 0)$ given by

$$\Delta_* F(q, x) = \left(F, \frac{\partial F}{\partial q_1}, \ldots, \frac{\partial F}{\partial q_k} \right)(q, x)$$

is not singular.

When F is a Morse family of hypersurfaces, the set-germ

$$\Sigma_F = \left\{ (q,x) \in (\mathbb{R}^k \times \mathbb{R}^n, 0) \mid F(q,x) = \frac{\partial F}{\partial q_1}(q,x) = \cdots = \frac{\partial F}{\partial q_k}(q,x) = 0 \right\}$$

is a germ of a smooth $(n-1)$-dimensional submanifold of $(\mathbb{R}^k \times \mathbb{R}^n, 0)$. Then we have the map-germ $\mathscr{L}(F) : (\Sigma_F, 0) \to PT^*\mathbb{R}^n$ defined by

$$\mathscr{L}(F)(q,x) = \left(x, \left[\frac{\partial F}{\partial x_1}(q,x) : \cdots : \frac{\partial F}{\partial x_n}(q,x) \right] \right).$$

By definition the local contact form for K is given by $\alpha = \sum_{i=1}^n \xi_i dx_i$, where $((x_1, \ldots, x_n), [\xi_1 : \cdots : \xi_n])$ are homogeneous coordinates of $PT^*(\mathbb{R}^n)$. Therefore, we have

$$\mathscr{L}(F)^*\alpha = \sum_{i=1}^n \frac{\partial F}{\partial x_i} dx_i|_{\Sigma_F} = dF|_{\Sigma_F} = d(F|_{\Sigma_F}) = 0.$$

We can show that $\mathscr{L}(F)$ is an immersion-germ, so that it is a Legendrian immersion-germ.

The family of function-germs F is called the *generating family* of the germ of the Legendrian submanifold $\mathscr{L}(F)(\Sigma_F)$ of $PT^*\mathbb{R}^n$.

Theorem 5.9. *Let $\mathscr{L}$ be a germ of a Legendrian submanifold of $PT^*\mathbb{R}^n$. Then there exists a germ of a Morse family of hypersurfaces F such that $\mathscr{L}(F)(\Sigma_F) = \mathscr{L}$.*

Proof. See [Arnol'd, Guseĭn-Zade and Varchenko (1985)] and [Zakalyukin (1976)]. □

The wavefront of the germ of the Legendrian map $\pi \circ \mathscr{L}(F) : (\Sigma_F, 0) \to \mathbb{R}^n$ is the set-germ

$$W(\mathscr{L}(F)) = \left\{ x \in (\mathbb{R}^n, 0) \mid \exists q \in (\mathbb{R}^k, 0) \text{ such that } (q,x) \in (\Sigma_F, 0) \right\}.$$

This is precisely the discriminant of the family F (see Chapter 3).

Let $i : (\mathscr{L}, x) \to (PT^*\mathbb{R}^n, p)$ and $i' : (\mathscr{L}', x') \to (PT^*\mathbb{R}^n, p')$ be two germs of Legendrian immersions. We say that i and i' (or, the Legendrian map-germs $\pi \circ i$ and $\pi \circ i'$) are *Legendrian equivalent* if there exists a germ of a Legendrian diffeomorphism $\Psi : (PT^*\mathbb{R}^n, p) \to (PT^*\mathbb{R}^n, p')$ such that $\Psi(i(\mathscr{L})) = i'(\mathscr{L}')$.

A germ of a Legendrian immersion $i : (\mathscr{L}, x) \to (PT^*\mathbb{R}^n, p)$ (or, the Legendrian map-germ $\pi \circ i$) is said to be *Legendrian stable* if for any representative $\bar{i} : V \to PT^*\mathbb{R}^n$ of i, there exist a neighbourhood W of $\bar{i}$ (in the Whitney C^∞-topology on the space of Legendrian immersions considered as a subspace of $C^\infty(\mathbb{R}^{n-1}, PT^*\mathbb{R}^n))$ and a neighbourhood W of $\bar{i}$ with the following property: For any Legendrian immersion $\bar{j}$ in W, there exists $x' \in V$ such that $\pi \circ i$ and $\pi \circ j$ are Legendrian equivalent, where $j : (L, x') \to (PT^*\mathbb{R}^n, p')$ is the germ of $\bar{j}$ at x'.

Since the Legendrian lift $i : (\mathscr{L}, p) \to (PT^*\mathbb{R}^n, p)$ is uniquely determined on the regular part of the wavefront $W(i(\mathscr{L}))$, we have the following property of germs of Legendrian immersions [Zakalyukin (1984)].

Theorem 5.10. *Let $i_j : (\mathscr{L}_j, p_j) \longrightarrow (PT^*\mathbb{R}^n, p_\ell)$ be a germ of a Legendrian immersion, $j = 1, 2$. Suppose that there exist representatives $\bar{i}_j : U_j \to PT^*\mathbb{R}^n$ of the germs such that the set of the regular points of $\pi \circ \bar{i}_j$ is dense in U_j for each $j = 1, 2$. Then the following statements are equivalent.*

(1) *The germs of Legendrian immersions i_1 and i_2 are Legendrian equivalent.*

(2) *The germs of the wavefront sets $W(i_1(\mathscr{L}_1)), W(i_2(\mathscr{L}_2))$ are diffeomorphic.*

Proof. Statement (1) implies statement (2) by the definition of Legendrian equivalence. We now show that statement (2) implies statement (1). By the uniqueness of the contact lift of a local diffeomorphism on $\mathbb{R}^n$ to $PT^*\mathbb{R}^n$, we may assume that there are open neighbourhoods $V_j \subset U_j$ of p_j such that $\pi \circ \bar{i}_1(V_1) = \pi \circ \bar{i}_2(V_2)$. We may also assume that V_j is relatively compact and $\overline{V_j} \subset U_j$ for each $j = 1, 2$. Then, we have

$$\pi \circ \bar{i}_1(\overline{V_1}) = \overline{\pi \circ \bar{i}_1(V_1))} = \overline{\pi \circ \bar{i}_2(V_2)} = \pi \circ \bar{i}_2(\overline{V_2}).$$

By hypothesis, the set of regular points of $\pi \circ \bar{i}_j | \overline{V_j}$ is dense for each $j = 1, 2$.

We set $S = \pi \circ \bar{i}_1(\overline{V_1}) = \pi \circ \bar{i}_2(\overline{V_2})$, $Z_j = \{\pi \circ \bar{i}_j(u) \in S \mid u \in \overline{V_j}$ is a singular point of $\pi \circ \bar{i}_j\}$,$j = 1, 2$, $Z = Z_1 \cup Z_2$ and $R = S \setminus Z$. Then we show that $(\pi \circ \bar{i}_j)^{-1}(a)$ is a finite set for each $a \in S \setminus Z_j$. Otherwise, there exists a sequence $\{p_n\}$ such that $\pi \circ \bar{i}_j(p_n) = a$, $j = 1, 2$. By taking a subsequence, we may assume that $\{p_n\}$ converges to a point $p \in \overline{V_j}$. It follows that $\pi \circ \bar{i}_j(p) = a$. Since $a \in S \setminus Z_\ell$, there exists a neighbourhood V of p such that $\pi \circ \bar{i}_\ell | V$ is an embedding. For sufficient large n, we have $p_n \in V$ and $p_n \neq p$. However, we have $\pi \circ \bar{i}_\ell(p_n) = a = \pi \circ i_\ell(p)$. This contradicts the fact that $\pi \circ \bar{i}_\ell | V$ is an embedding. Therefore $(\pi \circ \bar{i}_\ell)^{-1}(a)$ is a finite set.

For any $a \in R$, we have

$$(\pi \circ \bar{i}_1)^{-1}(a) = \{p_1, \ldots, p_m\},\ (\pi \circ \bar{i}_2)^{-1}(a) = \{q_1, \ldots, q_l\}.$$

We have $PT^*\mathbb{R}^n \equiv \mathbb{R}^n \times (\mathbb{R}P^{n-1})^*$, so we get

$$\bar{i}_1(p_\ell) = (a, [\nu_\ell]) \text{ and } \bar{i}_2(q_k) = (a, [\xi_k]),$$

where $\nu_\ell, \xi_k \in (\mathbb{R}^n)^* = \mathrm{Hom}_{\mathbb{R}}(\mathbb{R}^n, \mathbb{R})$ and $[\nu_\ell], [\xi_k]$ denote the homogeneous coordinates of the dual projective space. Since $\bar{i}_j,\ j = 1,2$ are embeddings, $\nu_1, \ldots, \nu_m$ (respectively, $\xi_1, \ldots, \xi_l$) are mutually distinct. Here, a is a regular value of $\bar{i}_j$ $(j = 1,2)$, so that ν_ℓ (respectively, ξ_k) is considered to be the tangent hyperplane of one of the components of the hypersurface $\pi \circ \bar{i}_1(V_1)$ (respectively, $\pi \circ \bar{i}_2(V_2)$) through a. Since $\pi \circ \bar{i}_1(\overline{V_1}) = \pi \circ \bar{i}_2(\overline{V_2})$, we may conclude that $m = l$ and $\bar{i}_1(p_\ell) = \bar{i}_2(q_\ell)$ $(\ell = 1, \ldots, m)$. We set $W_j = (\pi \circ \bar{i}_j|\overline{V_j})^{-1}(R),\ j = 1,2$. It follows that $\bar{i}_1(W_1) = \bar{i}_2(W_2)$. By the continuity of $\bar{i}_j,\ j = 1,2$, we have $\bar{i}_1(\overline{W_1}) = \bar{i}_2(\overline{W_2})$. Therefore, it is enough to show that W_j is dense in $\overline{V_j}$. Suppose that $(\pi \circ i_j|\overline{V_j})^{-1}(Z)$ has an interior point. Since the set of regular points of $\pi \circ \bar{i}_j$ is dense in $\overline{V_j}$, there exists an open subset $O_\ell \subset V_j$ such that $\pi \circ \bar{i}_j(O_j) \subset Z$ and $\pi \circ \bar{i}_j|O_j$ is an immersion. For a point $q_j \in O_j$, let T_j be the tangent hyperplane of the regular hypersurface $\pi \circ \bar{i}_j(O_j)$ at q_j. It follows that we have a local diffeomorphism

$$\Phi_j : T_\ell \to \pi \circ L_j(O_j)$$

around q_j. Since Z_j is the critical value set of $\pi \circ \bar{i}_j$, $\Phi_j(Z_j)$ is the critical value set of $\Phi_j \circ \pi \circ \bar{i}_j,\ j = 1,2$. By Sard's theorem, $\Phi_j(Z)$ is a measure zero set. However, we have $\Phi_j \circ \pi \circ \bar{i}_j(O_j) \subset \Phi_j(Z)$. This is a contradiction, therefore $(\pi \circ \bar{i}_j|V_j)^{-1}(Z)$ does not have interior points. Since

$$\begin{aligned}\overline{Vj} &= (\pi \circ \bar{i}j|V_j)^{-1}(S) = (\pi \circ \bar{i}_j|\overline{V_j})^{-1}(R \cup Z)\\ &= (\pi \circ \bar{i}_j|\overline{V_j})^{-1}(R) \cup (\pi \circ \bar{i}_j|\overline{V_j})^{-1}(Z),\end{aligned}$$

we get that $W_j = (\pi \circ \bar{i}_j|\overline{V_j})^{-1}(R)$ is dense in $\overline{V_j}$. □

We remark that in the original proof in [Zakalyukin (1984)] it is assumed that $\pi \circ \bar{i}_j|_U$ are proper mappings for $j = 1,2$. This assumption is not needed. The idea of the above proof for removing this assumption is given in [Kokubu, Rossman, Saji, Umehara, and Yamada (2005)].

The assumption on the density of the regular sets of $\pi \circ \bar{i}$ in Theorem 5.10 is a generic condition on i. If i is Legendrian stable, then there exist a neighbourhood W of a representative $\bar{i} : V \to PT^*\mathbb{R}^n$ of i (in the Whitney C^∞-topology) such that for any Legendrian immersion $\bar{j} \in W$, there exists a point $x' \in V$ such that the germ $j : (V, x') \to PT^*\mathbb{R}^n$ is Legendrian equivalent to i. Since the above property is generic, there exists a Legendrian immersion $\bar{j} : V \to PT^*\mathbb{R}^n \in W$ and $x' \in V$ such that the germ

$j : (V, x') \to PT^*\mathbb{R}^n$ is Legendrian equivalent to i. Therefore, i satisfies this property.

The Legendrian equivalence can be interpreted using the generating families.

Theorem 5.11. (i) *Let* $F : (\mathbb{R}^k \times \mathbb{R}^n, 0) \to (\mathbb{R}, 0)$ *and* $G : (\mathbb{R}^l \times \mathbb{R}^n, 0) \to (\mathbb{R}, 0)$ *be two Morse families of hypersurfaces. Then the Legendrian map-germs* $\pi \circ \mathscr{L}(F)$ *and* $\pi \circ \mathscr{L}(G)$ *are Legendrian equivalent if and only if* F *and* G *are stably* P-$\mathcal{K}$*-equivalent.*

(ii) *The Legendrian map-germ* $\pi \circ \mathscr{L}(F)$ *is Legendrian stable if and only if* F *is a* $\mathcal{K}$*-versal deformation of* $f(q) = F(q, 0)$.

Proof. See [Arnol'd, Guseĭn-Zade and Varchenko (1985)] and [Zakalyukin (1976)]. □

If $k = l$ in Theorem 5.11 (i), we replace stably P-$\mathcal{K}$-equivalence by P-$\mathcal{K}$-equivalence.

For a given map-germ $f : (\mathbb{R}^n, 0) \to (\mathbb{R}^p, 0)$, we have the local ring of f given by

$$Q_r(f) = \frac{\mathcal{E}_n}{f^*(\mathcal{M}_p)\mathcal{E}_n + \mathcal{M}_n^{r+1}}.$$

Proposition 5.2. *Let* $F, G : (\mathbb{R}^k \times \mathbb{R}^n, (0,0)) \to (\mathbb{R}, 0)$ *be two Morse families of hypersurfaces. Suppose that* $\pi \circ \mathscr{L}(F)$ *and* $\pi \circ \mathscr{L}(G)$ *are Legendrian stable. Then the following statements are equivalent.*

(1) $(W_{\mathscr{L}(F)}, 0)$ *and* $(W_{\mathscr{L}(G)}, 0)$ *are diffeomorphic as germs.*
(2) $\mathscr{L}(F)$ *and* $\mathscr{L}(G)$ *are Legendrian equivalent.*
(3) $Q_{n+1}(f)$ *and* $Q_{n+1}(g)$ *are isomorphic as* $\mathbb{R}$*-algebras, where* $f = F|_{\mathbb{R}^k \times \{0\}}$ *and* $g = G|_{\mathbb{R}^k \times \{0\}}$.

Proof. By Theorem 5.11, F and G are $\mathcal{K}$-versal deformations of f and g respectively. Since $\pi \circ \mathscr{L}(F)$ and $\pi \circ \mathscr{L}(G)$ are Legendrian stable, they satisfy the generic condition of Theorem 5.10, so statements (1) and (2) are equivalent. If F and G are $\mathcal{K}$-versal deformations of f and g respectively, then f and g are $(n+1)$-$\mathcal{K}$-determined (see [Martinet (1982); Mather (1969b)]) so statement (3) implies that f and g are $\mathcal{K}$-equivalent. It follows by the uniqueness of the $\mathcal{K}$-versal deformation of a germ of a function that F and G are P-$\mathcal{K}$-equivalent. This shows that statement (3) implies (2). Now statement (2) implies statement (3) by Theorem 5.11. □

The classification of singularities of germs of functions and of their $\mathcal{K}$-versal unfoldings (Table 3.1 and Theorem 3.15) can now be used to obtain a list the normal forms of the singularities of Legendrian map-germs in low dimensions.

Theorem 5.12. *Let* $F : (\mathbb{R}^k \times \mathbb{R}^n, 0) \to (\mathbb{R}, 0)$ *be a Morse family of hypersurfaces. Suppose that* $\mathscr{L}(F) : (\Sigma_F, 0) \to PT^*\mathbb{R}^n$ *is Legendrian stable and* $n \le 4$. *Then* $\mathscr{L}(F)$ *is Legendrian equivalent to a Legendrian submanifold germ whose generating family* $G(q_1, \dots, q_k, x_1, \dots, x_n)$ *is one of the following germs, where* $Q(q_r, \dots, q_k) = \pm q_r^2 \pm \cdots \pm q_k^2$.

(1) $Q(q_1, \dots, q_k) + x_1$.
(2) $Q(q_2, \dots, q_k) + q_1^3 + x_1 + x_2 q_1$.
(3) $Q(q_2, \dots, q_k) + q_1^4 + x_1 + x_2 q_1 + x_3 q_1^2$.
(4) $Q(q_2, \dots, q_k) + q_1^5 + x_1 + x_2 q_1 + x_3 q_1^2 + x_4 q_1^3$.
(5) $Q(q_3, \dots, q_k) + q_1^3 - q_1 q_2^2 + x_1 + x_2 q_1 + x_3 q_2 + x_4(q_1^2 + q_2^2)$.
(6) $Q(q_3, \dots, q_k) + q_1^3 + q_2^3 + x_1 + x_2 q_1 + x_3 q_2 + x_4 q_1 q_2$.

Proof. By the assumption and Theorem 5.11 (2), F is a $\mathcal{K}$-versal deformation of $f = F|_{\mathbb{R}^k \times \{0\}}$. Since $n \le 4$, $\mathrm{cod}_e(f, \mathcal{K}) \le 4$. By Theorem 3.15, F is P-$\mathcal{K}$-equivalent to one of the germs in the above list. □

5.3 Graph-like Legendrian submanifolds

We consider some special Legendrian submanifolds in $PT^*(\mathbb{R}^n \times \mathbb{R})$, namely the so-called *graph-like Legendrian submanifolds.* These were introduced in [Izumiya (1993)] and have also application in geometrical optics. Recall from Example 5.8 that the manifold of 1-jets of functions $J^1(\mathbb{R}^n, \mathbb{R})$ can be considered as a coordinate neighbourhood of $PT^*(\mathbb{R}^n \times \mathbb{R})$.

A Legendrian immersion $i : \mathscr{L} \to PT^*(\mathbb{R}^n \times \mathbb{R})$ is said to be a *graph-like Legendrian immersion* if $i(\mathscr{L}) \subset J^1(\mathbb{R}^n, \mathbb{R}) \subset PT^*(\mathbb{R}^n \times \mathbb{R})$.

The manifold $J^1(\mathbb{R}^n, \mathbb{R})$ can be identified canonically with $T^*\mathbb{R}^n \times \mathbb{R}$ and we have the canonical projection

$$\widetilde{\pi} : J^1(\mathbb{R}^n, \mathbb{R}) \to T^*\mathbb{R}^n$$

defined by $\widetilde{\pi}(x, p, y) = (x, p)$.

In what follows $\pi : PT^*(\mathbb{R}^n \times \mathbb{R}) \to \mathbb{R}^n \times \mathbb{R}$ and $\rho : T^*\mathbb{R}^n \to \mathbb{R}^n$ are the bundle projections.

Proposition 5.3. *Let* $i : \mathscr{L} \to PT^*(\mathbb{R}^n \times \mathbb{R})$ *be a graph-like Legendrian immersion. Then* $\widetilde{\pi} \circ i : \mathscr{L} \to T^*\mathbb{R}^n$ *is a Lagrangian immersion.*

Proof. Given a tangent vector $v \in T_z i(\mathscr{L})$, there exist λ_ℓ, η_ℓ, $\ell = 1, \ldots, n$, and μ in $\mathbb{R}$, such that

$$v = \sum_{\ell=1}^{n} \lambda_\ell \frac{\partial}{\partial x_\ell} + \sum_{\ell=1}^{n} \eta_\ell \frac{\partial}{\partial p_\ell} + \mu \frac{\partial}{\partial y}$$

as a tangent vector of $J^1(\mathbb{R}^n, \mathbb{R})$. Then

$$d\widetilde{\pi}_z(v) = \sum_{\ell=1}^{n} \lambda_i \ell \frac{\partial}{\partial x_\ell} + \sum_{\ell=1}^{n} \eta_\ell \frac{\partial}{\partial p_\ell}.$$

Since $i(\mathscr{L})$ is a Legendrian submanifold, $\mu = \sum_{\ell=1}^{n} p_\ell \lambda_\ell$. If $d\widetilde{\pi}_z(v) = 0$ then $\lambda_i = \eta_\ell = 0$, $i = \ell, \ldots, n$, which implies $\mu = 0$. Therefore, $d\widetilde{\pi}_z(v) = 0$ implies $v = 0$. This proves that $\widetilde{\pi}_{|i(\mathscr{L})}$ is an immersion, so $\widetilde{\pi} \circ i : \mathscr{L} \to T^*\mathbb{R}^n$ is an immersion.

Let α denote the canonical contact 1-form in $PT^*(\mathbb{R}^n \times \mathbb{R})$. The condition $\alpha|_{i(\mathscr{L})} = 0$ implies $\omega = -d\alpha = 0$ on $\widetilde{\pi} \circ i(\mathscr{L})$, where ω is the canonical symplectic form in $T^*\mathbb{R}^n$. This shows that $\widetilde{\pi} \circ i(\mathscr{L})$ is a Lagrangian submanifold of $T^*\mathbb{R}^n$. □

We call $\rho \circ \widetilde{\pi} \circ i$ the *induced Lagrangian map* from the graph-like Legendrian immersion $i : \mathscr{L} \to PT^*(\mathbb{R}^n \times \mathbb{R})$.

Proposition 5.4. *Let $i : \mathscr{L} \to PT^*(\mathbb{R}^n \times \mathbb{R})$ be a graph-like Legendrian immersion. Then the Legendrian map $\pi \circ i : \mathscr{L} \to \mathbb{R}^n \times \mathbb{R}$ is singular at $x \in \mathscr{L}$ if and only if the Lagrangian map $\rho \circ \widetilde{\pi} \circ i : \mathscr{L} \to \mathbb{R}^n$ is singular at $x \in \mathscr{L}$.*

Proof. We consider the canonical projection $\pi_1 : \mathbb{R}^n \times \mathbb{R} \to \mathbb{R}^n$ and remark that

$$\pi_1 \circ \pi|_{J^1(\mathbb{R}^n, \mathbb{R})}(x, p, y) = \pi_1(p, y) = p = \rho(x, p) = \rho \circ \widetilde{\pi}(x, p, y).$$

Therefore, it is enough to show that $\pi|_{i(\mathscr{L})}$ is singular at $z = (x, p, y)$ if and only if $\pi_1 \circ \pi|_{i(\mathscr{L})}$ is singular at z.

Suppose that $\pi|_{i(\mathscr{L})}$ is not singular at $z \in i(\mathscr{L})$. Then $d\pi_z|_{T_z i(\mathscr{L})}$ is a monomorphism. Any $v \in T_x i(\mathscr{L})$ can be written in the form

$$v = \sum_{\ell=1}^{n} \lambda_i \frac{\partial}{\partial x_\ell} + \sum_{\ell=1}^{n} \eta_i \frac{\partial}{\partial p_j} + \mu \frac{\partial}{\partial y}$$

for some $\lambda_\ell, \eta_\ell, \mu \in \mathbb{R}$, $\ell = 1, \ldots, n$. Then

$$d(\pi_1 \circ \pi)_z(v) = \sum_{\ell=1}^{n} \lambda_i \frac{\partial}{\partial x_\ell}.$$

Suppose that $d(\pi_1 \circ \pi)_z(v) = 0$. Then $\lambda_\ell = 0$, $\ell = 1, \dots, n$, and this implies $\mu = 0$ since $\mathscr{L}$ is a Legendrian submanifold (see details in the proof of Proposition 5.3). Therefore

$$d\pi_z(v) = \sum_{i\ell=1}^{n} \lambda_i \frac{\partial}{\partial x_\ell} + \mu \frac{\partial}{\partial y} = 0,$$

which implies $v = 0$ as we assumed $\pi|_{i(\mathscr{L})}$ to be non-singular at z. This means that $d(\pi_1 \circ \pi)_z|_{T_z i(\mathscr{L})}$ is a monomorphism, so $\pi_1 \circ \pi|_{i(\mathscr{L})}$ is not singular at z.

For the converse of the assertion, suppose that $\operatorname{rank}(d(\pi|_{i(\mathscr{L})})_z) < n$. Then we have $\operatorname{rank}(d(\rho \circ \widetilde{\pi}|_{i(\mathscr{L})})_z) = \operatorname{rank}(d(\pi_1 \circ \pi|_{i(\mathscr{L})})_z) < n$. This completes the proof. □

A Legendrian submanifold is given locally by a generating family (Theorem 5.9). We shall show the generating family of a graph-like Legendrian submanifold has a special form.

Let $\mathcal{F} : (\mathbb{R}^k \times (\mathbb{R}^n \times \mathbb{R}), 0) \to (\mathbb{R}, 0)$ be a Morse family of hypersurfaces and let (q, x, y) denote the parameters in $\mathbb{R}^k \times (\mathbb{R}^n \times \mathbb{R})$. We say that $\mathcal{F}$ is a *graph-like Morse family of hypersurfaces* if $\frac{\partial \mathcal{F}}{\partial y}(0) \neq 0$. In the homogeneous system of coordinates $((x_1, \dots, x_n, y), [\xi_1 : \cdots : \xi_n : \eta])$ of $PT^*(\mathbb{R}^n \times \mathbb{R})$, the germ $\mathscr{L}(\mathcal{F})(\Sigma_{\mathcal{F}})$ belongs to the affine coordinate neighbourhood

$$\mathbb{R}^n \times V_\eta = \{((x_1, \dots, x_n, y), [\xi_1 : \cdots : \xi_n : \eta]) \mid \eta \neq 0\} = J^1(\mathbb{R}^n, \mathbb{R}).$$

Therefore, the germ $\mathscr{L}(\mathcal{F})(\Sigma_{\mathcal{F}})$ of the Legendrian submanifold determined by the graph-like Morse family of hypersurfaces $\mathcal{F}$ is a graph-like Legendrian submanifold germ. We then say that $\mathcal{F}$ is a *graph-like generating family* of $\mathscr{L}(\mathcal{F})(\Sigma_{\mathcal{F}})$. In fact, any germ of a graph-like Legendrian submanifold can be identified with $\mathscr{L}(\mathcal{F})(\Sigma_{\mathcal{F}})$, for some graph-like Morse family of hypersurfaces $\mathcal{F}$.

Let $\mathcal{F}$ be a graph-like Morse family of hypersurfaces. By the implicit function theorem, there is a Morse family of functions $F : (\mathbb{R}^k \times \mathbb{R}^n, 0) \to (\mathbb{R}, 0)$ such that

$$\mathcal{F}^{-1}(0) = \{(q, x, F(q, x)) \mid (q, x) \in (\mathbb{R}^k \times \mathbb{R}^n, (0, 0))\ \}.$$

Since $\mathcal{F}^{-1}(0)$ is a regular hypersurface, we have

$$\mathcal{E}_{k+n+1} \cdot \{\mathcal{F}(q, x, t)\} = \mathcal{E}_{k+n+1} \cdot \{F(q, x) - y\}.$$

This means that there exists a germ of function $\lambda : (\mathbb{R}^k \times \mathbb{R}^n \times \mathbb{R}, (0, 0, 0)) \to \mathbb{R})$ with $\lambda(0, 0, 0) \neq 0$ such that

$$\lambda(q, x, t) \cdot \mathcal{F}(q, x, t) = F(q, x) - y.$$

The families $\mathcal{F}(q,x,y)$ and $F(q,x)-y$ define the same germ of a graph-like Legendrian submanifold. Therefore $F(q,x)-y$ is a graph-like generating family of $\mathscr{L}(\mathcal{F})(\Sigma_{\mathcal{F}})$. In this case,

$$\Sigma_{\mathcal{F}} = \{(q,x,F(q,x)) \in (\mathbb{R}^k \times (\mathbb{R}^n \times \mathbb{R}),0) \mid (q,x) \in C_F\}$$

and $\mathscr{L}(\mathcal{F}) : (\Sigma_{\mathcal{F}},0) \to J^1(\mathbb{R}^n,\mathbb{R})$ is given by

$$\mathscr{L}(\mathcal{F})(q,x,F(q,x)) = (L(F)(q,x),F(q,x)) \in J^1(\mathbb{R}^n,\mathbb{R}) \equiv T^*\mathbb{R}^n \times \mathbb{R}.$$

Define a map $\mathfrak{L}(F) : (C_F,0) \to J^1(\mathbb{R}^n,\mathbb{R})$ by

$$\mathfrak{L}(F)(q,x) = \left(x, \frac{\partial F}{\partial x_1}(q,x),\ldots,\frac{\partial F}{\partial x_n}(q,x),F(q,x)\right).$$

Then $\mathfrak{L}(F)(C_F) = \mathscr{L}(\mathcal{F})(\Sigma_{\mathcal{F}})$.

We call $W(\mathfrak{L}(F)) = \pi(\mathfrak{L}(F)(C_F))$ the *graph-like wavefront* of the germ of the graph-like Legendrian submanifold $\mathfrak{L}(F)(C_F)$ and call F a generating family*generating family* of the germ of the graph-like Legendrian submanifold $\mathfrak{L}(F)(C_F)$.

For a given Morse family of functions F, we set $\widetilde{F}(q,x,y) = F(q,x)-y$. We have, by definition, $\widetilde{\pi}(\mathscr{L}(\mathcal{F})(q,x,F(q,x))) = L(F)(q,x)$, so that F is a generating family of the germ of the Lagrangian submanifold $\widetilde{\pi}(\mathfrak{L}(F)(C_F))$.

Proposition 5.5. *Let $F : (\mathbb{R}^k \times \mathbb{R}^n,0) \to (\mathbb{R},0)$ and $G : (\mathbb{R}^l \times \mathbb{R}^n,0) \to (\mathbb{R},0)$ be two Morse families of functions. If $\widetilde{\pi}(\mathfrak{L}_F(C_F))$ and $\widetilde{\pi}(\mathfrak{L}_G(C_G))$ are Lagrangian equivalent, then $\mathfrak{L}(F)(C_F)$ and $\mathfrak{L}(G)(C_G)$ are Legendrian equivalent.*

Proof. By the observations following Theorems 5.4 and 5.11, it is enough to consider the case when $k=l$. By Theorem 5.4, if $\widetilde{\pi}(\mathfrak{L}_F(C_F))$ and $\widetilde{\pi}(\mathfrak{L}_G(C_G))$ are Lagrangian equivalent then F and G are P-$\mathcal{R}^+$-equivalent. Then there exist a germ of a diffeomorphism $\Phi : (\mathbb{R}^k \times \mathbb{R}^n,0) \to (\mathbb{R}^k \times \mathbb{R}^n,0)$ of the form $\Phi(q,x) = (\phi_1(q,x),\phi_2(x))$ and a germ of a function $\alpha : (\mathbb{R}^n,0) \to \mathbb{R}$ such that $F \circ \Phi(q,x) = G(q,x) + \alpha(x)$. We define the germ of a diffeomorphism $\widetilde{\Phi} : (\mathbb{R}^k \times \mathbb{R}^n \times \mathbb{R},0) \to (\mathbb{R}^k \times \mathbb{R}^n \times \mathbb{R},0)$ by

$$\widetilde{\Phi}(q,x,y) = (\phi_1(q,x),\phi_2(x),y+\alpha(x)).$$

We have

$$\widetilde{F} \circ \widetilde{\Phi}(q,x,y) = F \circ \Phi(q,x) - y - \alpha(x) = G(q,x) - y = \widetilde{G}(q,x,y).$$

It follows by Theorem 5.11, that $\mathfrak{L}_F(C_F)$ and $\mathfrak{L}_G(C_G)$ are Legendrian equivalent. □

We have the following result as a consequence of the classification theorem of Lagrangian singularities (Theorem 5.5).

Theorem 5.13. *Let $F : (\mathbb{R}^k \times \mathbb{R}^n, 0) \to (\mathbb{R}, 0)$ be a Morse family of functions. Suppose that $\widetilde{\pi} \circ \mathfrak{L}_F : C_F \to T^*\mathbb{R}^n$ is Lagrangian stable and $n \leq 4$. Then $\widetilde{\pi} \circ \mathfrak{L}(F)$ is Lagrangian equivalent to $\widetilde{\pi} \circ \mathfrak{L}(G)$ where $G(q_1, \ldots, q_k, x_1, \ldots, x_n)$ is one of the following germs, where $Q(q_r, \ldots, q_k) = \pm q_r^2 \pm \cdots \pm q_k^2$.*

(1) $Q(q_2, \ldots, q_k) + q_1^3 + x_1 q_1$.
(2) $Q(q_2, \ldots, q_k) \pm q_1^4 + x_1 q_1 + x_2 q_1^2$.
(3) $Q(q_2, \ldots, q_k) + q_1^5 + x_1 q_1 + x_2 q_1^2 + x_3 q_1^3$.
(4) $Q(q_2, \ldots, q_k) \pm q_1^6 + x_1 q_1 + x_2 q_1^2 + x_3 q_1^3 + x_4 q_1^4 + x_5 q_1^5$.
(5) $Q(q_3, \ldots, q_k) + q_1^3 - q_1 q_2^2 + x_1 q_1 + x_2 q_2 + x_3 (q_1^2 + q_2^2)$.
(6) $Q(q_3, \ldots, q_k) + q_1^3 + q_2^3 + x_1 q_1 + x_2 q_2 + x_3 q_1 q_2$.
(7) $Q(q_3, \ldots, q_k) + q_1^2 q_2 + q_2^4 + x_1 q_1 + x_2 q_2 + x_3 q_1^2 + x_4 q_2^2$.

As a consequence, the germ of the graph-like Legendrian submanifold $\mathfrak{L}_F$ is Legendrian equivalent to $\mathfrak{L}_G$, where G is one of the above germs.

In the list of the above theorem, the germ of the Lagrangian map corresponding to the germ (1) is the fold map-germ and the germ of the corresponding graph-like wavefront is the cuspidal edge. In the case of germ (2) the Lagrangian map-germ is the Whitney cusp and the graph-like wavefront is the swallowtail.

5.4 Versal unfoldings and Morse families of functions

We summarise here the results in the previous sections which will be used to describe generic properties of submanifolds in Euclidean and Minkowski spaces. We start with the following observation.

Proposition 5.6. *If $F : (\mathbb{R}^k \times \mathbb{R}^n, 0) \to (\mathbb{R}, (0,0))$ is an $\mathcal{R}^+$-versal deformation of the germ $f = F|_{\mathbb{R}^k \times \{0\}}$, then F is a Morse family of functions.*

Proof. By Thom's splitting lemma (Lemma 3.1), we may assume that $f \in \mathcal{M}_k^3$. By Theorem 3.12, we have

$$L\mathcal{R}_e \cdot f + \mathbb{R} \cdot \left\{1, \dot{F}_1, \ldots, \dot{F}_n\right\} = \mathcal{E}_k,$$

where $\dot{F}_i(q) = \frac{\partial F}{\partial q_i}(q, 0)$. By Theorem 3.1, f is $\mathcal{R}^+$-finitely determined so that there exists $r \in \mathbb{N}$ such that $\mathcal{M}_k^r \subset L\mathcal{R}_e \cdot f$. Then we have

$$\mathbb{R} \cdot \left\{j^{r-1}\dot{F}_1, \ldots, j^{r-1}\dot{F}_n\right\} = \mathcal{M}_k / (L\mathcal{R}_e \cdot f + \mathcal{M}_k^r).$$

Since $f \in \mathcal{M}_k^3$, we have $\mathcal{LR}_e \cdot f + \mathcal{M}_k^r \subset \mathcal{M}_k^2$. Therefore, we have

$$\mathcal{M}_k/(\mathcal{LR}_e \cdot f + \mathcal{M}_k^r) = (\mathcal{M}_k/\mathcal{M}_k^2) \oplus (\mathcal{M}_k^2/(\mathcal{LR}_e \cdot f + \mathcal{M}_k^r))$$

and

$$n \geq \operatorname{cod}_e(f, \mathcal{R}) - 1 = \dim_{\mathbb{R}} \mathcal{M}_k/(\mathcal{LR}_e \cdot f + \mathcal{M}_k^r) \geq \dim_{\mathbb{R}} \mathcal{M}_k/\mathcal{M}_k^2 = k.$$

On the other hand, the Jacobian matrix of ΔF at $(0,0)$ is

$$J_{\Delta F}(0,0) = \left(\frac{\partial^2 f}{\partial q_i \partial q_j}(0), \frac{\partial^2 F}{\partial q_i \partial x_\ell}(0,0)\right) = \left(0, \frac{\partial^2 F}{\partial q_i \partial x_\ell}(0,0)\right).$$

Here, we have

$$j^1 \dot{F}_\ell = \sum_{i=1}^{k} \frac{\partial^2 F}{\partial q_i \partial x_\ell}(0,0) q_i \in \mathcal{M}_k/\mathcal{M}_k^2.$$

It follows that $\left\{j^1\dot{F}_1, \ldots, j^1\dot{F}_n\right\}$ generates the $\mathbb{R}$-vector space $\mathcal{M}_k/\mathcal{M}_k^2$. This means that

$$\operatorname{rank} J_{\Delta F}(0,0) = \operatorname{rank}\left(\frac{\partial^2 F}{\partial q_i \partial x_\ell}(0,0)\right) = k.$$

Therefore F is a Morse family of functions. □

Let $F : (\mathbb{R}^k \times \mathbb{R}^n, (0,0)) \to (\mathbb{R}, 0)$ be an $\mathcal{R}^+$-versal unfolding of f. Then we have a Lagrangian submanifold of $T^*\mathbb{R}^n$ through the following Lagrangian embedding

$$\begin{aligned} L(F) : (C_F, 0) &\to T^*\mathbb{R}^n \\ (q, x) &\mapsto (q, dF_x(q, x))). \end{aligned}$$

The bifurcation set B_F of F coincides with the caustic of the Lagrangian submanifold $L(F)(C_F, 0)$. Then the catastrophe map $\pi_{C_F} : (C_F, 0) \mapsto \mathbb{R}^n$ can be identified with the Lagrangian map. Thus, the stable singularities of catastrophe map under the perturbations of F are the stable Lagrangian singularities.

Consider now the germ of the extended family of functions associated to F,

$$\begin{aligned} \widetilde{F} : (\mathbb{R}^k \times \mathbb{R}^n \times \mathbb{R}, 0) &\to (\mathbb{R}, 0) \\ (q, x, y) &\mapsto F(q, x) - y. \end{aligned}$$

We denote by G_F the set $\widetilde{F}^{-1}(0) = Graph(F)$.

Consider the contact manifold $J^1(\mathbb{R}^n, \mathbb{R})$ with its standard contact structure and the legendre fibration $\pi : J^1(\mathbb{R}^n, \mathbb{R}) \to \mathbb{R}^n \times \mathbb{R}$. By using the natural identification of $J^1(\mathbb{R}^n, \mathbb{R})$ with $T^*\mathbb{R}^n \times \mathbb{R}$ we have an immersion

$$\begin{array}{ll} \mathfrak{L}(F) : (C_F, 0) & \to T^*\mathbb{R}^n \times \mathbb{R} \\ \qquad (q, x) & \mapsto ((q, d_q F_x), F(q, x)), \end{array}$$

whose image is a graph-like Legendrian submanifold of $J^1(\mathbb{R}^n, \mathbb{R})$. Moreover, the discriminant (or level bifurcation set) D_F of F is its corresponding (graph-like) wavefront $W(\mathfrak{L}(F))$.

We define

$$\Sigma W(\mathfrak{L}(F)) = \{(x, z) \in W_F |\ \det \mathrm{Hess}(f_x)(q) = 0, (q, x) \in C_F\ \},$$

which is the singular set of $W(\mathfrak{L}(F))$. Observe that we have the following commutative diagram

$$\begin{array}{ccccc} \mathfrak{L}(F)(C_F) & \overset{i}{\subset} & T^*\mathbb{R}^n \times \mathbb{R} & \overset{\pi}{\longrightarrow} & \mathbb{R}^n \times \mathbb{R} \\ \mathfrak{L}(F) \uparrow & & \downarrow \bar{\pi} & & \downarrow \pi_1 \\ C_F & \overset{L(F)}{\longrightarrow} & T^*\mathbb{R}^n & \overset{\rho}{\longrightarrow} & \mathbb{R}^n \end{array}$$

where $i_{\mathcal{L}_F}$ is the inclusion, π and $\bar{\pi}$ are the canonical Legendre fibration, and $\tilde{\pi}_1$ and π_1' are the natural projections onto the first factor. Also note that $\pi_{C_F} = \rho \circ L(F)$. Therefore,

(a) $\pi(\mathfrak{L}(F)(C_F)) = W(\mathfrak{L}(F))$,

(b) $\pi_1(\Sigma(W(\mathfrak{L}(F))) = B_F$.

Figures 5.1, 5.2 and 5.3 illustrate the above relations for the fold (A_2) and cusp (A_3) singularities.

5.5 Families of functions on hypersurfaces in $\mathbb{R}^n$

In the subsequent section (unless stated otherwise), M denotes a hypersurface patch in the Euclidean space $\mathbb{R}^n$ parametrised by $\mathbf{x} : U \to \mathbb{R}^n$. We defined in Chapter 2 three families of functions on M which measure its contact with totally umbilic hypersurfaces in $\mathbb{R}^n$. These are

(1) The family of height functions $H : U \times S^{n-1} \to \mathbb{R}$ given by

$$H(u, \mathbf{v}) = \langle \mathbf{x}(u), \mathbf{v} \rangle.$$

(2) The extended family of height functions $\widetilde{H} : U \times (S^{n-1} \times \mathbb{R}) \to \mathbb{R}$ given by

$$\widetilde{H}(u, \mathbf{v}, r) = \langle \mathbf{x}(u), \mathbf{v} \rangle - r.$$

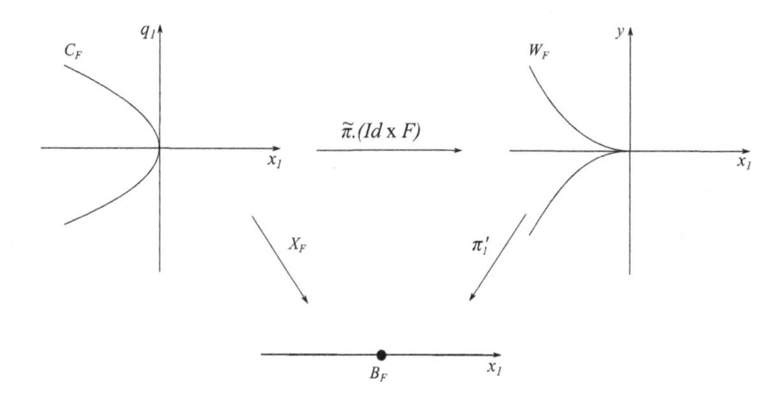

Fig. 5.1 Catastrophe set, wavefront and bifurcation set of a 1-parameter versal unfolding of the fold singularity.

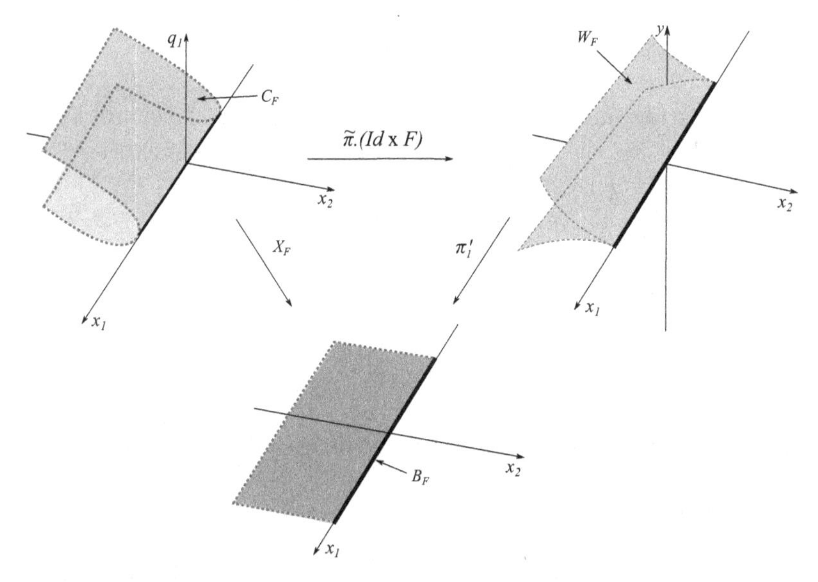

Fig. 5.2 Catastrophe set, wavefront and bifurcation set of a 2-parameter versal unfolding of the fold singularity.

(3) The family of distance squared functions $D : U \times \mathbb{R}^n \to \mathbb{R}$, given by

$$D(u, a) = \|\mathbf{x}(u) - a\|^2.$$

It follows from Chapter 4 that the germs of the families of height and distance squared functions associated to any immersion lying in a residual subset of $Imm(\mathbb{R}^{n-1}, \mathbb{R}^n)$ (referred to as generic immersion) can be seen as

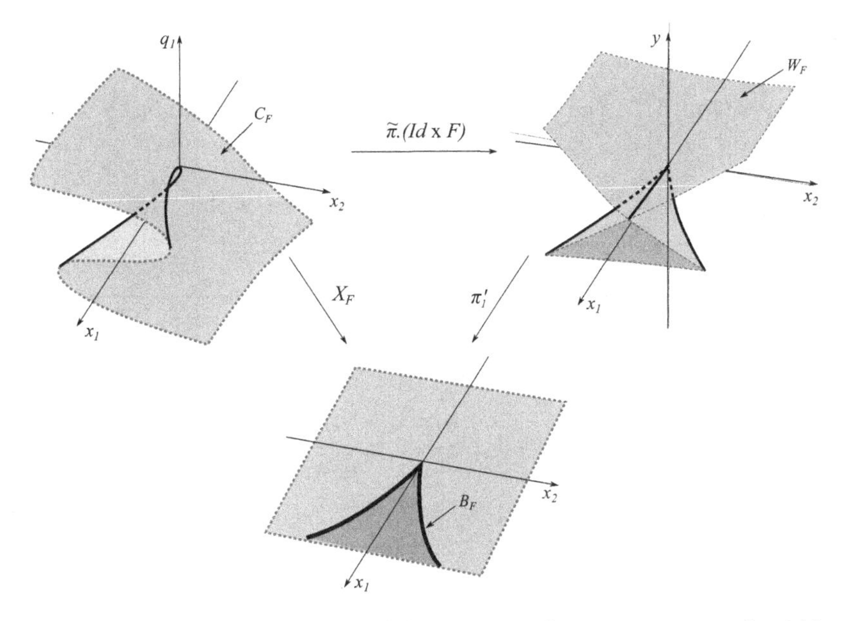

Fig. 5.3 Catastrophe set, wavefront and bifurcation set of a 2-parameter versal unfolding of the cusp singularity.

versal unfoldings of the germs of the corresponding function at each one of their points. Clearly, in such case, they must be Morse families of functions. We prove in the subsections below that the families of height functions and of distance squared functions are Morse families of functions at each point in their domain, regardless of the embeddings that induce them be generic or not. They are thus generating families of some Lagrangian manifolds (at least locally). We show that their associated caustics are geometric objects related to the hypersurface M. In the case of generic embeddings, these Lagrangian manifolds correspond to stable Lagrangian maps and the local behaviour of their caustics is well known. We also prove that the extended family of height functions is a Morse family of hypersurfaces at each point in their domain. It is therefore a generating family of a Legendrian manifold (at least locally), and the associated wavefront is also a geometric object related to the hypersurface M.

5.5.1 *The family of height functions*

Proposition 5.7. *The germ of the family of height functions H on M at each point $(u, \mathbf{v}) \in U \times S^{n-1}$ is a Morse family of functions.*

Proof. Let $\mathbf{v} = (v_1, \dots, v_n) \in S^{n-1}$ and assume, without loss of generality, that $v_n > 0$. Then $v_n = \sqrt{1 - v_1^2 - \dots - v_{n-1}^2}$ and

$$H(u, \mathbf{v}) = \mathbf{x}_1(u)v_1 + \ldots + \mathbf{x}_{n-1}(u)v_{n-1} + \mathbf{x}_n(u)\sqrt{1 - v_1^2 - \dots - v_{n-1}^2}.$$

We need to prove that the mapping

$$\Delta H(u, \mathbf{v}) = \left(\frac{\partial H}{\partial u_1}, \dots, \frac{\partial H}{\partial u_{n-1}}\right)(u, \mathbf{v})$$

is not singular at any point $(u, \mathbf{v}) \in C_H$. Its Jacobian matrix is given by

$$\begin{pmatrix} \langle \mathbf{x}_{u_1 u_1}(u), \mathbf{v}\rangle & \cdots & \langle \mathbf{x}_{u_1 u_{n-1}}(u), \mathbf{v}\rangle & \\ \vdots & \vdots & \vdots & A \\ \langle \mathbf{x}_{u_{n-1} u_1})(u), \mathbf{v}\rangle & \cdots & \langle \mathbf{x}_{u_{n-1} u_{n-1}}(u), \mathbf{v}\rangle & \end{pmatrix},$$

with

$$A = \begin{pmatrix} (\mathbf{x}_1)_{u_1}(u) - (\mathbf{x}_n)_{u_1}(u)\dfrac{v_1}{v_n} & \cdots & (\mathbf{x}_{n-1})_{u_1}(u) - (\mathbf{x}_n)_{u_1}(u)\dfrac{v_{n-1}}{v_n} \\ \vdots & \vdots & \vdots \\ (\mathbf{x}_1)_{u_{n-1}}(u) - (\mathbf{x}_n)_{u_{n-1}}(u)\dfrac{v_1}{v_n} & \cdots & (\mathbf{x}_{n-1})_{u_{n-1}}(u) - (\mathbf{x}_n)_{u_{n-1}}(u)\dfrac{v_{n-1}}{v_n} \end{pmatrix}$$

We show that the rank of the matrix A is $n - 1$ at $(u, \mathbf{v}) \in C(H)$. We set

$$a_i = \begin{pmatrix} (\mathbf{x}_i)_{u_1}(u) \\ \vdots \\ (\mathbf{x}_i)_{u_{n-1}}(u) \end{pmatrix} \quad \text{for } i = 0, \dots, n$$

and rewrite the matrix A in the form

$$A = \left(a_1 - a_n\frac{v_1}{v_n}, \dots, a_{n-1} - a_n\frac{v_{n-1}}{v_n}\right).$$

By Proposition 2.4, we have

$$C_H = \{(u, \pm N(u)) \mid u \in U\}.$$

Therefore,

$$\begin{aligned}\det A &= (-1)^{n+1}\frac{v_1}{v_n}\det(a_2,\ldots,a_n) + \cdots + (-1)^{2n}\frac{v_n}{v_n}\det(a_1,\ldots,a_{n-1}) \\ &= (-1)^{n-1}\left\langle\left(\frac{v_1}{v_n},\ldots,\frac{v_n}{v_n}\right), \mathbf{x}_{u_1}\times\cdots\times\mathbf{x}_{u_{n-1}}\right\rangle \\ &= \frac{(-1)^{n-1}}{v_n}\langle\pm N, \mathbf{x}_{u_1}\times\cdots\times\mathbf{x}_{u_{n-1}}\rangle \\ &= \pm\frac{(-1)^{n-1}}{v_n}\|\mathbf{x}_{u_1}\times\cdots\times\mathbf{x}_{u_{n-1}}\|\end{aligned}$$

and this is not zero for any $(u,\mathbf{v}) \in C_H$. This proves that H is a Morse family of functions. □

As a consequence of Proposition 5.7, we can define a germ of a Lagrangian immersion whose generating family is the family of height functions on the hypersurface patch M (Theorem 5.3).

We consider the standard coordinate neighbourhoods

$$U_i^+ = \{\mathbf{v} = (v_1,\ldots,v_n) \in S^{n-1} \mid v_i > 0\},\ i = 1,\ldots,n$$

and

$$U_i^- = \{\mathbf{v} = (v_1,\ldots,v_n) \in S^{n-1} \mid v_i < 0\},\ i = 1,\ldots,n$$

of the $(n-1)$-sphere S^{n-1}. Since $T^*S^{n-1}|U_i^\pm$ is a trivial bundle, we define the maps

$$L_i^\pm(H) : C_H \to T^*S^{n-1}|U_i^\pm,\ \ i = 1,\ldots,n$$

by

$$\begin{aligned}&L_i^\pm(H)(u,\pm N(u)) \\ &= \left(\pm N(u), \mathbf{x}_1(u) - \mathbf{x}_i(u)\tfrac{v_1}{v_i},\ldots,\widehat{\mathbf{x}_i(u) - \mathbf{x}_i(u)}\tfrac{v_i}{v_i},\ldots,\mathbf{x}_n(u) - \mathbf{x}_i(u)\tfrac{v_n}{v_i}\right),\end{aligned}$$

where $(x_1,\ldots,\hat{x}_i,\ldots,x_n)$ denotes the point in $\mathbb{R}^{n-1}$ obtained by removing the i-th component of $(x_1,\ldots,x_i,\ldots,x_n)$ in $\mathbb{R}^n$.

We have the following consequence of Proposition 5.7.

Corollary 5.1. *The maps $L_i^\pm(H)$, $i = 1,\ldots,n$, are Lagrangian immersions and the family of height functions H is their generating family at each point of $U \times U_i^\pm$.*

Proof. Since we have $v_i = \pm\sqrt{1 - \sum_{j \neq i} v_j^2}$ on $U_i^{\pm}$,

$$\frac{\partial H}{\partial v_j}(u, \mathbf{v}) = \mathbf{x}_j(u) - \mathbf{x}_i(u)\frac{v_j}{v_i}, \ (j \neq i),$$

at $(u, \mathbf{v}) \in U \times U_i^{\pm}$. By the construction of the germ of the Lagrangian immersion from the generating family in Section 5.1.2, $L_i^{\pm}(H)$ is a Lagrangian immersion. □

Remark 5.2. 1. The subset $C_H = \{(u, \pm N(u)) \ |u \in U \ \}$ is a double cover of the hypersurface M. It is not difficult to see that C_H is the catastrophe set of H and the corresponding catastrophe map can be identified with the normal Gauss map (up to a sign) on the hypersurface M. Thus, for a generic embedding $\mathbf{x}$, the Gauss map of $M = \mathbf{x}(U)$ behaves as a stable Lagrangian map.

2. We can apply the same construction to a submanifold M of codimension higher than one. In this case, the catastrophe set C_H is the unit normal bundle and it can be identified with a canal hypersurface around M in $\mathbb{R}^n$. The catastrophe map coincides with the generalised Gauss map on C_H.

5.5.2 *The extended family of height functions*

Proposition 5.8. *The extended family of height functions $\widetilde{H}$ on M is a graph-like Morse family of hypersurfaces.*

Proof. Since the height function H is a Morse family of fucntions at each point, the extended height function $\widetilde{H}(u, \mathbf{v}, r) = H(u, \mathbf{v}) - r$ is a graph-like Morse family of hypersurface at each point in $U \times (S^{n-1} \times \mathbb{R})$ (see §5.3 and the discussion before Proposition 5.5). □

We can define now a graph-like Legendrian immersion whose generating family at each point in the domain is the germ of the extended family of height functions $\widetilde{H}$.

We consider the contact manifold $\mathbb{R}^n \times S^{n-1}$ whose contact structure is given by the 1-form $\theta = \langle \mathbf{v}, dx \rangle|_{\{\|\mathbf{v}\|=1\}}$, where $(x, \mathbf{v}) \in \mathbb{R}^n \times S^{n-1}$. Since the tangent bundle $T\mathbb{R}^n$ is a trivial bundle, the above contact manifold is identified canonically with the unit tangent bundle $T_1\mathbb{R}^n$. The contact structure on $T_1\mathbb{R}^n$ is given by the above 1-form ([Blair (1976)]).

We consider the projection $\phi : \mathbb{R}^n \times S^{n-1} \to S^{n-1} \times \mathbb{R}$ definded by

$$\phi(x, \mathbf{v}) = (\mathbf{v}, \langle x, \mathbf{v} \rangle).$$

Since $\langle \mathbf{v}, dx\rangle|_{\{\|\mathbf{v}\|=1\}} = -\langle d\mathbf{v}, x\rangle|_{\{\|\mathbf{v}\|=1\}}$, $\theta = 0$ on the set $\mathbf{v} =$ constant, so π is a Legendrian fibration. We now define the map $\Phi : \mathbb{R}^n \times S^{n-1} \to T^*S^{n-1} \times \mathbb{R}$ by

$$\Phi(\mathbf{v}, x) = (\mathbf{v}, d_v\langle x, \mathbf{v}\rangle, \langle x, \mathbf{v}\rangle),$$

where $d_v\langle x, \mathbf{v}\rangle$ is defined as follows: For any $\mathbf{v} \in S^{n-1}$, we may consider that $T_vS^{n-1} \subset T_v\mathbb{R}^n$. Define a linear mapping $d_v\langle x, \mathbf{v}\rangle : T_vS^{n-1} \to \mathbb{R}$ by

$$d_v\langle x, \mathbf{v}\rangle(\eta) = \sum_{i=1}^{n} \frac{\partial\langle x, \mathbf{v}\rangle}{\partial v_i}\eta_i = \sum_{i=1}^{n} x_i\eta_i$$

where $x = (x_1, \ldots, x_n)$, $\mathbf{v} = (v_1, \ldots, v_n)$ and $\eta = (\eta_1, \ldots, \eta_n)$ in the basis $\{\partial/\partial v_1, \ldots, \partial/\partial v_n\}$ of $T_v\mathbb{R}^n$.

We have the projection $\widetilde{\pi} : T^*S^{n-1} \to S^{n-1}$ of the cotangent bundle. By the canonical identification $T^*S^{n-1} \times \mathbb{R} \equiv J^1(S^{n-1}, \mathbb{R})$, $T^*S^{n-1} \times \mathbb{R}$ is a contact manifold whose contact form is $dy - \lambda$, where λ is the Liouville form of T^*S^{n-1}. We consider the standard coordinate neighbourhoods

$$U_i^+ = \{\mathbf{v} = (v_1, \ldots, v_n) \in S^{n-1} \mid v_i > 0\ \},\ i = 1, \ldots, n$$

and

$$U_i^- = \{\mathbf{v} = (v_1, \ldots, v_n) \in S^{n-1} \mid v_i < 0\ \},\ i = 1, \ldots, n$$

of the $(n-1)$-sphere S^{n-1}.

Proposition 5.9. *The map $\Phi|_{\mathbb{R}^n \times U_i^\pm} : \mathbb{R}^n \times U_i^\pm \to T^*S^{n-1}|_{U_i^\pm} \times \mathbb{R}$ is a Legendrian diffeomorphism for $i = 1, \ldots, n$.*

Proof. We prove the proposition on the local coordinate neighbourhood

$$U_1^+ = \{\mathbf{v} = (v_1, \ldots, v_n) \in S^{n-1} \mid v_1 > 0\}.$$

On U_1^+, we have

$$d_v\langle x, \mathbf{v}\rangle = \langle x, d\mathbf{v}\rangle = \sum_{i=1}^{n} x_i dv_i||_{\{\|\mathbf{v}\|=1\}}.$$

Since $v_1 = \sqrt{1 - \sum_{j=2}^{n} v_j^2}$, we have $dv_1 = \sum_{i=2}^{n} -(v_i/v_1)dv_i$, so that

$$\langle x, d\mathbf{v}\rangle|_{U_1^+} = \sum_{j=2}^{n}\left(x_i - \frac{v_i}{v_1}x_1\right)dv_j.$$

Therefore, the local representation of Φ is given as

$$\Phi(x, \mathbf{v}) = \left(\mathbf{v}, \sum_{j=2}^{n}\left(x_j - \frac{v_j}{v_1}x_1\right)dv_j, \langle x, \mathbf{v}\rangle\right).$$

It follows that

$$\begin{aligned}\Phi^*(dy - \sum_{j=2}^{n} p_j dv_j) &= d\langle x, \mathbf{v}\rangle|_{U_1^+} - \sum_{j=1}^{n}\left(x_j - \frac{v_j}{v_1}x_1\right)dv_j \\ &= \langle dx, \mathbf{v}\rangle|_{U_1^+} + \langle x, d\mathbf{v}\rangle|_{U_1^+} - \langle x, d\mathbf{v}\rangle|_{U_1^+} \\ &= \langle dx, \mathbf{v}\rangle|_{U_1^+} = \theta|_{\mathbb{R}^n\times U_1^+}.\end{aligned}$$

Let $\Psi : (T^*S^{n-1}|_{U_1^+}) \times \mathbb{R} \to \mathbb{R}^n \times U_1^+$ be defined by

$$\Psi(\mathbf{v}, p, y) = (v_1(y - \sum_{j=2}^{n} p_j v_j), \dots, v_n(y - \sum_{j=2}^{n} p_j v_j), \mathbf{v}).$$

Then we have $\Psi \circ \Phi|_{\mathbb{R}^n\times U_1^+} = 1_{\mathbb{R}^n\times U_1^+}$ and $\Phi|_{\mathbb{R}^n\times U_1^+} \circ \Psi = 1_{(T^*S^{n-1}|_{U_1^+})\times\mathbb{R}}$. We proceed similarly for the other coordinate neighbourhoods. □

We define a map $\mathscr{L}_M : U \to \mathbb{R}^n \times S^{n-1}$ by $\mathscr{L}_M(u) = (\mathbf{x}(u), N(u))$. Then we have $\mathbf{x}^*(u) = \phi(\mathscr{L}_M(u))$.

Let $U_i^\pm$, $i = 1, \dots, n$, be the standard coordinate neighbourhoods of S^{n-1}. Since $PT^*(S^{n-1}\times\mathbb{R})_{|(U_i^\pm\times\mathbb{R})}$ is a trivial bundle, we define for $i = 1, \dots, n$, the map

$$\mathfrak{L}_{H,i} : C_H|_{U\times U_i^\pm} \to T^*S^{n-1}\times\mathbb{R}|_{U_i^\pm\times\mathbb{R}} \equiv J^1(U_i^\pm, \mathbb{R})$$

by $\mathfrak{L}_{H,i}(u, N(u)) = (N(u), \boldsymbol{\xi}(u), \langle N(u), \mathbf{x}(u)\rangle)$, where

$$\boldsymbol{\xi}(u) = \left(\mathbf{x}_1(u) - \mathbf{x}_i(u)\frac{v_1}{v_i} : \cdots : \widehat{\mathbf{x}_i(u) - \mathbf{x}_i(u)\frac{v_i}{v_i}} : \cdots : \mathbf{x}_n(u) - \mathbf{x}_i(u)\frac{v_n}{v_i}\right)$$

and $N(u) = (v_1, \dots. v_n)$. It follows from Corollary 5.1 that $\mathfrak{L}_{H,i}$ is a graph-like Legendrian immersion. We have

$$\Phi \circ \mathscr{L}_M(u) = \mathfrak{L}_{H,i}(u, N(u)) \text{ for } u \in \{u \in U\ |N(u) \in U_i^\pm\}.$$

We remark that $J^1(S^{n-1}, \mathbb{R}) = \cup_{i=1}^n J^1(U_i^\pm, \mathbb{R})$. Since Φ is an immersion, $\Phi \circ \mathscr{L}_M : U \to J^1(S^{n-1}, \mathbb{R})$ is a graph-like Legendrian immersion.

Theorem 5.14. *Let M be a hypersurface patch in $\mathbb{R}^n$. The map $\Phi \circ \mathscr{L}_M$ is a graph-like Legendrian immersion and the germ of the extended family of height functions $\widetilde{H}$ is its graph-like generating family at each point. The graph-like wavefront $W(\mathscr{L}_M)$ of $\Phi\circ\mathscr{L}_M$ is the cylindrical pedal $M^* = \mathbf{x}^*(U)$ of M. Moreover, the Gauss map $N : M \to S^{n-1}$ is the corresponding Lagrangian map. (As a consequence, the cylindrical pedal of a hypersurface in $\mathbb{R}^n$ has Legendrian singularities.)*

5.5.3 *The family of distance squared functions*

Proposition 5.10. *The germ defined by the family of the distance squared functions D on M at each point $(u_0, a_0) \in U \times \mathbb{R}^n$ is a Morse family of functions.*

Proof. We have

$$D(u, a) = \sum_{i=1}^{n} (\mathbf{x}_i(u) - a_i)^2,$$

where $\mathbf{x}(u) = (\mathbf{x}_1(u), \dots, \mathbf{x}_n(u))$, $u = (u_1, \dots, u_{n-1}) \in U$ and $a = (a_1 \dots, a_n) \in \mathbb{R}^n$. We shall prove that the mapping

$$\Delta D(u, a) = \left(\frac{\partial D}{\partial u_1}, \dots, \frac{\partial D}{\partial u_{n-1}} \right) (u, a)$$

is not singular at any point (u, a). Its Jacobian matrix is given by

$$\begin{pmatrix} A_{11} & \cdots & A_{1(n-1)} & -2(\mathbf{x}_1)_{u_1}(u) & \cdots & -2(\mathbf{x}_n)_{u_1}(u) \\ \vdots & \vdots & \vdots & \vdots & \vdots & \vdots \\ A_{(n-1)1} & \cdots & A_{(n-1)(n-1)} & -2(\mathbf{x}_1)_{u_{n-1}}(u) & \cdots & -2(\mathbf{x}_n)_{u_{n-1}}(u) \end{pmatrix},$$

where $A_{ij} = 2(\langle \mathbf{x}_{u_i u_j}(u), \mathbf{x}(u) - a \rangle + \langle \mathbf{x}_{u_i}(u), \mathbf{x}_{u_j}(u) \rangle)$. Since $\mathbf{x} : U \to \mathbb{R}^n$ is an embedding, the rank of the matrix

$$\mathbf{x} = \begin{pmatrix} -2(\mathbf{x}_1)_{u_1}(u) & \cdots & -2(\mathbf{x}_n)_{u_1}(u) \\ \vdots & \vdots & \vdots \\ -2(\mathbf{x}_1)_{u_{n-1}}(u) & \cdots & -2(\mathbf{x}_n)_{u_{n-1}}(u) \end{pmatrix}$$

is $n-1$ at each point u_0 in U. Therefore, the rank of the Jacobian matrix of ΔD is $n-1$, and this proves that the germ of D at (u_0, a_0) is a Morse family of functions. □

As a consequence of Proposition 5.10, we can define a germ of a Lagrangian immersion whose generating family is the family distance squared functions on the hypersurface patch M (Theorem 5.3).

Consider the set,

$$C_D = \{(u, \mathbf{x}(u) + \lambda N(u)) \mid \lambda \in \mathbb{R}, u \in U\}$$

and define the smooth mapping $L(D) : C_D \to T^*\mathbb{R}^n$ by

$$L(D)(u, \mathbf{x}(u) + \lambda N(u)) = (\mathbf{x}(u) + \lambda N(u), 2\lambda N(u)).$$

Corollary 5.2. *The map $L(D)$ is a Lagrangian immersion and the family of distance squared functions D is its generating family at each point in $U \times \mathbb{R}^n$. The caustic of the Lagrangian map $\pi \circ L(D)$ is precisely the focal set (or evolute) of M. (As a consequence, the focal set of a hypersurface in $\mathbb{R}^n$ has Lagrangian singularities.)*

Proof. By definition,

$$\frac{\partial D}{\partial a_i}(u,a) = -2(x_i(u) - a_i),$$

so that

$$\left(\frac{\partial D}{\partial a_1}(u,\mathbf{x}(u)+\lambda N(u)),\ldots,\frac{\partial D}{\partial a_n}(u,\mathbf{x}(u)+\lambda N(u))\right) = 2\lambda N(u).$$

By the construction of the Lagrangian immersion from the generating family in §5.1, we have a Lagrangian immersion

$$L(D)(u,\mathbf{x}(u)+\lambda N(u)) = (\mathbf{x}(u)+\lambda N(u), 2\lambda N(u)).$$

Therefore the Lagrangian map is given by

$$\pi \circ L(D)(u,\mathbf{x}(u)+\lambda N(u)) = \mathbf{x}(u)+\lambda N(u).$$

□

5.6 Contact from the viewpoint of Lagrangian and Legendrian singularities

We considered in Chapter 4 the theory of contact between two submanifolds and showed that the height (resp. distance squared) functions on a hypersurface in the Euclidean space measures the contact between the hypersurface and hyperplanes (resp. hyperspheres). In this section we apply the Lagrangian and Legendrian singularity theory to the study of this contact.

5.6.1 *Contact of hypersurfaces with hyperplanes*

First we consider the contact of hypersurfaces with hyperplanes. Let $\mathbf{x}_i^* : (U,u_i) \to (S^{n-1}\times\mathbb{R}, (\mathbf{v}_i, r_i))$, $i = 1,2$, be two germs of cylindrical pedal hypersurface patches M_i parametrised by $\mathbf{x}_i : (U,u_i) \to (\mathbb{R}^n, p_i)$.

Suppose that the regular set of $M_i^* = \mathbf{x}_i^*(U)$,$i = 1,2$, is dense in some neighbourhood V_i of u_i in U. Then, by Theorem 5.10, the germs $(M_1^*, (\mathbf{v}_1, r_1))$ and $(M_2^*, (\mathbf{v}_2, r_2))$ are diffeomorphic if and only if the corresponding Legendrian immersion germs $\mathscr{L}_{M_1} : (U,u_1) \to (\mathbb{R}^n\times S^{n-1}(p_1,\mathbf{v}_1))$ and $\mathscr{L}_{M_2} : (U,u_2) \to (\mathbb{R}^n \times S^{n-1}, (p_2,\mathbf{v}_2))$ are Legendrian equivalent. Since the germ of the extended family of height functions $\widetilde{H}_i$ on M_i, $i = 1,2$, is a generating family of the germ of the graph-like Legendrian immersion $\Phi \circ \mathscr{L}_{M_i}$, it follows by Theorem 5.11 that the condition that $(\mathscr{L}_{M_1}(U), (p_1,\mathbf{v}_1))$ and $(\mathscr{L}_{M_2}(U), (p_2,\mathbf{v}_2))$ are Legendrian equivalent is equivalent to the two generating families $\widetilde{H}_1$ and $\widetilde{H}_2$ being P-$\mathcal{K}$-equivalent.

We have the (universal) family of height functions $\mathcal{H} : \mathbb{R}^n \times S^{n-1} \to \mathbb{R}$ given by $\mathcal{H}(y, \mathbf{v}) = \langle y, \mathbf{v} \rangle = \mathfrak{h}_v(y)$, and an affine hyperplane $H(\mathbf{v}, r) = \mathfrak{h}_v^{-1}(r)$.

We consider the case when $\mathbf{v}_i = N_i(u_i)$ and $r_i = \langle \mathbf{x}_i(u_i), N_i(u_i) \rangle$. We write $\widetilde{h}_{i,(v_i,r_i)}(u) = \widetilde{H}_i(u, \mathbf{v}_i, r_i) = \mathfrak{h}_{v_i} \circ \mathbf{x}_i(u) - r_i$.

By Proposition 2.4, $\partial(\mathfrak{h}_{v_i} \circ \mathbf{x}_i)/\partial u_j(u_i) = 0$ for $j = 1, \ldots, n-1$ and $\mathbf{v}_i = N_i(u_i)$. This means that $M_i = \mathbf{x}_i(U)$ is tangent to the affine hypersurface $H(\mathbf{v}_i, r_i)$ at $p_i = \mathbf{x}_i(u_i)$. We call $H(\mathbf{v}_i, r_i)$ the *tangent affine hyperplane* of M_i at p_i and denote it by $T(M_i)_p$.

By Theorem 4.1,

$$K(M_1, T(M_1)_{p_1}, p_1) = K(M_2, T(M_2)_{p_2}, p_2)$$

if and only if $\widetilde{h}_{1,(v_1,r_1)}$ and $\widetilde{h}_{1,(v_2,r_2)}$ are $\mathcal{K}$-equivalent. Therefore, we can apply the arguments in Section 5.2.2.

We denote by $Q_r(\mathbf{x}(U), u_0)$ the local ring of the germ $\widetilde{h}_{v_0,r_0} : (U, u_0) \to \mathbb{R}$, where $(\mathbf{v}_0, r_0) = \mathrm{CPe}_M(u_0)$. We have

$$Q_r(\mathbf{x}(U), u_0) = \frac{C^\infty_{u_0}(U)}{\langle \langle \mathbf{x}(u), N(u_0) \rangle - r_0 \rangle_{C^\infty_{u_0}(U)} + \mathcal{M}^{r+1}_{u_0}},$$

where $r_0 = \langle \mathbf{x}(u_0), N(u_0) \rangle$ and $C^\infty_{u_0}(U)$ is the local ring of function germs at u_0 with the unique maximal ideal $\mathcal{M}_{u_0}$.

Theorem 5.15. *Let $\mathbf{x}_i : (U, u_i) \to (\mathbb{R}^n, p_i)$, $i = 1, 2$, be local parametrisations of hypersurfaces germs M_i such that the germs of Legendrian immersions*

$$\mathscr{L}_{M_i} : (U, u_i) \to (\mathbb{R}^n \times S^{n-1}, (p_i, \mathbf{v}_i)), \ i = 1, 2,$$

are Legendrian stable. Then the following statements are equivalent:

(1) *The germs of the cylindrical pedals $(M_1^*, (\mathbf{v}_1, r_1))$ and $(M_2^*, (\mathbf{v}_2, r_2))$ are diffeomorphic.*
(2) *The germs of the Legendrian immersions $\mathscr{L}_{M_1}$ and $\mathscr{L}_{M_2}$ are Legendrian equivalent.*
(3) *$\widetilde{H}_1$ and $\widetilde{H}_2$ are P-$\mathcal{K}$-equivalent.*
(4) *$\widetilde{h}_{1,(v_1,r_1)}$ and $\widetilde{h}_{1,(v_2,r_2)}$ are $\mathcal{K}$-equivalent, where $(\mathbf{v}_i, r_i) = \mathrm{CPe}_{M_i}(u_i)$.*
(5) *$K(M_1, T(M_1)_{p_1}, p_1) = K(M_2, T(M_2)_{p_2}, p_2)$.*
(6) *$Q_{n+2}(\mathbf{x}_1, u_1)$ and $Q_{n+2}(\mathbf{x}_2, u_2)$ are isomorphic as $\mathbb{R}$-algebras.*

Proof. Statements (4) and (5) are equivalent by Theorem 4.1. The equivalence of the other statements follow from Proposition 5.2 and Theorem 5.11. □

Proposition 5.11. *Let $\mathbf{x}_i : (U, u_i) \to (\mathbb{R}^n, p_i)$, $i = 1, 2$, be parametrisations of of hypersurfaces germs M_i such that their parabolic sets have no interior points in U. If the germs of the cylindrical pedals $(M_1^*, (\mathbf{v}_1, r_1))$ and $(M_2^*, (\mathbf{v}_2, r_2))$ are diffeomorphic, then*

$$K(M_1, T(M_1)_{p_1}, p_1) = K(M_2, T(M_2)_{p_2}, p_2).$$

In this case, $(\mathbf{x}_1^{-1}(T(M_1)_{p_1}), u_1)$ and $(\mathbf{x}_2^{-1}(T(M_2)_{p_2}), u_2)$ are diffeomorphic.

Proof. By Proposition 5.4 and Theorem 5.14, the singular set of the cylindrical pedal coincides with the singular set of the Gauss map, which is the parabolic set. Thus, the corresponding Legendrian lifts $\mathscr{L}_{M_i}$ satisfy the hypothesis of Theorem 5.10. If the germs of the cylindrical pedals $(M_1^*, (\mathbf{v}_1, r_1))$ and $M_2^*, (\mathbf{v}_2, r_2))$ are diffeomorphic, then $\mathscr{L}_{M_1}$ and $\mathscr{L}_{M_2}$ are Legendrian equivalent, so $\widetilde{H}_1, \widetilde{H}_2$ are P-$\mathcal{K}$-equivalent. Therefore, $\widetilde{h}_{1,(v_1,r_1)}, \widetilde{h}_{1,(v_2,r_2)}$ are $\mathcal{K}$-equivalent, where $r_i = \langle \mathbf{x}_i(u_i), N_i(u_i)\rangle$. By Theorem 4.1, this condition is equivalent to $K(M_1, T(M_1)_{p_1}, p_1) = K(M_2, T(M_2)_{p_2}, p_2)$.

On the other hand, we have

$$(\mathbf{x}_i^{-1}(T(M_i)_{p_i}), u_i) = (\widetilde{h}_{i,(v_i,r_i)}^{-1}(0), u_i).$$

Since the $\mathcal{K}$-equivalence preserves the zero level sets, the germs $(\mathbf{x}_1^{-1}(T(M_1)_{p_1}), u_1)$, $(\mathbf{x}_2^{-1}(T(M_2)_{p_2}), u_2)$ are diffeomorphic. □

We call $(\mathbf{x}^{-1}(T(M)_{p_0}), u_0)$ the *germ of tangent indicatrix* of M at u_0 (or, $p_0 = \mathbf{x}(u_0)$), which is denoted by $TI(M, p_0)$. By Proposition 5.11, the diffeomorphism type of the germ of the tangent indicatrix is an invariant of the diffeomorphism type of the germ of the cylindrical pedal $\mathrm{CPe}_M(U)$ of M.

We can make use of some basic invariants of germs of functions. The local ring of a germ of a function is a complete $\mathcal{K}$-invariant for generic germs, but it is not a numerical invariant. The $\mathcal{K}_e$-codimension of the germ is a numerical $\mathcal{K}$-invariant ([Martinet (1982)]). We denote that $\text{T-ord}(\mathbf{x}(U), u_0) = \mathrm{cod}_e(\widetilde{h}_{(v_0,r_0)}, \mathcal{K})$. By definition,

$$\text{T-ord}(\mathbf{x}(U), u_0) = \dim \frac{C_{u_0}^\infty(U)}{\langle \langle \mathbf{x}(u), N(u_0)\rangle - r_0, \langle \mathbf{x}_{u_i}(u), N(u_0)\rangle \rangle_{C_{u_0}^\infty}},$$

where $r_0 = \langle \mathbf{x}(u_0), N(u_0)\rangle$. Usually $\text{T-ord}(\mathbf{x}(U), u_0)$ is called the $\mathcal{K}_e$-codimension of $\widetilde{h}_{(v_0,r_0)}$. We call it the *order of contact of M with its tangent hyperplane* at $p_0 = \mathbf{x}(u_0)$.

We denote the corank of $\mathbf{x}$ at u_0 by

$$\text{T-corank}(\mathbf{x}(U), u_0) = (n-1) - \operatorname{rank} H(h_{v_0})(u_0),$$

where $\mathbf{v}_0 = N(u_0)$.

By Proposition 2.5, p_0 is a parabolic point if and only if T-corank$(\mathbf{x}(U), u_0) \geq 1$. Moreover p_0 is a flat point if and only if T-corank$(\mathbf{x}(U), u_0) = n - 1$.

By Thom's splitting lemma (Lemma 3.1), if T-corank$(\mathbf{x}(U), u_0) = 1$, then generically the height function h_{v_0} has the A_k-type singularity at u_0. In this case we have T-ord$(\mathbf{x}(U), u_0) = k$. For curves in the plane (i.e., $n = 2$), this number is equal to the order of contact of the curve with the tangent line in the classical sense (see §4.1 in Chapter 4 and [Bruce and Giblin (1992)]). This is the reason why we call T-ord$(\mathbf{x}(U), u_0)$ the order of contact of M with its tangent hyperplane at p_0.

We now consider the contact of hypersurfaces with families of hyperplanes. Let $\mathbf{x}_i : (U, u_i) \to (\mathbb{R}^n, p_i)$, $i = 1, 2$, be germs of parametrisations of hypersurfaces M_i. We consider height functions $H_i : (U \times S^{n-1}, (u_i, \mathbf{v}_i)) \to \mathbb{R}$ on M_i, where $\mathbf{v}_i = N(u_i)$. We denote that $h_{i,v_i}(u) = H_i(u, \mathbf{v}_i)$, then we have $h_{i,v_i}(u) = \mathfrak{h}_{v_i} \circ \mathbf{x}_i(u)$. We also consider that the germ of the foliation $\mathcal{F}_{\mathfrak{h}_{v_i}}$ defined in §4.1 for each $i = 1, 2$. We call $\mathcal{F}_{\mathfrak{h}_{v_i}}$ a *tangent family of affine hyperplanes* of M_i at p_i wich is denoted by $\mathcal{TFH}(M_i, p_i)$.

Theorem 5.16. *Let $\mathbf{x}_i : (U, u_i) \to (\mathbb{R}^n.p_i)$, $i = 1, 2$, be germs of hypersurfaces such that the corresponding germs of Lagrangian immersions*

$$L(H_i) : (C_{H_i}, (u_i, \mathbf{v}_i)) \to T^*S^{n-1}$$

are Lagrangian stable, where $\mathbf{v}_i = N(u_i)$. Then the following statements are equivalent:

(1) $K(M_1, \mathcal{TFH}(M_1, p_1); p_1) = K(M_2, \mathcal{TFH}(M_2, p_2); p_2)$.
(2) *h_{1,v_1} and h_{2,v_2} are $\mathcal{R}^+$- equivalent.*
(3) *H_1 and H_2 are P-$\mathcal{R}^+$- equivalent.*
(4) *$L(H_1)$ and $L(H_2)$ are Lagrangian equivalent.*
(5) *The ranks and the signatures of $H(h_{1,v_1})(u_1)$ and $H(h_{2,v_2})(u_2)$ are equal and there is an isomorphism $\gamma : \mathcal{R}_2(h_{1,v_1}) \to \mathcal{R}_2(h_{2,v_2})$ such that $\gamma(\overline{h_{1,v_1}}) = \overline{h_{2,v_2}}$.*

Proof. By Proposition 4.1, statement (1) is equivalent to statement (2). Statements (2), (3), (4), (5) are equivalent by Theorem 5.6. □

We remark that if $L(H_1)$ and $L(H_2)$ are Lagrangian equivalent, then the corresponding Lagrangian map-germs $\pi \circ L(H_1)$ and $\pi \circ L(H_1)$ are $\mathcal{A}$-equivalent. The Gauss map of a hypersurface $M = \mathbf{x}(U)$ is considered to be the Lagrangian map-germ of $L(H)$ (or, the catastrophe map-germ of H_1).

Moreover, if $h_{1.v_1}$ and h_{2,v_2} are $\mathcal{R}^+$-equivalent then the level set germs of function germs $h_{1.v_1}$ and h_{2,v_2} are diffeomorphic. For a hypersurface germ $\mathbf{x} : (U, u_0) \to (\mathbb{R}^n.p_0)$, the set germ $(\mathbf{x}^{-1}(\mathcal{F}_{\mathfrak{h}_{v_0}}), u_0)$ is a singular foliation germ at $u_0 \in U$, where $\mathbf{v}_0 = N(p_0)$. We call it the *tangential Dupin foliation* of $M = \mathbf{x}(\mathbf{U})$ at u_0, which is denoted by $\mathcal{DF}(\mathbf{x}(U), u_0)$. We have the following corollary.

Corollary 5.3. *If one of the statements in Theorem 5.16 is satisfied, then*

(1) *The Gauss map-germs G_1, G_2 are $\mathcal{A}$-equivalent.*
(2) *The germs of the Dupin foliations $\mathcal{DF}(\mathbf{x}_1(U), u_1)$, $\mathcal{DF}(\mathbf{x}_2(U), u_2)$ are diffeomorphic.*
(3) *The germs of Legendrian immersions $\mathscr{L}_{M_1}$ and $\mathscr{L}_{M_2}$ are Legednrian equivalent. This is equivalent to the germs of the cylindrical pedals $(M_1^*, (\mathbf{v}_1, r_1))$ and $(M_2^*, (\mathbf{v}_2, r_2))$ being diffeomorphic.*

5.6.2 Contact of hypersurfaces with hyperspheres

Let $p_i = \mathbf{x}_i(u_i)$, $i = 1, 2,$. We consider the following point on the evolute of M_i:

$$a_i = \mathrm{Ev}_{\kappa_j(u_i)}(u_i) = \mathbf{x}(u_i) + \frac{1}{\kappa_j(u_i)} N(u_i),$$

where $\kappa_j(u_i)$ is one of the principal curvatures of M_i at p_i. Let $D_i : (U \times \mathbb{R}^n, (u_i, a_i)) \to \mathbb{R}$ be the germ of the family of distance squared functions and denote by $d_{i,a}$ its restriction to a fixed a, so $d_{i,a}(u) = D_i(u, a)$.

We have the (*universal*) *family of distance squared functions* $\mathcal{D} : \mathbb{R}^n \times \mathbb{R}^n \to \mathbb{R}$ by $\mathcal{D}(y, a) = \|y - a\|^2 = \mathfrak{d}_a(y)$, so that $d_{i,a_i}(u) = \mathfrak{d}_{a_i} \circ \mathbf{x}_i(u)$.

By Proposition 2.6, we have $\partial(\mathfrak{d}_{a_i} \circ \mathbf{x}_i)/\partial u_\ell(u_i) = 0$ for $\ell = 1, \ldots, n-1$. This means that the hypersphere $S^{n-1}(a_i, 1/|\kappa_j(u_i)|) = \mathfrak{d}_{a_i}^{-1}(1/\kappa_j^2(u_i))$ is tangent to M_i at $p_i = \mathbf{x}_i(u_i)$. We have the foliation $\mathcal{F}_{\mathfrak{d}_{x_i}}$ (see §4.1) given by the family of hyperspheres with centres a_i which contains the osculating hypersphere $S^{n-1}(a_i, 1/|\kappa_j(u_i)|)$. We call this foliation the *osculating family of hyperspheres* of M_i at p_i with respect to $(a_i, \kappa_j(u_i))$. We write $\mathcal{F}_{\mathfrak{d}_{x_i}} = \mathcal{FS}(M_i, (a_i, \kappa_j(u_i)), p_i)$.

Theorem 5.17. *Let $\mathbf{x}_i : (U, u_i) \to (\mathbb{R}^n, p_i)$, $i = 1, 2$, be parametrisations of two germs of hypersurfaces M_i such that the corresponding germs of Lagrangian immersions $L(D_i) : (C_{D_i}, (u_i, a_i)) \to T^*\mathbb{R}^n$ are Lagrangian stable, where a_i are centres of the osculating hyperspheres of M_i. Then the following statements are equivalent.*

(1) $K(M_1, \mathcal{FS}(M_1, (a_1, \kappa_j(u_1)), p_1); p_1) =$
$K(M_2, \mathcal{FS}(M_2, (a_2, \kappa_j(u_2)), p_2); p_2)$.
(2) d_{1,x_1} *and* d_{2,x_2} *are* $\mathcal{R}^+$*-equivalent.*
(3) D_1 *and* D_2 *are* P-$\mathcal{R}^+$*-equivalent.*
(4) $L(D_1)$ *and* $L(D_2)$ *are Lagrangian equivalent.*
(5) *The rank and signature of the* $H(d_{1,x_1})(u_1)$ *and* $H(d_{2,x_2})(u_2)$ *are equal, and there is an isomorphism* $\gamma : \mathcal{R}^{(2)}(d_{1,x_1}) \to \mathcal{R}^{(2)}(d_{2,x_2})$ *such that* $\gamma(\overline{d_{1,x_1}}) = \overline{d_{2,x_2}}$.

Here $H(d_{i,x_i})(u_i)$ *denotes the Hessian matrix of* d_{i,x_i} *at* u_i.

Proof. Statements (1) and (2) are equivalent by Proposition 4.1. and statements (2)–(5) are equivalent by Theorem 5.6. □

We remark that if $L(D_1)$ and $L(D_2)$ are Lagrangian equivalent, then the corresponding evolutes are diffeomorphic. Since the evolute of a hypersurface M is the caustic of $L(D)$, Theorem 5.17 provides an interpretation for the contact of hypersurfaces with the family of hyperspheres from the view point of Lagrangian singularities.

For a germ of a parametrisation $\mathbf{x} : (U, u_0) \to (\mathbb{R}^n.p_0)$ of M, the set-germ $(\mathbf{x}^{-1}(\mathcal{F}_{\mathfrak{d}_{a_0}}), u_0)$ is a singular foliation-germ at $u_0 \in U$, where $\boldsymbol{a}_0 = \mathrm{Ev}_{\kappa_j(u_0)}(u_0)$. We call it a *spherical Dupin foliation* of $M = \mathbf{x}(U)$ at u_0, and denote it by S-$\mathcal{DF}(\mathbf{x}(U), u_0)$. We consider here the germ of the osculating hyperspherical foliation.

Corollary 5.4. *Under the hypotheses of Theorem 5.17 and if one of the statement there is satisfied, then*

(1) *The germs of the images of evolutes* $(\mathrm{Ev}_{M_1}(U), a_1)$ *and* $(\mathrm{Ev}_{M_2}(U), a_2)$ *are diffeomorphic.*
(2) *The germs of spherical Dupin foliations* S-$\mathcal{DF}(\mathbf{x}_1(U), u_1)$ *and* S-$\mathcal{DF}(\mathbf{x}_2(U), u_2)$ *are diffeomorphic.*

Proof. Statement (1) follows from the fact that Lagrangian equivalences among germs of Lagrangian immersions preserves their caustics. By definition, the germ of the spherical Dupin foliation S-$\mathcal{DF}(\mathbf{x}_i(U), u_i)$ is the level-set foliation $\mathcal{F}_{d_{i,x_i}}$ of d_{i,x_i}. Statement (2) follows from the fact that $\mathcal{R}^+$-equivalence sends the level sets of one germ to another. □

5.6.3 *Contact of submanifolds with hyperplanes*

Let M be a patch of a submanifold of $\mathbb{R}^n$ of dimension s parametrised by $\mathbf{x} : U \to \mathbb{R}^n$. We set $k = n - s$ (the codimension of M).

We defined in §2.2.3 of Chapter 2 the canal hypersurface $CM(\varepsilon)$ of M. We choose an orthonormal frame $\{\nu_1(u), \dots, \nu_k(u)\}$ of N_pM at $p = \mathbf{x}(u)$. Let S^{k-1} be the unit sphere in $\mathbb{R}^k$ and denote the coordinate of its points by $\mu = (\mu_1, \dots, \mu_k)$. Then , for ε sufficiently small, the map $\mathbf{y} : U \times S^{k-1} \to \mathbb{R}^n$ given by

$$\mathbf{y}(u, \mu) = \mathbf{x}(u) + \varepsilon N(u, \mu).$$

with $N(u, \mu) = \sum_{i=1}^{r} \mu_i \nu_i(u) \in N_pM$ is a parametrisation of $CM(\varepsilon)$ (see the proof of Theorem 2.7).

We consider the family of height functions $H : U \times S^{n-1} \to \mathbb{R}$ on M given, as usual, by $H(u, \mathbf{v}) = \langle \mathbf{x}(u), \mathbf{v} \rangle$. We also have the extended family of height functions $\widetilde{H} : U \times S^{n-1} \times \mathbb{R} \to \mathbb{R}$ defined by $\widetilde{H}(u, \mathbf{v}, r) = H(u, \mathbf{v}) - r$.

We denote that $h_v(u) = H(u, \mathbf{v})$ and $\widetilde{h}_{v,r}(u) = \widetilde{H}(u, \mathbf{v}, r)$ for any $(\mathbf{v}, r) \in S^{n-1} \times \mathbb{R}$.

Proposition 5.12. *With notation as above*

(i) $\partial h_{v_0}/\partial u_i(u) = 0, i = 1, \dots, s,$ *if and only if* $\mathbf{v}_0 = N(u, \mu)$ *for some* $\mu \in S^{k-1}$.

(ii) $\widetilde{h}_{v_0}(u) = \partial \widetilde{h}_{v_0}/\partial u_i(u) = 0, i = 1, \dots, s,$ *if and only if*

$$\mathbf{v}_0 = N(u, \mu) \text{ and } r_0 = \langle \mathbf{x}(u), \mathbf{v}_0 \rangle.$$

Proof. (i) There exist real numbers $\lambda_i, i = 1, \dots, s$ and $\mu_j, j = 1, \dots k$, such that

$$\mathbf{v}_0 = \sum_{i=1}^{s} \lambda_i \mathbf{x}_{u_i}(u) + \sum_{j=1}^{k} \mu_j \nu_j(u).$$

Now $\partial h_{\mathbf{v}_0}/\partial u_l(u) = \langle \mathbf{x}_{u_l}, \mathbf{v}_0 \rangle = 0$ if and only if

$$\sum_{i=1}^{s} \lambda_i \langle \mathbf{x}_{u_l}, \mathbf{x}_{u_i} \rangle = 0.$$

The matrix $(\langle \mathbf{x}_{u_l}, \mathbf{x}_{u_i} \rangle)$ is that of the first fundamental form of M, so is not singular. Therefore, $\lambda_i = 0, i = 1, \dots s$.

(ii) This follows from the fact that $\partial \widetilde{h}_{v_0}/\partial u_i(u) = \partial h_{v_0}/\partial u_l(u)$. □

Definition 5.7. Let M as above and let $X^* : U \times S^{k-1} \to S^{n-1} \times \mathbb{R}$ be defined by

$$X^*(u, \mu) = (N(u, \mu), \langle \mathbf{x}(u), N(u, \mu) \rangle).$$

We call the map X^* : the *canal cylindrical pedal* of M.

The canal cylindrical pedal X^* : depends on the choice of the orthonormal frame $\{\nu_1(u), \dots, \nu_k(u)\}$ of $N_p(M)$, $p = \mathbf{x}(u)$. However, its image does not.

Proposition 5.13. *The image of the canal cylindrical pedal of M is independent of the choice of the orthonormal frame in NM. We call it the* canal cylindrical pedal hypersurface of M.

Proof. Let $\{\overline{\nu}_1(u), \dots, \overline{\nu}_k(u)\}$ be another orthonormal frame of N_pM. Then $\nu_i = \sum_{j=1}^k \lambda^i_j \overline{\nu}_j$, with $\lambda^i_j = \langle \nu_i(u), \overline{\nu}_j(u)\rangle$.

The map $\phi : U \times S^{k-1} \to U \times S^{k-1}$ given by

$$\phi(u, \mu) = (u, (\sum_{j=1}^{k} \lambda^1_j(u)\mu_j, \dots, \sum_{j=1}^{k} \lambda^k_j(u)\mu_j)),$$

is a diffeomorphism. If we set $\overline{N}(u, \mu) = \sum_{i=1}^k \mu_i \overline{\nu}_i(u)$, then $N(u, \mu) = \overline{N} \circ \phi(u, \mu)$. Therefore,

$$X^*(u, \mu) = \overline{X}^* \circ \phi(u, \mu),$$

where $\overline{X}^*(u, \mu) = \mathbf{x}(u) + \overline{N}(u, \mu)$. This means that the image of $\overline{X}^*$ is the same as that of X^*. □

It follows from Proposition 5.12, the canal cylindrical pedal hypersurface is the discriminant set of the extended family of height functions of M. The following result follows by similar arguments to those in the proof of Proposition 5.7.

Proposition 5.14. *The germ of the family of height functions H on the submanifold M is a Morse family of functions at each point in the domain. Therefore, the germ family of the family of extended height functions $\widetilde{H}$ is a germ of the graph-like Morse family of hypersurfaces at each point in the domain.*

Proposition 5.15. *The canal cylindrical pedal hypersurface is a graph-like wavefront.*

Proof. Let $\mathbf{y} : U \times S^{k-1} \to \mathbb{R}^n$ be the parametrisation of the canal hypersurface $CM(\varepsilon)$. We define a mapping $\Psi_\varepsilon : J^1(S^{n-1}, \mathbb{R}) \to J^1(S^{n-1}, \mathbb{R})$ by $\Psi_\varepsilon(x, p, y) = (x, p, y - \varepsilon)$. Since $\Psi^*_\varepsilon(dy - \lambda) = d(y - \varepsilon) - \lambda = dy - \lambda$, Ψ_ε is a Legendrian diffeomorphism.

We have the Legendrian immersion $\mathscr{L}_{CM(\varepsilon)} : U \times S^{k-1} \to \mathbb{R}^n \times S^{n-1}$ corresponding to the canal hypersurface.

By Theorem 5.14, $\Phi \circ \mathscr{L}_{CM(\varepsilon)} : U \times S^{k-1} \to J^1(S^{n-1}, \mathbb{R})$ is a graph-like Legendrian immersion and the extended family of height function of $CM(\varepsilon)$ is its graph-like generating family at each point in $U \times S^{k-1}$. Here, $\Phi : \mathbb{R}^n \times S^{n-1} \to J^1(S^{n-1}, \mathbb{R})$ is a local Lagrangian diffeomorphism defined in the proof of Propositon 5.8. It follows that $\Psi_\varepsilon \circ \Phi \circ \mathscr{L}_{CM(\varepsilon)} : U \times S^{k-1} \to J^1(S^{n-1}, \mathbb{R})$ is the graph-like Legendrian immersion. The graph-like wavefront of $\Phi \circ \mathscr{L}_{CM(\varepsilon)}(U \times S^{k-1})$ is the cylindrial pedal hypersurface $CM(\varepsilon)^* = \mathbf{y}^*(U \times S^{k-1})$ of the canal hypersurface $CM(\varepsilon)$. The Legendrian diffeomorphism Ψ_ε induces a diffeomorphism $\widehat{\Psi}_\varepsilon : S^{n-1} \times \mathbb{R} \to S^{n-1} \times \mathbb{R}$ defined by $\widehat{\Psi}_\varepsilon(x, y) = (x, y - \varepsilon)$. Then we have $\widehat{\Psi}_\varepsilon \circ \mathbf{y}^* = X^*$. □

Following the notation in §5.6.1, we denote by $T(M_i; \mathbf{v}_i)_{p_i}$ the tangent affine hyperplane of M_i at p_i with respect to $\mathbf{v}_i = N(u_i, \mu_i)$.

Theorem 5.18. *Suppose that the germs of Legendrian immersions*

$$\mathscr{L}_{CM_i(\varepsilon)} : (U \times S^{k-1}, (u_i, \mu_i)) \to (\mathbb{R}^n \times S^{n-1}, (p_i, \mathbf{v}_i)), \; i = 1, 2,$$

are Legendrian stable. Then the following statements are equivalent:

(1) *The germs of the canal cylindrical pedals*

$$(X_1^*(U \times S^{k-1}), (\mathbf{v}_1, r_1)) \text{ and } (X_2^*(U \times S^{k-1}), (\mathbf{v}_2, r_2))$$

are diffeomorphic.

(2) $\mathscr{L}_{CM_{1\varepsilon}}$ *and* $\mathscr{L}_{CM_{2\varepsilon}}$ *are Legendrian equivalent.*

(3) $\widetilde{H}_1$ *and* $\widetilde{H}_2$ *are* P-$\mathcal{K}$*-equivalent.*

(4) $\widetilde{h}_{(\mathbf{v}_1, r_1)}$ *and* $\widetilde{h}_{(\mathbf{v}_2, r_2)}$ *are* $\mathcal{K}$*-equivalent.*

(5) $K(M_1, T(M_1; \mathbf{v}_1)_{p_1}, p_1) = K(\mathbf{x}_2(U), T(M_2; \mathbf{v}_2)_{p_2}, p_2)$.

(6) *The germs of cylindrical pedals of the canal hypersurfaces*

$$(CM_1(\varepsilon)^*, p_1 + \varepsilon \mathbf{v}_1) \text{ and } (CM_2(\varepsilon)^*, p_2 + \varepsilon \mathbf{v}_2)$$

are diffeomorphic.

(7) $K(CM_1(\varepsilon), TCM_1(\varepsilon)_{p_1 + \varepsilon v_1}, p_1 + \varepsilon \mathbf{v}_1) =$
$K(CM_2(\varepsilon), TCM_2(\varepsilon)_{p_2 + \varepsilon v_2}, p_2 + \varepsilon \mathbf{v}_2)$.

Proof. We remark that the germs of the extended families of height functions $\widetilde{H}_i$ are the graph-like generating families of the Legendrian submanifold covering $(X_i^*(U \times S^{k-1}), (u_i, \mu_i))$. By the relation $\widehat{\Psi}_\varepsilon \circ \mathbf{y}_i^* = X_i^*$ in the proof of Proposition 5.15, such a Legendrian submanifold is Legendrian equivalent to $\mathscr{L}_{CM_{i\varepsilon}}$.

Statements (1)–(4) are equivalent by Theorem 5.11 and Proposition 5.2. It also follows from the relation $\widehat{\Psi}_\varepsilon(CM(\varepsilon)_i^* = \mathbf{y}_i^*(U \times S^{k-1})$ that the statements (1) and (6) are equivalent.

Statements (6) and (7) are equivalent by Theorem 5.15 for hypersurfaces.

Statements (4) and (5) are equivalent by Theorem 4.4 and the fact that $\mathfrak{h}_{v_i}^{-1}(r_i) = T(M_i; \mathbf{v}_i)_{p_i}$. □

Chapter 6

Surfaces in the Euclidean 3-space

In the previous chapters, we considered the general framework of Lagrangian and Legendrian singularities as well as that of contact between submanifolds in $\mathbb{R}^n$. We shall apply here the results from those chapters to study the extrinsic geometry of surfaces embedded in the Euclidean 3-space $\mathbb{R}^3$.

There are some features on a surface in $\mathbb{R}^3$ that determine its local shape. The features of interest are those that can be followed if the surface is deformed. These are given the name of *robust features* by Ian Porteous, and proved to be of importance in computer vision and shape recognition ([Koenderink (1990); Porteous (2001); Siddiqi and Pizer (2008)]). An example of a robust feature of a surface is its parabolic set. The parabolic set is captured by the contact of the surface with planes and lines and can be detected, on some surfaces, by the naked eye (see for example the parabolic curves on the bust of Apollo in Figure 204 in [Hilbert and Cohen-Vossen (1932)], and Figure 6.1).

We start by recalling some basic concepts of the differential geometry of a surface M in $\mathbb{R}^3$ (§6.1). We then consider the contact of M with planes, lines and spheres and study the singularities of respectively the height functions, orthogonal projections and distance squared functions on M (§6.3, §6.4, §6.5). Some types of singularities of these mappings occur on curves on the surface. These curves are robust features of M and have geometric meanings. We deal with them in some detail in §6.6.

6.1 First and second fundamental forms

In Chapter 2 we considered hypersurfaces in $\mathbb{R}^{n+1}$. We deal here with the case of surfaces in the Euclidean 3-space $\mathbb{R}^3$.

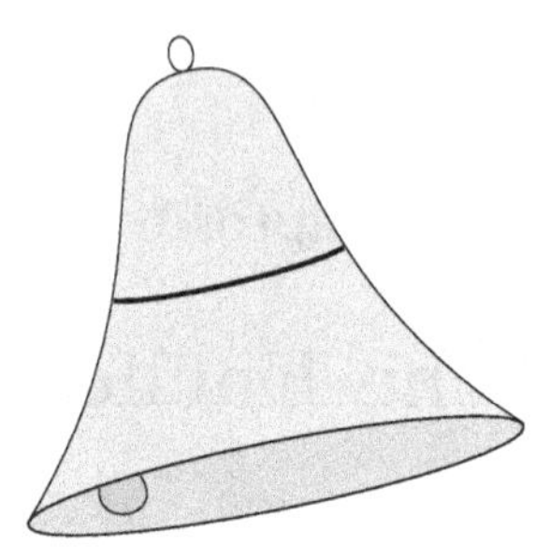

Fig. 6.1 Visible parabolic curves (in thick) on the surface of a bell.

Let $\mathbf{x} : U \to \mathbb{R}^3$ be a parametrisation of a surface patch M where U is an open subset of $\mathbb{R}^2$. The first fundamental form (or the metric) of $M = \mathbf{x}(U)$ at $p \in M$ is given by $\mathrm{I}_p(\mathbf{w}) = \langle w, w \rangle$, for any $w \in T_pM$. If $\mathbf{w} = a\mathbf{x}_{u_1} + b\mathbf{x}_{u_2}$ then

$$\mathrm{I}_p(\mathbf{w}) = a^2E + 2abF + b^2G$$

where

$$E = \langle \mathbf{x}_{u_1}, \mathbf{x}_{u_1} \rangle, \quad F = \langle \mathbf{x}_{u_1}, \mathbf{x}_{u_2} \rangle, \quad G = \langle \mathbf{x}_{u_2}, \mathbf{x}_{u_2} \rangle$$

are the coefficients of the first fundamental form.

Observe that $(EG - F^2)(u_1, u_2) > 0$ for all $(u_1, u_2) \in U$. Also, I_p is the Riemannian metric on M induced from the Euclidean scaler product. In the notation of Chapter 2, $E = g_{11}$, $F = g_{12} = g_{21}$ and $G = g_{22}$.

The surface patch M is given an orientation by choosing at each point $p \in M$ the unit normal vector $N(p)$ which together with the orientation of M gives the positive orientation of the ambient space $\mathbb{R}^3$ at p. By varying p on M, we obtain the Gauss map

$$N : M \to S^2.$$

The Weingarten map $W_p : T_pM \to T_pM$ is given by $W_p = -dN_p$. The second fundamental form of M at p is the bilinear symmetric form given by

$$\mathrm{II}_p(\mathbf{w}, \mathbf{v}) = \langle W_p(\mathbf{w}), \mathbf{v} \rangle = \langle W_p(\mathbf{v}), \mathbf{w} \rangle$$

for any $\mathbf{v}, \mathbf{w} \in T_pM$. We still denote by II_p the quadratic form associated to II_p and write

$$\mathrm{II}_p(\mathbf{w}) = \mathrm{II}_p(\mathbf{w}, \mathbf{w}) = la^2 + 2mab + nb^2,$$

where

$$
\begin{aligned}
l &= \langle -N_{u_1}, \mathbf{x}_{u_1} \rangle = \langle N, \mathbf{x}_{u_1 u_1} \rangle, \\
m &= \langle -N_{u_1}, \mathbf{x}_{u_2} \rangle = \langle N, \mathbf{x}_{u_1 u_2} \rangle, \\
n &= \langle -N_{u_2}, \mathbf{x}_{u_2} \rangle = \langle N, \mathbf{x}_{u_2 u_2} \rangle
\end{aligned}
$$

are the coefficients of the second fundamental form of M. Following the notation in Chapter 2, we have

$$l = h_{11},\ m = h_{12} = h_{21},\ n = h_{22}.$$

At each point on M, there are two principal curvatures κ_1 and κ_2 which are the eigenvalues of W_p. The Gaussian curvature K and the mean curvature H of M are given by

$$K = \kappa_1 \kappa_2, \qquad H = \frac{1}{2}(\kappa_1 + \kappa_2).$$

These can be expressed in terms of the coefficients of the first and second fundamental forms as follows (see for example [do Carmo (1976)] and Corollary 2.1)

$$K = \frac{ln - m^2}{EG - F^2}$$

and

$$H = \frac{lG - 2mF + nE}{2(EG - F^2)}.$$

The principal curvatures κ_1 and κ_2 are the solutions of the quadratic equation

$$\kappa^2 - 2H\kappa + K = 0,$$

so, one is equal to $H + \sqrt{H^2 - K}$ and the other to $H - \sqrt{H^2 - K}$.

Recall from Chapter 1 that the Gaussian curvature K is an intrinsic invariant of the surface M. However, the principal curvatures κ_1 and κ_2 and the mean curvature H are not intrinsic invariants of M. They provide information about the geometry of M as a surface in the ambient space $\mathbb{R}^3$, so they capture some extrinsic properties of M.

If p is not an umbilic point, that is, if $\kappa_1 \neq \kappa_2$, there are two orthogonal principal directions at p denoted by $\mathbf{v}_1$ and $\mathbf{v}_2$ which are parallel to the eigenvectors of W_p. At umbilic points , every tangent direction is considered a principal direction. A curve on M whose tangent direction at all points is a principal direction is called *a line of principal curvature*. The lines of principal curvature are the solution curves of the binary differential equation

$$(Fn - mG)du_2^2 + (En - lG)du_1 du_2 + (Em - lF)du_1^2 = 0, \tag{6.1}$$

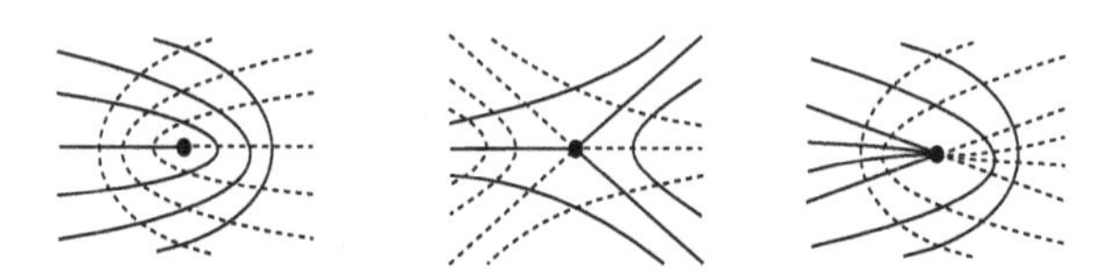

Fig. 6.2 Generic configurations of the lines of principal curvature at umbilic points: Lemon (left), Star (centre), Monstar (right).

see [do Carmo (1976)] for a proof. Equation (6.1) can also be written in the following determinant form

$$\begin{vmatrix} du_2^2 & -du_1du_2 & du_1^2 \\ E & F & G \\ l & m & n \end{vmatrix} = 0.$$

The lines of principal curvature form a pair of orthogonal foliations Their generic topological configurations at umbilic points are determined in [Sotomayor and Gutierrez (1982); Bruce and Fidal (1989)]; see Figure 6.2 and [Sotomayor (2004)] for historical notes. Porteous gave the colour blue to one foliation of the lines of curvature and the colour red to the other. In Figure 6.2, one foliation associated to the curvature κ_i is drawn in solid lines and the other associated to the curvature κ_j, $j \neq i$, is drawn in dashed lines.

The second fundamental form measures how the surface M bends in $\mathbb{R}^3$. Let $\alpha : (-\epsilon, \epsilon) \to M$ be a smooth curve on M with $\alpha(0) = p$. Suppose that α is parametrised by arc length and denote by s the arc length parameter. The acceleration vector α'' is orthogonal to the tangent vector $\mathbf{t} = \alpha'$, so it lies in the plane with orthonormal basis $\{\mathbf{t} \times N, N\}$, where $N(s) = N(\alpha(s))$ is the normal vector to M at $\alpha(s)$. We can write

$$\alpha'' = \kappa_g \, \mathbf{t} \times N + \kappa_n N,$$

for some scalars κ_g (called the *geodesic curvature*) and κ_n (called the *normal curvature*). The component $\kappa_g \, \mathbf{t} \times N$ of α'' at s lies in the tangent plane to M at $\alpha(s)$ and the component $\kappa_n N$ is parallel to the normal of the surface at $\alpha(s)$.

If the geodesic curvature $\kappa_g = \langle \alpha'', \mathbf{t} \times N \rangle$ of the curve α vanishes at some point s, we say that the curve α has a *geodesic inflection* at s. In general, at a geodesic inflection, the curve α is locally on both sides of the plane at $\alpha(s)$ parallel to $N(s)$ and $\alpha'(s)$ (so the projection of α to the tangent plane $T_{\alpha(s)}M$ is a plane curve with an inflection).

As its name suggests, the geodesic curvature has something to do with geodesic curves. There is a great amount of study of such curves as they minimise (at least locally) the length of the distance between points on the surface. However, we will not pursue them in this book.

If α is not parametrised by arc length, it can be shown by using a re-parametrisation by arc length that its geodesic curvature is given by

$$\kappa_g = \frac{1}{||\dot{\alpha}||^3}\langle \ddot{\alpha}, \dot{\alpha} \times N \rangle, \tag{6.2}$$

where $\dot{\alpha} = d\alpha/dt$ for arbitrary parameter t. We turn now to the normal curvature. Differentiating the identity $\langle N(s), \alpha'(s) \rangle = 0$ leads to

$$\langle N(s), \alpha''(s) \rangle = - \langle N'(s), \alpha'(s) \rangle .$$

Therefore,

$$\begin{aligned} \mathrm{II}_p(\alpha'(0)) &= \langle -dN_p(\alpha'(0)), \alpha'(0) \rangle , \\ &= - \langle N'(0), \alpha'(0) \rangle , \\ &= \langle N(0), \alpha''(0) \rangle , \\ &= \kappa_n . \end{aligned}$$

Observe that the minus sign in the Weingarten map $W_p = -dN_p$ is there so that we get $\mathrm{II}_p(\alpha'(0))$ is equal to κ_n and not to $-\kappa_n$. The equality $\mathrm{II}_p(\alpha'(0)) = \kappa_n$ shows that the normal curvature at p depends only on the unit tangent direction of the curve α at p and not on the curve itself, so we have the following result.

Proposition 6.1. *All curves on a surface M with the same tangent line at a point p on M have the same normal curvature at p.*

If α is not parametrised by arc length, the normal curvature at p is defined as

$$\kappa_n = \mathrm{II}_p(\frac{\dot{\alpha}(0)}{|\dot{\alpha}(0)|}) = \frac{\mathrm{II}_p(\dot{\alpha}(0))}{\mathrm{I}_p(\dot{\alpha}(0))} = \frac{\mathrm{II}_p(\dot{\alpha}(0))}{||\dot{\alpha}(0)||^2}.$$

If $\dot{\alpha}(0) = a\mathbf{x}_{u_1} + b\mathbf{x}_{u_2}$, then

$$\kappa_n = \frac{a^2 l + 2abm + b^2 n}{a^2 E + 2abF + b^2 G}.$$

Given a tangent direction $\mathbf{w} \in T_pM$, the *normal section* of M at p along $\mathbf{w}$ is the plane curve obtained by intersecting M with the plane through p generated by $N(p)$ and $\mathbf{w}$. By Proposition 6.1, the normal curvature κ_n at p along $\mathbf{w}$ is the curvature of the normal section at p along $\mathbf{w}$.

If p is not an umbilic point, we can write $\mathbf{w} = \cos\theta\mathbf{v}_1 + \sin\theta\mathbf{v}_2$ for any unit tangent vector in T_pM, where $\mathbf{v}_1$ and $\mathbf{v}_2$ are the orthonormal principal directions at p. Then

$$\begin{aligned}\kappa_n(\theta) = \mathrm{II}_p(\mathbf{w}) &= \langle W_p(\cos\theta\mathbf{v}_1 + \sin\theta\mathbf{v}_2), \cos\theta\mathbf{v}_1 + \sin\theta\mathbf{v}_2\rangle \\ &= \langle \cos\theta\kappa_1\mathbf{v}_1 + \sin\theta\kappa_2\mathbf{v}_2, \cos\theta\mathbf{v}_1 + \sin\theta\mathbf{v}_2\rangle \\ &= \kappa_1\cos^2\theta + \kappa_2\sin^2\theta.\end{aligned}$$

Therefore, the principal curvatures κ_1 and κ_2 are the extrema of the normal curvature $\kappa_n(\theta)$ at p when θ varies in $[0,\pi]$.

A direction along which the normal curvature is zero is called an *asymptotic direction*. Thus, a direction $\mathbf{w} \in T_pM$ is asymptotic if and only if $\mathrm{II}_p(\mathbf{w}) = 0$. We observe that the normal section along an asymptotic direction has an inflection and that there are two asymptotic directions at a hyperbolic point, one at a parabolic point and none at an elliptic point (see §2.1.4).

A curve on M whose tangent direction at each point is an asymptotic direction is called an *asymptotic curve*. The asymptotic curves are the solutions of the binary differential equation

$$n du_2^2 + 2m du_1 du_2 + l du_1^2 = 0. \tag{6.3}$$

Definition 6.1. Two directions $\mathbf{w}_1$ and $\mathbf{w}_2$ in T_pM are said to be conjugate (with respect to the second fundamental form) if and only if

$$\langle W_p(\mathbf{w}_1), \mathbf{w}_2\rangle = 0.$$

A conjugate direction to $\mathbf{w}$ is denoted by $\bar{\mathbf{w}}$.

Proposition 6.2. (i) *A conjugate direction to* $\mathbf{v} = a\mathbf{x}_{u_1} + b\mathbf{x}_{u_2}$ *is* $\bar{\mathbf{v}} = -(am+bn)\mathbf{x}_{u_1} + (al+bm)\mathbf{x}_{u_2}$.

(ii) $\mathrm{II}_p(\bar{\mathbf{v}}) = (nl - m^2)\mathrm{II}_p(\mathbf{v})$.

(iii) $\kappa_n(\bar{\mathbf{v}}) = \kappa_n(\mathbf{v})(nl - m^2)||\mathbf{v}||^2/||\bar{\mathbf{v}}||^2$.

(iv) $||\mathbf{v}||^2||\bar{\mathbf{v}}||^2 - \langle\bar{\mathbf{v}},\mathbf{v}\rangle^2 = (EG - F^2)\mathrm{II}_p(\mathbf{v})^2$.

Proof. (i) Write $\bar{\mathbf{v}} = \xi_1\mathbf{x}_{u_1} + \xi_2\mathbf{x}_{u_2}$. Then $\langle W_p(\bar{\mathbf{v}}), \mathbf{v}\rangle = 0$ is equivalent to

$$a\xi_1 l + (a\xi_2 + b\xi_1)m + b\xi_2 n = 0.$$

Rearranging the above expression gives

$$(al+bm)\xi_1 + (am+bn)\xi_2 = 0,$$

so we can take $\xi_1 = -(am+bn)$ and $\xi_2 = (al+bm)$.

(ii) We have

$$\begin{aligned}\mathrm{II}_p(\bar{\mathbf{v}}) &= (am+bn)^2l - 2(am+bn)(al+bm)m + (al+bm)^2n\\ &= (nl-m^2)(a^2l+2abm+b^2n)\\ &= (nl-m^2)\mathrm{II}_p(\mathbf{v}).\end{aligned}$$

(iii) It follows from (ii) that

$$\begin{aligned}\kappa_n(\bar{\mathbf{v}}) &= \frac{1}{||\bar{\mathbf{v}}||^2}\mathrm{II}_p(\bar{\mathbf{v}})\\ &= \frac{1}{||\bar{\mathbf{v}}||^2}(nl-m^2)\mathrm{II}_p(\mathbf{v})\\ &= \frac{||\mathbf{v}||^2}{||\bar{\mathbf{v}}||^2}(nl-m^2)\kappa_n(\mathbf{v}).\end{aligned}$$

(iv) We have

$$\begin{aligned}&||\mathbf{v}||^2||\bar{\mathbf{v}}||^2 - \langle\bar{\mathbf{v}},\mathbf{v}\rangle^2 =\\ &(a^2E+2abF+b^2G)((am+bn)^2E - 2(am+bn)(al+bm)F + (al+bm)^2G)\\ &-\left(a(am+bn)E + (a(al+bm)-b(am+bn))F + b(al+bm)G\right)^2\\ &= (EG-F^2)(a^2l+2abm+b^2n)^2\\ &= (EG-F^2)\mathrm{II}_p(\mathbf{v})^2.\end{aligned}$$

□

Example 6.1. An asymptotic direction can be defined as a direction which coincides with its conjugate direction, equivalently, as a direction which forms a zero angle with its conjugate direction. The conjugate to the principal direction $\mathbf{v}_1$ is the principal direction $\mathbf{v}_2$ and vice-versa. In fact, the angle between $\mathbf{w}$ and $\bar{\mathbf{w}}$ is $\pi/2$ if and only if $\mathbf{w}$ is a principal direction. Fletcher considered in [Fletcher (1996)] directions in T_pM which form a fixed oriented angle α with their conjugate directions. It turns out that, away from umbilic and parabolic points, there are at most two of these directions at each point on M. Fletcher called the integral curves of these directions the *conjugate curve congruence* and denoted them by $\mathcal{C}_\alpha$. The conjugate curve congruences are solutions of a binary differential equation and the family $\mathcal{C}_\alpha$ interpolates between the asymptotic curves ($\alpha = 0$) and the lines of principal curvature ($\alpha = \pi/2$). More work on the family C_α and other families can be found in [Bruce, Fletcher and Tari (2004); Bruce and Tari (2005); Nabarro and Tari (2009); Oliver (2010, 2011); Tari (2009)].

6.2 Surfaces in Monge form

At each point p on the surface M, we can choose a coordinate system in $\mathbb{R}^3$ so that p is the origin, T_pM is the plane $z = 0$ and the surface M is locally the graph of some function $z = f(x, y)$, with (x, y) in a open subset U of $\mathbb{R}^2$ containing the origin. Then we have the Monge form parametrisation $\phi(x, y) = (x, y, f(x, y))$, $(x, y) \in U$, of M at p. We shall say that M is parametrised locally in Monge form $z = f(x, y)$ at the origin p. Note that the Taylor expansion of f at the origin has no constant or linear terms.

Proposition 6.3. *Let M be parametrised locally in Monge form $z = f(x, y)$ at the origin p. Then,*

(i) $E = 1 + f_x^2$, $F = f_x f_y$, $G = 1 + f_y^2$.

(ii) $N = \dfrac{1}{\sqrt{1 + f_x^2 + f_y^2}}(-f_x, -f_y, 1)$.

(iii) $l = \dfrac{f_{xx}}{\sqrt{1 + f_x^2 + f_y^2}}$, $m = \dfrac{f_{xy}}{\sqrt{1 + f_x^2 + f_y^2}}$, $n = \dfrac{f_{yy}}{\sqrt{1 + f_x^2 + f_y^2}}$.

Proof. We have $\phi_x = (1, 0, f_x)$ and $\phi_y = (0, 1, f_y)$, so

$$\begin{aligned} E &= \langle \phi_x, \phi_x \rangle = 1 + f_x^2, \\ F &= \langle \phi_x, \phi_y \rangle = f_x f_y, \\ G &= \langle \phi_y, \phi_y \rangle = 1 + f_y^2. \end{aligned}$$

The Gauss map at (x, y) is given by

$$N = \frac{\phi_x \times \phi_y}{||\phi_x \times \phi_y||} = \frac{1}{\sqrt{1 + f_x^2 + f_y^2}}(-f_x, -f_y, 1).$$

The coefficients of the second fundamental form are given by

$$\begin{aligned} l &= \langle N, \phi_{xx} \rangle = f_{xx}/\sqrt{1 + f_x^2 + f_y^2}, \\ m &= \langle N, \phi_{xy} \rangle = f_{xy}/\sqrt{1 + f_x^2 + f_y^2}, \\ n &= \langle N, \phi_{yy} \rangle = f_{yy}/\sqrt{1 + f_x^2 + f_y^2}. \end{aligned}$$

□

For a surface given in Monge form, we write the homogeneous part of degree k of the Taylor expansion of f at the origin in $\mathbb{R}^2$ in the form

$$\sum_{i=0}^{k} a_{ki} x^{k-i} y^i.$$

6.3 Contact with planes

Let $\mathbf{x} : U \to \mathbb{R}^3$ be a parametrisation of a surface patch M. The contact of M with planes is measured by the singularities of the height functions on M. Recall that the family of height functions $H : U \times S^2 \to \mathbb{R}$ on M is given by

$$H(u, \mathbf{v}) = \langle \mathbf{x}(u), \mathbf{v} \rangle$$

and the extended family of height functions $\tilde{H} : U \times S^2 \times \mathbb{R} \to \mathbb{R}$ is given by

$$\tilde{H}(u, \mathbf{v}, r) = \langle \mathbf{x}(u), \mathbf{v} \rangle - r.$$

For $\mathbf{v} \in S^2$ fixed, the height function $h_{\mathbf{v}}$ along the direction $\mathbf{v}$ is the function $h_{\mathbf{v}}(u) = H(u, \mathbf{v})$. Similarly, for $\mathbf{v}$ and r fixed, $\tilde{h}_{\mathbf{v},r}$ is the function $\tilde{h}_{\mathbf{v},r}(u) = \tilde{H}(u, \mathbf{v}, r)$.

Theorem 6.1. *There is an open and dense set $\mathcal{O}_1$ of proper immersions $\mathbf{x} : U \to \mathbb{R}^3$, such that for any $\mathbf{x} \in \mathcal{O}_1$, the surface $M = \mathbf{x}(U)$ has the following properties. For any $\mathbf{v} \in S^2$, the height function $h_{\mathbf{v}}$ (resp. $\tilde{h}_{\mathbf{v},\langle p,\mathbf{v}\rangle}$) along the normal direction $\mathbf{v}$ at any point p on M has only local singularities of type A_1, A_2 or A_3 which are $\mathcal{R}^+$ (resp. $\mathcal{K}$) versally unfolded by the family H (resp. $\tilde{H}$).*

Proof. The result follows from Theorems 4.3, 4.5 and 4.7. □

Definition 6.2. A surface is called (*locally*) *height function generic* if any of its local parametrisations belongs to the set $\mathcal{O}_1$ in Theorem 6.1.

We shall derive geometric information about the surface M from the family of height functions. We start with the A_1-singularity. Given a direction $\mathbf{v}$ in S^2, the functions $h_{\mathbf{v}}$ and $\tilde{h}_{\mathbf{v},\langle p,v\rangle}$ have the same $\mathcal{R}$ or $\mathcal{K}$-type singularity at any point p on M.

Proposition 6.4. *The height functions $h_{\mathbf{v}}$ is singular at $p \in M$ if and only if $\mathbf{v}$ is a normal direction to M at p. The singularity of $h_{\mathbf{v}}$ at p is of type A_1^+ if and only if p is an elliptic point, A_1^- if and only if p is a hyperbolic point, and $A_{\geq 2}$ if and only if p is a parabolic point.*

Proof. We only need to characterise geometrically the A_1^+ and A_1^- singularities as the remaining part of the proposition is a particular case of Propositions 2.4 and 2.5. These singularities are distinguished by the sign of the determinant of the Hessian matrix of $h_{\mathbf{v}}$, positive for A_1^+ and negative for A_1^-.

It follows from the Weingarten formula Theorem 2.2 and Corollary 2.1 that

$$\det \mathcal{H}(h_{\mathbf{v}})(u) = K(u)(EG - F^2)(u).$$

As $(EG - F^2)(u) > 0$, $\det \mathcal{H}(h_{\mathbf{v}})(u)$ and $K(u)$ have the same sign, and the result follows. □

When the surface is taken in Monge form $z = f(x, y)$, the conditions for the height function $h_{N_0}(x, y) = f(x, y)$ along the normal direction $N_0 = (0, 0, 1)$ to have a given singularity at the origin p and for the family of height functions to be a $\mathcal{R}^+$-versal unfolding of these singularities can be expressed in terms of the coefficients of the Taylor expansion of f at p. We write

$$f(x,y) = a_{20}x^2 + a_{21}xy + a_{22}y^2 + a_{30}x^3 + a_{31}x^2y + a_{32}xy^2 + a_{33}y^3 + O(4). \tag{6.4}$$

We can rotate the coordinate axes in the tangent plane T_pM and set a chosen direction to be along the y-axis. The chosen direction could be, for example, an asymptotic direction if p is not an elliptic point or a principal direction.

Proposition 6.5. *The following hold for M in Monge form $z = f(x, y)$ at the origin p with f as in (6.4) and $\mathbf{v} = (0, 1, 0) \in T_pM$.*

(i) *The point p is a parabolic point if and only if $a_{21}^2 - 4a_{20}a_{22} = 0$.*
(ii) *The direction $\mathbf{v}$ is asymptotic at p if and only if $a_{22} = 0$.*
(iii) *If $\mathbf{v}$ is an asymptotic direction at p and p is a parabolic point but not a flat umbilic point, the parabolic set is a smooth curve at p if and only if $a_{32} \neq 0$ or $a_{33} \neq 0$.*

Proof. For f as in (6.4), the 1-jets at the origin of the coefficients of the second fundamental form for the Monge form setting in Proposition 6.3 are given by

$$\begin{aligned} j^1 l &= 2a_{20} + 6a_{30}x + 2a_{31}y, \\ j^1 m &= a_{21} + 2a_{31}x + 2a_{32}y, \\ j^1 n &= 2a_{22} + 2a_{32}x + 6a_{33}y. \end{aligned}$$

It follows that

$$\begin{aligned} j^1(ln - m^2)(x,y) = {} & 4a_{20}a_{22} - a_{21}^2 + 4(a_{20}a_{32} + 3a_{30}a_{22} - a_{21}a_{31})x + \\ & 4(3a_{20}a_{33} + a_{31}a_{22} - a_{21}a_{32})y. \end{aligned}$$

(i) In particular, the origin is a parabolic point if and only if

$$(ln - m^2)(0,0) = 4a_{20}a_{22} - a_{21}^2 = 0.$$

(ii) At the origin p, $l = 2a_{20}$, $m = a_{21}$ and $n = 2a_{22}$ so a direction $a\phi_x(0,0) + b\phi_y(0,0) = (a, b, 0)$ is asymptotic at p if and only if

$$a_{20}a^2 + a_{21}ab + a_{22}b^2 = 0$$

(see equation (6.3)). Therefore, $v = (0,1,0)$ is an asymptotic direction at p if and only if $a_{22} = 0$.

(iii) When $\mathbf{v}$ is an asymptotic direction and p is a parabolic point,

$$j^1(ln - m^2)(x,y) = 4a_{20}a_{32}x + 12a_{20}a_{33}y.$$

As p is not a flat umbilic point, $a_{20} \neq 0$, so $ln - m^2$ is not singular at the origin if and only if $a_{32} \neq 0$ or $a_{33} \neq 0$. □

Theorem 6.2. *The following hold for M in Monge form $z = f(x,y)$ at the origin p with f as in (6.4) and $\mathbf{v} = (0,1,0) \in T_pM$. Assume that p is not a flat umbilic point.*

(i) *For p a parabolic point and $\mathbf{v}$ an asymptotic direction at p, the height function H_{N_0} has a singularity at p of type*

$$\begin{array}{lcl} A_2 & \Longleftrightarrow & a_{33} \neq 0, \\ A_3 & \Longleftrightarrow & a_{33} = 0 \text{ and } a_{32}^2 - 4a_{20}a_{44} \neq 0. \end{array}$$

(ii) *The family of height functions on M is a $\mathcal{R}^+$-versal unfolding of the A_k-singularities, $k = 1,2,3$, of H_{N_0} if and only if*

A_1 : always
A_2 : always
A_3 : $a_{32} \neq 0$, equivalently, the parabolic set is a smooth curve.

Proof. (i) The height function along the normal direction $N_0 = (0,0,1)$ at the origin p is given by $h_{N_0}(x,y) = f(x,y)$. By Proposition 6.5, if p is a parabolic point and $\mathbf{v}$ is an asymptotic direction at p, then $a_{22} = a_{21} = 0$ so that

$$f(x,y) = a_{20}x^2 + a_{30}x^3 + a_{31}x^2y + a_{32}xy^2 + a_{33}y^3 + \Sigma_{i=0}^4 a_{4i}x^{4-i}y^i + O(5).$$

(We chose the unique asymptotic direction at p to be along $\mathbf{v}$ in order to have $j^2f = a_{20}x^2$. This makes the task of recognizing the singularities of h_{N_0} much easier.) The hypothesis on p not being a flat umbilic point is then equivalent to $a_{20} \neq 0$. The height function f has an $A_{\geq 2}$-singularity at the origin if and only if $a_{33} = 0$. Then the singularity is of type A_3 if and only if $a_{20}x^2 + a_{32}xy^2 + a_{44}y^4$ is not a perfect square, equivalently, $a_{32}^2 - 4a_{20}a_{44} \neq 0$.

(ii) Observe that the family of height functions $H : \mathbb{R}^2 \times S^2, (0, N_0) \to \mathbb{R}$ is P-$\mathcal{R}^+$-equivalent to the modified family, that we still denote by H : $\mathbb{R}^2 \times \mathbb{R}^2, (0,0) \to \mathbb{R}$, which is given by

$$H(x,y,\alpha,\beta) = \alpha x + \beta y + f(x,y)$$

(all we did here is to parametrise the sphere S^2 near $(0,0,1)$ by $(\alpha,\beta,1)$).

Recall from Theorem 3.12 that H is an $\mathcal{R}^+$-versal unfolding if and only if

$$L\mathcal{R}_e\cdot f+\mathbb{R}.\left\{1,\dot{H}_1,\dot{H}_2\right\}=\mathcal{E}_2. \tag{6.5}$$

We have

$$\begin{aligned}\dot{H}_1(x,y)&=H_\alpha(x,y,0,0)=x,\\ \dot{H}_2(x,y)&=H_\beta(x,y,0,0)=y.\end{aligned}$$

For f as in (6.4),

$$\begin{aligned}j^3f_x&=2a_{20}x+a_{21}y+3a_{30}x^2+2a_{31}xy+a_{32}y^2\\ &\quad+4a_{40}x^3+3a_{41}x^2y+2a_{42}xy^2+a_{43}y^3\\ j^3f_y&=a_{21}x+2a_{22}y+a_{31}x^2+2a_{32}xy+3a_{33}y^2\\ &\quad+a_{41}x^3+2a_{42}x^2y+3a_{43}xy^2+4a_{44}y^3.\end{aligned}$$

The A_1-singularity is 2-$\mathcal{R}$-determined, that is $\mathcal{M}_2^3\subset L\mathcal{R}\cdot f$, so for H to be an $\mathcal{R}^+$-versal unfolding it is enough to show that equality (6.5) holds modulo $\mathcal{M}_2^3$, that is

$$j^2(L\mathcal{R}_e\cdot f+\mathbb{R}.\left\{\dot{H}_1,\dot{H}_2\right\}+\langle 1\rangle_{\mathbb{R}})=J^2(2,1). \tag{6.6}$$

We view $J^2(2,1)$ as an $\mathbb{R}$-vector space with basis $1,x,y,x^2,xy,y^2$. To show that equality (6.6) holds, it is enough to show that the elements of the basis of $J^2(2,1)$ are in $j^2(L\mathcal{R}_e\cdot f+\mathbb{R}.\left\{1,\dot{H}_1,\dot{H}_2\right\})$. Clearly, 1, $x=\dot{H}_1(x,y)$ and $y=\dot{H}_2(x,y)$ are in there.

Let $\xi_1=j^2(yf_x)=2a_{20}xy+a_{21}y^2$ and $\xi_2=j^2(yf_y)=a_{21}xy+2a_{22}y^2$ be two vectors in $j^2(L\mathcal{R}_e\cdot f)$. Then

$$\begin{pmatrix}\xi_1\\ \xi_2\end{pmatrix}=A\begin{pmatrix}xy\\ y^2\end{pmatrix},$$

with

$$A=\begin{pmatrix}2a_{20}&a_{21}\\ a_{21}&2a_{22}\end{pmatrix}.$$

The matrix A is invertible as its determinant is $4a_{20}a_{22}-a_{21}^2\neq 0$. Hence

$$\begin{pmatrix}xy\\ y^2\end{pmatrix}=A^{-1}\begin{pmatrix}\xi_1\\ \xi_2\end{pmatrix},$$

which shows that $xy,y^2\in j^2(L\mathcal{R}_e\cdot f)$. From this we get $x^2=(j^2(xf_x)-a_{21}xy)/(2a_{20})\in j^2(L\mathcal{R}_e\cdot f)$. Therefore (6.6) always holds at an A_1-singularity.

We set now $a_{21} = a_{22} = 0$. For the A_2-singularity which is 3-$\mathcal{R}$-determined, it is enough to show that

$$j^3(L\mathcal{R}_e \cdot f + \mathbb{R}.\left\{1, \dot{H}_1, \dot{H}_2\right\}) = J^3(2,1) \tag{6.7}$$

for H to be an $\mathcal{R}^+$-versal unfolding.

Let P be a monomial of degree 2. Then the vector $j^3(Pf_x) \in j^3(L\mathcal{R}_e \cdot f)$, and this shows that all the monomials of degree 3 divisible by x, i.e., x^3, x^2y, xy^2 are in $j^3(L\mathcal{R}_e \cdot f)$. Then $y^3 = (j^3(yf_y) - a_{31}x^2y - 2a_{32}xy^2)/3a_{33} \in j^3(L\mathcal{R}_e \cdot f)$ (we do have $a_{33} \neq 0$ at an A_2-singularity of the height function f). We proceed similarly to show that all the degree 2 monomials are in $j^3(L\mathcal{R}_e \cdot f)$. This means that equality (6.7) always holds at an A_2-singularity.

For the A_3-singularity, we set $a_{21} = a_{22} = a_{33} = 0$. As this singularity is 4-$\mathcal{R}$-determined, we work modulo $\mathcal{M}_2^5$. Using vectors of the form $j^4(Pf_x)$, we get all monomials of degree 4 divisible by x in $j^4(L\mathcal{R}_e \cdot f)$, then those of degree 3 divisible by x^2. Working modulo these monomials we have

$$\begin{aligned} \eta_1 &= j^4(y^2f_x) \equiv 2a_{20}xy^2 + a_{32}y^4 \\ \eta_2 &= j^4(yf_y) \equiv 2a_{32}xy^2 + 4a_{44}y^4 \end{aligned}$$

that is

$$\begin{pmatrix} \eta_1 \\ \eta_2 \end{pmatrix} = A \begin{pmatrix} xy^2 \\ y^4 \end{pmatrix},$$

with

$$A = \begin{pmatrix} 2a_{20} & a_{32} \\ 2a_{32} & 4a_{44} \end{pmatrix}.$$

The determinant $2(4a_{20}a_{44} - a_{32}^2)$ of matrix A is distinct from zero as the height function has an A_3-singularity. Therefore xy^2 and y^4 are in $j^4(L\mathcal{R}_e \cdot f)$. Using now $j^4(xf_x)$ and $j^4(yf_y)$ modulo the monomials that we have shown are in $j^4(L\mathcal{R}_e \cdot f)$, we get $x^2, xy \in j^4(L\mathcal{R}_e \cdot f)$. We also get y^3 and xy using $\eta_1 = j^4(yf_x)$ and $\eta_2 = j^4(f_y)$ (we get the same matrix A as above).

The only way to get y^2 in $j^4(L\mathcal{R}_e \cdot f)$ is by using $j^4 f_x \equiv a_{32}y^2$, modulo the monomials that are already in $j^4(L\mathcal{R}_e \cdot f)$. Therefore, H is a P-$\mathcal{R}^+$-versal unfolding of the A_3-singularity of h_{N_0} if and only if $a_{32} \neq 0$. The geometric interpretation of this condition is given in Proposition 6.5(iii).□

Several geometric properties of M can be deduced from Theorems 6.1 and 6.2.

The intersection of the surface M with its tangent plane at p is called *the tangent Dupin indicatrix of* $\in M$ *at* p.

Proposition 6.6. *The tangent Dupin indicatrix of M at p consists locally of an isolated point if p is an elliptic point, a pair of transverse crossing curves if p is a hyperbolic point, a curve with a cusp singularity if p is an ordinary parabolic point (i.e., an A_2-singularity of the height function), an isolated point or a tacnode (a pair of tangential curves) if p is an A_3-singularity of the height function. See* Figure 6.3.

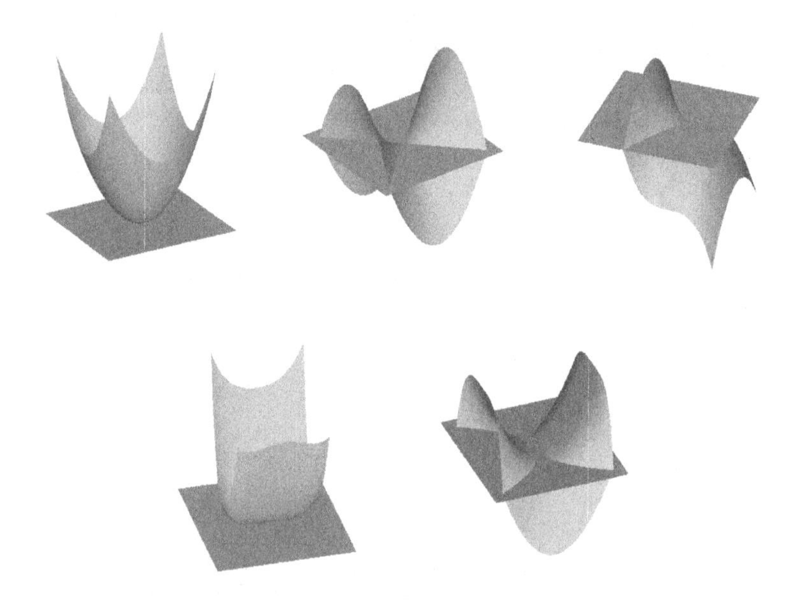

Fig. 6.3 Intersection of a surface with its tangent plane, top figures at an elliptic, hyperbolic and parabolic point respectively, and bottom figures at an A_3^+ and A_3^--singularities of the height function.

Proof. If we take the surface in Monge form $z = f(x, y)$, then the height function H_{N_0} along the normal $N_0 = (0, 0, 1)$ is just the function f. The intersection of the surface with its tangent plane $z = 0$ is the zero set of the function f, which is diffeomorphic to the zero set of the $\mathcal{R}$-normal form of the singularity of $h_{N_0} = f$. The zero sets of the normal forms (taken up to a sign $\pm$) are as follows

Name	$\mathcal{R}$-Model	Zero set
A_1^+	$x^2 + y^2$	$\{(0,0)\}$
A_1^-	$x^2 - y^2$	$\{(x,y) \in \mathbb{R}^2, 0 : x = y$ or $x = -y\}$
A_2	$x^2 + y^3$	$\{(t^3, -t^2), t \in \mathbb{R}, 0\}$
A_3^+	$x^2 + y^4$	$\{(0,0)\}$
A_3^-	$x^2 - y^4$	$\{(x,y) \in \mathbb{R}^2, 0 : x = y^2\} \cup \{(x,y) \in \mathbb{R}^2, 0 : x = -y^2\}$

□

Proposition 6.6 can be viewed as a local classification, up to diffeomorphisms, of the tangent Dupin indicatrices of a height function generic surface. We can also consider the *tangential Dupin foliation* of M at p, which is obtained by intersecting the surface M with planes parallel to T_pM.

Proposition 6.7. *The tangential Dupin foliation of a surface M at a point p is locally diffeomorphic to the foliation given by the level sets of the $\mathcal{R}$-model of the singularity of the height function on M at p along its normal direction. See* Figure 6.4.

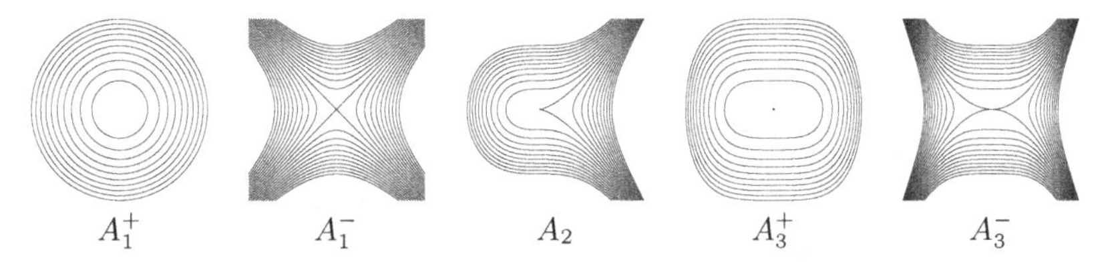

Fig. 6.4 The models, up to diffeomorphisms, of the tangential Dupin foliation of a height function generic surface.

Proof. We consider the same setting as in the proof of Proposition 6.6, so that $h_{N_0} = f$. The result follows by observing that the diffeomorphism in the $\mathcal{R}$-equivalence between two germs maps the level sets of one germ to the level sets of the other. □

The cylindrical pedal of the surface M is the image of the map $\mathbf{x}^* : U \to S^2 \times \mathbb{R}$, given by

$$\mathbf{x}^*(u) = (N(u), \langle \mathbf{x}(u), N(u) \rangle),$$

(see Chapters 2 and 5). The cylindrical pedal can be viewed as the dual surface of M.

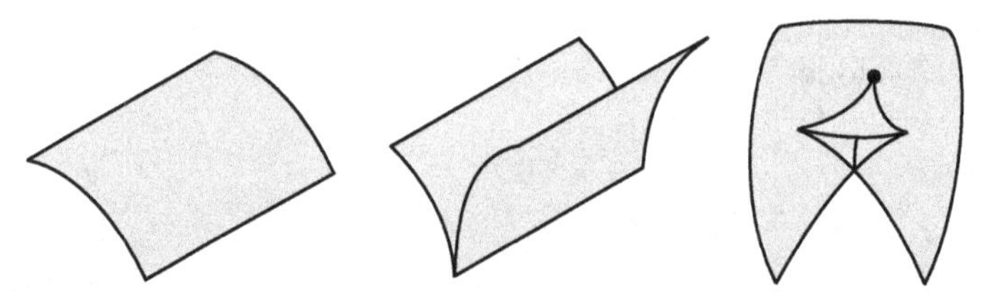

Fig. 6.5 The dual surface of M at the A_k-singularities of the height function: A_1 left, A_2 centre and A_3 right.

Proposition 6.8. *The dual surface of a height function generic surface M, or its cylindrical pedal, is locally diffeomorphic to*

(i) *a smooth surface when h_{N_0} has an A_1-singularity at p,*
(ii) *a cuspidal edge surface when h_{N_0} has an A_2-singularity at p,*
(iii) *a swallowtail surface when h_{N_0} has an A_3-singularity at p,*

where N_0 is the normal to M at p. See Figure 6.5.

Proof. The cylindrical pedal of M is a wavefront with $\tilde{H}$ its graph-like generating family (Theorem 5.14). Thus, it has Legendrian singularities. For a height function generic M, the family $\tilde{H}$ is $\mathcal{K}$-versal, so the cylindrical surface is diffeomorphic to the discriminant of a model of a $\mathcal{K}$-versal unfolding (with three parametres) of the singularities that occur in a given height function. These are A_1, A_2 and A_3-singularities (Theorem 6.1). The discriminants of the $\mathcal{K}$-versal unfoldings with three parameters of these singularities are as stated in the proposition. □

The stable singularities of the Gauss map are given in [Looijenga (1974)] and [Bleeker and Wilson (1978)]. These singularities are intimately related to those of the height functions on M.

Proposition 6.9. *The Gauss map $N : M \to S^2$ of a height function generic surface M is either a local diffeomorphism or has singularities of type fold or cusp. The Gauss map is singular at p if and only if p is a parabolic point. Its fold singularities occur on the parabolic set at the A_2-singularities of the height function and its cusp singularities occur at the A_3-singularities of the height functions.*

Proof. The family H is the generating family of the Lagrangian submanifold $L(H)(C_H) \subset T^*S^2$ and the Gauss map N is the corresponding catastrophe map $\chi_H : C_H \to S^2$. For a height function generic surface M, the family H is an $\mathcal{R}^+$-versal unfolding, so the catastrophe map $N = \chi_H$ is $\mathcal{A}$-equivalent to the catastrophe map of a model of an $\mathcal{R}^+$-versal unfolding (with two parametres) of the singularities that occur in a given height function. The catastrophe map of an $\mathcal{R}^+$-versal 2-parameter family of the A_1, A_2 or A_3 singularities are respectively the map-germs of a diffeomorphism, a fold or a cusp. □

Corollary 6.1. *For a height function generic surface M, the Gauss map at $p \in M$ is $\mathcal{A}$-equivalent to:*

(i) *a fold map-germ if and only if the dual surface of M at p is locally diffeomorphic to a cuspidal edge surface;*
(ii) *a cusp map-germ if and only if the dual surface of M at p is locally diffeomorphic to a swallowtail surface.*

It follows from Proposition 6.9 that the image of the parabolic set by the Gauss map is a curve with a cusp singularity at an A_3-singularity of the height function on M. For this reason, we have the following definition.

Definition 6.3. A point on the parabolic set where the height function along the normal direction to the surface has an A_3-singularity is called a *cusp of Gauss.* A cusp of Gauss p is elliptic if the height function along the normal direction at p has an A_3^+-singularity at p and hyperbolic if it has an A_3^--singularity.

We need the following lemma for some characterisations of the cusps of Gauss.

Lemma 6.1. *Let M be a surface parametrised locally by $\mathbf{x} : U \to \mathbb{R}^3$ and let $p_1 = \boldsymbol{x}(u_1)$ and $p_2 = \boldsymbol{x}(u_2)$ be two distinct points on M. Then,*

(1) *$N(u_1) = N(u_2)$ if and only if the tangent affine planes of M at p_1 and p_2 are parallel.*

(2) *$\boldsymbol{x}^*(u_1) = \boldsymbol{x}^*(u_2)$ if and only if the tangent affine planes of M at p_1 and p_2 are the same.*

Proof. The tangent affine plane of M at p_i, $i = 1, 2$, is given by $T_{p_i}(M; \mathbf{v}_i)_p = h_{\mathbf{v}_i}^{-1}(r_i)$, where $\mathbf{v}_i = N(u_i)$ and $r_i = H(p_i, \mathbf{v}_i)$. The condition $N(u_1) = N(u_2)$ means that the normal vectors to $T_{p_1}(M; \mathbf{v}_1)$ and $T_{p_2}(M; \mathbf{v}_2)$ are linearly dependent, so the two planes are parallel.

We have $\boldsymbol{x}^*(u_1) = \boldsymbol{x}^*(u_2)$ if and only if $N(u_1)$ and $N(u_2)$ are parallel and $r_1 = r_2$. This means that $T_{p_1}(M; \mathbf{v}_1) = T_{p_2}(M; \mathbf{v}_2)$. □

Cusps of Gauss can be identified geometrically in many ways ([Banchoff, Gaffney and McCrory (1982)]). We have the following result, the main part of which is given in [Banchoff, Gaffney and McCrory (1982)]. We include some additional information from the view point of the Legendrian singularity framework.

Theorem 6.3. *Let M be a height function generic surface parametrised locally by $\mathbf{x} : U \to \mathbb{R}^3$, and let $p_0 = \mathbf{x}(u_0)$ be a point on M. Then the following statements are equivalent.*

(i) *The point p_0 is a cusp of Gauss.*
(ii) *The dual surface of M is a swallowtail surface at p_0.*
(iii) *H_{N_0} has an A_3-singularity at p_0.*
(iv) *The order of contact of M with its tangent plane at p_0 is equal to 3 (i.e., $\text{T-ord}(\mathbf{x}(U), u_0) = 3$).*
(vi) *For any $\epsilon > 0$, there exist three distinct non-parabolic points $p_i = \mathbf{x}(u_i)$ such that $||u_0 - u_i|| < \epsilon$ for $i = 1, 2, 3$, and the tangent planes of M at p_1, p_2, p_3 are parallel.*
(vii) *For any $\epsilon > 0$, there exist two distinct non-parabolic points $p_i = \mathbf{x}(u_i)$ such that $||u_0 - u_i|| < \epsilon$ for $i = 1, 2$, and the tangent planes to M at p_1 and p_2 are equal.*

Proof. Statements (i) and (ii) are equivalent by Corollary 6.1.

It follows from Proposition 6.8 that statement (iii) is equivalent to statement (ii).

As the $\mathcal{K}_e$-codimention of the A_k-singularity is k, by definition of the order of contact of M with its tangent plane, statements (iii) and (iv) are equivalent.

Since M is a height function generic surface, the Gauss map has only fold or cusp singularities. If u_0 is a fold singularity, there is a neighbourhood of u_0 on which the Gauss map is 2 to 1 except on the parabolic curve (i.e, on the curve of fold singularities of the Gauss map). Of course, if u_0 is not a parabolic point, the Gauss map is one to one locally at u_0. By Lemma 6.1, the negation of statement (i) implies the negation of statement (vi). If u_0 is a cusp singularity, the image of the discriminant of the Gauss map has an ordinary cusp at N_0. By considering the $\mathcal{A}$-normal form $(x, xy + y^3)$ of the cusp singularity, it can be shown that a point has 3, 2 or 1 pre-image points by the Gauss map. By Lemma 6.1, this means that statement (vi) holds. Therefore, statement (i) implies statement (vi).

The dual surface of a height function generic surface M has singularities

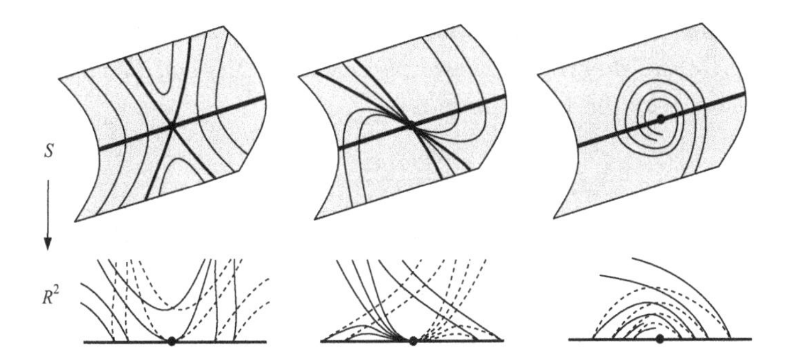

Fig. 6.6 Generic configurations of the asymptotic curves at a cusp of Gauss.

of type cuspidal edge or swallowtail. If p_0 is a cuspidal edge point, the dual surface has no self-intersections around the image of p_0, so by Lemma 6.1, the negation of statement (vii) implies the negation of statement (ii).

At a swallowtail point p_0, there is a curve on M containing p_0 which corresponds to the points of self-intersections of the dual surface of M. On this curve, there are two distinct points $p_1 = \boldsymbol{x}(u_1)$ and $p_1 = \boldsymbol{x}(u_2)$ such that $\boldsymbol{x}^*(u_1) = \boldsymbol{x}^*(u_2)$. By Lemma 6.1, this means that the tangent affine planes to M at points p_1 and p_2 are equal. Since the singularities are either cuspidal edges or swallowtails, statement (vii) characterises a swallowtail point of the dual surface. □

Remark 6.1. Cusps of Gauss can also be captured by the singularities of the asymptotic curves. The way these are studied is by lifting the bi-valued direction field in the plane determined by equation (6.3) to a single vector field ξ on a smooth surface S. The vector field ξ is singular at a point q if and only if q is the lift of a cusp of Gauss p. The singularity of ξ is of type saddle (has index -1) if and only if p is a hyperbolic cusp of Gauss and of type node or focus (has index $+1$) if and only if p is an elliptic cusp of Gauss. The generic topological configurations of the asymptotic curves at the cusp of Gauss are as in Figure 6.6 ([Banchoff, Gaffney and McCrory (1982); Banchoff and Thom (1980); Dara (1975); Davydov (1994); Kergosien and Thom (1980); Thom (1972)]). For more results on cusps of Gauss see [Bruce and Tari (2000); Banchoff, Gaffney and McCrory (1982); Oliver (2011); Uribe-Vargas (2006); Izumiya and Takahashi (2011)].

Proposition 6.10. *Let M be given in Monge form $z = f(x, y)$ at p with f as in* (6.4) *and let* $\mathbf{v} = (0, 1, 0) \in T_pM$.

If $\mathbf{v}$ *is an asymptotic direction at* p *and* p *is a parabolic point but not a cusp of Gauss, then the geodesic curvature at* $N(p)$ *of the image of the parabolic curve by the Gauss map is*

$$\kappa_g = -\frac{3a_{31}a_{33} - a_{32}^2}{24a_{20}^3 a_{33}}.$$

Proof. From the proof of Proposition 6.5(iii), when p is an A_2-singularity of the height function $a_{33} \neq 0$, so the zero set of $ln - m^2$ can be parametrised by

$$\alpha(x) = (x, -\frac{a_{32}}{3a_{33}}x + \eta x^2 + g(x))$$

with

$$\eta = -\frac{a_{42}}{3a_{33}} + \frac{a_{32}a_{43}}{3a_{33}^2} - \frac{2a_{32}^2 a_{44}}{9a_{33}^3} + \frac{(a_{32}^2 - 3a_{31}a_{33})^2}{27a_{33}^3 a_{20}},$$

and g is a germ, at the origin, of a smooth function with a zero 2-jet. (The constant η will not contribute to the geodesic curvature we are seeking here and is included for completeness. It contributes to the geodesic curvature of the parabolic set at p.) The parabolic set is the image by ϕ of the curve α and we have

$$j^2\phi(\alpha(x)) = (x, -\frac{a_{32}}{3a_{32}}x + \eta x^2, a_{20}x^2).$$

The Gauss map is given by

$$N = \frac{1}{(1 + (\frac{\partial f}{\partial x})^2 + (\frac{\partial f}{\partial y})^2)^{\frac{1}{2}}}(-\frac{\partial f}{\partial x}, -\frac{\partial f}{\partial y}, 1),$$

see Proposition 6.3. Its 2-jet at the origin is given by

$$(-2a_{20}x - 3a_{30}x^2 - 2a_{31}xy - a_{32}y^2, -a_{31}x^2 - 2a_{32}xy - 3a_{33}y^2, 1 - 2a_{20}^2x^2).$$

Let $\beta(x) = N \circ \alpha(x)$ be a local parametrisation of the image of the parabolic set by the Gauss map. Then

$$j^2\beta(x) = (-2a_{20}x - (3a_{30} - \frac{2a_{31}a_{32}}{3a_{33}} + \frac{a_{32}^3}{9a_{33}^2})x^2, -\frac{3a_{31}a_{33} - a_{32}^2}{3a_{33}}x^2, 1 - 2a_{20}^2x^2).$$

The geodesic curvature of β at $N(p)$ is (see formula (6.2))

$$\kappa_g = \frac{1}{||\beta'(0)||^3}\langle \beta''(0), \beta'(t) \times N(\alpha(0))\rangle = -\frac{3a_{31}a_{33} - a_{32}^2}{24a_{20}^3 a_{33}}.$$

□

6.4 Contact with lines

As seen in §4.6 of Chapter 4, the family of orthogonal projections $P : U \times S^2 \to TS^2$ on M is given by

$$P(u, \mathbf{v}) = (\mathbf{v}, \mathbf{x}(u) - \langle \mathbf{x}(u), \mathbf{v} \rangle \mathbf{v}).$$

We denote the second component of P by $P_{\mathbf{v}}$ and consider $P_{\mathbf{v}}$ as the orthogonal projection of M along the fixed direction $\mathbf{v}$. We have, for $i = 1, 2$,

$$\frac{\partial P_{\mathbf{v}}}{\partial u_i}(u_1, u_2) = \mathbf{x}_{u_i}(u) - \langle \mathbf{x}_{u_i}(u), \mathbf{v} \rangle \mathbf{v}.$$

The map $P_{\mathbf{v}}$ is singular at $p = \mathbf{x}(u)$ if and only if the vectors $\partial P_{\mathbf{v}} / \partial u_1(u)$ and $\partial P_{\mathbf{v}} / \partial u_2(u)$ are linearly dependent, and this occurs if and only if $\mathbf{v}$ is tangent to M at p.

We denote by $\Sigma_{\mathbf{v}}$ the set of critical points of $P_{\mathbf{v}}$ (these are the points on M where $\mathbf{v}$ is tangent to M) and by $\Delta_{\mathbf{v}}$ the image of $\Sigma_{\mathbf{v}}$ by $P_{\mathbf{v}}$, so $\Delta_{\mathbf{v}} = P_{\mathbf{v}}(\Sigma_{\mathbf{v}})$. The set $\Delta_{\mathbf{v}}$ is called *the profile* or *apparent contour* of M along the direction $\mathbf{v}$ and $\Sigma_{\mathbf{v}}$ is called the *contour generator* of M along the direction $\mathbf{v}$ (see Figure 6.7).

The $\mathcal{K}$-singularities of $P_{\mathbf{v}}$ measure the contact of M with lines parallel to $\mathbf{v}$. In Chapter 4, we reviewed the general framework of contact between submanifolds. When the dimension of the target of these map-germs is greater than 1, the group $\mathcal{K}$ is too large to give significant geometric information. This is why we use its subgroup $\mathcal{A}$ instead. (Montaldi's results are still valid for the group $\mathcal{A}$, see Chapter 4.)

Another justification for using the group $\mathcal{A}$ is the following. Our aim is to understand the singularities of the apparent contour and how they bifurcate as the direction of projection varies in S^2. We can suppose that the point $p \in M$ is the image of $0 \in \mathbb{R}^2$ by the parametrisation $\mathbf{x}$, identify the plane of projection $T_{\mathbf{v}}S^2$ with $\mathbb{R}^2$ and suppose that $P_{\mathbf{v}}(p)$ is also the origin in $\mathbb{R}^2$. We have the diagram

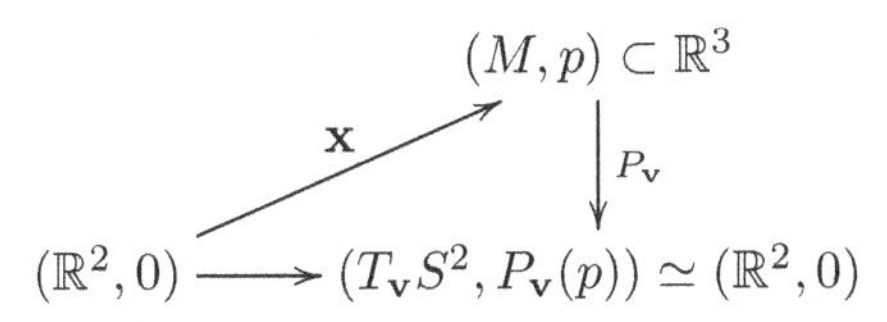

and still denote the composite map-germ $P_{\mathbf{v}} \circ \mathbf{x} : (\mathbb{R}^2, 0) \to (\mathbb{R}^2, 0)$ by $P_{\mathbf{v}}$. The singularities of the apparent contour of interest are those which are

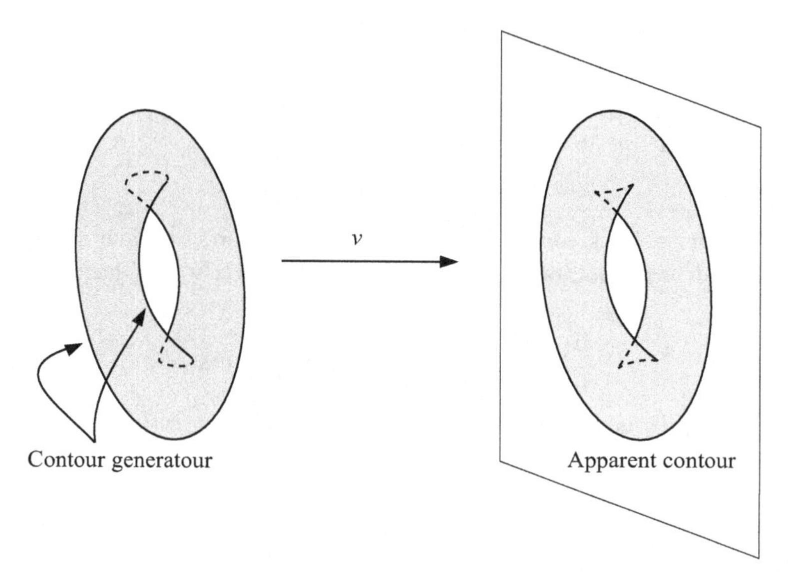

Fig. 6.7 An apparent contour of a torus: the dashed parts represent the invisible part of the contour generator and apparent contour when the surface is not transparent.

not altered by a re-parametrisation of the surface (given by germs of diffeomorphisms h in the source $(\mathbb{R}^2, 0)$) nor by smooth changes of coordinates in the plane of projection (given by germs of diffeomorphisms k in the target $(\mathbb{R}^2, 0)$), as shown in the diagram below

$$\begin{array}{ccc} (\mathbb{R}^2,0) & \xrightarrow{P_{\mathbf{v}}} & (\mathbb{R}^2,0) \\ \Big\downarrow h & & \Big\downarrow k \\ (\mathbb{R}^2,0) & \xrightarrow{k\circ P_{\mathbf{v}}\circ h^{-1}} & (\mathbb{R}^2,0) \end{array}$$

It is thus appropriate to use the group $\mathcal{A}$ for studying the singularities of the germs of orthogonal projections of M.

6.4.1 *Contour generators and apparent contours*

Proposition 6.11. (i) *The contour generator $\Sigma_{\mathbf{v}}$ is a singular curve at p if and only if p is a parabolic point and $\mathbf{v}$ is the unique asymptotic direction at p.*

(ii) *The apparent contour $\Delta_{\mathbf{v}}$ is a smooth curve at $P_{\mathbf{v}}(p)$ if and only if $\mathbf{v}$ is not an asymptotic direction at p.*

Proof. Let $\mathbf{x} : U \to \mathbb{R}^3$ be a local parametrisation of M.

(i) A point $\mathbf{x}(u)$ is in $\Sigma_{\mathbf{v}}$ if and only if

$$[v, \mathbf{x}_{u_1}(u), \mathbf{x}_{u_2}(u)] = 0,$$

where $[.,.,.]$ denotes the determinant of the matrix formed by three vectors in $\mathbb{R}^3$. The contour generator $\Sigma_{\mathbf{v}}$ is singular at u if and only if

$$[\mathbf{v}, \mathbf{x}_{u_1u_1}(u), \mathbf{x}_{u_2}(u)] + [\mathbf{v}, \mathbf{x}_{u_1}(u), \mathbf{x}_{u_1u_2}(u)] = 0, \tag{6.8}$$

$$[\mathbf{v}, \mathbf{x}_{u_1u_2}(u), \mathbf{x}_{u_2}(u)] + [\mathbf{v}, \mathbf{x}_{u_1}(u), \mathbf{x}_{u_2u_2}(u)] = 0. \tag{6.9}$$

We drop the argument u and write $\mathbf{v} = a\mathbf{x}_{u_1} + b\mathbf{x}_{u_2}$. The vectors $\mathbf{x}_{u_iu_j}$ can be written, with respect to the basis $\{\mathbf{x}_{u_1}, \mathbf{x}_{u_2}, N\}$ of $\mathbb{R}^3$, in the form

$$\begin{aligned}
\mathbf{x}_{u_1u_1} &= \Gamma_{11}^1\mathbf{x}_{u_1} + \Gamma_{11}^2\mathbf{x}_{u_2} + lN \\
\mathbf{x}_{u_1u_2} &= \Gamma_{12}^1\mathbf{x}_{u_1} + \Gamma_{12}^2\mathbf{x}_{u_2} + mN \\
\mathbf{x}_{u_2u_2} &= \Gamma_{22}^1\mathbf{x}_{u_1} + \Gamma_{22}^2\mathbf{x}_{u_2} + nN
\end{aligned}$$

where Γ_{ij}^k are the Christoffel symbols (see for example [do Carmo (1976)]). Then equation (6.8) becomes

$$\begin{aligned}
&[a\mathbf{x}_{u_1} + b\mathbf{x}_{u_2}, \Gamma_{11}^1\mathbf{x}_{u_1} + \Gamma_{11}^2\mathbf{x}_{u_2} + lN, \mathbf{x}_{u_2}] \\
&+[a\mathbf{x}_{u_1} + b\mathbf{x}_{u_2}, \mathbf{x}_{u_1}, \Gamma_{12}^1\mathbf{x}_{u_1} + \Gamma_{12}^2\mathbf{x}_{u_2} + mN] = 0,
\end{aligned}$$

which is equivalent to

$$al + bm = 0$$

as $[\mathbf{x}_{u_1}, \mathbf{x}_{u_2}, N] \neq 0$. Similarly, equation (6.9) is equivalent to $am + bn = 0$. Therefore, the contour generator is singular if and only if

$$\begin{cases} al + bm = 0 \\ am + bn = 0 \end{cases}$$

equivalently, $nl - m^2 = 0$ and $\mathbf{v} = -n\mathbf{x}_{u_1} + m\mathbf{x}_{u_2}$ at $u = (u_1, u_2)$. These are precisely the conditions for p to be a parabolic point and for $\mathbf{v}$ to be the unique asymptotic direction at p.

(ii) The contour generator $\Sigma_{\mathbf{v}}$ must be a smooth curve at p otherwise $\Delta_{\mathbf{v}}$ is singular at $P_{\mathbf{v}}(p)$. We take, without loss of generality, $am + bn \neq 0$ at p so that $\Sigma_{\mathbf{v}}$ can be parametrised by $u_1 \mapsto (u_1, u_2(u_1))$ for some smooth function $u_2(u_1)$ with

$$u_2' = -\frac{al + bm}{am + bn}. \tag{6.10}$$

The apparent contour $\Delta_{\mathbf{v}}$ is then parametrised by

$$P_v(u_1) = \mathbf{x}(u_1, u_2(u_1)) - \langle \mathbf{x}(u_1, u_2(u_1)), v\rangle v.$$

We have

$$\begin{aligned}
P'_{\mathbf{v}} &= \mathbf{x}_{u_1} + u'_2\mathbf{x}_{u_2} - \langle \mathbf{x}_{u_1} + u'_2\mathbf{x}_{u_2}, v\rangle v \\
&= \mathbf{x}_{u_1} + u'_2\mathbf{x}_{u_2} - \langle \mathbf{x}_{u_1} + u'_2\mathbf{x}_{u_2}, a\mathbf{x}_{u_1} + b\mathbf{x}_{u_2}\rangle (a\mathbf{x}_{u_1} + b\mathbf{x}_{u_2}) \\
&= (1 - a(aE + (u'_2 a + b)F + u'_2 bG))\,\mathbf{x}_{u_1} \\
&\quad + (u'_2 - b(aE + (u'_2 a + b)F + u'_2 bG))\,\mathbf{x}_{u_2}
\end{aligned}$$

Therefore, $\Delta_{\mathbf{v}}$ is singular if and only if

$$1 - a(aE + (u'_2 a + b)F + u'_2 bG) = 0,$$
$$u'_2 - b(aE + (u'_2 a + b)F + u'_2 bG) = 0.$$

The above two equations can be written, after substituting u'_2 by its expression in (6.10), in the form

$$am + bn = a\left(a^2(mE - lF) + ab(nE - lG) + b^2(nF - mG)\right), \quad (6.11)$$

$$al + bm = -b\left(a^2(mE - lF) + ab(nE - lG) + b^2(nF - mG)\right). \quad (6.12)$$

Observe that $a^2(mE - lF) + ab(nE - lG) + b^2(nF - mG) = 0$ means that $\mathbf{v}$ is a solution of equation (6.1), that is, $\mathbf{v}$ is a principal direction. If this is the case, the system of equations (6.11) and (6.12) becomes

$$\begin{cases} al + bm = 0 \\ am + bn = 0 \end{cases}$$

equivalently, p is a parabolic point and $\mathbf{v}$ the unique asymptotic direction at p. But this is excluded as it implies that $\Sigma_{\mathbf{v}}$, and hence $\Delta_{\mathbf{v}}$, is singular.

Suppose then that $a^2(mE - lF) + ab(nE - lG) + b^2(nF - mG) \neq 0$. Dividing side by side equation (6.11) by equation (6.12) yields

$$\frac{am + bn}{al + bm} = -\frac{a}{b}.$$

Rearranging the above equality gives

$$a^2 l + 2abm + b^2 n = 0.$$

This implies that $\mathbf{v}$ is a solution of equation (6.3), that is, $\mathbf{v}$ is an asymptotic direction.

Conversely, let $\mathbf{v}$ be a unit asymptotic direction (we project along $\mathbf{v}$ so we require $v \in S^2$). Then $a^2l + 2abm + b^2n = 0$, so $al + bm = -(am + bn)b/a$. The right hand side of equation (6.11) can then be rearranged to get

$$\begin{aligned}
&a\left(a^2(mE - lF) + ab(nE - lG) + b^2(nF - mG)\right) \\
&= a\left((a^2m + abn)E + (b^2n - a^2l)F - (abl + mb^2)G\right) \\
&= a\left(a(am + bn)E + 2b(am + bn)F + \frac{b^2}{a}(am + bn)G\right) \\
&= (am + bn)(a^2E + 2abF + b^2G) \\
&= am + bn.
\end{aligned}$$

Similarly, rearranging the right hand side of equation (6.12) gives

$$\begin{aligned}
&-b\left(a^2(mE-lF)+ab(nE-lG)+b^2(nF-mG)\right)\\
&=-\frac{b}{a}a\left((a^2m+abn)E+(b^2n-a^2l)F-(abl+mb^2)G\right)\\
&=-\frac{b}{a}(am+bn)\\
&=al+bm.
\end{aligned}$$

Therefore, equations (6.11) and (6.12) are satisfied. □

Proposition 6.12. *Suppose that the critical set $\Sigma_{\mathbf{v}}$ is a smooth curve at p. Then its tangent direction at p is the conjugate direction to $\mathbf{v}$ with respect to the second fundamental form at p.*

Proof. Following the proof of Proposition 6.11, the tangent direction to the contour generator is parallel to $\mathbf{w}=-(am+bn)\mathbf{x}_{u_1}+(al+bm)\mathbf{x}_{u_2}$. This is precisely the conjugate direction to $\mathbf{v}$ as $\langle W_p(\mathbf{w}),\mathbf{v}\rangle=0$. □

We give below an alternative proof to Koenderink's Theorem [Koenderink (1984)].

Theorem 6.4 (Koenderink's Theorem). *Suppose that the apparent contour $\Delta_{\mathbf{v}}$ is a smooth curve at $P_{\mathbf{v}}(p)$. Then the Gaussian curvature of M at p is equal to the product of the curvature of the apparent contour together with the curvature of the normal section of M at p along the direction $\mathbf{v}$.*

Proof. The curvature of the normal section is just the normal curvature of the surface M at p along the tangent direction $\mathbf{v}$ and is given by

$$\kappa_n(\mathbf{v})=\frac{\mathrm{II}_p(\mathbf{v})}{||v||^2}.$$

Let $\alpha(s)$ be an arc length parametrisation of the contour generator $\Sigma_{\mathbf{v}}$. Then, by Proposition 6.12,

$$\begin{aligned}
\alpha'&=\mathbf{t}=\frac{\bar{\mathbf{v}}}{||\bar{\mathbf{v}}||}\\
\alpha''&=\kappa_g\,\mathbf{t}\times N+\kappa_n(\bar{\mathbf{v}})N.
\end{aligned}$$

The apparent contour $\Delta_{\mathbf{v}}$ is parametrised by $\beta(s)=P_{\mathbf{v}}(\alpha(s))$. The projection $P_{\mathbf{v}}$ for $\mathbf{v}$ fixed is a linear map $\mathbb{R}^3\to T_{\mathbf{v}}S^2$, so

$$\begin{aligned}
\beta'&=dP_{\mathbf{v}}(\alpha')=\alpha'-\langle\alpha',\mathbf{v}\rangle\mathbf{v},\\
\beta''&=dP_{\mathbf{v}}(\alpha'')=\alpha''-\langle\alpha'',\mathbf{v}\rangle\mathbf{v}.
\end{aligned}$$

The vector $N(s)$ is a unit normal vector to the curve $\Delta_{\mathbf{v}}$ at $\beta(s)$, and $\beta'(s), N(s)$ form a positively oriented frame. If we denote $\beta'^{\perp}$ the vector obtained by rotating β' anti-clockwise by $\pi/2$, then $\beta'^{\perp} = ||\beta'||N$, so the curvature $\kappa(\Delta_{\mathbf{v}})$ of the apparent contour is given by

$$\kappa(\Delta_{\mathbf{v}}) = \frac{\langle \beta'^{\perp}, \beta''\rangle}{||\beta'||^3} = \frac{\langle ||\beta'||N, \beta''\rangle}{||\beta'||^3} = \frac{\langle N, \alpha''\rangle}{||\beta'||^2} = \frac{\kappa_n(\bar{v})}{||\beta'||^2}.$$

We have

$$\beta' = dP_{\mathbf{v}}(\frac{\bar{\mathbf{v}}}{||\bar{\mathbf{v}}||}) = \frac{\bar{\mathbf{v}}}{||\bar{\mathbf{v}}||} - \langle \frac{\bar{\mathbf{v}}}{||\bar{\mathbf{v}}||}, \frac{\mathbf{v}}{||\mathbf{v}||}\rangle \frac{\mathbf{v}}{||\mathbf{v}||} = \frac{1}{||\mathbf{v}||^2||\bar{\mathbf{v}}||}(||\mathbf{v}||^2\bar{\mathbf{v}} - \langle \bar{\mathbf{v}}, \mathbf{v}\rangle \mathbf{v}),$$

so that

$$\begin{aligned} ||\beta'||^2 &= \frac{1}{||\mathbf{v}||^2||\bar{\mathbf{v}}||^2}(||\mathbf{v}||^2||\bar{\mathbf{v}}||^2 - \langle \bar{\mathbf{v}}, \mathbf{v}\rangle^2) \\ &= \frac{1}{||\mathbf{v}||^2||\bar{\mathbf{v}}||^2}(EG - F^2)\mathrm{II}_p(\mathbf{v})^2 \text{ (by Proposition 6.2(iv))} \\ &= \frac{||\mathbf{v}||^2}{||\bar{\mathbf{v}}||^2}(EG - F^2)\left(\frac{\mathrm{II}_p(\mathbf{v})^2}{||\mathbf{v}||^4}\right) \\ &= \frac{||\mathbf{v}||^2}{||\bar{\mathbf{v}}||^2}(EG - F^2)\kappa_n(\mathbf{v})^2. \end{aligned}$$

Therefore,

$$\kappa(\Delta_{\mathbf{v}}) = \frac{||\bar{\mathbf{v}}||^2\kappa_n(\bar{\mathbf{v}})}{||\mathbf{v}||^2(EG - F^2)\kappa_n(\mathbf{v})^2}.$$

We obtain by substituting $\kappa_n(\bar{\mathbf{v}})$ by its expression from Proposition 6.2 that

$$\kappa(\Delta_{\mathbf{v}}) = \frac{||\bar{\mathbf{v}}||^2}{||\mathbf{v}||^2(EG - F^2)\kappa_n(\mathbf{v})^2}\left(\frac{||\mathbf{v}||^2}{||\bar{\mathbf{v}}||^2}(nl - m^2)\kappa_n(\mathbf{v})\right) = \frac{K(p)}{\kappa_n(\mathbf{v})}.$$

Consequently,

$$K(p) = \kappa_n(\mathbf{v})\kappa(\Delta_{\mathbf{v}}).$$

□

We have the following consequences of Koenderink's theorem.

Corollary 6.2. (i) *The direction of the projection* $\mathbf{v}$ *is a principal direction if and only if the curvature of the apparent contour is equal to the principal curvature associated to the other principal direction.*

(ii) *The apparent contour* $\Delta_{\mathbf{v}}$ *has an inflection if and only if the corresponding point on the surface is a parabolic point and* $\mathbf{v}$ *is not an asymptotic direction.*

6.4.2 The generic singularities of orthogonal projections

Theorem 6.5. *There is an open and dense set $\mathcal{O}_2$ of immersions $\mathbf{x}: U \to \mathbb{R}^3$ such that for any $\mathbf{x} \in \mathcal{O}_2$, the surface $M = \mathbf{x}(U)$ has the following properties. For any $\mathbf{v} \in S^2$, the orthogonal projection $P_\mathbf{v}$ along $\mathbf{v}$ has only $\mathcal{A}$-singularities of $\mathcal{A}_e$-codimension* ≤ 2 (Table 6.1) *at any point p on M. Furthermore, these singularities are $\mathcal{A}_e$-versally unfolded by the family P.*

Proof. The proof is a consequence of Theorems 4.4 and 4.12. □

Definition 6.4. A surface is called (*locally*) *projection generic* if any of its local parametrisations belongs to the set $\mathcal{O}_2$ in Theorem 6.5.

We shall identify geometrically the conditions for a given local generic singularity of $P_\mathbf{v}$ to occur. The conditions for having a given singularity have simpler expressions when considering the following setting. We take the surface M in Monge-form $\phi(x,y) = (x, y, f(x,y))$ at a point p considered to be the origin in $\mathbb{R}^3$. We can rotate the coordinate axes if necessary and set $v = (0,1,0)$. Observe that there is no loss of generality with this setting as the choice of a coordinate system in $\mathbb{R}^3$ is arbitrary. With the above setting, we have

$$P_\mathbf{v}(x,y) = (x, f(x,y)).$$

When considering the family of orthogonal projections P, the computations also simplify considerably if we modify P as follows. We parametrise the directions near $v = (0,1,0)$ by $(\alpha, 1, \beta)$, with α and β close to zero, and project to the fixed plane $\pi : y = 0$. The projection of the point $(x, y, f(x,y))$ along $(\alpha, 1, \beta)$ is the point $(x, y, f(x,y)) + \lambda(\alpha, 1, \beta) \in \pi$, which implies that $\lambda = -y$. We obtain the family of (germs of) projections $\bar{P} : (\mathbb{R}^2 \times \mathbb{R}^2, (0,0)) \to (\mathbb{R}^2, 0)$, given by $\bar{P}(x,y,\alpha,\beta) = (x - \alpha y, f(x,y) - \beta y)$. The change of coordinates $(x,y) \mapsto (x - \alpha y, y)$ transforms the family $\bar{P}$ to the following family

$$\tilde{P}(x,y,\alpha,\beta) = (x, f(x+\alpha y, y) - \beta y) \tag{6.13}$$

which we call the *modified family of projections.* It is clear from the transformations carried out above that the family P of orthogonal projections on M is $\mathcal{A}$-equivalent to the modified family of projections $\tilde{P}$.

We need the following concept.

Definition 6.5. A point p on M is a *flecnodal point* if there is a tangent line through p which has at least 4-point contact with M at p. Equivalently,

p is a flecnodal point if it is in the closure of the set of points where the projection along an asymptotic direction has a swallowtail singularity, i.e., it is $\mathcal{A}$-equivalent to $(x, xy+y^4)$. The *flecnodal set* of M is the set flecnodal points.

Theorem 6.6. *The following holds for M in Monge form $z = f(x,y)$ at p with f as in (6.4) and $v = (0,1,0)$.*

(i) *The flecnodal set (in the parameter space) consists locally of points (x,y) for which there exist α and β near zero such that*

$$\begin{gathered}(\alpha f_x + f_y)(x+\alpha y, y) - \beta = 0,\\ (\alpha^2 f_{xx} + 2\alpha f_{xy} + f_{yy})(x+\alpha y, y) = 0,\\ (\alpha^3 f_{xxx} + 3\alpha^2 f_{xxy} + 3\alpha f_{xyy} + f_{yyy})(x+\alpha y, y) = 0.\end{gathered}$$

In particular, the origin p is a flecnodal point if and only if

$$f_y = f_{yy} = f_{yyy} = 0,\ f_{xy} f_{yyyy} \neq 0 \ \text{ at } p,$$

that is,

$$a_{22} = a_{33} = 0 \text{ and } a_{21}a_{44} \neq 0.$$

(ii) *If p is a flecnodal point, then the 1-jet of the equation of the flecnodal set is given by*

$$6(a_{43} - \frac{a_{32}^2}{a_{21}})x + 24a_{44}y.$$

Proof. (i) The swallowtail singularities of the projection $P_{\mathbf{v}}$ are $\mathcal{A}$-invariant, so we can use the modified family of projections (6.13).

For α, β fixed, we have a map-germ $\tilde{P}_{(\alpha,\beta)}(x,y) = (x, f(x+\alpha y, y) - \beta y)$ from the plane to the plane. We can now use Saji's criteria for recognition of the swallowtail singularities of $\tilde{P}_{(\alpha,\beta)}$ ([Saji (2010)]). The advantage of using the modified family of projections is that the critical set $\Sigma_{\alpha,\beta}$ of $\tilde{P}_{(\alpha,\beta)}$ is given by

$$\frac{\partial}{\partial y}(f(x+\alpha y, y) - \beta y) = 0,$$

and the kernel of $d\tilde{P}_{(\alpha,\beta)}$ at the singular points is along the constant direction $(0,1)$. Thus, by Theorem 3 in [Saji (2010)], (x,y) is a swallowtail singularity of $\tilde{P}_{(\alpha,\beta)}$ if and only if its critical set $\Sigma_{\alpha,\beta}$ is a smooth curve, that is,

$$(f_{xy}(0,0), f_{yy}(0,0)) = (a_{21}, 2a_{22}) \neq (0,0),$$

and has 4-point contact at (x, y) with the kernel of $d\tilde{P}_{(\alpha,\beta)}$, that is,

$$\frac{\partial}{\partial y}(f(x+\alpha y, y) - \beta y) = 0,$$

$$\frac{\partial^2}{\partial y^2}(f(x+\alpha y, y) - \beta y) = 0,$$

$$\frac{\partial^3}{\partial y^3}(f(x+\alpha y, y) - \beta y) = 0,$$

$$\frac{\partial^4}{\partial y^4}(f(x+\alpha y, y) - \beta y) \neq 0.$$

The last inequality is satisfied locally at p if and only if

$$\frac{\partial^4}{\partial y^4}(f(x+\alpha y, y) - \beta y)|_{(0,0,0,0)} = f_{yyyy}(0,0) = 24a_{44} \neq 0.$$

The first three equations are equivalent to

$$(\alpha f_x + f_y)(x + \alpha y, y) - \beta = 0, \tag{6.14}$$

$$(\alpha^2 f_{xx} + 2\alpha f_{xy} + f_{yy})(x+\alpha y, y) = 0, \tag{6.15}$$

$$(\alpha^3 f_{xxx} + 3\alpha^2 f_{xxy} + 3\alpha f_{xyy} + f_{yyy})(x+\alpha y, y) = 0. \tag{6.16}$$

(ii) The above three equations determine the flecnodal set. Evaluating at $\alpha = \beta = 0$, we get that the origin p is a flecnodal point if and only if $f_y = f_{yy} = f_{yyy} = 0$, at $(0,0)$, equivalently, $a_{22} = a_{33} = 0$, and $a_{21} \neq 0$, $a_{44} \neq 0$. (As $a_{22} = 0$, the condition for $\Sigma_{\alpha,\beta}$ to be a smooth curve becomes $a_{21} \neq 0$.)

We can use equation (6.14) to obtain β as a function of (x, y, α). Substituting in equation (6.15), we can solve for α as $f_{xy}(0,0) = a_{21} \neq 0$. We obtain α as a function of (x, y) and substituting in equation (6.16), we obtain an equation for the flecnodal set. The 1-jet of this equation is $6(a_{43} - \frac{a_{32}^2}{a_{21}})x + 24a_{44}y$. □

Theorem 6.7. *Suppose that M is given in Monge form $z = f(x, y)$ at the origin p with f as in (6.4) and suppose that the direction of projection is parallel to $v = (0, 1, 0)$. Then the conditions for $P_{\mathbf{v}}$ to have a generic singularity at p are given in Table 6.1 in terms of the coefficients of the Taylor expansion of f at p.*

The geometric characterisation of the generic singularities of $P_{\mathbf{v}}$ are as in Table 6.2. *(In* Table 6.2 *is given only the geometric meaning of the conditions in* Table 6.1 *that are equal to zero.)*

Table 6.1 Algebraic conditions for the local singularities of $P_{\mathbf{v}}$.

Name	Normal form	Algebraic conditions
Fold	(x, y^2)	$a_{22} \neq 0$
Cusp	$(x, xy + y^3)$	$a_{22} = 0,\ a_{21} \neq 0,\ a_{33} \neq 0$
Lips/beaks	$(x, y^3 \pm x^2 y)$	$a_{22} = 0,\ a_{21} = 0,\ a_{33} \neq 0,\ a_{32}^2 - 3a_{31}a_{33} \neq 0$
Goose	$(x, y^3 + x^3 y)$	$a_{22} = 0,\ a_{21} = 0,\ a_{32}^2 - 3a_{31}a_{33} = 0,\ a_{33} \neq 0,$ $27a_{41}a_{33}^3 - 18a_{42}a_{32}a_{33}^2 + 9a_{43}a_{32}^2 a_{33} - 4a_{44}a_{32}^3 \neq 0$
Swallowtail	$(x, xy + y^4)$	$a_{22} = 0,\ a_{33} = 0,\ a_{21} \neq 0,\ a_{44} \neq 0$
Butterfly	$(x, xy + y^5 \pm y^7)$	$a_{22} = 0,\ a_{33} = 0,\ a_{44} = 0,\ a_{21} \neq 0,\ a_{55} \neq 0,$ $(8a_{55}a_{77} - 5a_{66}^2)a_{21}^2 + 2a_{55}(a_{32}a_{66} - 20a_{43}a_{55})a_{21}+$ $35a_{32}^2 a_{55}^2 \neq 0$
Gulls	$(x, xy^2 + y^4 + y^5)$	$a_{22} = 0,\ a_{21} = 0,\ a_{33} = 0,\ a_{32} \neq 0,\ a_{44} \neq 0,$ $a_{55}a_{32}^2 - 2a_{43}a_{44}a_{32} + 4a_{31}a_{44}^2 \neq 0$

Table 6.2 Geometric characterisation of the local singularities of $P_{\mathbf{v}}$ ([Gaffney (1983)]).

Name	Geometric characterisation
Fold	$\mathbf{v}$ tangent to M at p
Cusp	p a hyperbolic point, $\mathbf{v}$ an asymptotic direction at p
Lips/beaks	p a parabolic point, $\mathbf{v}$ an asymptotic direction at p
Goose	p a parabolic point, $\mathbf{v}$ an asymptotic direction at p, the Gauss image of the parabolic set has a geodesic inflection
Swallowtail	p a flecnodal point, $\mathbf{v}$ an asymptotic direction at p
Butterfly	p a flecnodal point, $\mathbf{v}$ an asymptotic direction at p and tangent to the flecnodal curve at p
Gulls	p a cusp of Gauss, $\mathbf{v}$ an asymptotic direction

Proof. If $v \notin T_pM$, $P_{\mathbf{v}}$ is a local diffeomorphism, so is locally $\mathcal{A}$-equivalent to the germ (x, y). We consider from now on the case when $v \in T_pM$ and take M in Monge form $z = f(x, y)$ at p with f as in (6.4) and $v = (0, 1, 0)$. Then $P_{\mathbf{v}}(x, y) = (x, f(x, y))$.

The $\mathcal{A}$-classification of map-germs $(\mathbb{R}^2, 0) \to (\mathbb{R}^2, 0)$ of $\mathcal{A}_e$-codimension ≤ 2 is given in [Gaffney (1983)]. This is extended in [Rieger (1987)] to cover the germs of $\mathcal{A}_e$-codimension ≤ 6. In [Rieger (1987)], the germs are also taken in the form $(x, f(x, y))$ and the work there is about determining the $\mathcal{A}$-classes of such germs. The problem here is a recognition one: given a germ of the form $P_{\mathbf{v}}(x, y) = (x, f(x, y))$, find the conditions on the coefficients of the Taylor expansion of f for $P_{\mathbf{v}}$ to be $\mathcal{A}$-equivalent to one of the germs in Table 6.1. We shall use of course the fact that when the singularity of $P_{\mathbf{v}}$ is k-$\mathcal{A}$-determined we can work with the k-jet of $P_{\mathbf{v}}$ and ignore higher degree terms in the Taylor expansion of f.

All monomials of the form $(0, a_{i0}x^i)$ $i = 0, \ldots, k$ in $j^kP_{\mathbf{v}}$ can be eliminated by a change of coordinates in the target of the form $(U, V) \mapsto (U, V - a_{i0}U^i)$, so $j^kP_{\mathbf{v}}(x, y) \sim_{\mathcal{A}} (x, j^kf(x, y) - j^kf(x, 0))$. To avoid repetition, we shall carry out these changes of coordinates without mentioning them.

The singular set $\Sigma_{\mathbf{v}}$ of $P_{\mathbf{v}}$ is the zero set of

$$f_y(x, y) = a_{21}x + 2a_{22}y + a_{31}x^2 + 2a_{32}xy + 3a_{33}y^2 + O(3).$$

It is a smooth curve at the origin if and only $a_{21} \neq 0$ or $a_{22} \neq 0$.

• *Fold*

The fold singularity (x, y^2) is 2-$\mathcal{A}$-determined, so consider $j^2P_{\mathbf{v}}(x, y) \sim_{\mathcal{A}^{(2)}} (x, a_{21}xy + a_{22}y^2)$.

If $a_{22} \neq 0$, the change of coordinates $(x, y) \mapsto (x, y - a_{21}/(2a_{22})x)$ removes the term $a_{21}xy$ in the second component of $j^2P_{\mathbf{v}}$. Re-scaling gives $j^2P_{\mathbf{v}} \sim_{\mathcal{A}^{(2)}} (x, y^2)$, hence $P_{\mathbf{v}} \sim_{\mathcal{A}} (x, y^2)$.

If $a_{22} = 0$, then $j^2P_{\mathbf{v}}$ is $\mathcal{A}^{(2)}$-equivalent to (x, xy) if $a_{21} \neq 0$ and to $(x, 0)$ if $a_{21} = 0$. In both cases $j^2P_{\mathbf{v}}$ cannot be $\mathcal{A}$-equivalent to the fold singularity. Therefore, the singularity of $P_{\mathbf{v}}$ at p is a fold if and only if $a_{22} \neq 0$, equivalently, $\mathbf{v}$ is transverse to $\Sigma_{\mathbf{v}}$ at p.

• *Cusp*

We have $j^2P_{\mathbf{v}} \sim_{\mathcal{A}^{(2)}} (x, xy)$ if and only if $a_{22} = 0$ and $a_{21} \neq 0$, equivalently, $\mathbf{v}$ is an asymptotic direction at p and p is not a parabolic point (so it must be a hyperbolic point). If furthermore $a_{33} \neq 0$, then $j^3P_{\mathbf{v}}$ and hence $P_{\mathbf{v}}$ is $\mathcal{A}$-equivalent to the cusp singularity $(x, xy + y^3)$. The condition $a_{33} \neq 0$ means that $P_{\mathbf{v}} \sim_{\mathcal{K}} (x, y^3)$, that is $\mathbf{v}$ has 2-point contact with $\Sigma_{\mathbf{v}}$.

• *Swallowtail*

If $a_{22} = a_{33} = 0$ and $a_{21} \neq 0$, we have

$$j^4P_{\mathbf{v}} \sim_{\mathcal{A}^{(4)}} (x, a_{21}x(y + g(x, y)) + a_{44}y^4),$$

where g has no constant or linear terms. The change of coordinates $(x, y) \mapsto (x, y + g(x, y))$ gives $j^4P_{\mathbf{v}} \sim_{\mathcal{A}^{(4)}} (x, a_{21}xy + a_{44}y^4)$. If $a_{44} \neq 0$, then $j^4P_{\mathbf{v}}$ and hence is $\mathcal{A}$-equivalent to the swallowtail singularity $(x, xy + y^4)$. In particular, p is a flecnodal point (Definition 6.5).

• *Butterfly*

Suppose now that $a_{22} = a_{33} = a_{44} = 0$ and $a_{21} \neq 0$. By Theorem 6.6(ii), this means that the direction $\mathbf{v}$ is tangent to the flecnodal set when this is a smooth curve (which is the case for a generic surface, see Theorem 6.8 and Table 6.3). Following the calculation for the Swallowtail singularity, $j^7P_{\mathbf{v}}$ is $\mathcal{A}^{(7)}$-equivalent to

$$F_1(x, y) = (x, a_{21}x(y + g(x, y)) + a_{55}y^5 + a_{66}y^6 + a_{77}y^7),$$

where g is a polynomial with no constant or linear terms. If $a_{55} \neq 0$, then $F_1 \sim_{\mathcal{A}} (x, xy + y^5 \pm y^7)$ or to $(x, xy + y^5)$. We need to make appropriate changes of coordinates and find the condition for F_1, and hence for $P_{\mathbf{v}}$, to be $\mathcal{A}$-equivalent to $(xy + y^5 \pm y^7)$.

The calculation can be carried out with the help of a computer algebra package (such as Maple or Matematica). We first eliminate the term $a_{21}xg(x, y)$ in the second component of F_1 by a change of coordinate of the form $(x, y) \mapsto (x, y + h(x, y))$, where h is a polynomial of degree 6 with no linear terms. To obtain h we proceed as follows. We write $h(x, y) = h_2(x, y) + \ldots + h_6(x, y)$, with

$$h_i(x, y) = \sum_{j=0}^{i} h_{ij}x^{i-j}y^j$$

and consider $F_1(x, y + h(x, y))$. The 3-jet of $F_1(x, y + h(x, y))$ is given by

$$(x, a_{21}xy + a_{21}h_{20}x^3 + (a_{21}h_{21} + a_{31})x^2y + (a_{21}h_{22} + a_{32})xy^2)$$

and equating the coefficients of $(0, x^3)$, $(0, x^2y)$ and $(0, xy^2)$ to zero gives $h_{20} = 0$, $h_{21} = -a_{31}/a_{21}$ and $h_{22} = -a_{32}/a_{21}$. This determines completely the quadratic part $h_2(x, y)$ of $h(x, y)$.

We proceed inductively on the jet level of $F_1(x, y+h(x, y))$ to determine completely h, and this gives $F_1(x, y + h(x, y))$, and hence $j^7P_{\mathbf{v}}(x, y)$, $\mathcal{A}^{(7)}$-equivalent to

$$F_2(x, y) = (x, a_{21}xy + a_{55}y^5 + \alpha_1 y^6 + \alpha_2 y^7),$$

with

$$\begin{aligned} \alpha_1 &= \tfrac{1}{a_{21}}(a_{21}a_{66} - 5a_{55}a_{32}), \\ \beta_1 &= \tfrac{1}{a_{21}^2}(a_{77}a_{21}^2 - 6a_{66}a_{32}a_{21} - 5a_{21}a_{55}a_{43} + 20a_{55}a_{32}^2). \end{aligned}$$

We need to eliminate the term $(0, y^6)$ in F_2. For this we make the change of coordinate $(U, V) \mapsto k(U, V) = (U - \lambda V, V)$ in the target and set $X = x - \lambda(a_{21}xy + a_{55}y^5 + \alpha_1 y^6 + \beta_1 y^7)$, so that $x = (X + \lambda(a_{55}y^5 + \alpha_1 y^6 + \beta_1 y^7))/(1 - \lambda a_{21}y)$. This gives $j^7(k \circ F_2)(X, y)$ equivalent to a germ of the form

$$j^7(k \circ F_2) \sim_{\mathcal{A}^{(7)}} (X, a_{21}X(y + G(y))) + a_{55}y^5 + \beta_2 y^6 + \beta_3 y^7).$$

We now make a change of coordinate in the source of the form $H(X, y) = (X, y + h_2y^2 + h_3y^3 + h_4y^4 + h_5y^5 + h_6y^6)$. The coefficients of $(0, Xy^2)$ and $(0, y^6)$ in $j^7(k \circ F_2 \circ H)$ are, respectively, $a_{21}(\lambda a_{21} - h_2)$ and $-a_{55}a_{21}\lambda +$

$5a_{55}h_2 + (a_{21}a_{66} - 5a_{55}a_{32})/a_{21}$. Setting these to be zero and solving the linear system in λ and h_2 gives

$$\lambda = -\frac{a_{21}a_{66} - 5a_{55}a_{32}}{4a_{55}a_{21}^2}, \qquad h_2 = -\frac{a_{21}a_{66} - 5a_{55}a_{32}}{4a_{55}a_{21}}.$$

We then equate the coefficients of $(0, Xy^i)$ in $j^7(k \circ F_2 \circ H)$ to zero to get $h_3, \ldots, h_6$. As a result, $j^7(k \circ F_2 \circ H)$, and hence $j^7 P_{\mathbf{v}}$, is $\mathcal{A}^{(7)}$-equivalent to $(x, a_{21}xy + a_{55}y^5 + \Lambda y^7)$ with

$$\Lambda = \frac{(8a_{55}a_{77} - 5a_{66}^2)a_{21}^2 + 2a_{55}(a_{32}a_{66} - 20a_{43}a_{55})a_{21} + 35a_{32}^2a_{55}^2}{8a_{21}^2a_{55}}.$$

The singularity of $P_{\mathbf{v}}$ at p is a Butterfly if and only if $\Lambda \neq 0$.

- *Lips/beaks*

For the lips/beaks singularity which is 3-$\mathcal{A}$-determined, we consider

$$j^3 P_{\mathbf{v}} \sim_{\mathcal{A}^{(3)}} (x, a_{21}xy + a_{22}y^2 + a_{31}x^2y + a_{32}xy^2 + a_{33}y^3).$$

For $P_{\mathbf{v}}$ to have a lips/beaks singularity it is necessary that $a_{21} = a_{22} = 0$ and $a_{33} \neq 0$. From Proposition 6.5 and Theorem 6.2, these conditions mean that $\mathbf{v}$ is an asymptotic direction and p is a parabolic point but not a cusp of Gauss. Then the change of coordinates $(x, y) \mapsto (x, y - \frac{a_{32}}{3a_{33}}x)$ in the source and a change of coordinate in the target yields

$$j^3 P_{\mathbf{v}} \sim_{\mathcal{A}^{(3)}} (x, \frac{3a_{31}a_{33} - a_{32}^2}{3a_{33}}x^2y + a_{33}y^3).$$

This is equivalent to a lips/beaks singularity if and only if $3a_{31}a_{33} - a_{32}^2 \neq 0$. By Proposition 6.10, $3a_{31}a_{33} - a_{32}^2 = 0$ if and only if the geodesic curvature of the image of the parabolic curve by the Gauss map vanishes at $N(p)$.

- *Goose*

Suppose that $a_{21} = a_{22} = 3a_{31}a_{33} - a_{32}^2 = 0$ and $a_{33} \neq 0$. Thus, p is a parabolic point, $\mathbf{v}$ is an asymptotic direction and the geodesic curvature of the image of the parabolic curve by the Gauss map is zero at $N(p)$. The coordinate changes for the lips/beaks give

$$j^4 P_{\mathbf{v}} \sim_{\mathcal{A}^{(4)}} F_3 = (x, a_{33}y^3 + \alpha_1 x^3 y + \alpha_2 x^2 y^2 + \alpha_3 x y^3 + \alpha_4 y^4)$$

with

$$\alpha_1 = \frac{1}{27a_{33}^3}(27a_{41}a_{33}^3 - 18a_{42}a_{32}a_{33}^2 + 9a_{43}a_{32}^2a_{33} - 4a_{44}a_{32}^3)$$

and $\alpha_2, \alpha_3, \alpha_4$ are constants depending on the coefficients a_{ij}. The terms $(0, \alpha_2x^2y^2 + \alpha_3xy^3 + \alpha_4y^4)$ in F_3 can be eliminated by a change of coordinate of the form $(x, y) \mapsto (x, y + g(x, y))$ with g a polynomial with no constant or linear terms. This change of coordinate does not alter α_1, so $j^4P_{\mathbf{v}}$ is $\mathcal{A}^{(4)}$-equivalent $(x, a_{33}y^3 + \alpha_1 x^3 y)$, which is 4-$\mathcal{A}$-determined and is the Goose singularity if and only if $\alpha_1 \neq 0$.

• *Gulls*

Suppose that $a_{21} = a_{22} = a_{33} = 0$ and $a_{32} \neq 0$. Thus, p is a cusp of Gauss and $\mathbf{v}$ is an asymptotic direction. (The condition $a_{32} \neq 0$ means that the geodesic curvature of the image of the parabolic curve by the Gauss map is not zero at $N(p)$, see Proposition 6.10).

The Gulls singularity is 5-$\mathcal{A}$-determined, so we shall restrict to the 5-jet space and make changes of coordinate to put the 3-jet, the 4-jet and then the 5-jet of $P_{\mathbf{v}}$ to a simple form. Observe, as before, that the terms of the form $(0, \alpha_k x^k)$ in the Taylor expansion of $P_{\mathbf{v}}$ can be eliminated by changes of coordinates in the target and will be ignored.

With the above conditions on the coefficients of f, $j^3P_{\mathbf{v}} \sim_{\mathcal{A}^{(3)}} (x, x(a_{31}xy + a_{32}y^2))$. The change of coordinates $(x,y) \mapsto (x, y - a_{31}/(2a_{32})x)$ in the source eliminates the term $(0, a_{31}x^2y)$ in the Taylor expansion of $j^5P_{\mathbf{v}}$. We get

$$\begin{aligned} j^5P_{\mathbf{v}} &\sim_{\mathcal{A}^{(5)}} (x, a_{32}xy^2 + \alpha_1 x^3 y + \alpha_2 x^2 y^2 + \alpha_3 x y^3 + x g_1(x,y) + a_{44}y^4 + a_{55}y^5) \\ &\sim_{\mathcal{A}^{(5)}} (x, x(a_{32}y^2 + y(\alpha_1 x^2 + \alpha_2 xy + \alpha_3 y^2)) + x g_1(x,y) + a_{44}y^4 + a_{55}y^5) \end{aligned}$$

with α_i, $i = 1,2,3$ depending on the coefficients of f and $g_1(x,y)$ is a homogeneous polynomial of degree 4. The change of coordinates $(x,y) \mapsto (x, y - \frac{1}{2}(\alpha_1 x^2 + \alpha_2 xy + \alpha_3 y^2))$ shows that $j^5P_{\mathbf{v}}$ is $\mathcal{A}^{(5)}$-equivalent to

$$(x, a_{32}xy^2 + a_{44}y^4 + \frac{1}{a_{32}^2}(a_{55}a_{32}^2 - 2a_{43}a_{44}a_{32} + 4a_{31}a_{44}^2)y^5 + x g_2(x,y))$$

with $g_2(x,y)$ a homogeneous polynomial of degree 4. Similar change of coordinates carried out to clear the 4-jet can be done to get rid of $(0, x g_2(x,y))$ and this change of coordinates does not alter the coefficients of $(0, y^5)$. Therefore,

$$j^5P_{\mathbf{v}} \sim_{\mathcal{A}^{(5)}} (x, a_{32}xy^2 + a_{44}y^4 + \frac{1}{a_{32}^2}(a_{55}a_{32}^2 - 2a_{43}a_{44}a_{32} + 4a_{31}a_{44}^2)y^5),$$

and it is 5-determined and is a Gulls singularity if and only if $a_{44} \neq 0$ and $a_{55}a_{32}^2 - 2a_{43}a_{44}a_{32} + 4a_{31}a_{44}^2 \neq 0$. □

Theorem 6.7 is about the singularities of the orthogonal projection $P_{\mathbf{v}}$ along the fixed direction $\mathbf{v}$. If the family of orthogonal projection P is an $\mathcal{A}_e$-versal unfolding of the singularity of $P_{\mathbf{v}}$, one can describe completely the bifurcations in the apparent contour $\Delta_{\mathbf{v}}$ as $\mathbf{v}$ varies locally in S^2.

Theorem 6.8. *Let M be given in Monge form $z = f(x, y)$ at the origin p with f as in* (6.4) *and $v = (0, 1, 0) \in T_pM$. Then the conditions for the family of orthogonal projections to be an $\mathcal{A}_e$-versal unfolding of the singularities in* Table 6.1 *are as in* Table 6.3 (*where it is assumed that the algebraic conditions in* Table 6.1 *are satisfied for each given singularity type*).

Table 6.3 The conditions for the family of projections to be a versal unfolding.

Name	Algebraic conditions	Geometric interpretation
Fold	–	Always
Cusp	–	Always
Lips/beaks	–	Always
Swallowtail	–	Always
Goose	$a_{20} \neq 0$	p is not a flat umbilic
Gulls	$a_{32}^2 - 4a_{20}a_{44} \neq 0$	The image of the parabolic set by the Gauss map is a curve with a $(2, 3)$-cusp at $N(p)$
Butterfly	$a_{32}^2 - a_{21}a_{43} \neq 0$	The flecnodal curve is not singular

Proof. We consider the modified family of projections (6.13) given by

$$\tilde{P}(x, y, \alpha, \beta) = (x, f(x + \alpha y, y) - \beta y)$$

and denote by $\tilde{P}_0$ the map-germ $\tilde{P}_0(x, y) = \tilde{P}(x, y, 0, 0) = P_{\mathbf{v}}(x, y)$. The conditions for P and $\tilde{P}$ to be $\mathcal{A}_e$-versal families of a given singularity of $\tilde{P}_0$ are the same as the two families $\mathcal{A}$-equivalent.

According to Theorem 3.6, the family $\tilde{P}$ is an $\mathcal{A}_e$-versal unfolding of the singularity of $\tilde{P}_0$ if and only if

$$L\mathcal{A}_e \cdot \tilde{P}_0 + \mathbb{R}.\left\{\dot{\tilde{P}}_1, \dot{\tilde{P}}_2\right\} = \mathcal{E}(2, 2),$$

where

$$\begin{aligned}\dot{\tilde{P}}_1(x, y) &= \tilde{P}_\alpha(x, y, 0, 0) = (0, y f_x(x, y)),\\ \dot{\tilde{P}}_2(x, y) &= \tilde{P}_\beta(x, y, 0, 0) = (0, -y).\end{aligned}$$

The fold and cusp are stable singularities so are versally unfolded by any family. We consider the remaining singularities in Table 6.1. As these singularities are finitely k-$\mathcal{A}$-determined for some k, we have $\mathcal{M}_2^{k+1}.\mathcal{E}(2, 2) \subset L\mathcal{A} \cdot \tilde{P}_0$, so to prove that $\tilde{P}$ is an $\mathcal{A}_e$-versal unfolding of the singularity of $\tilde{P}_0$ we only need to show that

$$j^k\left(L\mathcal{A}_e \cdot \tilde{P}_0 + \mathbb{R}.\left\{\dot{\tilde{P}}_1, \dot{\tilde{P}}_2\right\}\right) = J^k(2, 2). \tag{6.17}$$

We shall work downwards on jet levels (as in the proof of Theorem 6.2) and start by showing that all monomials $(x^iy^j, 0)$ and $(0, x^iy^j)$ of degree k are in the left hand side of (6.17). We have $\frac{\partial \tilde{P}_0}{\partial x}(x,y) = (1, f_x(x,y))$, thus $j^k(Q\frac{\partial \tilde{P}_0}{\partial x}) = (Q(x,y), 0)$ is in the left hand side of (6.17) for any monomial $Q(x,y)$ of degree k. Therefore, we only need to consider the monomials $(0, x^iy^j)$ of degree k. Once we got these, we can work modulo $\mathcal{M}_2^k.\mathcal{E}(2,2)$ and consider the monomials of degree $k-1$. Again we only need to consider monomials of the form $(0, Q(x,y))$. We observe that, using the first component of $\tilde{P}$, the monomials $(x^i, 0)$ and $(0, x^i)$ are in $L\mathcal{L}_e \cdot \tilde{P}_0$ and hence in the left hand side of (6.17). We also get the monomial $(0, y)$ from $\dot{\tilde{P}}_2$.

• *Swallowtail*

The swallowtail singularity is 4-$\mathcal{A}$-determined, so we need to show that (6.17) holds for $k = 4$. We have

$$\begin{aligned} j^3(\tfrac{\partial \tilde{P}_0}{\partial x}) &= (1, 2a_{20}x + a_{21}y + 3a_{30}x^2 + 2a_{31}xy + a_{32}y^2 \\ &\qquad +4a_{40}x^3 + 3a_{41}x^2y + 2a_{42}xy^2 + a_{43}y^3) \\ j^3(\tfrac{\partial \tilde{P}_0}{\partial y}) &= (0, a_{21}x + a_{31}x^2 + 2a_{32}xy \\ &\qquad +a_{41}x^3 + 2a_{42}x^2y + 3a_{43}xy^2 + 4a_{44}y^3) \end{aligned}$$

We get all the monomial of degree 4 of the form $(0, xQ_1(x,y))$ using $j^4(Q_1\frac{\partial \tilde{P}_0}{\partial y})$, where Q_1 is of degree 3. From this we also get the degree 3 monomial $(0, x^2y)$ using $j^4(x^2\frac{\partial \tilde{P}_0}{\partial y})$.

Working modulo these monomials, we have

$$\begin{aligned} \xi_1 &= j^3(y\tfrac{\partial \tilde{P}_0}{\partial y}) \equiv (0, a_{21}xy + a_{32}xy^2) \\ \xi_2 &= j^3(0, f(x,y)) \equiv (0, a_{21}xy + a_{32}xy^2 + a_{44}y^4) \end{aligned}$$

so that $\xi_2 - \xi_1 \equiv (0, a_{44}y^4)$. As $a_{44} \neq 0$, we get $(0, y^4)$.

Now $j^3(y^2\frac{\partial \tilde{P}_0}{\partial y}) \equiv (0, a_{21}xy^2)$ which gives $(0, xy^2)$ as $a_{21} \neq 0$. Similarly, $j^3(x\frac{\partial \tilde{P}_0}{\partial x}) \equiv (0, a_{21}xy)$ which gives $(0, xy)$.

We get $(0, y^3)$ from $j^3(\frac{\partial \tilde{P}_0}{\partial y})$ and $(0, y^2)$ from $\dot{\tilde{P}}_1$. Therefore (6.17) holds without additional conditions on the coefficients of f. Thus P is always an $\mathcal{A}_e$-versal unfolding of the swallowtail singularity.

• *Lips/beaks*

The lips/beaks singularity is 3-$\mathcal{A}$-determined, so we need to show that (6.17) holds for $k = 3$. In this case we have

$$\begin{aligned} j^2(\tfrac{\partial \tilde{P}_0}{\partial x}(x,y)) &= (1, 2a_{20}x + 3a_{30}x^2 + 2a_{31}xy + a_{32}y^2) \\ j^2(\tfrac{\partial \tilde{P}_0}{\partial y}(x,y)) &= (0, a_{31}x^2 + 2a_{32}xy + 3a_{33}y^2). \end{aligned}$$

We write the two vectors $\xi_1 = j^3(x\frac{\partial \tilde{P}_0}{\partial x}) \equiv (0, 2a_{31}x^2y + a_{32}xy^2)$ and $\xi_2 = j^3(x\frac{\partial \tilde{P}_0}{\partial y}) \equiv (0, 2a_{32}x^2y + 3a_{33}xy^2)$ in matrix form

$$\begin{pmatrix} \xi_1 \\ \xi_2 \end{pmatrix} = \begin{pmatrix} 2a_{31} & a_{32} \\ 2a_{32} & 3a_{33} \end{pmatrix} \begin{pmatrix} (0, x^2y) \\ (0, xy^2) \end{pmatrix}.$$

The determinant of the 2×2 matrix in the above equality is $-2(a_{32}^2 - 3a_{31}a_{33})$ and does not vanish as the singularity is a lips/beaks (see Table 6.1). Therefore we get $(0, x^2y)$ and $(0, xy^2)$. We can work now modulo these monomials to get $(0, y^3)$ from $j^3(y\frac{\partial \tilde{P}_0}{\partial y})$ as $a_{33} \neq 0$.

A similar argument to the above shows that we can get $(0, xy)$ and $(0, y^2)$ from the two vectors $j^3(\frac{\partial \tilde{P}_0}{\partial x}) \equiv (0, 2a_{31}xy + a_{32}y^2)$ and $j^3(\frac{\partial \tilde{P}_0}{\partial y}) \equiv (0, 2a_{32}xy + 3a_{33}y^2)$. Therefore (6.17) holds without additional conditions on the coefficients of f, and this shows that P is always an $\mathcal{A}_e$-versal unfolding of the lips/beaks singularity.

- *Goose*

The goose singularity is 4-$\mathcal{A}$-determined, so we need to show that (6.17) holds for $k = 4$. We have

$$\begin{aligned} j^3(\tfrac{\partial \tilde{P}_0}{\partial x}(x,y)) &= (1, 2a_{20}x + 3a_{30}x^2 + 2a_{31}xy + a_{32}y^2 \\ &\quad +4a_{40}x^3 + 3a_{41}x^2y + 2a_{42}xy^2 + a_{43}y^3) \\ j^3(\tfrac{\partial \tilde{P}_0}{\partial y}(x,y)) &= (0, a_{31}x^2 + 2a_{32}xy + 3a_{33}y^2 \\ &\quad +a_{41}x^3 + 2a_{42}x^2y + 3a_{43}xy^2 + 4a_{44}y^3) \end{aligned}$$

with $a_{31} = a_{32}^2/(3a_{33})$.

The following vectors

$$\begin{aligned} j^4(x^2\tfrac{\partial \tilde{P}_0}{\partial y}) &\equiv (0, 2a_{32}x^3y + 3a_{33}x^2y^2) \\ j^4(xy\tfrac{\partial \tilde{P}_0}{\partial y}) &\equiv (0, a_{31}x^3y + 2a_{32}x^2y^2 + 3a_{33}xy^3) \\ j^4(a_{32}x\tfrac{\partial \tilde{P}_0}{\partial x} - a_{31}x\tfrac{\partial \tilde{P}_0}{\partial y}) &\equiv (0, (3a_{41}a_{32} - 2a_{42}a_{31})x^3y \\ &\quad +(2a_{42}a_{32} - 3a_{43}a_{31})x^2y^2 + (a_{32}a_{43} - 4a_{44}a_{32})xy^3) \end{aligned}$$

generate a 3×3 matrix with determinant

$$\frac{1}{a_{33}}a_{32}(27a_{41}a_{33}^3 - 18a_{42}a_{32}a_{33}^2 + 9a_{43}a_{32}^2a_{33} - 4a_{44}a_{32}^3).$$

The third component of the above determinant does not vanish as the singularity is a goose (Table 6.1). If $a_{32} \neq 0$, the determinant does not vanish and we get $(0, x^3y)$, $(0, x^2y^2)$, $(0, xy^3)$, and using $j^4(y^2\frac{\partial \tilde{P}_0}{\partial y})$ we also get $(0, y^4)$ as $a_{33} \neq 0$. If $a_{32} = 0$, then $a_{31} = 0$ (condition for a Goose singularity; see Table 6.1) and we get $(0, x^2y^2)$, $(0, xy^3)$ and $(0, y^4)$ from respectively $j^4(x^2\frac{\partial \tilde{P}_0}{\partial y})$, $j^4(xy\frac{\partial \tilde{P}_0}{\partial y})$ and $j^4(y^2\frac{\partial \tilde{P}_0}{\partial y})$. We can get now the monomial $(0,x^3y)$ from $j^4(x\frac{\partial \tilde{P}_0}{\partial x})$ as $a_{41}\neq 0$ (from the Goose condition). Therefore we get all the monomials of degree 4 without imposing any extra condition on the coefficients of f.

The calculation for the degree 3 monomials follows similarly. Using the vectors $j^3(x\frac{\partial \tilde{P}_0}{\partial y})$, $j^3(y\frac{\partial \tilde{P}_0}{\partial y})$ and $j^3(a_{32}\frac{\partial \tilde{P}_0}{\partial x}-a_{31}\frac{\partial \tilde{P}_0}{\partial y})$ we get the monomials $(0,x^2y)$, $(0,xy^2)$, $(0,y^3)$ if $a_{32}\neq 0$. If $a_{32}=0$, we get $(0,xy^2)$ and $(0,y^3)$ from respectively $j^3(x\frac{\partial \tilde{P}_0}{\partial y})$ and $j^3(y\frac{\partial \tilde{P}_0}{\partial y})$. We subsequently get $(0,x^2y)$ from $j^3(\frac{\partial \tilde{P}_0}{\partial x})$. Therefore we get all the monomial of degree 3 without imposing any extra condition on the coefficients of f.

For the monomials of degree 2, the vectors $j^2(\frac{\partial \tilde{P}_0}{\partial x})\simeq(0,2a_{31}xy+a_{32}y^2)$ and $j^2(\frac{\partial \tilde{P}_0}{\partial y})\simeq(0,2a_{32}xy+3a_{33}y^2)$ are linearly dependent and the only other possibility to get monomials of degree 2 is from the vector $j^2(\dot{\tilde{P}}_1)=(0,2a_{20}xy)$, which gives the monomial $(0,xy)$ if and only if $a_{20}\neq 0$. We can then get $(0,y^2)$ from $j^2(\frac{\partial \tilde{P}_0}{\partial y})$ as $a_{33}\neq 0$. Therefore, the family P is an $\mathcal{A}_e$-versal unfolding of the Goose singularity if and only if $a_{20}\neq 0$. As $a_{21}=a_{22}=0$, the condition $a_{20}=0$ means that f has an identically zero quadratic part, that is, the point p is a flat umbilic point.

- *Gulls*

Here we take $k=5$ in (6.17) as the gulls singularity is 5-$\mathcal{A}$-determined. We have

$$\begin{aligned}
j^4(\tfrac{\partial \tilde{P}_0}{\partial x}(x,y)) = (1,\, & 2a_{20}x+3a_{30}x^2+2a_{31}xy+a_{32}y^2\\
&+4a_{40}x^3+3a_{41}x^2y+2a_{42}xy^2+a_{43}y^3\\
&+5a_{50}x^4+4a_{51}x^3y+3a_{52}x^2y^2+2a_{43}xy^3+a_{54}y^4)\\
j^4(\tfrac{\partial \tilde{P}_0}{\partial y}(x,y)) = (0,\, & a_{31}x^2+2a_{32}xy\\
&+a_{41}x^3+2a_{42}x^2y+3a_{43}xy^2+4a_{44}y^3\\
&+a_{51}x^4+2a_{52}x^3y+3a_{43}x^2y^2+4a_{54}xy^3+5a_{55}y^4)
\end{aligned}$$

The two vectors $j^5(x^3\frac{\partial \tilde{P}_0}{\partial x})\equiv(0,2a_{31}x^4y+a_{32}x^3y^2)$ and $j^5(x^2y\frac{\partial \tilde{P}_0}{\partial y})\equiv(0,a_{31}x^4y+2a_{32}x^3y^2)$ give $(0,x^3y^2)$ and $(0,x^4y)$ if $a_{31}\neq 0$. Then using $\frac{\partial \tilde{P}_0}{\partial y}$ we get the remaining monomials of degree 5 of the form $(0,xyQ(x,y))$. If $a_{31}=0$, we get all the monomials of degree 5 of the form $(0,xyQ(x,y))$ using $j^5(Q\frac{\partial \tilde{P}_0}{\partial y})$ as $a_{32}\neq 0$. We can get similarly the monomials $(0,x^3y)$

and $(0,x^2y^2)$. The vectors

$$
\begin{aligned}
j^5(y^2\tfrac{\partial\tilde{P}_0}{\partial y}) &\equiv (0, 2a_{32}xy^3+4a_{44}y^5)\\
j^5(y\tfrac{\partial\tilde{P}_0}{\partial y}) &\equiv (0, a_{31}x^2y+2a_{32}xy^2+3a_{43}xy^3+4a_{44}y^4+5a_{55}y^5)\\
j^5(x\tfrac{\partial\tilde{P}_0}{\partial y}) &\equiv (0, 2a_{32}x^2y+4a_{44}xy^3)\\
j^5(x\tfrac{\partial\tilde{P}_0}{\partial x}) &\equiv (0, 2a_{31}x^2y+a_{32}xy^2+a_{43}xy^3)\\
j^5(0,f(x,y)) &\equiv (0, a_{31}x^2y+a_{32}xy^2+a_{43}xy^3+a_{44}y^4+a_{55}y^5)
\end{aligned}
$$

generate a 5×5 matrix with determinant

$$4a_{32}a_{44}(a_{55}a_{32}^2-2a_{43}a_{44}a_{32}+4a_{31}a_{44}^2)$$

which does not vanish at a gulls singularity. Therefore we get the monomials $(0,x^2y)$, $(0,xy^2)$, $(0,xy^3)$, $(0,y^4)$ and $(0,y^5)$. This shows that all monomials of degree 5 and 4 are in the left hand side of (6.17) and the only monomial of degree 3 still to get is $(0,y^3)$. The only way to get this monomial is by using the following vectors

$$
\begin{aligned}
j^3(\tfrac{\partial\tilde{P}_0}{\partial y}) &\equiv (0, 2a_{32}xy+4a_{44}y^3)\\
j^3(\tfrac{\partial\tilde{P}_0}{\partial x}) &\equiv (0, 2a_{31}xy+a_{32}y^2+a_{43}y^3)\\
j^3(\tilde{P}_1) &\equiv (0, 2a_{20}xy+a_{32}y^3)
\end{aligned}
$$

which generate a matrix with determinant $2a_{32}(a_{32}^2-4a_{20}a_{44})$. Thus, we require the additional condition $a_{32}^2-4a_{20}a_{44}\neq 0$ for P to be an $\mathcal{A}_e$-versal unfolding of the gulls singularity. From Theorem 6.2, this is the condition for the height function along the normal direction to have precisely an A_3-singularity at p, equivalently, the image of the parabolic set by the Gauss map has a cusp singularity $\mathcal{A}$-equivalent to (t^2,t^3).

- *Butterfly*

This singularity is 7-$\mathcal{A}$-determined, so we take $k=7$ in (6.17).

For Q any monomial of degree 6, $j^7(Q\frac{\partial\tilde{P}_0}{\partial y})=(0,a_{21}xQ(x,y))$, so we get all monomials of degree 7 of the form $(0,xQ(x,y))$. Using $\frac{\partial\tilde{P}_0}{\partial y}$ again and working modulo these monomials we get all monomials of the form $(0,xQ(x,y))$ of degree 6 and 5 as well as the monomials $(0,x^3y)$ and $(0,x^2y^2)$.

The following vectors

$$
\begin{aligned}
j^7(y^3\tfrac{\partial\tilde{P}_0}{\partial y}) &\equiv (0, a_{21}xy^3+5a_{55}y^7)\\
j^7(y^2\tfrac{\partial\tilde{P}_0}{\partial y}) &\equiv (0, a_{21}xy^2+2a_{32}xy^3+5a_{55}y^6+6a_{66}y^7)\\
j^7(y\tfrac{\partial\tilde{P}_0}{\partial y}) &\equiv (0, a_{21}xy+2a_{32}xy^2+3a_{43}xy^3+5a_{55}y^5+6a_{66}y^6+7a_{77}y^7)\\
j^7(x\tfrac{\partial\tilde{P}_0}{\partial x}) &\equiv (0, a_{21}xy+a_{32}xy^2+a_{43}xy^3)\\
j^7(Q\tfrac{\partial\tilde{P}_0}{\partial x}-(f,0)) &\equiv (0, 2a_{32}a_{21}xy^3+a_{21}^2xy^2+a_{55}a_{21}y^6+(a_{66}a_{21}+a_{55}a_{32})y^7)\\
j^7(0,f) &\equiv (0, a_{21}xy+a_{32}xy^2+a_{43}xy^3+a_{55}y^5+a_{66}y^6+a_{77}y^7)
\end{aligned}
$$

generate a matrix with determinant

$$a_{55}a_{21}^2\left((8a_{55}a_{77}-5a_{66}^2)a_{21}^2+2a_{55}(a_{32}a_{66}-20a_{43}a_{55})a_{21}+35a_{32}^2a_{55}^2\right),$$

which does not vanish as the singularity is of type butterfly. Therefore we get $(0,xy)$, $(0,xy^2)$, $(0,xy^3)$, $(0,y^5)$, $(0,y^6)$ and $(0,y^7)$.

Now $j^7(\frac{\partial \tilde{P}_0}{\partial y}) \equiv (0, 5a_{55}y^4)$ so we get $(0,y^4)$ as $a_{55}\neq 0$. The only way to get $(0,y^2)$ and $(0,y^3)$ is by using the following two vectors

$$j^7(\dot{\tilde{P}}_1) \equiv (0, a_{21}y^2+a_{32}y^3),$$
$$j^7(\tfrac{\partial \tilde{P}_0}{\partial x}-\dot{\tilde{P}}_2) \equiv (0, a_{32}xy^2+a_{43}y^3).$$

These vectors generate a matrix with determinant $a_{21}a_{43}-a_{32}^2$. Therefore, the family P is an $\mathcal{A}_e$-versal unfolding of the butterfly singularity if and only if $a_{21}a_{43}-a_{32}^2\neq 0$. As $a_{44}=0$, it follows from Theorem 6.6(ii) that $a_{21}a_{43}-a_{32}^2\neq 0$ if and only if the flecnodal curve is not singular at p. □

Remark 6.2. When the family P is a versal unfolding, the bifurcations in the apparent contour can be described by considering the bifurcations in the discriminants of the model families in Table 4.1. The figures for these bifurcations can be found in [Gibson, Hawes and Hobbs (1994)].

6.5 Contact with spheres

Recall that the family of distance squared functions $D: U\times\mathbb{R}^3\to\mathbb{R}$ is given by

$$D(u,a)=\langle \mathbf{x}(u)-a, \mathbf{x}(u)-a\rangle$$

and the extended family of distance squared functions $\tilde{D}: U\times\mathbb{R}^3\times\mathbb{R}\to\mathbb{R}$ is given by

$$\tilde{D}(u,a,r)=\langle \mathbf{x}(u)-a, \mathbf{x}(u)-a\rangle - r^2.$$

The distance squared function d_{a_0}, for a_0 fixed, measures the contact of M with spheres of centre a_0. Observe that the functions d_{a_0} and $\tilde{d}_{a_0,r}$ have respectively the same $\mathcal{R}$ or $\mathcal{K}$-singularity type.

Theorem 6.9. *There is a residual set $\mathcal{O}_3$ of immersions $\mathbf{x}: U\to\mathbb{R}^3$ such that for any $\mathbf{x}\in\mathcal{O}_3$, the surface $M=\mathbf{x}(U)$ has the following properties.*

For any $a_0\in\mathbb{R}^3$, the distance squared function d_{a_0} (resp $\tilde{d}_{a_0}$) at any point p on M has only local singularities of type A_1, A_2, A_3, A_4 or D_4. Furthermore, the singularities of d_{a_0} (resp. $\tilde{d}_{a_0}$) are $\mathcal{R}^+$ (resp. $\mathcal{K}$) versally unfolded by the family D (resp. $\tilde{D}$).

Proof. The result is a consequence of Theorems 4.4 and 4.8. □

Definition 6.6. A surface is called *(locally) distance squared function generic* if any of its local parametrisations belongs to the set $\mathcal{O}_3$ in Theorem 6.9.

When M is given in Monge form $z = f(x, y)$, the conditions for d_{a_0} to have one of the singularities in Proposition 6.9 and for the family D to be a versal deformation of these singularities can be expressed in terms of the coefficients of the Taylor expansion of f at the origin p. If $p \in M$ is not an umbilic point, we can take the coordinate axes parallel to the principal directions of M at p. We choose the x-axis in the direction of the principal direction $\mathbf{v}_1$ and the y-axis in the direction of $\mathbf{v}_2$. We can also rescale the coordinates in the source if necessary, i.e., make a change of coordinates of the form $(x, y) \mapsto (\alpha x, \beta y)$, so that the coefficients of the first fundamental form at p are given by

$$E(0,0) = 1,\ F(0,0) = 0,\ G(0,0) = 1.$$

A principal direction $(a, b, 0)$ at p is a solution of the equation

$$\begin{vmatrix} b^2 & -ab & a^2 \\ E(0,0) & F(0,0) & G(0,0) \\ l(0,0) & m(0,0) & n(0,0) \end{vmatrix} = 0,$$

that is,

$$-m(0,0)a^2 + (n(0,0) - l(0,0))ab + m(0,0)b^2 = 0.$$

As the principal directions at p are taken to be along $(1,0,0)$ and $(0,1,0)$, it follows that $m(0,0) = 0$, so $f_{xy}(0,0) = 0$. We have

$$l(0,0) = f_{xx}(0,0),\ n(0,0) = f_{yy}(0,0),$$

so the principal curvatures at the origin p are given by

$$\kappa_1(p) = f_{xx}(0,0),\ \kappa_2(p) = f_{yy}(0,0).$$

Thus, at the origin p and with the above setting,

$$f(x,y) = a_{20}x^2 + a_{22}y^2 + a_{30}x^3 + a_{31}x^2y + a_{32}xy^2 + a_{33}y^3 + O(4) \quad (6.18)$$

with $\kappa_1(p) = 2a_{20}$ and $\kappa_2(p) = 2a_{22}$.

It follows from Proposition 2.6 that d_{a_0} is singular at the origin p if and only if $a_0 = (0, 0, a_3)$ with a_3 some non-zero real number. That is, a is on the normal line to M at p. The singularity is of type A_1 if and only if a is not on the focal set of M, that is, $a_3 \neq 1/\kappa_i(p)$, $i = 1, 2$.

We consider now the degenerate singularities d_{a_0} associated to the sphere of centre $a_0 = (0, 0, 1/\kappa_1(p))$ on the focal set $\mathcal{F}_1$ (we assume that $\kappa_1(p) \neq 0$). The case when $a_0 = (0, 0, 1/\kappa_2(p))$ follows similarly. (It requires appropriate changes of indices of the coefficients of the Taylor expansion f.)

Theorem 6.10. *Let M be given in Monge form $z = f(x, y)$ at the origin p with f as in* (6.18) *and $c_0 = (0, 0, 1/\kappa_1(p))$.*

(i) *The conditions for the distance squared function d_{a_0} to have one of the degenerate singularities in* Theorem 6.9 *are as follows.*

A_2: $a_{30} \neq 0$;

A_3: $a_{30} = 0,\ a_{31}^2 - 4(a_{22} - a_{20})(a_{40} - a_{20}^3) \neq 0$;

A_4: $a_{30} = 0,\ a_{31}^2 - 4(a_{22} - a_{20})(a_{40} - a_{20}^3) = 0,$
$a_{32}a_{31}^2 - 2a_{41}a_{31}(a_{22} - a_{20}) + 4a_{50}(a_{20} - a_{22})^2 \neq 0$;

D_4: $a_{20} = a_{22} \neq 0,$
$a_{31}^2 a_{32}^2 - 4a_{30}a_{32}^3 - 4a_{33}a_{31}^3 - 27a_{30}^2 a_{33}^2 + 18a_{30}a_{31}a_{32}a_{33} \neq 0.$

(ii) *The family of distance squared functions on M is a $\mathcal{R}^+$ versal unfolding of the singularities of d_{a_0} in* (i) *if and only if*

A_2: always;

A_3: always;

A_4: $8a_{31}a_{20}^4 - 8a_{31}a_{22}a_{20}^3 - 4a_{51}a_{20}^2 + 4(2a_{51}a_{22} - a_{31}a_{42})a_{20}$
$-3a_{31}^2 a_{33} + 4a_{31}a_{42}a_{22} - 4a_{51}a_{22}^2 \neq 0$;

D_4: $2a_{32}^2 - a_{31}^2 + 3a_{31}a_{33} \neq 0$.

Proof. The proof follows the same steps as those of Theorem 6.2. We consider only the singularities A_4 and D_4.

We start with the A_4-singularity. We denote by d_{ij} the coefficient of $x^i y^j$ in the Taylor expansion of d_{a_0} at the origin. Observe that

$$d_{02} = \frac{\kappa_1(p) - \kappa_2(p)}{2\kappa_1(p)} = \frac{a_{22} - a_{20}}{2a_{22}} \neq 0$$

as p is not an umbilic point. The singularity is more degenerate than an A_3, so $a_{30} = 0$ and $a_{31}^2 - 4(a_{22} - a_{20})(a_{40} - a_{20}^3) = 0$. Then $L = d_{02}y^2 + d_{21}x^2 y + d_{40}x^4$ is a perfect square.

We make the change of coordinates $(x, Y) = (x, y - d_{21}/2(d_{02})x^2)$. Then $L(x, Y) = d_{02}Y^2$ and the singularity of d_{a_0} is of type A_4 if and only if the coefficient of x^5 in the Taylor expansion of $D_{a_0}(x, Y)$ is not zero. That coefficient is not zero when the expression in the statement of the theorem is not zero.

The family D, with $a = (a_1, a_2, a_3)$ varying near a_0, is an $\mathcal{R}^+$- versal deformation of the A_4-singularity of d_{a_0} if and only if

$$j^5\left(L\mathcal{R}_e \cdot D_{a_0} + \mathbb{R}.\left\{\dot{D}_1, \dot{D}_2, \dot{D}_3\right\} + \langle 1\rangle_{\mathbb{R}}\right) = J^5(2,1), \tag{6.19}$$

where

$$\begin{aligned}
\dot{D}_1(x,y) &= D_{a_1}(x,y,0,0,1/\kappa_1(p)) = -2x,\\
\dot{D}_2(x,y) &= D_{a_2}(x,y,0,0,1/\kappa_1(p)) = -2y,\\
\dot{D}_3(x,y) &= D_{a_3}(x,y,0,0,1/\kappa_1(p)) = -2f(x,y).
\end{aligned}$$

We consider the family $D(x, Y, a_1, a_2, a_3)$ with Y as above. Denote by $\tilde{d}_{ij}$ the coefficient of x^iY^j in the Taylor expansion of $D_{a_0}(x, Y)$. Then

$$\begin{aligned}
j^5 D_{a_0}(x,Y) = \tfrac{1}{\kappa_1(p)^2} &+ \tilde{d}_{22}Y^2 + \tilde{d}_{32}xY^2 + \tilde{d}_{33}Y^3\\
&+\tilde{d}_{41}x^3Y + \tilde{d}_{42}x^2Y^2 + \tilde{d}_{43}xY^3 + \tilde{d}_{44}Y^4\\
&+\tilde{d}_{50}x^5 + \tilde{d}_{51}x^4Y + \tilde{d}_{52}x^3Y^2 + \tilde{d}_{53}x^2Y^3 + \tilde{d}_{54}xY^4 + \tilde{d}_{55}Y^5.
\end{aligned}$$

The arguments for showing that the family D is a versal deformation are similar to those in the proof of Theorem 6.2. We use $(D_{a_0})_x$ and $(D_{a_0})_Y$ to show that all the monomials of degree 5 and 4 are in the 5-jet of the left hand side of (6.19). We also get xy in there using $(D_{a_0})_Y$. Now x^3 appears only in $(D_{a_0})_Y$, and we need its coefficient $\tilde{d}_{41}$ to be non-zero for it to be in the 5-jet of the left hand side of (6.19). The coefficient $\tilde{d}_{41}$ is a non-zero multiple of the expression in the statement of the theorem. For degree 2 and 1 monomials, we use $(D_{a_0})_Y$, $\dot{D}_1$, $\dot{D}_2$, and $\dot{D}_3$ to show that they are all in the 5-jet of the left hand side of (6.19). Thus, the family D is a versal deformation if and only if $\tilde{d}_{41} \neq 0$.

We turn now to the D_4-singularity of d_{a_0}. The 2-jet of d_{a_0} vanishes identically if and only if $\kappa_1(p) = \kappa_2(p) = \kappa$, equivalently, p is an umbilic point. Then

$$j^3 D_{a_0}(x,y) = \frac{1}{\kappa^2} - \frac{2}{\kappa}(a_{30}x^3 + a_{31}x^2y + a_{32}xy^2 + a_{33}y^3).$$

The singularity of d_{a_0} is of type D_4 if and only if the cubic form $a_{30}x^3 + a_{31}x^2y + a_{32}xy^2 + a_{33}y^3$ has no repeated root, equivalently, when its discriminant is not zero. The discriminant of this cubic form is as in the statement of the theorem.

Following the same steps as above we obtain all the monomials of degree 3 in the 3-jet of the left hand side of (6.19) using $(D_{a_0})_x$ and $(D_{a_0})_y$. For the monomials of degree 2 (and working modulo monomials of degree 3), we have

$$\begin{pmatrix} j^2(-\frac{1}{\kappa}\dot{D}_3)\\ j^2(-\frac{\kappa}{2}D_{a_0})_x\\ j^2(-\frac{\kappa}{2}D_{a_0})_y \end{pmatrix} = M\begin{pmatrix} x^2\\ xy\\ y^2 \end{pmatrix},$$

with

$$M = \begin{pmatrix} 1 & 0 & 1 \\ 3a_{30} & 2a_{31} & a_{32} \\ a_{31} & 2a_{32} & 3a_{33} \end{pmatrix}.$$

The monomial of degree 2 are in the 3-jet of the left hand side of (6.19) if and only if the determinant of the matrix M is not zero, that is, $2a_{32}^2 - a_{31}^2 + 3a_{31}a_{33} \neq 0$. We then use $\dot{D}_1$ and $\dot{D}_2$ to get x, y in the 3-jet of the left hand side of (6.19). Therefore, the family D is a versal deformation of the D_4-singularity of d_{a_0} if and only if $2a_{32}^2 - a_{31}^2 + 3a_{31}a_{33} \neq 0$. □

Remark 6.3. The geometric interpretation of the algebraic conditions in Theorem 6.10 are given in §6.6.3.

Geometric objects can be derived from the family of distance squared functions on the surface M, such as the focal set of M and the spherical Dupin foliation defined below. For a distance squared function generic surface, the local structure of these objects up to diffomorphisms is determined by the $\mathcal{R}$-singularity type of the distance squared function d_{a_0}. It turns out that the diffeomorphism models of these objects determine the $\mathcal{R}$-singularity type of d_{a_0}. We start with the focal set.

Theorem 6.11. *Away from umbilic points, the focal set $\mathcal{F}_1$ and $\mathcal{F}_2$ of M are disjoint surfaces, and for a distance squared function generic M they are diffeomorphic to*

(i) *a smooth surface if and only if d_{a_0} has an A_2-singularity at p;*
(ii) *a cuspidal edge surface if and only if d_{a_0} has an A_3-singularity at p;*
(iii) *a swallowtail surface if and only if d_{a_0} has an A_4-singularity at p, where $a_0 = p + 1/\kappa_i N(p)$, $i = 1$ or 2.*

At an umbilic point the focal set $\mathcal{F}_1 \cup \mathcal{F}_2$ is diffeomorphic to a pyramid (resp. purse) if and only if d_c has a D_4^- (resp. D_4^+)-singularity at p.

Proof. ([Arnol'd, Guseĭn-Zade and Varchenko (1985)]) The focal set of M is a caustic with D the generating family of the Lagrangian submanifold $L(D)(C_D) \subset T^*\mathbb{R}^3$. Thus, it has Lagrangian singularities. For a distance squared function generic surface M, the family D is $\mathcal{R}^+$-versal, so the focal set is diffeomorphic to the bifurcation set of a model $\mathcal{R}^+$-versal unfolding (with two parameters) of the singularities that occur in a given distance squared function. These are A_1, A_2, A_3, A_4 and D_4-singularities (Theorem 6.9). The bifurcation set of an $\mathcal{R}^+$-versal 2-parameter family of these singularities are as stated in the proposition (see also §3.9.2) and are as in Figure 6.8. □

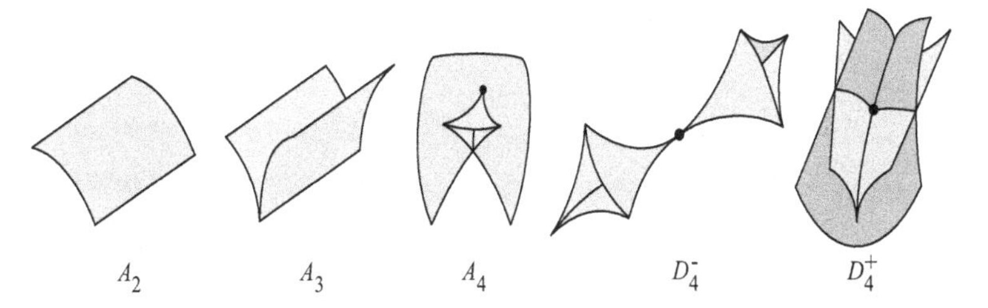

Fig. 6.8 Models of the focal set of a generic surface in $\mathbb{R}^3$. The fourth and fifth figures model the focal sets $\mathcal{F}_1$ and $\mathcal{F}_2$ joining at an umbilic point. The first three figures model the focal sets $\mathcal{F}_1$ or $\mathcal{F}_2$. One can have the following generic combinations for the pairs $(\mathcal{F}_1, \mathcal{F}_2)$ or $(\mathcal{F}_2, \mathcal{F}_1)$: (A_2, A_2), (A_2, A_3), (A_2, A_4), (A_3, A_3).

Definition 6.7. The spherical Dupin foliation of M at p_0 associated to the tangential sphere of centre c_0 with radius r_0 is the family of curves obtained by intersecting M with the 1-parameter family of spheres of centre c_0 and radius r varying near r_0.

Theorem 6.12. *The spherical Dupin foliation of M at p_0 associated to the tangential sphere of centre c_0 are diffeomorphic to the level sets of the germs of the following functions (see also* Figure 6.9)*:*

$$\begin{array}{lcl}
\pm x_1^2 + x_2^2 & \Longleftrightarrow & d_{a_0} \textit{ has an } A_1\textit{-singularity at } p_0 \\
x_1^3 + x_2^2 & \Longleftrightarrow & d_{a_0} \textit{ has an } A_2\textit{-singularity at } p_0 \\
\pm x_1^4 + x_2^2 & \Longleftrightarrow & d_{a_0} \textit{ has an } A_3\textit{-singularity at } p_0 \\
x_1^5 + x_2^2 & \Longleftrightarrow & d_{a_0} \textit{ has an } A_4\textit{-singularity at } p_0 \\
x_1^3 - x_1 x_2^2 & \Longleftrightarrow & d_{a_0} \textit{ has an } D_4^+\textit{-singularity at } p_0 \\
x_1^3 + x_2^3 & \Longleftrightarrow & d_{a_0} \textit{ has an } D_4^-\textit{-singularity at } p_0
\end{array}$$

Proof. The function d_{a_0} has A_2-singularity at p_0 if and only if is $\mathcal{R}$-equivalent to $f(x_1, x_2) = x_1^3 + x_2^2$. Since two $\mathcal{R}$-equivalent germs of functions have diffeomorphic the level sets, the spherical Dupin foliation is diffeomorphic to the level set of germ of $f(x_1, x_2) = x_1^3 + x_2^2$ (see Figure 6.9). The other cases follow by a similar argument. □

6.6 Robust features of surfaces

We have already encountered several special curves on a surface M in $\mathbb{R}^3$, such as its lines of principal curvatures and asymptotic curves. These form

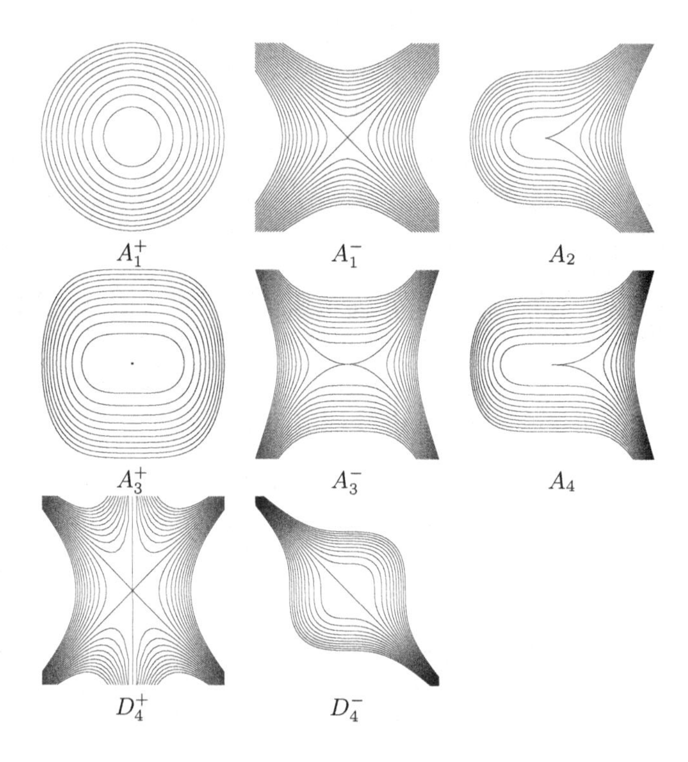

Fig. 6.9 Dupin foliations of M associated to tangential spheres.

pairs of foliations on the surface and one cannot trace in general an individual curve when the surface is deformed. There may of course exist lines of curvature or asymptotic curves which are homeomorphic to a circle, i.e., limit cycles. These are indeed robust features of the surface. Such curves are studied in [Sotomayor and Gutierrez (1982); Garcia and Sotomayor (1997)].

We consider below some special curves which are robust features on M, captured via the contact of the surface with planes, lines and spheres.

6.6.1 *The parabolic curve*

For a surface parametrised by $\mathbf{x} : U \to \mathbb{R}^3$, the parabolic set is the image by $\mathbf{x}$ of the zero set of the function $ln - m^2$. We start by establishing its generic structure.

Proposition 6.13. *The parabolic set of a height function generic surface is* *a smooth curve (when not empty). It consists of the A_2 and A_3-singularities of the height functions along the normal directions to the surface. The A_3-singularities of these functions (i.e., the cusps of Gauss) occur at isolated points on the parabolic set.*

Proof. The surface M is height function generic, so the height functions have only singularities of type A_1, A_2 or A_3 for any $v \in S^2$ and these are $\mathcal{K}$-versally unfolded by the extended family of height functions.

The map $g(u) = (u, N(u), \langle \mathbf{x}(u), N(u)\rangle)$ is an embedding from the surface to the domain of the Legendrian map $\mathcal{L}(\tilde{H}) : \Sigma(H) \to \mathbb{R}^3$. Therefore $\mathcal{L}(\tilde{H}) \circ g$ is a cuspidal edge map at an A_2-singularity of the height function and a swallowtail map at an A_3-singularity of this function. In particular, its singular set is diffeomorphic to that of the $\mathcal{A}$-model of the cuspidal edge or the swallowtail map in Definition 1.4. The result now follows by direct computation of the critical sets of these models. These are given by $t = 0$ for the model of the cuspidal edge map and $t + 6s^2 = 0$ for that of the swallowtail map. □

Remark 6.4. One can also use the arguments in the proof of Proposition 6.13 to show that there is a smooth curve on M of pairs of points where the surface admits a bi-tangent plane meeting tangentially the parabolic curve at a cusp of Gauss (Figure 6.10). These pairs of points are where the height function along their common normal direction has an A_1-singularity. The curve of bi-tangent planes corresponds to the curve of self-intersections of the dual surface of M (a swallowtail at the cusp of Gauss). The set is denoted the A_1A_1-set or A_1^2-set. For a height function generic surface, the A_1A_1-set is a smooth curve which has ordinary tangency with the parabolic curve at a cusp of Gauss. As in Proposition 6.13, to prove this it is enough to carry out the calculations for the $\mathcal{A}$-model of a swallowtail surface in Definition 1.4. The A_1A_1-set is modelled by the pair of (distinct) points (s_1, t_1) and (s_2, t_2) which have the same image by the map $\psi(s,t) = (3s^4 + s^2t, 4s^3 + 2st, t)$. Then $t_1 = t_2 = t$ and

$$3u_1^4 + s_1^2 t = 3s_2^4 + s_2^2 t, \quad 4s_1^3 + 2s_1 t = 4s_2^3 + 2s_2 t.$$

Equivalently,

$$(s_1^2 - s_2^2)(3s_1^2 + 3s_2^2 + t) = 0, \quad 2(s_1 - s_2)(2(s_1^2 + s_1 s_2 + s_2^2) + t) = 0.$$

The solution $s_1 = s_2$ is excluded as it gives $(s_1, t_1) = (s_2, t_2)$. The system of equations $3s_1^2 + 3s_2^2 + t = 0$ and $2(s_1^2 + s_1 s_2 + s_2^2) + t = 0$ also gives $s_1 = s_2$. Therefore the solution of the system is $s_2 = -s_1$ (from the first equation) and $t = -2s_1^2$ (substituting in the second equation). Thus, the curve of self-intersection is a smooth curve with equation $t + 2s^2 = 0$. It has an ordinary tangency with the singular set $t + 6s^2 = 0$.

Remark 6.5. The parabolic curve is also detected by the singularities of the orthogonal projections of the surface to planes. It follows from Theorem 6.7 that p is a parabolic point if and only if the singularity of $P_{\mathbf{v}}$ is a lips/beaks, goose or gulls, see Table 6.2.

6.6.2 *The flecnodal curve*

From Theorem 6.6, the flecnodal set of a generic surface is a smooth curve when not empty. It is in fact the locus of geodesic inflections of the asymptotic curves. Before showing this we need the following result from [Bruce and Tari (2000)].

Proposition 6.14. *An asymptotic curve on M has a geodesic inflection at p if and only if its projection to the tangent plane T_pM has an inflection at p.*

Proof. We take, without loss of generality, the surface M parametrised in Monge form $z = f(x, y)$ at the origin p. The tangent plane T_pM is the plane $z = 0$. Let $\gamma(t) = (x(t), y(t), f(x(t), y(t)))$ be a parametrisation of the asymptotic curve with $\gamma(0) = p$, so that its projection to T_pM along the normal direction $(0, 0, 1)$ to M at p is the curve $\alpha(t) = (x(t), y(t))$ (here we are taking the orthogonal projection of γ to T_pM, but any projection along a direction transverse to T_pM will do). As γ is an asymptotic curve, we have

$$f_{xx}(\alpha(t))x'(t)^2 + 2f_{xy}(\alpha(t))x'(t)y'(t) + f_{yy}(\alpha(t))y'(t)^2 = 0,$$

for all t near zero. The geodesic curvature of the asymptotic curve γ vanishes at t if and only if $[\gamma'(t), \gamma''(t), N(\alpha(t))] = 0$ (see (6.2)). We have,

$$[\gamma', \gamma'', N] = \frac{1}{\delta}\begin{vmatrix} x' & y' & x'f_x + y'f_y \\ x'' & y'' & f_{xx}x'^2 + 2f_{xy}x'y' + f_{yy}y'^2 + x''f_x + y''f_y \\ -f_x & -f_y & 1 \end{vmatrix}$$

$$= \frac{1}{\delta}\begin{vmatrix} x' & y' & x'f_x + y'f_y \\ x'' & y'' & x''f_x + y''f_y \\ -f_x & -f_y & 1 \end{vmatrix}$$

$$= (x'y'' - x''y')\delta,$$

where $\delta = (1 + f_x^2 + f_y^2)^{1/2}$ and all the partial derivatives of f are evaluated at $\alpha(t)$. Thus $[\gamma'(t), \gamma''(t), N(\alpha(t))] = 0$ if and only if $(x''y' - x'y'')(t) = 0$, which completes the proof. □

Proposition 6.15. *Let p be a hyperbolic point on M. Then p is a flecnodal point and corresponds to the swallowtail singularity of some projection $P_{\mathbf{v}}$ if and only if p is a geodesic inflection of the asymptotic curve through p with tangent direction $\mathbf{v}$ at p.*

Proof. We take M in Monge form $z = f(x, y)$ at p with f as in (6.4) and $v = (0, 1, 0)$. By Theorem 6.6, p is a flecnodal point if and only if $f_y = f_{yy} = f_{yyy} = 0$ and $f_{xy} f_{yyyy} \neq 0$ at $(0, 0)$.

Asymptotic curves satisfy the binary differential equation (6.3). As p is a hyperbolic point and $\mathbf{v}$ an asymptotic direction (then $f_{yy}(0,0) = 0$), the two asymptotic curves through p are smooth and the one tangent to $\mathbf{v}$ can be parametrised by $\gamma(t) = (g(t), t, f(g(t), t))$ for some germ of a smooth function g with $g'(0) = 0$. We get from equation (6.3)

$$f_{xx}(g(t), t) g'(t)^2 + f_{xy}(g(t), t) g'(t) + f_{yy}(g(t), t) = 0.$$

Differentiating the above identity and evaluating at $t = 0$ yields

$$2 f_{xy} g'' + f_{xxx} g'^3 + 3 f_{xxy} g'^2 + 3 f_{xyy} g' + f_{yyy} = 2 f_{xy} g'' + f_{yyy} = 0$$

at $t = 0$. We have $f_{xy}(0,0) \neq 0$, so $g''(0) = 0$ if and only if $f_{yyy}(0,0) = 0$, equivalently, p is a flecnodal point. By Proposition 6.14, the condition $g''(0) = 0$ means that the asymptotic curve has a geodesic inflection at $\gamma(0)$. □

Theorem 6.6 gives the equations of the flecnodal curve for M in Monge form. Using Proposition 6.15, we can obtain its equations in terms of the coefficients of the second fundamental form and their derivatives.

Theorem 6.13. *Let $\mathbf{x} : U \to \mathbb{R}^3$ be a local parametrisation of M. The flecnodal curve is the image by $\mathbf{x}$ of the set of points (u_1, u_2) for which the following system of equations in a, b*

$$\begin{gathered} la^2 + 2mab + nb^2 = 0, \\ l_{u_1} a^3 + (l_{u_2} + 2m_{u_1}) a^2 b + (2m_{u_2} + n_{u_1}) ab^2 + n_{u_2} b^3 = 0, \end{gathered}$$

has a solution, where l, m, n and their partial derivatives are evaluated at $u = (u_1, u_2)$.

Proof. According to Proposition 6.15, the flecnodal curve is the locus of geodesic inflections of the asymptotic curves. Let $\alpha(t) = (u_1(t), u_2(t))$ be a local parametrisation of an asymptotic curve and write $\alpha' = (a, b)$. Suppose that $b \neq 0$ (the case $b = 0$ follows similarly) and re-parametrise α in the form $(g(t), t)$, with $g' = a/b$. As α is an asymptotic curve,

$$l(g(t), t)g'(t)^2 + 2m(g(t), t)g'(t) + n(g(t), t) = 0.$$

Differentiating with respect to t and dropping the arguments, we get

$$l_{u_1}g'^3 + (l_{u_2} + 2m_{u_1})g' + (2m_{u_2} + n_{u_1})g' + n_{u_2} + 2(lg' + m)g'' = 0.$$

The point p is not parabolic, so $lg' + m \neq 0$ which implies that g'' vanishes at some t if and only if

$$l_{u_1}g'^3 + (l_{u_2} + 2m_{u_1})g' + (2m_{u_2} + n_{u_1})g' + n_{u_2} = 0,$$

equivalently,

$$l_{u_1}a^3 + (l_{u_2} + 2m_{u_1})a^2b + (2m_{u_2} + n_{u_1})ab^2 + n_{u_2}b^3 = 0.$$

□

Theorem 6.14. *For a height function and projection generic surface, the flecnodal set contains the cusps of Gauss, is a smooth curve at such points and has 2-point contact with the parabolic curve there.*

Proof. We take M in Monge form $z = f(x, y)$ at the origin p which we assume to be a cusp of Gauss. We choose $(0, 1, 0)$ to be the unique asymptotic direction at p, so

$$\begin{aligned} j^4f = {} & a_{20}x^2 + a_{30}x^3 + a_{31}x^2y + a_{32}xy^2 + \\ & a_{40}x^4 + a_{41}x^3y + a_{42}x^2y^2 + a_{43}xy^3 + a_{44}y^4, \end{aligned}$$

with $a_{20} \neq 0$ and $a_{32}^2 - 4a_{20}a_{44} \neq 0$ (Theorem 6.2). As the surface is height function generic, $a_{32} \neq 0$ (Theorem 6.2). Similar calculations to those in the proof of Proposition 6.5 show that the parabolic set is a smooth curve at p and can be parametrised locally in the form $x = x_1(y)$ with

$$j^2x_1(y) = \frac{a_{32}^2 - 6a_{20}a_{44}}{a_{20}a_{32}}y^2. \tag{6.20}$$

For the flecnodal set we need to solve the system of equations (6.14), (6.15) and (6.16). These equations are satisfied when all the parameters are set to zero as $f_y = f_{yy} = f_{yyy} = 0$ at p. Therefore, p is in the flecnodal set.

We can use (6.14) to obtain β as a function of (x, y, α). Substituting in (6.16) we get α as a function of (x, y) as $a_{32} \neq 0$. Now (6.15) becomes

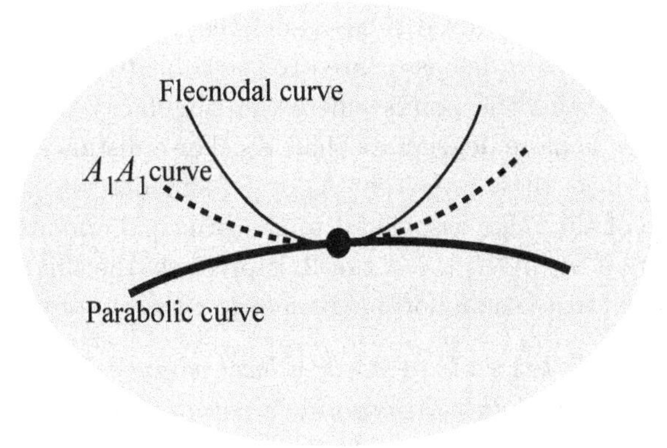

Fig. 6.10 A sketch of a configuration of the parabolic, flecnodal and A_1A_1 curves of a surface patch at a cusp of Gauss.

an equation in (x, y) and gives the flecnodal curve. This curve can be parametrised locally in the form $x = x_2(y)$ with

$$j^2x_2(y) = -\frac{a_{44}}{a_{32}}y^2. \tag{6.21}$$

It is clear from (6.20) and (6.21) that the flecnodal and parabolic curves are tangential at p. We have $j^2(x_1 - x_2)(y)/y^2 = a_{32}/a_{20} \neq 0$, so the two curves have 2-point contact at p. □

We sketch in Figure 6.10 the three robust features of the surface at a cusp of Gauss, captured by the contact of the surface with planes and lines. There are several configurations of the three curve at the cusp of Gauss ([Uribe-Vargas (2006)]). These curves are used in [Uribe-Vargas (2006)] to obtain a projective invariant of the surface at cusps of Gauss.

6.6.3 *The ridge curve*

Porteous introduced the following concept (see [Porteous (1987, 1983a, 2001)]).

Definition 6.8. A point on a surface M is called a *ridge point* if it is an A_3-singularity of some distance squared function on M. The closure of the set of the ridge points is called the *ridge* of the surface M.

The ridge inherits the colour of the principal directions. Following Porteous' colouring, we have a red ridge associated, say, to the principal curvature κ_1 and the blue ridge associated to the principal curvature κ_2.

The ridge contains the points where the singularity of some distance squared function is more degenerate than A_3. For a distance squared function generic surface, these are either A_4 or D_4-singularities.

The image of the ridge associated to the principal curvature κ_i by the map $\varepsilon_i(p) = p + 1/\kappa_i(p)N(p)$, $i = 1$ or 2, is precisely the singular set of the focal set, and captures the following geometric property of the surface.

Proposition 6.16. *Let $\mathbf{x} : U \to \mathbb{R}^3$ be a local parametrisation of M away from umbilic points such that the coordinate curves are the lines of principal curvature. Then the ridge is given by the set of points $\mathbf{x}(u_1, u_2)$ where*

$$\frac{\partial \kappa_i}{\partial u_i}(u_1, u_2) = 0,$$

for $i = 1$ or 2, that is, the ridge is the set of points where a principal curvature is extremal along its associated lines of principal curvature.

Proof. Consider, for example, the focal set associated to the principal curvature κ_1 which is parametrised by

$$\varepsilon_1(u_1, u_2) = (\mathbf{x} + \frac{1}{\kappa_1}N)(u_1, u_2).$$

The directions $\mathbf{x}_{u_1}$ and $\mathbf{x}_{u_2}$ are principal directions at all points in U so $N_{u_i} = -\kappa_i \mathbf{x}_{u_i}$, $i = 1, 2$. Differentiating ε_1 and dropping the argument, we get

$$\begin{aligned} \frac{\partial \varepsilon_1}{\partial u_1} &= -\frac{1}{\kappa_1^2}\frac{\partial k_1}{\partial u_1}N, \\ \frac{\partial \varepsilon_1}{\partial u_2} &= (1 - \frac{\kappa_2}{\kappa_1})\mathbf{x}_{u_2} - \frac{1}{\kappa_1^2}\frac{\partial k_1}{\partial u_2}N. \end{aligned}$$

A point $p = \mathbf{x}(u_1, u_2)$ is on the ridge if the focal set is singular at $\varepsilon_1(u_1, u_2)$. As p is not an umbilic point, the vectors $\partial \varepsilon_1/\partial u_1$ and $\partial \varepsilon_1/\partial u_2$ are linearly dependent at (u_1, u_2) if and only if

$$\frac{\partial k_1}{\partial u_1}(u_1, u_2) = 0.$$

□

Remark 6.6. It follows from the proof of Proposition 6.16 that, away from umbilic points, the normal direction to the focal set $\mathcal{F}_1$, which is given by $\partial \varepsilon_1/\partial u_1 \times \partial \varepsilon_1/\partial u_2$, is the principal direction $\mathbf{v}_1$ associated to the principal curvature κ_1. Similarly, the normal direction to the focal set $\mathcal{F}_2$ is the principal direction $\mathbf{v}_2$ associated to the principal curvature κ_2.

Proposition 6.17. *Suppose that M is a distance squared function generic surface patch without umbilic points. Then the ridge is a smooth curve (when not empty).*

Proof. The surface M is distance squared function generic, so the distance squared function D_a has only singularities of $\mathcal{R}$-type A_1, A_2, A_3, A_4 or D_4 for any $a \in \mathbb{R}^3$. The D_4-singularities occur at umbilic points and the surface patch is supposed to be umbilic free. Therefore, the ridge contains only A_3 or A_4 singular points of the distance squared function.

The map $h_i(u) = (u, \varepsilon_i(u))$, $i = 1$ or 2, is an embedding from the surface to the critical set of the Lagrangian map $L(D) : C(D) \to \mathbb{R}^3$. Thus, $\varepsilon_i = L(D) \circ h_i$ is a cuspidal edge map at an A_3-singularity of the distance squared function and a swallowtail map at an A_4-singularity of this function. The result now follows by direct computation of the critical sets of the $\mathcal{A}$-models of the cuspidal edge and swallowtail maps in Definition 1.4. These are given by $t = 0$ for the model of the cuspidal edge map and $t + 6s^2 = 0$ for that of the swallowtail map. □

Corollary 6.3. *The distance squared function has an A_3-singularity if and only if the ridge is transverse to its associated principal direction. The singularity is of type A_4 if and only if the ridge is tangent to its associated principal direction.*

Proof. According to [Saji, Umehara and Yamada (2009)], a map-germ $h : \mathbb{R}^2, 0 \to \mathbb{R}^3$ is a cuspidal edge map if and only if its critical set is a smooth curve and is transverse to the kernel of dh. It is a swallowtail map if and only if its critical set is a smooth curve and has an ordinary tangency with the kernel of dh at the swallowtail point.

The proof is a direct consequence of the above criteria applied to the map ε_i, $i = 1, 2$. The singular set of ε_i is the ridge curve and it follows from the proof of Proposition 6.16 that the kernel of $d\varepsilon_i$ is along the principal direction associated to the ridge. □

We can now characterise geometrically the generic singularities of the distance squared function on M.

Theorem 6.15. *Suppose that M is a distance squared function generic surface. Then, the singularities of the distance squared function d_c at $p \in M$ are characterised as follows.*

A_1: *c is on the normal line to M at p but is not a focal point.*
A_2: *c is a focal point and p is not a ridge point*
A_3: *c is a focal point, p is a ridge point and the ridge curve is transverse to its associated principal direction*
A_4: *c is a focal point, p is a ridge point and the ridge is tangent to its associated principal direction*
D_4: *c is a focal point and p is an umbilic point.*

One expects, by looking at the focal sets in Figure 6.8, one or three ridge curves arriving at an umbilic point. The structure of the closure of these curves is established in [Bruce (1984)]. At an umbilic point the normal curvatures in all tangent directions to the surface coincide and are equal to some constant κ. At such a point the surface can be taken locally in Monge-form $z = f(x, y)$ with

$$j^3 f(x,y) = \frac{\kappa}{2}(x^2 + y^2) + C(x,y),$$

where C is a cubic form in x, y (i.e., a homogeneous polynomial of degree 3 in x, y). The distance-squared function d_c with $c = (0, 0, 1/\kappa)$ is given locally by $C(x, y) + O(4)$. It has a D_4-singularity if and only if the cubic form C has three distinct roots. If the three roots are real, d_c has a D_4^--singularity (elliptic umbilic). If two of the roots are complex (conjugate), d_c has a D_4^+-singularity (hyperbolic umbilic).

Theorem 6.16 ([Bruce (1984)]). *The ridge set through an umbilic point of a generic surface has local model, up to diffeomorphism, $xy(x-y) = 0$ at a D_4^- and $x = 0$ at a D_4^+.*

Proof. The idea of the proof is as follows. At every point q on the surface, one can write the surface locally in Monge form $z = f_q(x, y)$. The 1-jet of f_q at the origin is identically zero. Let V_k denote the set of polynomials in (x, y) of degree d with $2 \le d \le k$. Define the Monge-Taylor map $j^k\phi : M \to V_k$, by $j^k\phi(q) = j^k f_q(0, 0)$.

Recall that an A_3-singularity is 4-$\mathcal{R}$-determined. The conditions for a distance squared function to have an A_3-singularity at p are expressed as algebraic conditions on the coefficients of $j^3 f_p$ (expressing the failure to have an A_2-singularity) together with an open semi-algebraic condition involving degree 4 terms in $j^4 f_p$ (expressing the condition to have an A_3-singularity and not a more degenerate one). The algebraic condition defines a variety in V_k ($k \ge 4$) labeled the A_3-stratum. The structure of the closure of the A_3-stratum is determined in [Bruce (1984)] and it is also shown there that the Monge-Taylor map is transverse to the D_4-stratum at an umbilic point. The ridge is the pre-image by $j^k\phi$ of the intersection of $j^k\phi(M)$ with the A_3-stratum and is diffeomorphic to the models in the statement of the theorem. See [Bruce (1984)] for more details. □

It is also known that ridges change colour at an umbilic [Porteous (2001)]. A consequence of Theorem 6.16 and of this remark is the following characterisation of the D_4-singularities of the distance squared function.

Corollary 6.4. *Let M be a distance squared function generic surface parametrised by $\mathbf{x} : U \to \mathbb{R}^3$. Suppose that $p_0 = \mathbf{x}(u_0)$ is an umbilic point. Then*

(i) *d_{a_0} has a D_4^--singularity if and only if for any $\epsilon > 0$, there exist six distinct points $p_i = \mathbf{x}(u_i)$ such that $||u_0 - u_i|| < \epsilon$ for $i = 1, \dots, 6$, and all the points p_i are ridge points.*

(ii) *d_{a_0} has a D_4^+-singularity if and only if for any $\epsilon > 0$, there exist two distinct points $p_i = \mathbf{x}(u_i)$ such that $||u_0 - u_i|| < \epsilon$ for $i = 1, 2$, and p_1 and p_2 are ridge points of different colours.*

At a hyperbolic umbilic there are two possible configurations of the ridges (Figure 6.11, left). The configuration depends only on the cubic form C (Proposition 6.18 below). Any cubic form can be written as $\mathrm{Re}(\alpha z^3 + \beta z^2 \bar{z})$, where α and β are complex numbers, $z = x + iy$ and $\mathrm{Re}(w)$ denotes the real part of the complex number w. Any such cubic form is $SO(2)$-equivalent to one of the form

$$\mathrm{Re}(z^3 + \beta z^2 \bar{z})$$

or is $SO(2)$-equivalent to $\mathrm{Re}(z^2\bar{z}) = x(x^2 + y^2)$. (The $SO(2)$-equivalence corresponds to a change of an orthonormal basis in T_pM and does not affect the 2-jet of $f(x, y)$.) Therefore, the set of cubic forms can be viewed as the set of points in the β-plane. There are the following three exceptional curves in this plane:

$\beta = 2e^{i\theta} + e^{-2i\theta}$: this is a hypocycloid in Figure 6.11 (right) and consists of the umbilics which are more degenerate than D_4, i.e., those cubic forms with repeated roots.

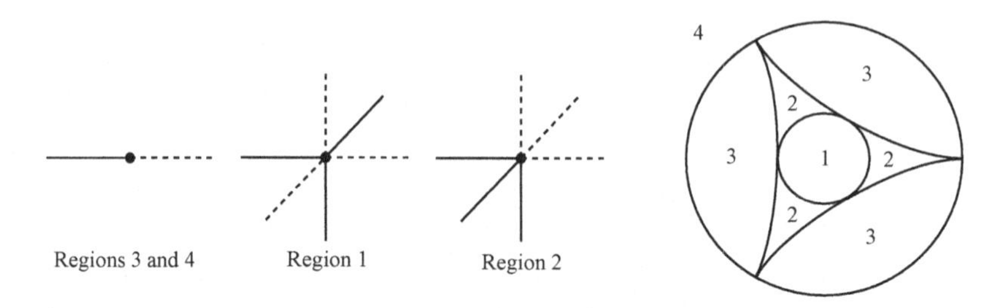

Fig. 6.11 Generic configurations of the ridge curves at umbilic points left and partition of the β-plane right. The ridge associated to one focal sheet is drawn in continuous line and the one associated to the other focal sheet is drawn in dashed line.

$|\beta| = 3$: this is the outer circle in Figure 6.11. It consists of the umbilics which are not versally unfolded by the family of distance-squared functions (this is equivalent to the algebraic condition given in Theorem 6.10(ii)).

$|\beta| = 1$: this is the inner circle in Figure 6.11. It consists of umbilics at which two of the ridge lines (or more) are tangential.

Proposition 6.18 ([Bruce (1984)]). *The configuration ridges at an umbilic point is completely determined by the cubic C. For β not on one of the three exceptional curves above, the configuration is as in Figure 6.11.*

Proof. See [Bruce (1984)]. □

6.6.4 *The sub-parabolic curve*

In [Bruce and Wilkinson (1991); Wilkinson (1991)], the authors considered the singularities of the folding maps on M. Let π be a plane in $\mathbb{R}^3$ and let $\mathbf{v}$ be a unit vector orthogonal to π. Denote by $d(p,\pi)$ the distance of a point $p \in \mathbb{R}^3$ to π. Then the folding map along π is the map $F_\pi : \mathbb{R}^3 \to \mathbb{R}^3$ given by $F_\pi(p) = P_{\mathbf{v}}(p) + d(p,\pi)^2 v$, where $P_{\mathbf{v}}$ is the orthogonal projection along $\mathbf{v}$ to π. For instance, the folding map along the plane $y = 0$ is the map $F_{y=0}(x, y, z) = (x, y^2, z)$. By varying π, one obtains a 3-parameter family of folding maps in the ambient space $\mathbb{R}^3$ ([Bruce and Wilkinson (1991)]). The restriction of this family to M is called the family of folding maps on M.

The authors in [Bruce and Wilkinson (1991)] proved that the bifurcation set of the family of folding maps is dual to the bifurcation set of the family of distance squared function (which is the focal set). Thus, the folding maps capture the geometry of the focal set obtained via its contact with planes. In particular, it captures its parabolic set. The set of points on the surface M which correspond to the parabolic set of its focal set is defined in [Bruce and Wilkinson (1991); Wilkinson (1991)] as the *sub-parabolic curve* of M. Before giving other geometric characterisations of the sub-parabolic curve, we require the following result.

Proposition 6.19. *Let $\mathbf{x} : U \to \mathbb{R}^3$ be a local parametrisation of M at a non-umbilic point p such that the coordinate curves are the lines of principal curvature. Suppose that p is not a ridge point. Then the coefficients of the first fundamental form of the focal set $\mathcal{F}_1$ are*

$$E_{\mathcal{F}_1} = \frac{1}{\kappa_1^4}(\frac{\partial k_1}{\partial u_1})^2, \quad F_{\mathcal{F}_1} = \frac{1}{\kappa_1^4}\frac{\partial k_1}{\partial u_1}\frac{\partial k_1}{\partial u_2}, \quad G_{\mathcal{F}_1} = (1 - \frac{\kappa_2}{\kappa_1})^2 + \frac{1}{\kappa_1^4}(\frac{\partial k_1}{\partial u_2})^2$$

and those of its second fundamental form are

$$l_{\mathcal{F}_1} = \frac{1}{\kappa_1}\frac{\partial k_1}{\partial u_1}\sqrt{E}, \quad m_{\mathcal{F}_1} = 0, \quad n_{\mathcal{F}_1} = \frac{1}{\sqrt{E}}(1 - \frac{\kappa_2}{\kappa_1})\langle \mathbf{x}_{u_1}, \mathbf{x}_{u_2u_2}\rangle.$$

For the focal set $\mathcal{F}_2$, the coefficients of its first fundamental form are

$$E_{\mathcal{F}_2} = \frac{1}{\kappa_2^4}(\frac{\partial k_2}{\partial u_2})^2, \quad F_{\mathcal{F}_2} = \frac{1}{\kappa_2^4}\frac{\partial k_2}{\partial u_1}\frac{\partial k_2}{\partial u_2}, \quad G_{\mathcal{F}_2} = (1 - \frac{\kappa_1}{\kappa_2})^2 + \frac{1}{\kappa_2^4}(\frac{\partial k_2}{\partial u_1})^2$$

and those of its second fundamental form are

$$l_{\mathcal{F}_2} = \frac{1}{\kappa_2}\frac{\partial k_2}{\partial u_2}\sqrt{G}, \quad m_{\mathcal{F}_2} = 0, \quad n_{\mathcal{F}_2} = \frac{1}{\sqrt{G}}(1 - \frac{\kappa_1}{\kappa_2})\langle \mathbf{x}_{u_2}, \mathbf{x}_{u_1u_1}\rangle.$$

Proof. The focal set $\mathcal{F}_1$ is parametrised by $\varepsilon_1(u) = \mathbf{x}(u) + \frac{1}{\kappa_1(u)}N(u)$. The coefficients of the first fundamental form of $\mathcal{F}_1$ are computed using the partial derivatives of the map ε_1 which are given in the proof of Proposition 6.16.

The Gauss map of the focal set $\mathcal{F}_1$ is $N_{\mathcal{F}_1} = (1/\sqrt{E})\mathbf{x}_{u_1}$, so the coefficients of the second fundamental form of $\mathcal{F}_1$ are

$$l_{\mathcal{F}_1} = \langle N_{\mathcal{F}_1}, \frac{\partial^2 \varepsilon_1}{\partial u_1^2}\rangle = \frac{1}{\kappa_1}\frac{\partial k_1}{\partial u_1}\sqrt{E}$$

$$m_{\mathcal{F}_1} = \langle N_{\mathcal{F}_1}, \frac{\partial^2 \varepsilon_1}{\partial u_1 u_2}\rangle = 0$$

$$n_{\mathcal{F}_1} = \langle N_{\mathcal{F}_1}, \frac{\partial^2 \varepsilon_1}{\partial u_2^2} \rangle = (1 - \frac{\kappa_2}{\kappa_1 \sqrt{E}}) \langle \mathbf{x}_{u_1}, \mathbf{x}_{u_2 u_2} \rangle$$

The coefficients of the first and second fundamental forms of $\mathcal{F}_2$ are computed in a similar way. □

Theorem 6.17 ([Morris (1996)]). *Let $\mathbf{x} : U \to \mathbb{R}^3$ be a local parametrisation of M at a non-umbilic point p such that the coordinate curves are the lines of principal curvature. Suppose that p is not a ridge point associated to $\mathcal{F}_1$. Then the sub-parabolic curve of M which corresponds to the parabolic set of the focal set $\mathcal{F}_1$ is the set of points $\mathbf{x}(u_1, u_2)$ where*

$$\frac{\partial \kappa_2}{\partial u_1}(u_1, u_2) = 0.$$

Similarly, the sub-parabolic curve which corresponds to the parabolic curve of the focal set $\mathcal{F}_2$ is given by

$$\frac{\partial \kappa_1}{\partial u_2}(u_1, u_2) = 0.$$

That is, the sub-parabolic curve is the set of points where a principal curvature is extremal along the other line of principal curvature.

Proof. We shall give a slightly different proof to that of Morris in [Morris (1996)] and consider only the case of the focal set $\mathcal{F}_1$. The case of the focal set $\mathcal{F}_2$ follows in a similar way.

The sub-parabolic curve associated to the focal set $\mathcal{F}_1$ is the set of points on M where the Gaussian curvature

$$K_{\mathcal{F}_1} = \frac{l_{\mathcal{F}_1} n_{\mathcal{F}_1} - m_{\mathcal{F}_1}^2}{E_{\mathcal{F}_1} G_{\mathcal{F}_1} - F_{\mathcal{F}_1}^2}$$

of $\mathcal{F}_1$ vanishes. It follows from Proposition 6.19 that $K_{\mathcal{F}_1} = 0$ if and only if

$$\langle \mathbf{x}_{u_1}, \mathbf{x}_{u_1 u_2} \rangle = 0 \tag{6.22}$$

The coordinate curves are the lines of principal curvature, so $\langle \mathbf{x}_{u_1}, \mathbf{x}_{u_2} \rangle = 0$ and we get by differentiating this identity with respect to u_2

$$\langle \mathbf{x}_{u_2}, \mathbf{x}_{u_1 u_2} \rangle = -\langle \mathbf{x}_{u_1}, \mathbf{x}_{u_2 u_2} \rangle. \tag{6.23}$$

Again, by the choice of the coordinate system, we have $N_{u_1} = \kappa_1 \mathbf{x}_{u_1}$ and $N_{u_2} = \kappa_2 \mathbf{x}_{u_2}$, so

$$\langle N_{u_1}, \mathbf{x}_{u_2} \rangle = \langle \kappa_1 \mathbf{x}_{u_1}, \mathbf{x}_{u_2} \rangle = 0.$$

Differentiating the above identity with respect to u_2 yields

$$\langle N_{u_1u_2}, \mathbf{x}_{u_2}\rangle + \langle N_{u_1}, \mathbf{x}_{u_1u_2}\rangle = 0.$$

Hence,

$$\langle N_{u_1u_2}, \mathbf{x}_{u_2}\rangle = -\langle N_{u_1}, \mathbf{x}_{u_1u_2}\rangle = -\kappa_1\langle \mathbf{x}_{u_1}, \mathbf{x}_{u_1u_2}\rangle. \tag{6.24}$$

Differentiating $N_{u_2} = \kappa_2\mathbf{x}_{u_2}$ with respect to u_1 gives

$$N_{u_1u_2} = \frac{\partial \kappa_2}{\partial u_1}\mathbf{x}_{u_2} + \kappa_2\mathbf{x}_{u_1u_2}.$$

so that

$$\langle N_{u_1u_2}, \mathbf{x}_{u_2}\rangle = \frac{\partial \kappa_2}{\partial u_1}\langle \mathbf{x}_{u_2}, \mathbf{x}_{u_2}\rangle + \kappa_2\langle \mathbf{x}_{u_2}, \mathbf{x}_{u_1u_2}\rangle.$$

Equivalently,

$$\langle N_{u_1u_2}, \mathbf{x}_{u_2}\rangle - \kappa_2\langle \mathbf{x}_{u_2}, \mathbf{x}_{u_1u_2}\rangle = \frac{\partial \kappa_2}{\partial u_1}G. \tag{6.25}$$

Using (6.24) and (6.23), equation (6.25) becomes

$$(\kappa_2 - \kappa_1)\langle \mathbf{x}_{u_1}, \mathbf{x}_{u_2u_2}\rangle = \frac{\partial \kappa_2}{\partial u_1}G. \tag{6.26}$$

As the surface patch is umbilic free, it follows from (6.26) that equation (6.22) is satisfied if and only if

$$\frac{\partial \kappa_2}{\partial u_1} = 0.$$

□

Proposition 6.20 ([Morris (1996)]). *The sub-parabolic curve corresponding to the focal set $\mathcal{F}_1$ (resp. $\mathcal{F}_2$) is the locus of geodesic inflections of the lines of curvature corresponding to the principal curvature κ_2 (resp. κ_1).*

Proof. We consider the sub-parabolic curve corresponding to the focal set $\mathcal{F}_1$, the other case follows similarly. With the parametrisation as in Theorem 6.17, a line of principal curvature corresponding to the principal curvature κ_2 is parametrised by $\alpha(u_2) = \mathbf{x}(a, u_2)$, for some constant a. By (6.2), its geodesic curvature is given by

$$\kappa_g = \frac{1}{||\alpha'||^3}\langle \alpha'', \alpha' \times N\rangle = \frac{1}{G^{\frac{3}{2}}}\langle \mathbf{x}_{u_2u_2}, \mathbf{x}_{u_2} \times N\rangle = \frac{1}{G^{\frac{3}{2}}}\langle \mathbf{x}_{u_1}, \mathbf{x}_{u_2u_2}\rangle.$$

Therefore, $\kappa_g = 0$ if and only if $\langle \mathbf{x}_{u_1}, \mathbf{x}_{u_2u_2}\rangle = 0$ and the result follows by (6.22) and (6.23). □

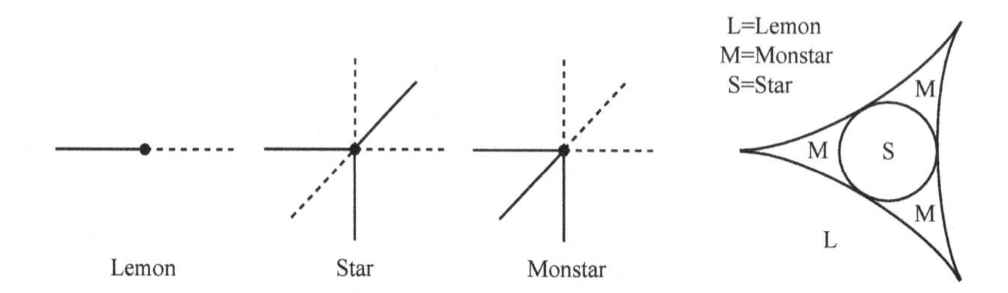

Fig. 6.12 The partition of the β-plane (right) and the generic configurations of the sub-parabolic curves at an umbilic point (first three left). The sub-parabolic curve associated to one focal sheet is drawn in continuous line and the one associated to the other focal sheet is drawn in dashed line.

The sub-parabolic curve of a generic surface is a smooth curve away from umbilic points. At an umbilic point it consists of three transverse curves or a single curve. The configurations of the sub-parabolic curves are closely related to those of the lines of principal curvature. If we write the surface in Monge-form $\frac{\kappa}{2}(x^2+y^2)+C(x,y)+O(4)$ and take $C(x,y) = \mathrm{Re}(z^3+\beta z^2\bar{z})$, then the configuration of the sub-parabolic curves depends only on the cubic C. There are two exceptional curves in the β plane:

$\beta = -3(2e^{i\theta}+e^{-2i\theta})$: this is the hypocycloid in Figure 6.12 (right), separating the lemon from the star and monstar umbilics (Figure 6.2).

$|\beta| = 3$: this is the circle in Figure 6.12 (right) separating the star and monstar umbilics.

The configurations of the sub-parabolic curves in the open regions of the β-plane delimited by the above two exceptional curves are as in Figure 6.12 ([Wilkinson (1991)]).

6.7 Notes

The contact of surfaces with planes and lines is an affine and projective property of the surface, so does not depend on the metric in the ambient space (see for example [Bruce, Giblin and Tari (1995); Shcherbak (1986)]). The results in this chapter are, in particular, valid for surfaces embedded in Minkowski 3-space. For such surfaces, additional geometric information can be obtained when considering projections along the lightlike directions in $\mathbb{R}^3_1$, see [Izumiya and Tari (2013)].

Extensive work was carry out on the singularities of orthogonal projections of surfaces to planes. We refer to the following (incomplete) list for related works: [Arnol'd (1983); Gaffney and Ruas (1979); Gaffney (1983); Kergosien (1981); Koenderink (1990); Koenderink and van Doorn (1976); Koenderink (1984); Platonova (1984); Rieger (1987); Landis (1981); Lyashko (1979); Goryunov (1981a,b)].

We considered here only the local singularities of the relevant germs of functions and mappings. Their multi-local singularities are also of importance and yield geometric information about the surface. For instance, the results in Theorems 6.1, 6.5, 6.9 can be extended, using multi-transversality arguments, to cover the multi-local singularities. See [Banchoff, Gaffney and McCrory (1982); Uribe-Vargas (2006)] for some geometric consequences of these at a cusp of Gauss and [Bruce, Giblin and Gibson (1985); Damon (2003, 2004, 2006); Giblin and Holtom (1999); Giblin and Janeczko (2012); Giblin and Zakalyukin (2005)] for work on symmetry sets and medial axis. Other works on multi-local singularities for other dimensions of the target can be found in [Atique (2000); Dreibelbis (2001, 2006)].

Orthogonal projections of non-smooth surfaces are also studied, see [Arnol'd (1979); Bruce and Giblin (1990); Goryunov (1990)] for projections of surfaces with boundary and [Tari (1991)] for those with creases and corners.

Suppose the direction of the orthogonal projection of a surface is changing along a curve. One question of great interest is how to reconstruct the surface from the resulting family of apparent contours. One can in fact reconstruct part of the surface (the visible part) from this family of apparent contours [Giblin and Weiss (1987); Siddiqi and Pizer (2008)].

We highlighted some properties of the parabolic, ridge and sub-parabolic curves. The complete catalogues of the bifurcations of these curves as the surface is deformed in generic 1-parameter families of surfaces are given in [Bruce, Giblin and Tari (1995, 1998, 1999)], see also [Uribe-Vargas (2001)].

We considered only smooth and regular surfaces. The techniques described in this book are also currently been used to study the geometry of singular submanifolds in $\mathbb{R}^n$ and $\mathbb{R}^n_1$, see for example [Bruce and West (1998); Dias and Nuño (2008); Oset Sinha and Tari (2010); Saji, Umehara and Yamada (2009); West (1995)]. See also Chapter 1 for a brief outline of the techniques and for references for the study of special singular surfaces such as rules and developable surfaces.

The families of orthogonal projections and height functions on a surface are related via a duality result between their bifurcation sets [Bruce and

Romero-Fuster (1991); Bruce (1994a); Bruce, Giblin and Tari (1998); Shcherbak (1986)]. The families of distance squared functions and of the folding maps are also related via a duality result between their bifurcation sets [Bruce and Wilkinson (1991); Bruce, Giblin and Tari (1999); Wilkinson (1991)]. Similar duality results in other dimensions can be found for instance in [Bruce and Nogueira (1998); Nabarro (2003)].

For results on the contact of the surface with circles see [Bruce (1994b); Montaldi (1986b)].

A line of research that we did not explore in detail in this book is on the pairs of foliations determined by the asymptotic and principal directions (Figure 6.2 and Figure 6.6). Their study also reveals a great deal of the geometry of the surface; see for example [Tari (2010)] for a survey article and references.

Chapter 7

Surfaces in the Euclidean 4-space

We consider in this chapter the extrinsic differential geometry of a surface M immersed in $\mathbb{R}^4$. The study of the second order geometry of M is considered in depth in [Little (1969)]. The work of Little inspired much of recent work on the subject.

Geometric properties of which are related to its second fundamental form depend only on the 2-jet of the immersion. These properties can be derived, at each point p on M, from an ellipse in the normal plane of M at p, called the curvature ellipse. Isometric invariants of the curvature ellipse, such as its area, are isometric invariants of the surface. Also, the position of the point p with respect to the curvature ellipse (outside, on, inside) gives an isometric invariant partition of the surface.

We consider the contact of the surface with flat objects (hyperplanes, planes and lines) and derive from it extrinsic properties of the surface. This contact is affine invariant, so the derived properties from it are also affine invariant. In fact, various properties of the surface derived from the isometric properties of the curvature ellipse can also be derived from the affine properties of this ellipse. A key observation is that the second fundamental form at each point p on the surface defines a pair of quadratic forms $(Q_1(x, y), Q_2(x, y))$. The action of $GL(2, \mathbb{R}) \times GL(2, \mathbb{R})$ on the set of these quadratic forms determines much of the second order affine properties of the surface at p.

As in the case of surfaces in $\mathbb{R}^3$, the contact viewpoint reveals new extrinsic geometric properties of the surface that depend on higher order jets of the immersion. We consider in this chapter in detail the contact of M with hyperplanes and lines and touch briefly on its contact with planes. Certain types of singularities of the map-germs that define the contact of the surface with a given model object occur on curves on the surface.

These curves are robust features of the surface and part of this chapter is devoted to determining and characterising them. Certain robust curves are related to the contact of the surface with two model objects. This suggest a (duality) relationship between the mappings that define the contact of the surface with the model objects. We explore briefly this relationship.

We recall in §7.1 some properties of the second fundamental form of M and consider some of its local intrinsic and affine invariants. We also define the asymptotic directions at a given point and give the expression of the binary differential equation of their integral curves. The coefficients of this equation define at each point on the surface a point in the projective plane. It turns out that its polar line with respect to the conic of degenerate quadratic forms represent the ν-shape operators. Exploring the polarity further gives us a way of choosing one special pair of ν-lines of curvature which we call the lines of curvature of M. We study the contact of the surface with hyperplanes (§7.6), lines (§7.7) and planes (§7.8), and derive robust features of the surface. The contact of a surface with hyperplanes is intimately related to that of its canal hypersurface with hyperplanes (§7.6.1). The contact of the surface with lines is captured by the singularities of orthogonal projections to 3-spaces. The image of the surface by a projection can be singular and the geometry of the singular projected surface gives geometric information about the surface itself. We discuss in the last section (§7.9) the contact of the surface with hyperspheres and give geometric characterisations of the generic singularities of the family of distance squared functions on the surface.

7.1 The curvature ellipse

Let $\mathbf{x} : U \to \mathbb{R}^4$ be a local parametrisation of M, where U is an open subset of $\mathbb{R}^2$. Let $\{\mathbf{e}_1, \mathbf{e}_2, \mathbf{e}_3, \mathbf{e}_4\}$ be a positively oriented orthonormal frame in $\mathbb{R}^4$ such that at any $u \in U$, $\{\mathbf{e}_1(u), \mathbf{e}_2(u)\}$ is a basis for the tangent plane T_pM and $\{\mathbf{e}_3(u), \mathbf{e}_3(u)\}$ is a basis for the normal plane N_pM at $p = \mathbf{x}(u)$. Associated to this frame is the dual basis of 1-forms $\{\omega_1, \omega_2, \omega_3, \omega_4\}$ in the dual space of $\mathbb{R}^4$.

In [Little (1969)] are given some properties of the surface which are invariant under rotations in T_pM and N_pM, that is, invariant under the action of $\mathrm{SO}(2) \times \mathrm{SO}(2)$ on $T_pM \times N_pM$. The expressions for the invariants are given with respect to an orthonormal frame $\{\mathbf{e}_1, \mathbf{e}_2, \mathbf{e}_3, \mathbf{e}_4\}$. We recall here some of them.

The vectors $\mathbf{e}_i$ and the 1-forms ω_1 can be extended to an open subset in $\mathbb{R}^4$, and we keep the same notation for their extensions. Define the 1-forms by ω_i and ω_{ij}, $i, j = 1, \ldots, 4$ by

$$\omega_i = \langle d\mathbf{x}, \mathbf{e}_i \rangle \quad \text{and} \quad \omega_{ij} = \langle d\mathbf{e}_i, \mathbf{e}_j \rangle,$$

where d is the exterior differential. It is worth observing that $\omega_3 = \omega_4 = 0$ and $\omega_{ij} = -\omega_{ji}$. The Maurer-Cartan structural equations (see [do Carmo (1976)]) are

$$d\omega_i = \Sigma_{j=1}^4 \omega_{ij} \wedge \omega_j \quad \text{and} \quad d\omega_{ij} = \Sigma_{k=1}^4 \omega_{ik} \wedge \omega_{kj}. \tag{7.1}$$

We have

$$\begin{aligned} 0 = d\omega_3 = \omega_{31} \wedge \omega_1 + \omega_{32} \wedge \omega_2, \\ 0 = d\omega_4 = \omega_{41} \wedge \omega_1 + \omega_{42} \wedge \omega_2. \end{aligned}$$

It follows by a lemma of Cartan (see [do Carmo (1976)]) that there exist a, b, c, e, f, g, such that

$$\begin{aligned} \omega_{13} &= a\omega_1 + b\omega_2, \qquad & \omega_{14} &= e\omega_1 + f\omega_2, \\ \omega_{23} &= b\omega_1 + c\omega_2, \qquad & \omega_{24} &= f\omega_1 + g\omega_2. \end{aligned}$$

The second fundamental form of M is the vector valued quadratic form associated to the normal component of the second derivative $d^2\mathbf{x}$ of $\mathbf{x}$ at p (see Chapter 2), that is,

$$\mathrm{II}_p = \langle d^2\mathbf{x}, \mathbf{e}_3 \rangle \mathbf{e}_3 + \langle d^2\mathbf{x}, \mathbf{e}_4 \rangle \mathbf{e}_4.$$

We have

$$\begin{aligned} \langle d^2\mathbf{x}, \mathbf{e}_3 \rangle &= -\langle d\mathbf{x}, d\mathbf{e}_3 \rangle \\ &= -\langle \omega_1 \mathbf{e}_1 + \omega_2 \mathbf{e}_2, d\mathbf{e}_3 \rangle \\ &= -\omega_1 \langle \mathbf{e}_1, d\mathbf{e}_3 \rangle - \omega_2 \langle \mathbf{e}_2, d\mathbf{e}_3 \rangle \\ &= -\omega_1 \omega_{31} - \omega_2 \omega_{32} \\ &= \omega_1 \omega_{13} + \omega_2 \omega_{23} \\ &= a\omega_1^2 + 2b\omega_1\omega_2 + c\omega_2^2. \end{aligned}$$

Similarly, $\langle d^2\mathbf{x}, \mathbf{e}_4 \rangle = e\omega_1^2 + 2f\omega_1\omega_2 + g\omega_2^2$, so that

$$\mathrm{II}_p = (a\omega_1^2 + 2b\omega_1\omega_2 + c\omega_2^2)\mathbf{e}_3 + (e\omega_1^2 + 2f\omega_1\omega_2 + g\omega_2^2)\mathbf{e}_4.$$

The matrix $\alpha = \begin{pmatrix} a & b & c \\ e & f & g \end{pmatrix}$ is called the matrix of the second fundamental form with respect to the orthonormal frame $\{\mathbf{e}_1, \mathbf{e}_2, \mathbf{e}_3, \mathbf{e}_4\}$. Its entries are called the coefficients of the second fundamental form with respect to that frame.

The 1-form ω_{12} is the connection form in the tangent bundle of M and ω_{34} is the connection form in the normal bundle of M, and $d\omega_{12}$ and $d\omega_{34}$ are the respective curvature forms in those bundles. Following Little, the forms ω_1, ω_2 and ω_{12} can be regarded as forms in the tangent bundle and depend only on the metric $\omega_1^2+\omega_2^2$ on the surface. The Gaussian curvature of the surface can be found using

$$d\omega_{12} = -K\omega_1 \wedge \omega_2.$$

Little defines another scalar invariant N, called the normal curvature, by $d\omega_{34} = -N\omega_1 \wedge \omega_2$. The Gaussian and normal curvatures K and N can be expressed as follows in terms of the coefficients of the second fundamental form:

$$K = (ac - b^2) + (eg - f^2) \quad \text{and} \quad N = (a-c)f - (e-g)b.$$

Let α_1 and α_2 be the symmetric matrices associated to the quadratic forms $\langle d^2\mathbf{x}, \mathbf{e}_3\rangle$ and $\langle d^2\mathbf{x}, \mathbf{e}_4\rangle$ respectively, so that

$$\alpha_1 = \begin{pmatrix} a & b \\ b & c \end{pmatrix} \quad \text{and} \quad \alpha_2 = \begin{pmatrix} e & f \\ f & g \end{pmatrix}.$$

Then

$$K = K_1 + K_2 \quad \text{with} \quad K_1 = \det \alpha_1, \; K_2 = \det \alpha_2.$$

In fact, there is a more general result explaining the above relation (see [do Carmo (1976); Basto-Gonçalves (2013)]).

Theorem 7.1. *The Gaussian curvature K of M at a point p is the sum of the curvatures K_1 and K_2 of the images M_1 and M_2 of the surface M by orthogonal projections along any two orthogonal normal directions $\mathbf{n}_1$ and $\mathbf{n}_2$ in N_pM respectively.*

Let S^1 be the unit circle in T_pM parametrised by θ, and denote by γ_θ the normal section of M in the direction $\mathbf{u} = \cos\theta\mathbf{e}_1 + \sin\theta\mathbf{e}_2$. We parametrise γ_θ by arc length and denote by $\eta(\theta)$ its curvature vector at p. Thus, $\eta(\theta)$ is the projection of γ_θ'' to N_pM, equivalently, $\eta(\theta) = \mathrm{II_p}(\mathbf{u})$. We have then, at each point p on M, a map $\eta : S^1 \subset T_pM \to N_pM$, given by

$$\begin{aligned} \eta(\theta) = &\, (a\cos^2\theta + 2b\cos\theta\sin\theta + c\sin^2\theta)\mathbf{e}_3 \\ &+ (e\cos^2\theta + 2f\cos\theta\sin\theta + g\sin^2\theta)\mathbf{e}_4. \end{aligned} \tag{7.2}$$

Using the trigonometric identities for double angles, expression (7.2) can be written in the form

$$\eta(\theta) = \mathbf{H} + \cos(2\theta)\mathbf{B} + \sin(2\theta)\mathbf{C},$$

with

$$\begin{aligned}\mathbf{H} &= \tfrac{1}{2}(a+c)\mathbf{e_3} + \tfrac{1}{2}(e+g)\mathbf{e_4},\\ \mathbf{B} &= \tfrac{1}{2}(a-c)\mathbf{e_3} + \tfrac{1}{2}(e-g)\mathbf{e_4},\\ \mathbf{C} &= b\mathbf{e_3} + f\mathbf{e_4}.\end{aligned}$$

The normal field $\mathbf{H}$ is called the *mean curvature vector* of M at p. The map η can also be written in matrix form

$$\eta(\theta) - \mathbf{H} = \mathbf{A}\begin{pmatrix}\cos 2\theta\\ \sin 2\theta\end{pmatrix}, \tag{7.3}$$

with

$$\mathbf{A} = \begin{pmatrix}\frac{1}{2}(a-c) & b\\ \frac{1}{2}(e-g) & f\end{pmatrix}.$$

It follows from (7.3) that the image of the map $\eta : S^1 \subset T_pM \to N_pM$ is an ellipse in the normal plane N_pM with centre $\mathbf{H}$ and principal axes along the vectors $\mathbf{B}$ and $\mathbf{C}$.

Definition 7.1. The *curvature ellipse* of M at p is defined as the image of the map $\eta : S^1 \subset T_pM \to N_pM$.

Rotations in the tangent plane T_pM leave invariant the curvature ellipse and rotations in the normal plane result in rotating the curvature ellipse (see Propositions 7.2 and 7.3). Thus any isometric scalar invariant of the curvature ellipse is an isometric scalar invariant of the surface (under the action of $\mathrm{SO}(2) \times \mathrm{SO}(2)$ on $T_pM \times N_pM$). For instance, the area of the curvature ellipse (which is $\pi||N||/2$) and the length $||\mathbf{H}||$ of the mean curvature vector are scalar invariants of the surface. Little defined another scalar invariant of the surface, namely the resultant

$$\begin{aligned}\Delta &= \frac{1}{4}\begin{vmatrix} a & b & c & 0\\ e & f & g & 0\\ 0 & a & b & c\\ 0 & e & f & g\end{vmatrix} \qquad (7.4)\\ &= \frac{1}{4}\left(4(af-eb)(bg-fc) - (ag-ec)^2\right)\end{aligned}$$

of the two polynomials $ax^2+2bxy+cy^2$ and $ex^2+2fxy+gy^2$. The function Δ has the following property. The curvature ellipse passes through the point p if and only if $\eta(\theta) = 0$ for some θ. It follows from (7.2) that this happens if and only if the above two polynomials have a common root. Therefore, p is a point on the curvature ellipse if and only if $\Delta(p) = 0$. The point p

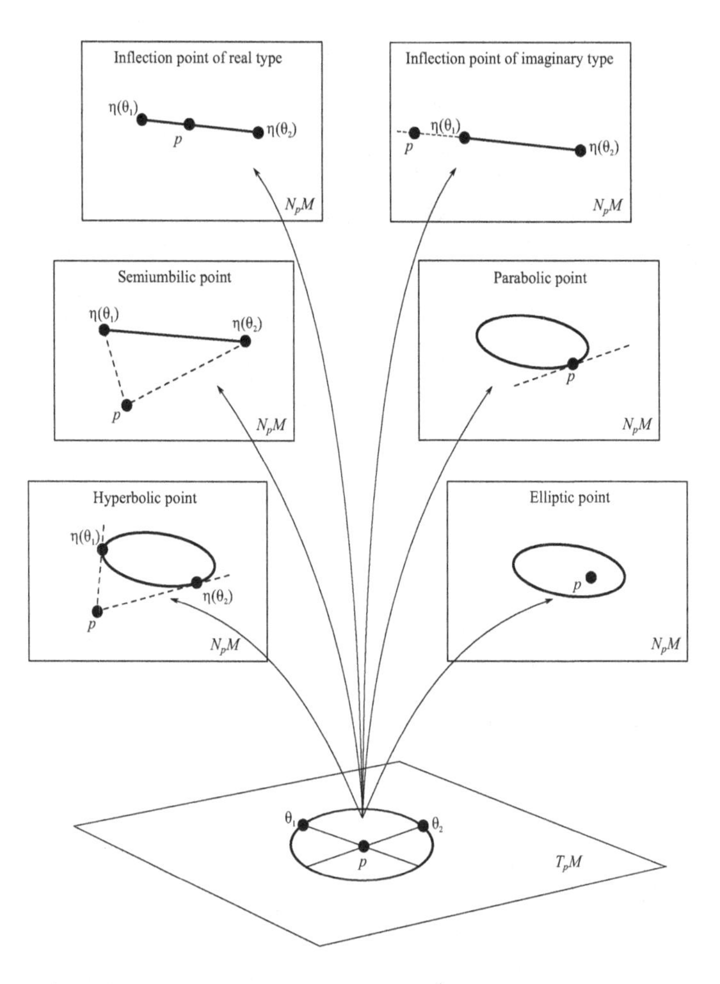

Fig. 7.1 Classification of points on a surface in $\mathbb{R}^4$ in terms of the curvature ellipse.

being inside or outside the curvature ellipse can also be determined by the sign of $\Delta(p)$, positive for inside and negative for outside.

Points on the surface are classified in terms of the curvature ellipse.

Definition 7.2. A point $p \in M$ is called *semiumbilic* if the curvature ellipse degenerates to a line segment that contains p. If the curvature ellipse is a radial segment, the point p is called an *inflection point*. An inflection point is of *real type*, (resp. *imaginary type*, *flat*) if p is an interior point of the radial segment, (resp. does not belong to it, is one of its end points).

When the curvature ellipse reduces to the point p, then p is said to be a *flat umbilic*. A non inflection point $p \in M$ is called *elliptic* (resp. *hyperbolic*, *parabolic*) when it lies inside (resp. outside, on) the curvature ellipse. See Figure 7.1.

The notation hyperbolic, elliptic and parabolic point in Definition 7.2 has nothing to do with the sign of the Gaussian curvature. They are introduced in [Mond (1982)] in an analogy to the behaviour of the asymptotic directions and curves on surfaces in $\mathbb{R}^3$ (see §7.3).

7.2 Second order affine properties

The invariants in §7.1 are computed using an orthonormal frame $\{\mathbf{e}_1, \mathbf{e}_2, \mathbf{e}_3, \mathbf{e}_4\}$ with $\{\mathbf{e}_1, \mathbf{e}_2\}$ a basis of T_pM and $\{\mathbf{e}_3, \mathbf{e}_4\}$ is a basis of N_pM at each point p on the surface. Those invariants are under the action of $\mathrm{SO}(2) \times \mathrm{SO}(2)$ on $T_pM \times N_pM$.

Given a local parametrisation $\mathbf{x} : U \to \mathbb{R}^4$ of the surface, a practical basis of the tangent plane T_pM to work with is $\{\mathbf{f}_1, \mathbf{f}_2\} = \{\mathbf{x}_{u_1}, \mathbf{x}_{u_2}\}$. Of course, in general it is not an orthonormal basis of T_pM. One can complete it at each point p on M to obtain a basis $\{\mathbf{f}_1, \mathbf{f}_2, \mathbf{f}_3, \mathbf{f}_4\}$ of $\mathbb{R}^4$ varying smoothly with p, with $\{\mathbf{f}_3, \mathbf{f}_4\}$ a basis of N_pM (which is also not necessarily orthonormal). The vector fields $\mathbf{f}_i$, $i = 1, \dots, 4$ defined on M can be extended locally to an open subset of $\mathbb{R}^4$.

We are interested in the contact of the surface with hyperplanes, planes and lines and this contact is affine invariant ([Bruce, Giblin and Tari (1995)]). Thus, it is natural to consider (affine) invariants properties of the surface under the action of $\mathrm{GL}(2, \mathbb{R}) \times \mathrm{GL}(2, \mathbb{R})$ on $T_pM \times N_pM$, where $\mathrm{GL}(2, \mathbb{R})$ denotes the general linear group. Given an orthonormal basis $\{\mathbf{e}_1, \mathbf{e}_2, \mathbf{e}_3, \mathbf{e}_4\}$ in $T_pM \times N_pM$, the above action can be viewed as a change of basis in T_pM and N_pM.

We analyse the effect of this change of basis on the coefficients of the second fundamental form and on the curvature ellipse. We assume that we are given a frame $\{\mathbf{e}_1, \mathbf{e}_2, \mathbf{e}_3, \mathbf{e}_4\}$ of M near p. We start with the $\mathrm{GL}(2, \mathbb{R})$ action on T_pM. Let $\Omega = \begin{pmatrix} \alpha_1 & \alpha_2 \\ \beta_1 & \beta_2 \end{pmatrix}$ represent an element in $\mathrm{GL}(2, \mathbb{R})$ and write

$$\begin{aligned} \mathbf{f}_1 &= \alpha_1 \mathbf{e}_1 + \beta_1 \mathbf{e}_2, \\ \mathbf{f}_2 &= \alpha_2 \mathbf{e}_1 + \beta_2 \mathbf{e}_2. \end{aligned}$$

Then we have the following.

Proposition 7.1. *Denote by a, b, c, e, f, g the coefficients of the second fundamental form with respect to the basis $\{\mathbf{e}_1, \mathbf{e}_2, \mathbf{e}_3, \mathbf{e}_4\}$ and by $l_1, m_1, n_1, l_2, m_2, n_2$ its coefficients with respect to the basis $\{\mathbf{f}_1, \mathbf{f}_2, \mathbf{e}_3, \mathbf{e}_4\}$. Then*

$$\begin{pmatrix} l_1 \\ m_1 \\ n_1 \end{pmatrix} = \mathbf{\Lambda} \begin{pmatrix} a \\ b \\ c \end{pmatrix} \quad \textit{and} \quad \begin{pmatrix} l_2 \\ m_2 \\ n_2 \end{pmatrix} = \mathbf{\Lambda} \begin{pmatrix} e \\ f \\ g \end{pmatrix}$$

with

$$\mathbf{\Lambda} = \begin{pmatrix} \alpha_1^2 & 2\alpha_1\beta_1 & \beta_1^2 \\ \alpha_1\alpha_2 & \alpha_1\beta_2 + \alpha_2\beta_1 & \beta_1\beta_2 \\ \alpha_2^2 & 2\alpha_2\beta_2 & \beta_2^2 \end{pmatrix} \quad \textit{and} \quad \det \mathbf{\Lambda} = (\det \Omega)^3.$$

Proof. The proof is an elementary linear algebra exercise . Let $\mathbf{w} \in T_pM$ and denote by (w_1, w_2) (resp. $(\bar{w}_1, \bar{w}_2)$) its coordinates with respect to the basis $\{\mathbf{e}_1, \mathbf{e}_2\}$ (resp. $\{\mathbf{f}_1, \mathbf{f}_2\}$) of T_pM. We have $w_1 = \alpha_1\bar{w}_1 + \alpha_2\bar{w}_2$ and $w_2 = \beta_1\bar{w}_1 + \beta_2\bar{w}_2$ and $\mathrm{II}_p(w_1, w_2) = (aw_1^2 + 2bw_1w_2 + cw_2)\mathbf{e}_3 + (ew_1^2 + 2ew_1w_2 + gw_2)\mathbf{e}_4$ so that

$$\begin{aligned} \langle \mathrm{II}_p(w_1, w_2), \mathbf{e}_3 \rangle &= a(\alpha_1\bar{w}_1 + \alpha_2\bar{w}_2)^2 + 2b(\alpha_1\bar{w}_1 + \alpha_2\bar{w}_2)(\beta_1\bar{w}_1 + \beta_2\bar{w}_2) \\ &\quad + c(\beta_1\bar{w}_1 + \beta_2\bar{w}_2)^2 \\ &= (\alpha_1^2 a + 2\alpha_1\beta_1 b + \beta_1^2 c)\bar{w}_1^2 \\ &\quad + 2(\alpha_1\alpha_2 a + (\alpha_1\beta_2 + \alpha_2\beta_1)b + \beta_1\beta_2 c)\bar{w}_1\bar{w}_2 \\ &\quad + (\alpha_2^2 a + 2\alpha_2\beta_2 b + \beta_2^2 c)\bar{w}_2^2. \end{aligned}$$

It follows that

$$\begin{aligned} l_1 &= \alpha_1^2 a + 2\alpha_1\beta_1 b + \beta_1^2 c, \\ m_1 &= \alpha_1\alpha_2 a + (\alpha_1\beta_2 + \alpha_2\beta_1)b + \beta_1\beta_2 c, \\ n_1 &= \alpha_2^2 a + 2\alpha_2\beta_2 b + \beta_2^2 c. \end{aligned}$$

The coefficients l_2, m_2, n_2 can be obtained similarly using the scalar product or $\mathrm{II}_p(w_1, w_2)$ with $\mathbf{e}_4$. $\square$

Remark 7.1. (1). The change of basis in Proposition 7.1 leaves the curvature ellipse unchanged as a point set. It is parametrised as in (7.3) where a, b, c, e, f, g in $\mathbf{H}$ and $\mathbf{A}$ are replaced by their expressions in terms of $l_1, m_1, n_1, l_2, m_2, n_2$ obtained by inverting the matrix Λ in Proposition 7.1.

(2). If we take $\mathbf{f}_1 = \mathbf{x}_{u_1}$ and $\mathbf{f}_2 = \mathbf{x}_{u_2}$, then we can use the orthonormal basis $\{\mathbf{e}_1, \mathbf{e}_2\}$ in T_pM, with $\mathbf{f}_1 = \alpha_1\mathbf{e}_1$, $\mathbf{f}_2 = \alpha_2\mathbf{e}_1 + \beta_2\mathbf{e}_2$ and

$$\begin{pmatrix} \alpha_1 & \alpha_2 \\ 0 & \beta_2 \end{pmatrix} = \begin{pmatrix} \sqrt{E} & \frac{F}{E} \\ 0 & \frac{\sqrt{EG - F^2}}{\sqrt{E}} \end{pmatrix},$$

with E, F, G the coefficients of the first fundamental form with respect to the parametrisation $\mathbf{x}$.

We turn now to the action of $GL(2,\mathbb{R})$ on N_pM and still represent by $\Omega = \begin{pmatrix} \alpha_1 & \alpha_2 \\ \beta_1 & \beta_2 \end{pmatrix}$ and element in $\mathrm{GL}(2,\mathbb{R})$. We write

$$\mathbf{f}_3 = \alpha_1\mathbf{e}_3 + \beta_1\mathbf{e}_4,$$
$$\mathbf{f}_4 = \alpha_2\mathbf{e}_3 + \beta_2\mathbf{e}_4.$$

and denote by

$$a = \langle \mathbf{x}_{u_1u_1}, \mathbf{e}_3\rangle, \quad b = \langle \mathbf{x}_{u_1u_2}, \mathbf{e}_3\rangle, \quad c = \langle \mathbf{x}_{u_2u_2}, \mathbf{e}_3\rangle,$$
$$e = \langle \mathbf{x}_{u_1u_1}, \mathbf{e}_4\rangle, \quad f = \langle \mathbf{x}_{u_1u_2}, \mathbf{e}_4\rangle, \quad g = \langle \mathbf{x}_{u_2u_2}, \mathbf{e}_4\rangle$$

and

$$l_1 = \langle \mathbf{x}_{u_1u_1}, \mathbf{f}_3\rangle, \quad m_1 = \langle \mathbf{x}_{u_1u_2}, \mathbf{f}_3\rangle, \quad n_1 = \langle \mathbf{x}_{u_2u_2}, \mathbf{f}_3\rangle,$$
$$l_2 = \langle \mathbf{x}_{u_1u_1}, \mathbf{f}_4\rangle, \quad m_2 = \langle \mathbf{x}_{u_1u_2}, \mathbf{f}_4\rangle, \quad n_2 = \langle \mathbf{x}_{u_2u_2}, \mathbf{f}_4\rangle$$

the coefficients of the second fundamental form with respect to the basis $\{\mathbf{e}_3, \mathbf{e}_4\}$ and $\{\mathbf{f}_3, \mathbf{f}_4\}$ respectively. Also denote by

$$E_n = \langle \mathbf{f}_3, \mathbf{f}_3\rangle, \quad F_n = \langle \mathbf{f}_3, \mathbf{f}_4\rangle, \quad G_n = \langle \mathbf{f}_4, \mathbf{f}_4\rangle.$$

Proposition 7.2. *Denote by a, b, c, e, f, g the coefficients of the second fundamental form with respect to the basis $\{\mathbf{e}_1, \mathbf{e}_2, \mathbf{e}_3, \mathbf{e}_4\}$ and by $l_1, m_1, n_1, l_2, m_2, n_2$ its coefficients with respect to the basis $\{\mathbf{e}_1, \mathbf{e}_2, \mathbf{f}_3, \mathbf{f}_4\}$. Then*

$$\begin{pmatrix} a & b & c \\ e & f & g \end{pmatrix} = \begin{pmatrix} \alpha_1 & \alpha_2 \\ \beta_1 & \beta_2 \end{pmatrix} \begin{pmatrix} E_n & F_n \\ F_n & G_n \end{pmatrix}^{-1} \begin{pmatrix} l_1 & m_1 & n_1 \\ l_2 & m_2 & n_2 \end{pmatrix}.$$

Proof. We have

$$\mathrm{II}_p(w_1, w_2) = \mathbf{x}_{u_1u_1}w_1^2 + 2\mathbf{x}_{u_1u_2}w_1w_2 + \mathbf{x}_{u_2u_2}w_2^2 = \mu_1\mathbf{f}_3 + \mu_2\mathbf{f}_4.$$

The scalar product with $\mathbf{f}_3$ and $\mathbf{f}_4$ gives

$$\langle \mathrm{II}_p(w_1, w_2), \mathbf{f}_3\rangle = l_1w_1^2 + 2m_1w_1w_2 + n_1w_2^2 = \mu_1E_n + \mu_2F_n,$$
$$\langle \mathrm{II}_p(w_1, w_2), \mathbf{f}_4\rangle = l_2w_1^2 + 2m_2w_1w_2 + n_2w_2^2 = \mu_1F_n + \mu_2G_n$$

so that

$$\begin{pmatrix} \mu_1 \\ \mu_2 \end{pmatrix} = \begin{pmatrix} E_n & F_n \\ F_n & G_n \end{pmatrix}^{-1} \begin{pmatrix} l_1w_1^2 + 2m_1w_1w_2 + n_1w_2^2 \\ l_2w_1^2 + 2m_2w_1w_2 + n_2w_2^2 \end{pmatrix}.$$

We also have

$$\begin{aligned} \mathrm{II}_p(w_1, w_2) &= \mu_1\mathbf{f}_3 + \mu_2\mathbf{f}_4 \\ &= \mu_1(\alpha_1\mathbf{e}_3 + \beta_1\mathbf{e}_4) + \mu_2(\alpha_2\mathbf{e}_3 + \beta_2\mathbf{e}_4) \\ &= (\alpha_1\mu_1 + \alpha_2\mu_2)\mathbf{e}_3 + (\beta_1\mu_1 + \beta_2\mu_2)\mathbf{e}_4 \\ &= (aw_1^2 + 2bw_1w_2 + cw_2^2)\mathbf{e}_3 + (ew_1^2 + 2fw_1w_2 + gw_2^2)\mathbf{e}_4 \end{aligned}$$

and from it we get

$$\begin{pmatrix} aw_1^2 + 2bw_1w_2 + cw_2^2 \\ ew_1^2 + 2fw_1w_2 + gw_2^2 \end{pmatrix} = \begin{pmatrix} \alpha_1 & \alpha_2 \\ \beta_1 & \beta_2 \end{pmatrix} \begin{pmatrix} E_n & F_n \\ F_n & G_n \end{pmatrix}^{-1} \begin{pmatrix} l_1w_1^2 + 2m_1w_1w_2 + n_1w_2^2 \\ l_2w_1^2 + 2m_2w_1w_2 + n_2w_2^2 \end{pmatrix}$$

which gives

$$\begin{pmatrix} a & b & c \\ e & f & g \end{pmatrix} = \begin{pmatrix} \alpha_1 & \alpha_2 \\ \beta_1 & \beta_2 \end{pmatrix} \begin{pmatrix} E_n & F_n \\ F_n & G_n \end{pmatrix}^{-1} \begin{pmatrix} l_1 & m_1 & n_1 \\ l_2 & m_2 & n_2 \end{pmatrix}.$$

□

Proposition 7.3. *With notation as in Proposition 7.2, the image of the curvature ellipse by the affine transformation Ω is an ellipse parametrised by*

$$\eta(\theta) = \begin{pmatrix} \mu_1 \\ \mu_2 \end{pmatrix} = \begin{pmatrix} \alpha_1 & \alpha_2 \\ \beta_1 & \beta_2 \end{pmatrix}^{-1} \mathbf{H} + \begin{pmatrix} \alpha_1 & \alpha_2 \\ \beta_1 & \beta_2 \end{pmatrix}^{-1} \mathbf{A} \begin{pmatrix} \cos 2\theta \\ \sin 2\theta \end{pmatrix} \tag{7.5}$$

with respect to the basis $\{\mathbf{f}_3, \mathbf{f}_4\}$ of N_pM.

Proof. The curvature ellipse is parametrised by

$$\begin{aligned} \eta(\theta) &= \mathbf{x}_{u_1u_1} \cos^2\theta + 2\mathbf{x}_{u_1u_2} \cos\theta \sin\theta + \mathbf{x}_{u_2u_2} \sin^2\theta \\ &= \mu_1 \mathbf{f}_3 + \mu_2 \mathbf{f}_4. \end{aligned}$$

It follows from the proof of Proposition 7.2 that

$$\begin{aligned} \begin{pmatrix} E_n & F_n \\ F_n & G_n \end{pmatrix} \begin{pmatrix} \mu_1 \\ \mu_2 \end{pmatrix} &= \begin{pmatrix} l_1 \cos^2\theta + 2m_1 \cos\theta \sin\theta + n_1 \sin^2\theta \\ l_2 \cos^2\theta + 2m_2 \cos\theta \sin\theta + n_2 \sin^2\theta \end{pmatrix} \\ &= \bar{\mathbf{H}} + \bar{\mathbf{A}} \begin{pmatrix} \cos 2\theta \\ \sin 2\theta \end{pmatrix}, \end{aligned}$$

with

$$\bar{\mathbf{H}} = \begin{pmatrix} E_n & F_n \\ F_n & G_n \end{pmatrix} \begin{pmatrix} \alpha_1 & \alpha_2 \\ \beta_1 & \beta_2 \end{pmatrix}^{-1} \mathbf{H} \quad \text{and} \quad \bar{\mathbf{A}} = \begin{pmatrix} E_n & F_n \\ F_n & G_n \end{pmatrix} \begin{pmatrix} \alpha_1 & \alpha_2 \\ \beta_1 & \beta_2 \end{pmatrix}^{-1} \mathbf{A}.$$

Multiplying both sides with $\begin{pmatrix} E_n & F_n \\ F_n & G_n \end{pmatrix}^{-1}$ gives the parametrisation (7.5) of the image of curvature ellipse by the transformation Ω with respect to the basis $\{\mathbf{f}_3, \mathbf{f}_4\}$. □

Remark 7.2. It is clear from (7.5) that some properties of the curvature ellipse (Definition 7.2) remain invariant under the action of $GL(2, \mathbb{R})$ on N_pM. For instance, the position of the point p with respect to the curvature ellipse is an affine invariant property. The concept of an inflection point is also affine invariant. However, semiumbilicity is not affine invariant.

Table 7.1 The $GL(2,\mathbb{R}) \times GL(2,\mathbb{R})$-classes of pairs of quadratic forms.

$GL(2,\mathbb{R}) \times GL(2,\mathbb{R})$-class	Name
(x^2, y^2)	hyperbolic point
$(xy, x^2 - y^2)$	elliptic point
(x^2, xy)	parabolic point
$(x^2 \pm y^2, 0)$	inflection point
$(x^2, 0)$	degenerate inflection
$(0, 0)$	degenerate inflection

The position of p with respect to the curvature ellipse is determined by the sign of $\Delta(p)$. Given a parametrisation $\mathbf{x}$, denote by $l_1, m_1, n_1, l_2, m_2, n_2$ the coefficients of the second fundamental form with respect to any basis $\{\mathbf{x}_{u_1}, \mathbf{x}_{u_2}, \mathbf{f}_3, \mathbf{f}_4\}$ of $T_pM \times N_pM$, and define $\tilde{\Delta}$ by

$$\tilde{\Delta} = \frac{1}{4} \begin{vmatrix} l_1 & 2m_1 & n_1 & 0 \\ l_2 & 2m_2 & n_2 & 0 \\ 0 & l_1 & 2m_1 & n_1 \\ 0 & l_2 & 2m_2 & n_2 \end{vmatrix} \tag{7.6}$$

$$= \frac{1}{4}\left(4(l_1m_2 - l_2m_1)(m_1n_2 - m_2n_1) - (l_1n_2 - l_2n_1)^2\right).$$

It follows from Propositions 7.1 and 7.2 that $\tilde{\Delta} < 0$ (resp $= 0$, > 0) if and only if $\Delta < 0$ (resp $= 0$, > 0), as the two determinant differ by an even power of the determinant of Ω.

7.2.1 *Pencils of quadratic forms*

The second order affine invariants of M at p are described by the class of the second fundamental form

$$\mathrm{II}_p(w_1, w_2) = (l_1w_1^2 + 2m_1w_1w_2 + n_1w_2^2, l_2w_1^2 + 2m_2w_1w_2 + n_2w_2^2)$$

under the action of $GL(2,\mathbb{R}) \times GL(2,\mathbb{R})$. The second fundamental form at p is a pair of quadratic forms (Q_1, Q_2) and the action of $GL(2,\mathbb{R}) \times GL(2,\mathbb{R})$ on these pairs is as given in Propositions 7.1 and 7.2.

Theorem 7.2 ([Gibson (1979)]). *The orbits of the action of $GL(2,\mathbb{R}) \times GL(2,\mathbb{R})$ on the set of pairs of quadratic forms are as in* Table 7.1.

We have the following characterisation of the points of M in terms of the sign of the invariant Δ.

Proposition 7.4. *Let M be a smooth surface in $\mathbb{R}^4$ and p a point on M. Then*

(i) $\Delta(p) < 0$ *if and only if* p *is a hyperbolic point.*
(ii) $\Delta(p) > 0$ *if and only if* p *is a parabolic point.*
(iii) $\Delta(p) = 0$ *and* $\operatorname{rank}\alpha(p) = 2$ *if and only if* p *is an elliptic point.*
(iv) $\Delta(p) = 0$ *and* $\operatorname{rank}\alpha(p) < 2$ *if and only if* p *is an inflection point.*

Proof. All the properties in the statement of the proposition are invariant under the action of $GL(2,\mathbb{R})\times GL(2,\mathbb{R})$ on $T_pM\times N_pM$. Thus, it is enough to take (Q_1, Q_2) as one of the normal forms in Table 7.1. After that, the proof follows by straightforward calculations. □

Corollary 7.1. *The inflection points are the common solutions of the* 2×2 *minors of the matrix*

$$\alpha(u) = \begin{pmatrix} l_1 & m_1 & n_1 \\ l_2 & m_2 & n_2 \end{pmatrix}(u),$$

of the second fundamental form. That is, $p = \mathbf{x}(u)$ *is an inflection point if and only if*

$$\begin{aligned}(l_1m_2 - m_1l_2)(u) &= 0,\\ (l_1n_2 - n_1l_2)(u) &= 0,\\ (m_1n_2 - n_1m_2)(u) &= 0.\end{aligned}$$

A quadratic form $Aw_1^2 + 2Bw_1w_2 + Cw_2^2$ can be represented by the point $Q = (A : B : C)$ in the real projective plane $\mathbb{R}P^2$. Denote by $\Gamma = \{Q \mid B^2 - AC = 0\}$ the conic of degenerate quadratic forms in $\mathbb{R}P^2$.

The second order affine invariant of M at p can also be described via pencils of quadratic forms determined by (Q_1, Q_2) ([Bruce and Nogueira (1998)] and also [Mochida (1993)]).

Proposition 7.5. *With notation as above,*

(i) *If* Q_1, Q_2 *are distinct, they determine a line in* $\mathbb{R}P^2$ *which meets the conic* Γ *in* 0 *(resp.* 1, 2*) points according to* $\delta(p) < 0$ *(resp.* $= 0, > 0$*) with*

$$\delta(p) = (l_1n_2 - l_2n_1)^2 - 4(l_1m_2 - l_2m_1)(m_1n_2 - n_1m_2) = -4\tilde{\Delta}(p).$$

In particular, p *is a hyperbolic (resp. parabolic, elliptic) point if and only if the pencil intersect the conic* Γ *in* 2 *(resp.* 1, 0*) points.*

(ii) *If the quadratic forms* Q_1 *and* Q_2 *are linearly dependent but not identically zero, the pencil determines a point in* $\mathbb{R}P^2$ *which may lie inside, on or outside the conic* Γ.

Proof. (i) If Q_1 and Q_2 are distinct, they determine a line in $\mathbb{R}P^2$ parametrized by $\alpha Q_1 + \beta Q_2$, with $(\alpha : \beta) \in \mathbb{R}P^1$. The intersection points of this line with Γ are given by the solutions of the quadratic equation

$$(\alpha l_1 + \beta l_2)(\alpha n_1 + \beta n_2) - (\alpha m_1 + \beta m_2)^2 = 0,$$

equivalently,

$$(l_1 n_1 - m_1^2)\alpha^2 - (l_1 n_2 + l_2 n_1 - 2m_1 m_2)\alpha\beta + (l_2 n_2 - m_2^2)\beta^2 = 0.$$

The number of the real solutions of the above equation is determined by its discriminant

$$\begin{aligned}\delta(p) &= (l_1 n_2 + l_2 n_1 - 2m_1 m_2)^2 - 4(l_1 n_1 - m_1^2)(l_2 n_2 - m_2^2)\\ &= (l_1 n_2 - l_2 n_1)^2 - 4(l_1 m_2 - l_2 m_1)(m_1 n_2 - n_1 m_2)\\ &= -4\tilde{\Delta}(p).\end{aligned}$$

The rest follows by Proposition 7.4.

(ii) Here Q_2 and Q_1 determine the same point in the projective plane. The point is outside (resp. on, inside) Γ if the quadratic forms have two (resp. one, no) real roots. □

7.3 Asymptotic directions

Definition 7.3. Let $\eta(\theta)$ be the parametrisation (7.2) of the curvature ellipse. A tangent direction θ at a point $p \in M$ is an *asymptotic direction* at p if $\eta(\theta)$ and $\frac{d\eta}{d\theta}(\theta)$ are linearly dependent vectors in N_pM.

A curve on M whose tangent direction at each point is an asymptotic direction is called an *asymptotic curve.*

By definition, a direction θ is asymptotic if the line joining p and $\eta(\theta)$ is tangent to the curvature ellipse. This is invariant under the action of $GL(2, \mathbb{R})$ in the normal plane, so the concept of asymptotic direction is affine invariant (we will show this too using the contact of the surface with hyperplanes, planes and lines). This means that there are 2 asymptotic directions at a hyperbolic point, one at a parabolic point and none at an elliptic point.

Theorem 7.3. *Let* $\mathbf{x} : U \to \mathbb{R}^4$ *be a local parametrisation of a surface* M *and denote by* $l_1, m_1, n_1, l_2, m_2, n_2$ *the coefficients of its second fundamental form with respect to any frame* $\{\mathbf{x}_{u_1}, \mathbf{x}_{u_2}, \mathbf{f}_3, \mathbf{f}_4\}$ *of* $T_pM \times N_pM$ *which* *depends smoothly on* $p = \mathbf{x}(u_1, u_2)$*. Then the asymptotic curves of* M *are the solution curves of the binary differential equation*

$$(l_1m_2 - l_2m_1)du_1^2 + (l_1n_2 - l_2n_1)du_1du_2 + (m_1n_2 - m_2n_1)du_2^2 = 0, \quad (7.7)$$

which can also be written in the following determinant form

$$\begin{vmatrix} du_2^2 & -du_1du_2 & du_1^2 \\ l_1 & m_1 & n_1 \\ l_2 & m_2 & n_2 \end{vmatrix} = 0.$$

Proof. We first write the second fundamental form with respect to an orthonormal frame $\{\mathbf{e}_1, \mathbf{e}_2, \mathbf{e}_3, \mathbf{e}_4\}$ of $T_pM \times N_pM$ at p. By definition, a direction θ at $p \in M$ is asymptotic if and only if

$$\eta(\theta) = \lambda \frac{\partial \eta}{\partial \theta}(\theta),$$

with $\eta(\theta)$ as in (7.2). Differentiating the above equality gives

$$l_1\cos^2\theta + 2m_1\cos\theta\sin\theta + n_1\sin^2\theta = 2\lambda((n_1 - l_1)\cos\theta\sin\theta + m_1(\cos^2\theta - \sin^2\theta))$$
$$l_2\cos^2\theta + 2m_2\cos\theta\sin\theta + n_2\sin^2\theta = 2\lambda((n_2 - l_2)\cos\theta\sin\theta + m_2(\cos^2\theta - \sin^2\theta))$$

Eliminating λ gives

$$(af - be)\cos^2\theta + (ag - ce)\cos\theta\sin\theta + (bg - cf)\sin^2\theta = 0. \quad (7.8)$$

Consider now a frame $\{\mathbf{x}_{u_1}, \mathbf{x}_{u_2}, \mathbf{e}_3, \mathbf{e}_4\}$ of $T_pM \times N_pM$, write

$$\begin{pmatrix} \cos\theta \\ \sin\theta \end{pmatrix} = \Omega \begin{pmatrix} du_1 \\ du_2 \end{pmatrix} = \begin{pmatrix} \alpha_1 & \alpha_2 \\ \beta_1 & \beta_2 \end{pmatrix} \begin{pmatrix} du_1 \\ du_2 \end{pmatrix}$$

and use the relations in Proposition 7.1. Then equation (7.8) becomes

$$\frac{1}{\det\Omega}\left((l_1m_2 - l_2m_1)du_1^2 + (l_1n_2 - l_2n_1)du_1du_2 + (m_1n_2 - m_2n_1)du_2^2\right) = 0,$$

that is

$$(l_1m_2 - l_2m_1)du_1^2 + (l_1n_2 - l_2n_1)du_1du_2 + (m_1n_2 - m_2n_1)du_2^2 = 0.$$

Similarly, using the relations in Proposition 7.2, equation (7.8) becomes equation (7.7) with respect to a frame $\{\mathbf{e}_1, \mathbf{e}_2, \mathbf{f}_3, \mathbf{f}_4\}$ in $T_pM \times N_pM$. □

An immediate consequence of Theorem 7.3 is the following.

Corollary 7.2. *Let* M *be a smooth surface in* $\mathbb{R}^4$*.*

(i) *The coefficients of the binary differential equation of the asymptotic curves (7.7) all vanish at a point* p *if and only if* p *is an inflection point. At such points all tangent directions are asymptotic directions.*

(ii) *The discriminant of equation* (7.7) *is* $-4\tilde{\Delta}(u_1, u_2)$. *Consequently, there are* 2 (*resp.* 1, 0) *asymptotic directions at hyperbolic* (*resp. not inflection parabolic, elliptic*) *points.*

Remark 7.3. Asymptotic curves determine a pair of regular foliations on the hyperbolic region of the surface. At generic points on the parabolic curve, they form a family of cusps with the cusps tracing the parabolic set. At isolated parabolic points (see Remark 7.7), the configuration of the asymptotic curves is as in Figure 7.2 ([Bruce and Tari (2002)]). At inflection points, the parabolic set generically has a Morse singularity. The singularity is of type A_1^+ at an inflection point of imaginary type and of type A_1^- at an inflection point of real type. The generic configurations of the asymptotic curves at an inflection point of imaginary type are as in Figure 7.3 ([Garcia, Mochida, Romero Fuster and Ruas (2000)]) and those at an inflection point of real type are as in Figure 7.4 ([Bruce and Tari (2002)]).

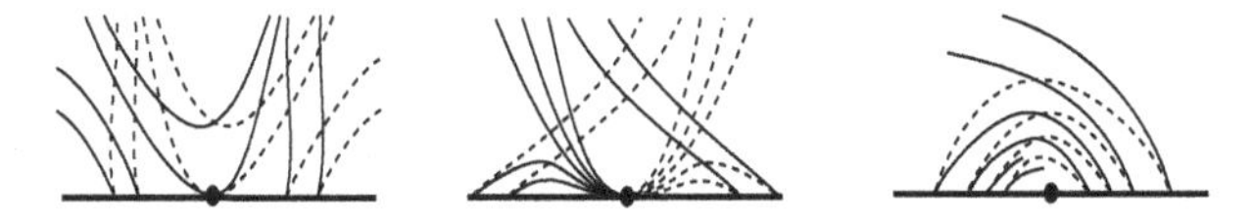

Fig. 7.2 Topological configurations of the asymptotic curves at special parabolic points.

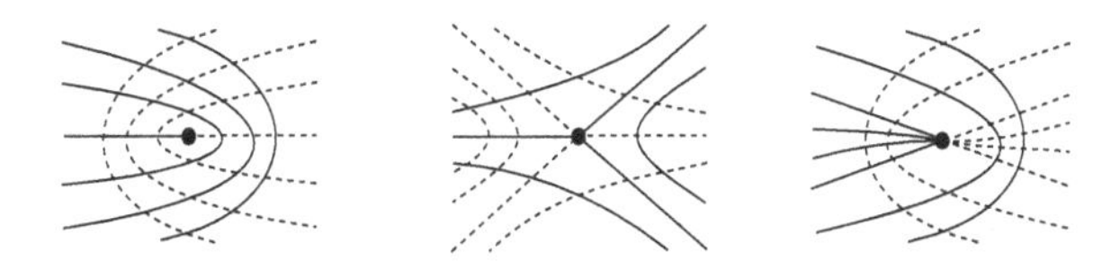

Fig. 7.3 Generic topological configurations of the asymptotic curves at an inflection point of imaginary type.

Equation (7.7) is, at each point on the surface, a quadratic form in du_1, du_2. As multiplying the equation by a nowhere vanishing function gives an equation with the same solutions as the original one, we can represent (7.7), at each point on the surface, by a point $A = (\alpha : \beta : \gamma)$ in the real projective plane $\mathbb{R}P^2$, with $\alpha = l_1 m_2 - l_2 m_1$, $\beta = l_1 n_2 - l_2 n_1$, $\gamma = m_1 n_2 - m_2 n_1$ the coefficients of (7.7).

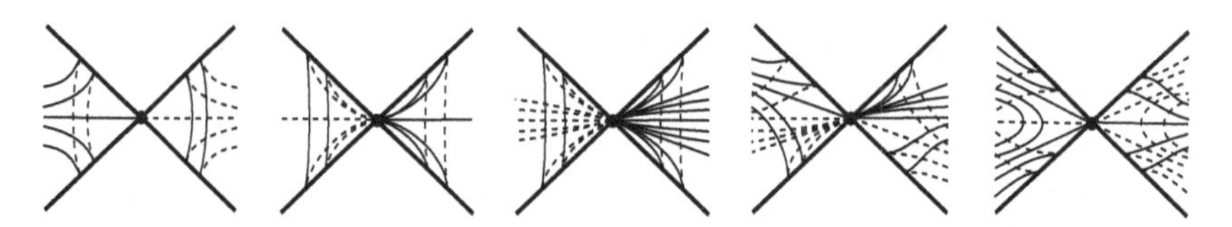

Fig. 7.4 Generic topological configurations of the asymptotic curves at an inflection point of real type.

The *polar line* $\widehat{Q}$ of a point Q in $\mathbb{R}P^2$ (with respect to the conic Γ of degenerate quadratic forms) is the line that contains all points Q such that Q and Q' are harmonic conjugate points with respect to the intersection points R_1 and R_2 of the conic Γ and a variable line through Q. Geometrically, if the polar line $\widehat{Q}$ meets Γ, then the tangents to Γ at the points of intersection meet at Q (Figure 7.5). Furthermore, if $Q = \omega_1\omega_2$, then ω_1^2 and ω_2^2 are the points of intersection of $\widehat{Q}$ with Γ. A point $(a' : 2b' : c')$ is on the polar line of a point $Q = (a : b : c)$, if and only if $ac' - bb' + ca' = 0$. Three points in the projective plane are said to form a *self-polar triangle* if the polar of any vertex of the triangle is the line through the remaining two points.

Let $\{\mathbf{x}_{u_1}, \mathbf{x}_{u_2}, \mathbf{e}_3, \mathbf{e}_4\}$ be a frame in $\mathbb{R}^4$ with $\{\mathbf{e}_3, \mathbf{e}_4\}$ and orthonormal basis of N_pM. Denote by $l_i = \langle \mathbf{x}_{u_1u_1}, \mathbf{e}_{2+i}\rangle$, $m_i = \langle \mathbf{x}_{u_1u_2}, \mathbf{e}_{2+i}\rangle$, $n_i = \langle \mathbf{x}_{u_2u_2}, \mathbf{e}_{2+i}\rangle$, $i = 1, 2$, the coefficient of the second fundamental form with respect to the above frame.

Given a vector $\mathbf{v} = \lambda_1\mathbf{e}_3 + \lambda_2\mathbf{e}_4$ in N_pM, consider the shape operator $W_p^\nu : T_pM \to T_pM$ along any normal vector field with $\nu(p) = \mathbf{v}$. Then the matrix of W_p^ν with respect to the basis $\{\mathbf{x}_{u_1}, \mathbf{x}_{u_2}\}$ is the symmetric matrix

$$\begin{pmatrix} \lambda_1 l_1 + \lambda_2 l_2 & \lambda_1 m_1 + \lambda_2 m_2 \\ \lambda_1 m_1 + \lambda_2 m_2 & \lambda_1 n_1 + \lambda_2 n_2 \end{pmatrix}.$$

We can represent W_p^ν by the point $(\lambda_1 l_1 + \lambda_2 l_2 : 2(\lambda_1 m_1 + \lambda_2 m_2) : \lambda_1 n_1 + \lambda_2 n_2)$ in $\mathbb{R}^4$. Varying $(\lambda_1 : \lambda_2) \in \mathbb{R}P^1$ gives a pencil in $\mathbb{R}P^1$, which we call the pencil of the ν-shape operators at p.

Theorem 7.4. *At each point p on a surface M, the polar line of the asymptotic curves BDE* (7.7) *is the pencil of the ν-shape operators at p; see* Figure 7.5.

Proof. A point $Q = (a' : 2b' : c')$ is on the pencil of the ν-shape operators at p if and only if there exists λ_1 and λ_2 such that

$$\lambda_1 l_1 + \lambda_2 l_2 = a', \quad \lambda_1 m_1 + \lambda_2 m_2 = b', \quad \lambda_1 n_1 + \lambda_2 n_2 = c'.$$

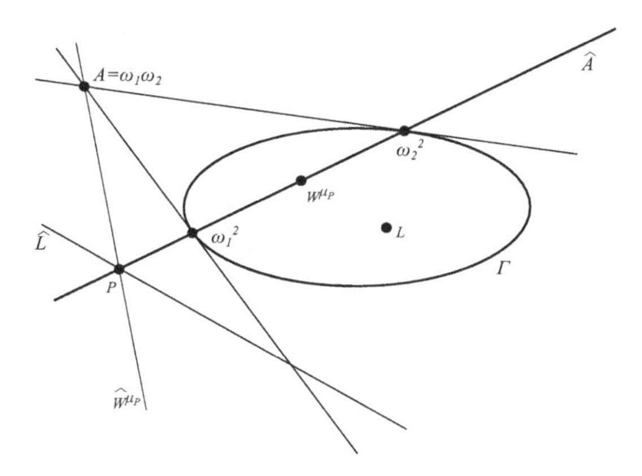

Fig. 7.5 Polar line of the asymptotic BDE tracing the pencil of ν-shape operators, and the determination of a special shape operator.

The above system of three equations in λ_1 and λ_2 has a solution if and only if

$$(l_1m_2 - l_2m_1)c' - (l_1n_2 - l_2n_1)b' + (m_1n_2 - m_2n_1)c' = 0$$

which is precisely the condition for the point Q to be on the polar line of asymptotic BDE A. □

Remark 7.4. (1) The induced metric on M can also be represented at each point by the point $L = (E : 2F : G)$ in $\mathbb{R}P^2$. The point L lies inside the conic Γ as the metric is Riemannian. Its polar line $\hat{L}$ represents BDE with orthogonal solutions at p and intersect the polar line $\hat{A}$ of the asymptotic BDE (7.7) at a unique point P (Figure 7.5). There is a unique point $W_p^{\nu_P}$ on $\hat{A}$ such that $A, C, W_p^{\nu_P}$ form a self-polar triangle (see Figure 7.5). In fact the solutions of P are the ν_P-principal directions at p. This construction gives a way of choosing a unique pair of orthogonal foliation on M coming from a ν-shape operator, which can be called the lines of principal curvature of M.

(2) The above construction is in fact inspired from and is analogous to that for surfaces in $\mathbb{R}^3$. In that case too the asymptotic curves are given by a *BDE* A and the lines of principal curvatures by a *BDE* P (see Chapter 6). There is a unique *BDE* C such that A, P, C form a self-polar triangle, which is the BDE of characteristic curves ([Bruce and Tari (2005)]). In fact using the metric L as above, the triple A, P, C is completely determined by A. The construction runs into difficulties for surfaces immersed in higher

dimensional Euclidean spaces. For surfaces in $\mathbb{R}^5$ the asymptotic curves are given a binary quintic differential equation ([Mochida, Romero Fuster and Ruas (1999); Romero Fuster, Ruas and Tari (2008)]). An attempt to choose special ν-lines of curvature is made in [Ruas and Tari (2012)] using invariants and covariants of binary forms (see for example [Olver (1999)]).

(3) More on binary differential equations and polarity in the projective plane with respect to Γ can be found in [Bruce and Tari (2005); Oliver (2010, 2011); Tari (2009)]. The case when the metric is Lorentzian (L is outside Γ) is treated in [Nabarro and Tari (2011)].

7.4 Surfaces in Monge form

At each point p on the surface M, we can choose a coordinate system $q = (x, y, z, w)$ in $\mathbb{R}^4$ so that p is the origin, T_pM is the plane $z = w = 0$ and the surface M is locally the graph of some smooth function $f : U \to \mathbb{R}^2$, with (x, y) in a open subset U of $\mathbb{R}^2$ containing the origin. Let (f_1, f_2) be the coordinate functions of f. Then we have the Monge form parametrisation

$$\phi(x, y) = (x, y, f_1(x, y), f_2(x, y)),$$

of M at p. The Taylor expansions of f_1 and f_2 at the origin have no constant or linear terms. We use the following notation for the Taylor polynomials of order k at the origin of f_1 and f_2

$$\begin{aligned} j^k f_1(x, y) &= a_{20}x^2 + a_{21}xy + a_{22}y^2 + \textstyle\sum_{i=3}^{i=k} \sum_{j=0}^{j=i} a_{ij}x^{i-j}y^j, \\ j^k f_2(x, y) &= b_{20}x^2 + b_{21}xy + b_{22}y^2 + \textstyle\sum_{i=3}^{i=k} \sum_{j=0}^{j=i} b_{ij}x^{i-j}y^j. \end{aligned}$$

The following is straightforward.

Lemma 7.1. *Let M be parametrised locally in Monge form $\phi(x, y) = (x, y, f_1(x, y), f_2(x, y))$ at the origin p. Then,*

(i) $E = 1 + f_{1x}^2 + f_{2x}^2$, $F = f_{1x}f_{1y} + f_{2x}f_{2y}$, $G = 1 + f_{1y}^2 + f_{2y}^2$.
(ii) *We can choose the frame $\mathbf{F} = \{\mathbf{f}_1, \mathbf{f}_2, \mathbf{f}_3, \mathbf{f}_4\}$ in U with*

$$\begin{aligned} \mathbf{f}_1 &= \phi_x = (1, 0, f_{1x}, f_{2x}) \\ \mathbf{f}_2 &= \phi_y = (0, 1, f_{1y}, f_{2y}) \\ \mathbf{f}_3 &= (-f_{1x}, -f_{1y}, 1, 0) \\ \mathbf{f}_4 &= (-f_{2x} + f_{1y}(f_{1x}f_{2y} - f_{1y}f_{2x}), -f_{2y} - f_{1x}(f_{1x}f_{2y} - f_{1y}f_{2x}), \\ &\quad -f_{1x}f_{2x} - f_{1y}f_{2y}, 1 + {f_{1x}}^2 + {f_{2x}}^2) \end{aligned}$$

The frame $\mathbf{F}$ at the origin is the standard basis of $\mathbb{R}^4$.

Proposition 7.6. *Let M be parametrised locally in Monge form $\phi(x,y) = (x, y, f_1(x,y), f_2(x,y))$ at the origin p. The second fundamental form at p with respect to the frame* **F** *in* Lemma 7.1 *is given by*

$$\mathrm{II_p}(\mathbf{w}) = 2(a_{20}w_1^2 + a_{21}w_1w_2 + a_{22}w_2^2, b_{20}w_1^2 + b_{21}w_1w_2 + b_{22}w_2^2), \quad (7.9)$$

with $\mathbf{w} = (w_1, w_2, 0, 0)$ *in* T_pM, *and*

$$\Delta(p) = 4\left((a_{20}b_{21} - b_{20}a_{21})(a_{21}b_{22} - b_{21}a_{22}) - (a_{20}b_{22} - b_{20}a_{22})^2\right).$$

Proof. The frame **F** at the origin is the standard basis of $\mathbb{R}^4$, so the coefficients of the second fundamental form of M at the origin are

$$l_1 = f_{1xx}(0) = 2a_{20},\ m_1 = f_{1xy}(0) = a_{21},\ n_1 = f_{1yy}(0) = 2a_{22},$$
$$l_2 = f_{2xx}(0) = 2b_{20},\ m_2 = f_{2xy}(0) = b_{21},\ n_2 = f_{2yy}(0) = 2b_{22}. \qquad \square$$

7.5 Examples of surfaces in $\mathbb{R}^4$

We present below three examples of orientable surfaces in 4-space, namely surfaces in S^3, complex curves and S^1-bundles, taken from [Garcia, Mochida, Romero Fuster and Ruas (2000)].

Example 7.1 (Surfaces in S^3). Examples of surfaces in the unit sphere $S^3 \subset \mathbb{R}^4$ can be obtained as images of surfaces in 3-space by the inverse of the stereographic projection $\varphi : \mathbb{R}^3 \to S^3 \subset \mathbb{R}^4$.

Let (x, y, z) and (X, Y, Z, W) denote the coordinates of $\mathbb{R}^3$ and $\mathbb{R}^4$, respectively. Then, $\varphi(x, y, z) = (X, Y, Z, W)$, with

$$X = \frac{2x}{1+\rho},\ Y = \frac{2y}{1+\rho},\ Z = \frac{2z}{1+\rho},\ W = \frac{-1+\rho}{1+\rho}$$

and $\rho = x^2 + y^2 + z^2$.

If M is a surface in $\mathbb{R}^3$ parametrised in Monge form by $\phi = (x, y, f(x,y))$ with (x, y) in an open set U of $\mathbb{R}^2$. The $\psi = \varphi \circ \phi : U \to \mathbb{R}^4$ is a parametrisation of $\varphi(M)$. The (X,Y)-plane and the (Z,W)-plane are respectively the tangent and normal planes to $\varphi(M)$ at the origin. We have by differentiating

$$\psi_{xx}(0) = (0, 0, 2f_{xx}(0), 2),$$
$$\psi_{xy}(0) = (0, 0, 2f_{xy}(0), 0),$$
$$\psi_{yy}(0) = (0, 0, 2f_{yy}(0), 2).$$

The matrix of the second fundamental form of $\varphi(M)$ at the origin p (with respect to the standard basis of $\mathbb{R}^4$) is

$$\alpha(p) = \begin{pmatrix} 2f_{xx}(0) & 2f_{xy}(0) & 2f_{yy}(0) \\ 2 & 0 & 2 \end{pmatrix}(0).$$

Since one of the rows of $\alpha(0)$ is a definite positive quadratic form, it follows that $\Delta(0) \leq 0$. Moreover, $\operatorname{rank} \alpha(0) = 1$ if and only $f_{xy}(0) = 0$ and $f_{xx}(0) = f_{yy}(0)$. It follows that $\varphi(M)$ has no elliptic points. It also follows by Theorem 4.11 that a point $p \in M$ is an umbilic point if and only if $\varphi(p)$ is an inflection point of imaginary type of $\varphi(M)$. In particular, the image by φ of any ellipsoid in $\mathbb{R}^3$ with three distinct axes is an example of a 2-sphere in $\mathbb{R}^4$ with 4 inflection points.

Sufaces in S^3 are also considered in [Nagai (2012)] using Legendrian dualities.

Example 7.2 (Complex curves). An example of a surface with boundary in $\mathbb{R}^4$ is given by an algebraic regular complex curve in $\mathbb{C}^2$ defined as a level set of a polynomial function $f : \mathbb{C}^2 \to \mathbb{C}$.

Let c be a regular value of f and let $p = (z_0, w_0)$ be a point on the level set $M = \{(z, w) | f(z, w) = c\}$. By the inverse function theorem $f(z, w) = c$ if and only if $z = g(w)$ in some neighbourhood of p for some holomorphic function g. Hence, $(g(w), w)$ parametrises locally the surface M in a neighbourhood U of w_0. Writing $w = x+iy$, $g(w) = g_1(w)+ig_2(w)$ and using the Cauchy-Riemann equations $\frac{\partial g_1}{\partial x} = \frac{\partial g_2}{\partial y}$, $\frac{\partial g_1}{\partial y} = -\frac{\partial g_2}{\partial x}$, we get the following matrix of the second fundamental form of M at w_0

$$\alpha(w_0) = \begin{pmatrix} \frac{\partial^2 g_1}{\partial x^2}(w_0) & -\frac{\partial^2 g_2}{\partial x^2}(w_0) & -\frac{\partial^2 g_1}{\partial x^2}(w_0) \\ \frac{\partial^2 g_2}{\partial x^2}(w_0) & \frac{\partial^2 g_1}{\partial x^2}(w_0) & -\frac{\partial^2 g_2}{\partial x^2}(w_0) \end{pmatrix}.$$

Hence,

$$\tilde{\Delta}(w_0) = \left(\left(\frac{\partial^2 g_1}{\partial x^2} \right)^2 + \left(\frac{\partial^2 g_1}{\partial x^2} \right)^2 \right)^2 (w_0) \geq 0.$$

We have $\tilde{\Delta}(w_0) = 0$ if and only if all second order derivatives of g_1 and g_2 at w_0 are zero. That is, the inflection points of M are degenerate and the non inflection points are all elliptic points. Moreover, one can show that $\mathbf{H}(w) = 0$ for all $w \in U$. Thus, M is a minimal surface in $\mathbb{R}^4$.

Example 7.3 (Torus embedded in $\mathbb{R}^4$). An example of a surface in $\mathbb{R}^4$ is an S^1-bundle over a closed curve in $\mathbb{R}^4$, with the fibres lying in the normal space of the curve at each point.

Let f_s be the one-parameter family of tori, parametrised by

$$f_s(x, y) = (f_1(x, y, s), f_2(x, y), f_3(x, y), f_4(x, y)),$$

with

$$
\begin{aligned}
f_1(x,y,s) &= (1-\frac{1}{10}\cos(y))\cos(x)+\frac{1}{10}\sin(x)\sin(y)+s\cos(y),\\
f_2(x,y) &= (1-\frac{1}{10}\cos(y))\sin(x)-\frac{1}{10}\cos(x)\sin(y),\\
f_3(x,y) &= (1-\frac{2}{5}\cos(y))\cos(2x)+\frac{4}{5}\sin(2x)\sin(y),\\
f_4(x,y) &= (1-\frac{2}{5}\cos(y))\sin(2x)-\frac{4}{5}\cos(2x)\sin(y).
\end{aligned}
$$

The inflection points on the image of f_s can be computed numerically for a given value of the parameter s. These are the solutions of any two of the following three equations

$$
\begin{aligned}
\alpha &= l_1m_2-m_1l_2 &&= 0,\\
\beta &= l_1n_2-n_1l_2 &&= 0,\\
\gamma &= m_1n_2-n_1m_2 &&= 0,
\end{aligned}
$$

where l_i, m_i, n_i, $i=1,2$ are the coefficients of the second fundamental form with respect to any frame.

Figure 7.6 left ($s=1/100$) and Figure 7.6 right ($s=1/20$) are computer plots of the curves $\Delta=0$, $\alpha=0$, $\beta=0$ and $\gamma=0$. The surface for $s=1/100$ has no inflection points. When $s=1/20$ the surface has 4 inflection points which can be depicted as the triple points formed by the intersection of the three curves $\alpha=0$, $\beta=0$ and $\gamma=0$. In fact, one can see from Figure 7.6 right that the parabolic curve does not pass through these points (it has a Morse singularity of type A_1^+ at such points), so the inflection points are of imaginary type.

7.6 Contact with hyperplanes

The contact of a smooth surface M in $\mathbb{R}^4$ with hyperplanes is measured by the singularities of the height functions on M. Given a local parametrisation $\mathbf{x}: U \to \mathbb{R}^4$ of M, the family of height functions $H: U\times S^3 \to \mathbb{R}$ on M is given by

$$H(u,\mathbf{v}) = \langle \mathbf{x}(u), \mathbf{v}\rangle.$$

For $\mathbf{v}$ fixed, we have the height function $h_{\mathbf{v}}$ on M given by $h_{\mathbf{v}}(u) = H(u,\mathbf{v})$. A point $p=\mathbf{x}(u)$ is a singular point of $h_{\mathbf{v}}$ if and only if $\mathbf{v}$ is a normal vector to M at p.

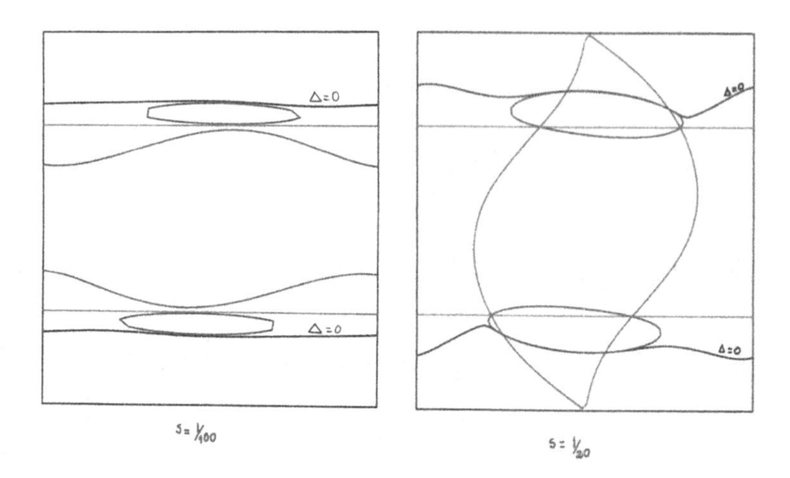

Fig. 7.6 No inflection points for $s = 1/100$ (left), and four inflection points for $s = 1/20$ (right) depicted as the points of intersection of the curves $\alpha = 0$, $\beta = 0$ and $\gamma = 0$.

Theorem 7.5. *There is an open and dense set $\mathcal{O}_H$ in $Imm(U, \mathbb{R}^4)$ such that for any $\mathbf{x} \in \mathcal{O}_H$, the surface $M = \mathbf{x}(U)$ has the following properties. For any $\mathbf{v} \in S^3$, the height function $h_{\mathbf{v}}$ along the normal direction $\mathbf{v}$ at any point p on M has only local singularities of $\mathcal{R}$ type A_1, A_2, A_3, A_4 or D_4. Furthermore, the singularities of $h_{\mathbf{v}}$ are $\mathcal{R}^+$-versally unfolded by the family H.*

Proof. The proof follows from Theorem 4.4. □

Definition 7.4. A hyperplane with orthogonal direction $\mathbf{v}$ is an *osculating hyperplane* of M at $p = \mathbf{x}(u)$ if it is tangent to M at p and $h_{\mathbf{v}}$ has a degenerate (i.e., non Morse) singularity at u. We call the direction $\mathbf{v}$ a *binormal direction* of M at p.

We parametrise M locally at p in Monge form $\phi(x, y) = (x, y, f_1(x, y), f_2(x, y))$, with f_1 and f_2 as in §7.4. For $\mathbf{v} = (v_1, v_2, v_3, v_4) \in S^3$, we have

$$h_{\mathbf{v}}(x, y) = v_1 x + v_2 y + v_3 f_1(x, y) + v_4 f_2(x, y)$$

so that $h_{\mathbf{v}}$ is singular at the origin p if and only if $v_1 = v_2 = 0$, that is $\mathbf{v} \in N_pM$. Denote by

$$Q_1(x, y) = j^2 f_1(x, y) = a_{20}x^1 + a_{21}xy + a_{22}y^2,$$
$$Q_2(x, y) = j^2 f_2(x, y) = b_{20}x^1 + b_{21}xy + c_{22}y^2.$$

Then $j^2h_{\mathbf{v}}(x,y)$ is the pencil $v_3Q_1(x,y)+v_4Q_2(x,y)$, and its $\mathcal{R}$-singularity type is completely determined by the $GL(2,\mathbb{R})\times GL(2,\mathbb{R})$-class of the pair (Q_1,Q_2). In particular, we can take (Q_1,Q_2) as in Table 7.1.

If p is a hyperbolic point, we take $(Q_1,Q_2) = (x^2,y^2)$ so that $j^2h_{\mathbf{v}}(x,y) = v_3x^2+v_4y^2$. There are exactly two binormal directions at p along $(0,0,1,0)$ and $(0,0,0,1)$.

If p is an elliptic point, we take $(Q_1,Q_2) = (xy,x^2-y^2)$ so that $j^2h_{\mathbf{v}}(x,y) = v_3xy+v_4(x^2-y^2)$. The discriminant of this quadratic form is $4v_3^2+v_4^2>0$, so the singularity of $h_{\mathbf{v}}$ at the origin is always of type A_1^-. Consequently, there are no binormal directions at p.

If p is a parabolic point, we take $(Q_1,Q_2)=(x^2,xy)$ so that $j^2h_{\mathbf{v}}(x,y) = v_3x^2+v_4xy$. There is one binormal direction at p along $(0,0,1,0)$.

Finally, if p is a non-degenerate inflection point and $(Q_1,Q_2)=(x^2\pm y^2,0)$, we get $j^2h_{\mathbf{v}}(x,y)=v_3(x^2\pm y^2)$. The direction $(0,0,0,1)$ is the unique binormal direction at p and the 2-jet of $j^2h_{\mathbf{v}}$ is identically zero along this direction. We have thus the following proposition.

Proposition 7.7. *Let M be a smooth surface immersed in $\mathbb{R}^4$ and let p be a point on M.*

(i) *If p is an elliptic point, then $h_{\mathbf{v}}$ has a non-degenerate singularity at p of type A_1^- for all $\mathbf{v}$ in N_pM.*
(ii) *If p is a hyperbolic point, then there are exactly two distinct binormal directions $\mathbf{v}_1,\mathbf{v}_2$ in N_pM.*
(iii) *If p is a parabolic point but not an inflection point, then there is a unique binormal direction $\mathbf{v}$ in N_pM.*
(iv) *If p is a non-degenerate inflection point, then there is a unique binormal direction $\mathbf{v}$ in N_pM and the 2-jet of $h_{\mathbf{v}}$ at p is identically zero.*

Proposition 7.8. *A normal direction $\mathbf{v}$ at $p=\mathbf{x}(u)$ is a binormal direction if and only if any tangent direction lying in the kernel of the Hessian of $h_{\mathbf{v}}$ at u is an asymptotic direction of M at p.*

Proof. Take the case p a hyperbolic point and $j^2h_{\mathbf{v}}(x,y)=v_3x^2+v_4y^2$. The Hessian matrix of $\mathcal{H}(h_{\mathbf{v}})$ at the origin is the diagonal matrix with entries $2v_3$ and $2v_4$. When $v_3=0$ (resp. $v_4=0$), the kernel of $\mathcal{H}(h_{\mathbf{v}})(0,0)$ is along $(1,0,0,0)$ (resp. $(0,1,0,0)$). The directions $(1,0,0,0)$ and $(0,1,0,0)$ are the asymptotic directions at p (see (7.7)). □

Remark 7.5. It is not difficult to check that when $\mathbf{v}$ is a binormal direction at p, its associated asymptotic direction $\mathbf{u}$ (i.e., a direction in the kernel of $\mathcal{H}(h_{\mathbf{v}})(p))$ is an eigenvector of the shape operator $W_p^{\mathbf{v}}$ along $\mathbf{v}$ and its associated eigenvalue is zero. That is, $\mathbf{u}$ is $\mathbf{v}$-principal direction.

Next, we characterise the degenerate singularity of $h_{\mathbf{v}}$ at the origin in terms of the coefficients in the Taylor expansion of the functions f_1 and f_2 giving M in Monge form. By a rotation at the origin in the normal plane, we can assume that $\mathbf{v}_0 = (0,0,0,1)$ is a binormal direction and take nearby directions in the form $\mathbf{v} = (v_1, v_2, v_3, 1)$, with v_i, $i = 1,2,3$ small enough. Then, the (modified) height function in the direction $\mathbf{v}$ is given by

$$h_{\mathbf{v}}(x,y) = v_1 x + v_2 y + v_3 f_1(x,y) + f_2(x,y).$$

Proposition 7.9. *With notation as in §7.4, the conditions for $h_{\mathbf{v}_0}$ to have one of the generic singularities at the origin are as follows*

$A_2 : b_{20} \neq 0, b_{33} \neq 0$
$A_3 : b_{20} \neq 0, b_{33} = 0, 4b_{20}b_{44} - b_{32}^2 \neq 0$
$A_4 : b_{20} \neq 0, b_{33} = 0, 4b_{20}b_{44} - b_{32}^2 = 0, 4{b_{20}}^2 b_{55} - 2b_{20}b_{32}b_{43} + b_{31}b_{32}^2 \neq 0$
$D_4 : b_{20} = 0,\ b_{30}x^3 + b_{31}x^2y + b_{32}xy^2 + b_{33}y^3$ *is non-degenerate.*

The above singularities are $\mathcal{R}^+$-versally unfolded by the family of height functions if and only if

A_2 : *always*
$A_3 : a_{22} \neq 0$ *or* $b_{32} \neq 0$
$A_4 : a_{22}(b_{20}b_{43} - b_{31}b_{32}) - b_{32}(b_{20}a_{33} - \frac{1}{2}a_{21}b_{32}) \neq 0$
$D_4 : 3b_{30}(a_{22}b_{32} - \frac{3}{2}a_{21}b_{33}) - b_{31}(a_{22}b_{31} - \frac{1}{2}a_{21}b_{32})$
$\quad + a_{20}(3b_{31}b_{33} - b_{32}^2) \neq 0.$

Proof. The height function on M along the normal direction $\mathbf{v}_0 = (0,0,0,1)$ is given by $h_{\mathbf{v}_0}(x,y) = f_2(x,y)$.

We first assume that $b_{20} \neq 0$. Then, the singularity of $h_{\mathbf{v}_0}$ at the origin is of type A_k. In this case, $h_{\mathbf{v}_0}$ has an A_2-singularity at the origin if and only if $b_{33} \neq 0$. If $b_{33} = 0$, we can make changes of coordinates in the source to reduce the 4-jet of f_2 to $j^4 f_2(x,y) = b_{20}x^2 + b_{32}xy^2 + b_{44}y^4$. Then, the singularity is of type A_3 if and only if $b_{33} = 0$ and $b_{20}x^2 + b_{32}xy^2 + b_{44}y^4$ is not a perfect square, equivalently, $b_{32}^2 - 4b_{20}b_{44} \neq 0$. If $b_{33} = 0$ and $b_{20}x^2 + b_{32}xy^2 + b_{44}y^4$ is a perfect square, we can make changes of coordinates in the source and show that $j^5 f_2 \sim_{\mathcal{R}^{(5)}} b_{20}x^2 + Dy^5$, with $D = a_{22}(b_{20}b_{43} - b_{31}b_{32}) - b_{32}(b_{20}a_{33} - \frac{1}{2}a_{21}b_{32})$. Hence, $h_{\mathbf{v}_0}$ has A_4 singularity if and only if $D \neq 0$.

If $b_{20} = 0$, the result follows from the characterisation of a D_4 singularity.

The modified family of height functions $H : \mathbb{R}^2 \times \mathbb{R}^3, (0, \mathbf{v}_0) \to \mathbb{R}$ is given by

$$H(x, y, v_1, v_2, v_3) = v_1 x + v_2 y + v_3 f_1(x, y) + f_2(x, y).$$

The family H is an $\mathcal{R}^+$-versal deformation if and only if

$$L\mathcal{R}_e \cdot f_2 + \mathbb{R}.\left\{\dot{H}_1, \dot{H}_2, \dot{H}_3\right\} + \langle 1 \rangle_{\mathbb{R}} = \mathcal{E}_2 \tag{7.10}$$

(see Theorem 3.12). We have

$$\begin{aligned}\dot{H}_1(x, y) &= H_{v_1}(x, y, 0, 0, 0) = x,\\ \dot{H}_2(x, y) &= H_{v_2}(x, y, 0, 0, 0) = y,\\ \dot{H}_3(x, y) &= H_{v_3}(x, y, 0, 0, 0) = f_1(x, y).\end{aligned}$$

We carry out the calculations for the A_3-singularity. We have

$$\begin{aligned}j^3 f_{2x} &= 2b_{20}x + 3b_{30}x^2 + 2b_{31}xy + b_{32}y^2 +\\ &\quad 4b_{40}x^3 + 3b_{41}x^2y + 2b_{42}xy^2 + b_{43}y^3.\\ j^3 f_{2y} &= b_{31}x^2 + 2b_{32}xy + 3b_{33}y^2 +\\ &\quad b_{41}x^3 + 2b_{42}x^2y + 3b_{43}xy^2 + 4b_{44}y^3.\end{aligned}$$

Since f_2 is 4-$\mathcal{R}$-determined, it is enough to verify that (7.10) holds in $J^4(2, 1)$. As $b_{20} \neq 0$, for any $d \geq 2$ and $k \geq 1$, we have $(x^{k-1}y^{d-k})j^3 f_{2x} \in L\mathcal{R}_e \cdot f_2$, hence

$$x^k y^{d-k} \in J^d(L\mathcal{R}_e \cdot f_2). \tag{7.11}$$

Then, when $d = 2$, if $b_{32} \neq 0$ or $a_{22} \neq 0$, we can solve (7.10) in $J^2(2, 1)$.

When $d = 3$ and 4, writing $y^i j^3 f_{2x}$, and $y^{i-1} j^3 f_{2y}$ $i = 1, 2$ and using (7.11) to eliminate the monomials of degree $2 + i$, $i = 1, 2$ divisible by x, we can write the equations

$$\begin{aligned}2b_{20}xy^i + b_{32}y^{2+i} &\in J^{2+i}(L\mathcal{R}_e \cdot f_2),\ i = 1, 2,\\ b_{31}x^2y^{i-1} + 2b_{32}xy^i + 4b_{44}y^{3+i-1} &\in J^{2+i}(L\mathcal{R}_e \cdot f_2),\ i = 1, 2,\\ b_{31}x^2y^{i-1} &\in J^{2+i}(L\mathcal{R}_e \cdot f_2).\end{aligned}$$

Since $4b_{20}b_{44} - b_{32}^2 \neq 0$, it follows that all monomials of degree 3 and 4 are in $J^4(L\mathcal{R}_e \cdot f_2)$. □

7.6.1 *The canal hypersurface*

The canal hypersurface of the surface $M = \mathbf{x}(U) \subset \mathbb{R}^4$ is the 3-manifold

$$CM(\varepsilon) = \{p + \varepsilon\mathbf{v} \in \mathbb{R}^4 | \, p \in M \text{ and } \mathbf{v} \in (N_pM)_1\}$$

where $(N_pM)_1$ denotes the unit sphere in N_pM and ε is a sufficiently small positive real number chosen so that $CM(\varepsilon)$ is a regular hypersurface (see

Theorem 2.7). We can consider $(N_pM)_1$ a subset of S^3 and identify $(p, \mathbf{v})$ and $p + \varepsilon\mathbf{v}$.

Let $\bar{H} : CM(\varepsilon) \times S^3 \to \mathbb{R}$ denote the family of height functions on $CM(\varepsilon)$. For $\mathbf{w} \in S^3$, let $\bar{h}_{\mathbf{w}} : CM(\varepsilon) \to \mathbb{R}$, be given by $\bar{h}_{\mathbf{w}}(p, \mathbf{v}) = \bar{H}((p, \mathbf{v}), \mathbf{w})$. Since $CM(\varepsilon)$ is a hypersurface in $\mathbb{R}^4$, we have the Gauss map $G : CM(\varepsilon) \to S^3$. By Proposition 2.12, it is given by $G(p, \mathbf{v}) = \mathbf{v}$. We call it the *generalised Gauss map* of M.

Proposition 7.10. *Let M be a smooth surface in $\mathbb{R}^4$ and let p be a point on M.*

(i) *The point p is a singular point of $h_{\mathbf{v}}$ if and only if $(p, \mathbf{v}) \in CM(\varepsilon)$ is a singular point of $\bar{h}_{\mathbf{v}}$. Furthermore, the $\mathcal{R}$-singularity type of $h_{\mathbf{v}}$ at p is the same as that of $\bar{h}_{\mathbf{v}}$ at $(p, \mathbf{v})$.*
(ii) *A point $(p, \mathbf{v})$ is a degenerate singular point of $\bar{h}_{\mathbf{v}}$ if and only if $(p, \mathbf{v})$ is a singular point of the generalised Gauss map G.*

Proof. The proof follows by Proposition 2.12 and Corollary 2.4. □

Theorem 7.6. *There is an open and dense set $\mathcal{O}_{\bar{H}}$ in $Imm(U, \mathbb{R}^4)$ such that for any $\mathbf{x} \in \mathcal{O}_{\bar{H}}$, the surface $M = \mathbf{x}(U)$ satisfies the conditions in* Theorem 7.5. *Moreover, the following properties hold.*

(i) *For any $\mathbf{v} \in S^3$, the height function $\bar{h}_{\mathbf{v}}$ on $CM(\varepsilon)$ has only local singularities of $\mathcal{R}$ type A_1, A_2, A_3, A_4 or D_4.*
(ii) *The singularities of $\bar{h}_{\mathbf{v}}$ are $\mathcal{R}^+$-versally unfolded by the family $\bar{H}$.*
(iii) *The generalised Gauss map G is stable as a Lagrangian map. The singularities of G are given in Table 7.2.*

Proof. By Proposition 5.14, $\bar{H}$ is a Morse family of functions and the generalised Gauss map is the Lagrangian map of $L(\bar{H})$. We apply Theorem 4.4 to obtain an open and dense set $\mathcal{O}'_{\bar{H}}$ of immersions for which $\bar{H}$ is an $\mathcal{R}^+$-versal unfolding of $\bar{h}_{\mathbf{v}}$. For any immersion $\mathbf{x}$ in this set, the associated generalised Gauss mapping $G : CM(\varepsilon) \to S^3$ is Lagrangian stable. We take $\mathcal{O}_{\bar{H}}$ as the intersection of $\mathcal{O}'_{\bar{H}}$ and the set $\mathcal{O}_H$ in Theorem 7.5. The singularities of the generalised Gauss maps appearing in Table 7.2 are the Lagrangian stable singularities when $n \leq 3$ from Table 5.1. □

Definition 7.5. A surface is called *height function generic* if any of its local parametrisation belongs to the set $\mathcal{O}_{\bar{H}}$ in Theorem 7.6.

Table 7.2 Lagrangian stable singularities of generalised Gauss maps of surfaces in $\mathbb{R}^4$.

$h_{\mathbf{v}}$ singularity type	G singularity type	Normal form
A_1	Immersion	(x, y, v_3)
A_2	Fold	(x, y^2, v_3)
A_3	Cuspidal edge	$(x, xy + y^3, v_3)$
A_4	Swallowtail	$(x, y^4 + xy + v_3y^2, v_3)$
D_4^+	Hyperbolic umbilic	$(x^2 + v_3y, y^2 + v_3x, v_3)$
D_4^-	Elliptic umbilic	$(x^2 - y^2 + v_3x, xy + v_3y, v_3)$

Proposition 7.11. *Let M be a height function generic surface in $\mathbb{R}^4$. Let $K_c : CM(\varepsilon) \to \mathbb{R}$ be the Gauss-Kronecker curvature function of $CM(\varepsilon)$. Then the following hold.*

(i) *The singular set of G is the parabolic set $K_c^{-1}(0)$ of $CM(\varepsilon)$.*
(ii) *$K_c^{-1}(0) := \{p + \varepsilon\mathbf{v} \in CM(\varepsilon) \mid h_{\mathbf{v}}$ has a degenerate singularity at $p\}$.*
(iii) *$K_c^{-1}(0)$ is a regular surface except at a finite number of singular points corresponding to the $D_4^{\pm}$-singularities of $\bar{h}_{\mathbf{v}}$.*

Proof. The proof of part (i) is analogous to the proof of Proposition 6.9, part (ii) follows from Proposition 7.10, and part (iii) follows from the normal forms (1), (2), (3), (5) and (6) in Theorem 5.5. The $D_4^{\pm}$ are the corank 2 singularities of G, so they correspond to the singular points of $K_c^{-1}(0)$. □

Remark 7.6. Let M be a height function generic surface, so that the generalised Gauss map G is a Lagrangian stable map. We use the Thom-Boardman symbols (see §3.4) to distinguish the singular points of G. Let $K_c^{-1}(0) = S_1(G) \cup S_2(G)$, where $S_1(G)$ are the singular points of corank 1 and $S_2(G)$ are the singular points of corank 2 of G. Since G is a Lagrangian stable map, the set $S_1(G)$ is a smooth surface which can be decomposed as $S_{1,0}(G) \cup S_{1,1,0}(G) \cup S_{1,1,1,0}(G)$. The points in $S_{1,0}(G)$ are the regular points of the restriction of the map G to $S_1(G)$; they are the fold points of G. The singular set of $G|_{S_1(G)}$ is the smooth curve $S_{1,1}(G) = S_{1,1,0}(G) \cup S_{1,1,1,0}(G)$. The restriction of G to this set is regular on $S_{1,1,0}(G)$ and singular on $S_{1,1,1,0}(G)$; they are respectively the cuspidal edge points and the swallowtail points of G. These geometric conditions are verified for the normal forms in Table 7.2.

At each point $(p, \mathbf{v}) \in K_c^{-1}(0) \setminus S_2(G)$, the unique kernel direction of dG_p is the principal direction of zero curvature of $CM(\varepsilon)$ at p. This direction coincides with the kernel of the Hessian quadratic form of the height function in the normal direction $\mathbf{v}$ on $CM(\varepsilon)$ (see the proof of Proposition

2.13). It follows from the definition of the sets $S_{1,0}(G), S_{1,1,0}(G), S_{1,1,1,0}(G)$ that this direction is transverse to the surface $S_{1,0}(G)$ and is tangent to the surface $K_c^{-1}(0)$ on the curve of points of type $S_{1,1}(G)$. It is transverse to this curve at general points and tangent to it at points of type $S_{1,1,1}(G)$.

Let $\xi : CM(\varepsilon) \to M$ be the projection of $CM(\varepsilon)$ to M, given by $\xi(p, \mathbf{v}) = p$. Then, it follows from part (ii) in Proposition 7.11 that the image of the parabolic set $K_c^{-1}(0)$ by ξ is the set $\Delta \leq 0$. More precisely, let $\bar{\xi}$ be the restriction of ξ to the regular surface $S_1(G) = K_c^{-1}(0) \setminus S_2(G)$.

We denote by $M_- = \{p \in M : \Delta(p) < 0\}$ the set of hyperbolic points on M and by $B = \{(p, \mathbf{v}) \in K_c^{-1}(0) : p \in M_-\}$.

Proposition 7.12. *With notation as above,*

(i) *the restriction $\bar{\xi}_{|B} : B \to M_-$ is a local diffeomorphism, more precisely, it is a double cover;*
(ii) *a point $p \in \Delta$ is not an inflection point if and only if there exists $\mathbf{v} \in S^3$ such that $(p, \mathbf{v})$ is a fold singularity of $\bar{\xi}$.*

Proof. We take M locally in Monge form with f_1 and f_2 as in Section 7.4 and choose the setting in the proof of Proposition 7.9. The modified height function on M along $\mathbf{v} = (v_1, v_2, v_3, 1)$ near $\mathbf{v}_0 = (0, 0, 0, 1)$ is given by

$$h_{\mathbf{v}}(x, y) = v_1 x + v_2 y + v_3 f_1(x, y) + f_2(x, y).$$

The function $h_{\mathbf{v}}$ has a singularity at a point (x, y) if and only if

$$v_1 = -v_3 \frac{\partial f_1}{\partial x}(x, y) - \frac{\partial f_2}{\partial x}(x, y) \quad \text{and} \quad v_2 = -v_3 \frac{\partial f_1}{\partial y}(x, y) - \frac{\partial f_2}{\partial y}(x, y).$$

The singularity $p = (x, y)$ is degenerate if furthermore $\det \mathcal{H}(h_{\mathbf{v}})(x, y) = 0$, equivalently $K_c(x, y, v_3) = 0$ (see Proposition 7.11). We have

$$\det\mathcal{H}(h_{\mathbf{v}})(x, y) = A_0(x, y) v_3^2 + A_1(x, y) v_3 + A_2(x, y) = 0,$$

where

$$\begin{aligned} A_0(x, y) &= \tfrac{\partial^2 f_1}{\partial x^2}(x, y) \tfrac{\partial^2 f_1}{\partial y^2}(x, y) - (\tfrac{\partial^2 f_1}{\partial x \partial y})^2(x, y) \\ A_1(x, y) &= \tfrac{\partial^2 f_1}{\partial x^2}(x, y) \tfrac{\partial^2 f_2}{\partial y^2}(x, y) + \tfrac{\partial^2 f_1}{\partial y^2}(x, y) \tfrac{\partial^2 f_2}{\partial x^2}(x, y) \\ &\quad -2 \tfrac{\partial^2 f_1}{\partial x \partial y}(x, y) \tfrac{\partial^2 f_2}{\partial x \partial y}(x, y) \\ A_2(x, y) &= \tfrac{\partial^2 f_2}{\partial x^2}(x, y) \tfrac{\partial^2 f_2}{\partial y^2}(x, y) - (\tfrac{\partial^2 f_2}{\partial x \partial y})^2(x, y). \end{aligned} \tag{7.12}$$

Write $\tilde{K}_c = \det(\mathcal{H}(h_{\mathbf{v}}))$ so that $K_c = \lambda \tilde{K}_c$ for some function λ with $\lambda(x, y, v_3) \neq 0$. Then, denoting by z any of the variables x, y, v_3, one can show inductively that

$$K_c = \frac{\partial K_c}{\partial z} = \ldots = \frac{\partial^m K_c}{\partial z^m} = 0 \text{ and } \frac{\partial^{m+1} K_c}{\partial z^{m+1}} \neq 0$$

at (x, y, v_3) if and only if

$$\tilde{K}_c = \frac{\partial \tilde{K}_c}{\partial z} = \ldots = \frac{\partial^m \tilde{K}_c}{\partial z^m} = 0 \text{ and } \frac{\partial^{m+1} \tilde{K}_c}{\partial z^{m+1}} \neq 0$$

at (x, y, v_3). For (i), we have $v_3 = 0$ is a simple root of $\tilde{K}_c(x, y, v_3) = 0$ if and only if $\frac{\partial \tilde{K}_c}{\partial v_3}(x, y, 0) \neq 0$. Equivalently, $K_c(x, y, v_3) = 0$ if and only if $\frac{\partial K_c}{\partial v_3}(x, y, 0) \neq 0$. That is, $\bar{\xi}$ is a local diffeomorphism.

For (ii), $v_3 = 0$ is a double root of $\tilde{K}_c(x, y, v_3) = 0$ if and only if $\frac{\partial \tilde{K}_c}{\partial v_3}(x, y, 0) = 0$ and $\frac{\partial^2 \tilde{K}_c}{\partial v_3^2}(x, y, 0) \neq 0$. Equivalently $K_c(x, y, v_3) = 0$, if and only if $\frac{\partial K_c}{\partial v_3}(x, y, 0) = 0$ and $\frac{\partial^2 K_c}{\partial v_3^2}(x, y, 0) \neq 0$. That is $(p, \mathbf{v})$ is a fold point for $\bar{\xi}$. □

Proposition 7.13. *The zero curvature principal directions on the parabolic set $K_c^{-1}(0) \setminus S_2(G)$ are mapped by $d\bar{\xi}$ to the asymptotic directions of M.*

Proof. We choose the setting as in the proof of Proposition 7.12 and take coordinates for M and $CM(\varepsilon)$ such that $p = (0, 0)$ and $\mathbf{v} = (0, 0, 0, 1)$. Hence $\tilde{K}_c(p, v_3) = A_0(x, y)v_3^2 + A_1(x, y)v_3 + A_2(x, y)$.

At $p = (0, 0)$, $\mathbf{v} = (0, 0, 0, 1)$, the singularity of the Gauss map $G : CM(\varepsilon) \to S^3$ is the singularity at the origin of the map-germ $g : (\mathbb{R}^3, 0) \to (\mathbb{R}^3, 0)$, given by

$$g(x, y, v_3) = (-v_3 \frac{\partial f_1}{\partial x}(x, y) - \frac{\partial f_2}{\partial x}(x, y), -v_3 \frac{\partial f_1}{\partial y}(x, y) - \frac{\partial f_2}{\partial y}(x, y), v_3).$$

The eigenvectors of the matrices dG_0 and dg_0 are the same. Thus, if $\mathbf{e}_2 = (0, 1, 0, 0) \in T_{(p,\mathbf{v})}CM(\varepsilon)$ is the zero curvature direction, it follows that $b_{11} = b_{02} = 0$. Then $T\xi(p, \mathbf{v})\mathbf{e}_2 = \mathbf{e}_2 \in T_pM$ is an asymptotic direction at p. □

7.6.2 *Characterisation of the singularities of the height function*

We characterise geometrically the degenerate singularities of generic height functions.

Denote by γ the normal section of M tangent to the asymptotic direction θ at p associated to the binormal direction $\mathbf{v}$.

The characterisation of the singularities of the height functions at hyperbolic points is given below.

Theorem 7.7. *Let p be a hyperbolic point on a height function generic surface M. Then p is an A_2 singularity of $h_\mathbf{v}$ if and only if γ has non* *vanishing normal torsion at p. If γ has a vanishing normal torsion at p, then:*

(i) *p is an A_3 singularity of $h_{\mathbf{v}}$ if and only if the direction θ is transversal to the curve $\bar{\xi}(S_{1,1}(G))$.*
(ii) *p is an A_4 singularity of $h_{\mathbf{v}}$ if and only if the direction θ is tangent to the curve $\bar{\xi}(S_{1,1}(G))$ with first order contact at p.*

Proof. We choose local coordinates as in Proposition 7.9. Let $\tilde{K}_c : \mathbb{R}^2 \times \mathbb{R}, 0$ given by

$$\tilde{K}_c(x, y, v_3) = A_0(x, y)v_3^2 + A_1(x, y)v_3 + A_2(x, y),$$

where A_0, A_1, A_2 are as in Equation (7.12).

The zero curvature direction of CM at $(p, \mathbf{v})$ in these coordinates is the y-axis. Then $d\bar{\xi}_{(p,\mathbf{v})}\mathbf{e}_2 = \mathbf{e}_2 = \theta$ (Proposition 7.13).

The normal section γ can be parametrised by

$$\gamma(s) = (0, s, a_{20}s^2 + ..., b_{33}s^3 + ...)$$

and it follows that γ has non-zero torsion if and only if $b_{33} \neq 0$. This is the case if and only if p is an A_2 singularity of $h_{\mathbf{v}}$.

Now, it follows from Proposition 7.10 and Theorem 7.6 that a hyperbolic point $p \in M(\epsilon)$ is a singularity of type A_2, A_3, A_4 of $h_{\mathbf{v}}$ if and only if $(p, \mathbf{v}) \in CM$ is respectively a fold, cusp or swallowtail singularity of G. Hence, if $b_{33} = 0$, the following follows from Remark 7.6.

(i) p is an A_3 singularity of $h_{\mathbf{v}}$ if and only if $(p, \mathbf{v}) \in S_{1,1,0}(G)$, if and only if the zero curvature direction is tangent to $K_c^{-1}(0)$ and transverse to the curve $S_{1,1,0}(G)$.
(ii) p is an A_4 singularity of $h_{\mathbf{v}}$ if and only if $(p, \mathbf{v}) \in S_{1,1,1,0}(G)$, and this happens if and only if the zero principal direction is tangent to $S_{1,1,0}(G)$, with first order contact.

Since $\bar{\xi} : B \to M_-$ is a local diffeomorphism, θ is transversal to $\bar{\xi}(S_{1,1,0}(G))$ in (i) and tangent to $\bar{\xi}(S_{1,1,0}(G))$ with first order contact in (ii). $\square$

Definition 7.6. By analogy with the case of surfaces in $\mathbb{R}^3$, we define the *flat rib set* of a surface M in $\mathbb{R}^4$ as the subset of points $(p, \mathbf{v}) \in CM(\epsilon)$ such that the height function $h_{\mathbf{v}}$ on M has a singularity of type $A_k, k \geq 3$ at p. We call its projection to M by ξ the *flat ridge set* of M.

We give now the characterisation of the singularities of the height functions at parabolic points.

Theorem 7.8. *Let M be a height function generic surface in $\mathbb{R}^4$ and $p \in M$.*

i) *Suppose p is a parabolic point, but not an inflection point. Then*
   (a) *p is an A_2-singularity of $h_{\mathbf{v}}$ if and only if θ is transversal to the parabolic curve Δ.*
   (b) *p is an A_3-singularity of $h_{\mathbf{v}}$ if and only if θ is tangent to the parabolic curve Δ with first order contact.*

ii) *The point p is a D_4-singularity of $h_{\mathbf{v}}$ if and only if p is an inflection point of M. Moreover,*
   (c) *p is a normal crossing point of Δ if and only if p is an inflection point of real type;*
   (d) *p is an isolated point of Δ if and only if p is an inflection point of imaginary type.*

Proof. (i) Observe first that the curves of points p in $M = \mathbf{x}(U)$ such that $h_{\mathbf{v}}$ has a singularity of type A_3 or more degenerate for some $\mathbf{v} \in N_pM$, cannot meet the curve Δ at a point of type A_4. This follows from Thom transversality theorem by observing that for a point $(x,y) \in U$ to be an A_4 point of $h_{\mathbf{v}}$, the 4-jet extension $j^4h : U \times S^3 \to J^4(U,\mathbb{R})$, must meet transversally the set of jets of type A_4 in $J^4(U,\mathbb{R})$. This set is an algebraic variety of codimension 5 in $J^4(U,\mathbb{R})$. The conditions on the 4-jet $j^4\phi$ to intercept this set transversally at the origin, where $\phi(x,y) = (x,y,f_1(x,y),f_2(x,y))$ is the Monge form embedding of M, are given in Proposition 7.9. Now, as the point also belongs to Δ, the image of j^4h meets another algebraic variety of codimension 1. Thus to have both conditions at the same time, the map j^4h meets the intersection of both varieties, which has codimension 6 in $J^4(M,\mathbb{R})$. But this can be avoided by a generic embedding of the surface.

We choose local coordinates as in Proposition 7.12. Let $\tilde{K}_c : \mathbb{R}^2 \times \mathbb{R}, 0$ given by

$$\tilde{K}_c(x,y,v_3) = A_0(x,y)v_3^2 + A_1(x,y)v_3 + A_2(x,y),$$

where A_0, A_1, A_2 are defined in (7.12).

Since $p \in \Delta$, $a_{22} = 0$ and there is a unique binormal direction $\mathbf{v}$ at p. The discriminant set

$$\{(x,y) \,|\, \exists\, v_3 : \tilde{K}_c(x,y,v_3) = 0 \,\text{and}\, \frac{\partial \tilde{K}_c}{\partial v_3}(x,y,v_3) = 0\}$$

is $A_0(x,y)A_2(x,y) - \frac{1}{4}A_1^2(x,y)$, which is exactly the set Δ. Then:

(a) p is an A_2 singularity of $h_{\mathbf{v}}$ if and only if $\frac{\partial A_2}{\partial y}(0,0) = 12b_{20}b_{33} \neq 0$ if and only if the asymptotic direction e_2 is transversal to the curve Δ.
(b) If p is an A_3 singularity of $h_{\mathbf{v}}$ then $4b_{20}b_{33} = 0$, $12b_{20}b_{32} \neq 0$. Then, the asymptotic direction e_2 is tangent to the curve Δ.

(ii) If the point p is a D_4-singularity of $h_{\mathbf{v}}$, then $b_{20} = 0$, and it follows that rank$\alpha(p) = 1$, which implies that p is an inflection point. We can verify conditions (c) and (d) by direct calculations. □

The following result is a corollary of the proof of the above theorem.

Corollary 7.3. *For a height function generic immersion, the flat ridge set is a regular curve in M which is tangent to the curve $\Delta^{-1}(0)$.*

7.7 Contact with lines

The family of orthogonal projections $P : M \times S^3 \to TS^3$ of M to 3-spaces is given by

$$P(p, \mathbf{v}) = (p, P_{\mathbf{v}}(p))$$

where $P_{\mathbf{v}}(p) = p - \langle p, \mathbf{v}\rangle \mathbf{v}$. For a given $\mathbf{v} \in S^3$, the map $P_{\mathbf{v}}$ is singular at p if and only if $\mathbf{v}$ is in T_pM.

The map $P_{\mathbf{v}}$ can be considered locally as a smooth map-germ $\mathbb{R}^2, 0 \to \mathbb{R}^3, 0$. Since P is a 3-parameter family, we expect the map $P_{\mathbf{v}}$ to have only simple singularities of $\mathcal{A}_e$-codimension ≤ 3 or non-simple singularities with the $\mathcal{A}_e$-codimension of the stratum ≤ 3. (The stratum is formed by the jets in some k-jet space which are $\mathcal{A}^{(k)}$-equivalent to a member of the family parametrised by the moduli). The expected singularities are extracted from the classification in [Mond (1985)] and are as in Table 7.3. The following result follows from Theorem 4.12.

Theorem 7.9. *There is an open and dense set $\mathcal{O}_P$ in Imm$(U, \mathbb{R}^4)$ such that for any $\mathbf{x} \in \mathcal{O}_P$, the surface $M = \mathbf{x}(U)$ has the following properties.*

(i) *Given any point $p \in M$, the projection $P_{\mathbf{v}}$ has local singularities of $\mathcal{A}$-type S_k, $k = 0, 1, 2, 3$, B_2, B_3, C_3, H_2, H_3 or $P_3(c)$ at p.*
(ii) *The simple singularities of $P_{\mathbf{v}}$ are $\mathcal{A}_e$-versally unfolded by the family P. The singularity $P_3(c_0)$, $c_0 \neq 0, \frac{1}{2}, 1, \frac{3}{2}$, is $\mathcal{A}_e$-versally unfolded by the family P with c near c_0 included as a parameter in the family.*

Table 7.3 Generic local singularities of $P_{\mathbf{v}}$.

Name	Normal form	$\mathcal{A}_e$-codimension
Immersion	$(x,y,0)$	0
Cross-cap	(x,xy,y^2)	0
S_1=B_1	$(x,y^2,y^3\pm x^2y)$	1
S_2	(x,y^2,y^3+x^3y)	2
S_3	$(x,y^2,y^3\pm x^4y)$	3
B_2	$(x,y^2,x^2y\pm y^5)$	2
B_3	$(x,y^2,x^2y\pm y^7)$	3
C_3	$(x,y^2,xy^3\pm x^3y)$	3
H_2	$(x,xy+y^5,y^3)$	2
H_3	$(x,xy+y^8,y^3)$	3
$P_3(c)$	$(x,xy+y^3,xy^2+cy^4)$, $c\neq 0,\frac{1}{2},1,\frac{3}{2}$	4(3*)

* The codimension of stratum P_3 is 3.

Definition 7.7. An immersion $\mathbf{x}: M \to \mathbb{R}^4$ is *projection P-generic* if $\mathbf{x}$ belongs to the set $\mathcal{O}_P$.

In §7.6, the second order geometry of a surface derived from its contact with hyperplanes is determined by a pencil of quadratic forms. This pencil also determines some of the second order geometry of the surface associated to its contact with lines.

We take the surface in Monge form $\phi(x,y) = (x,y,f_1(x,y),f_2(x,y))$, at the origin p with f_1 and f_2 as in §7.4 and project along a unit tangent direction $\mathbf{v} = (v_1,v_2,0,0) \in T_pM$ (so that $P_{\mathbf{v}}$ is singular at p). Then

$$\begin{aligned}
P_{\mathbf{v}}(x,y) &= \phi(x,y) - \langle\phi(x,y),\mathbf{v}\rangle\mathbf{v}\\
&= (x,y,f_1(x,y),f_2(x,y)) - (xv_1+yv_2)(v_1,v_2,0,0)\\
&= (x-(xv_1+yv_2)v_1, y-(xv_1+yv_2)v_2, f_1(x,y), f_2(x,y))\\
&= (v_2(v_2x-v_1y), -v_1(v_2x-v_1y), f_1(x,y), f_2(x,y)).
\end{aligned}$$

The image of $P_{\mathbf{v}}$ is in $\mathbf{v}^{\perp}$, the linear 3-space orthogonal to $\mathbf{v}$. If we denote by (X,Y,Z,W) the coordinates in $\mathbb{R}^4$, and if $v_2 \neq 0$, we can compose $P_{\mathbf{v}}$ with the projection $(X,Y,Z,W) \mapsto (X,Z,W)$ and obtain a map-germ $\mathcal{A}$-equivalent to $P_{\mathbf{v}}$ (the projection is a diffeomorphism restricted to $\mathbf{v}^{\perp}$). If $v_2 = 0$, we compose $P_{\mathbf{v}}$ with the projection $(X,Y,Z,W) \mapsto (Y,Z,W)$ instead. In both cases we still denote by $P_{\mathbf{v}}$ the composite map. Rescaling the first coordinate (or the second if $v_2 = 0$) gives

$$P_{\mathbf{v}}(x,y) = (v_2x - v_1y, f_1(x,y), f_2(x,y)). \qquad (7.13)$$

We have $j^2P_{\mathbf{v}}(x,y) = (v_2x - v_1y, Q_1(x,y), Q_2(x,y))$, with Q_1 and Q_2 representing j^2f_1 and j^2f_2 at the origin, respectively. We are interested in the $\mathcal{A}$-singularities of $P_{\mathbf{v}}$. Then one of the conditions that determines the

$\mathcal{A}$-type of $P_{\mathbf{v}}$ at the origin is whether or not $v_2x - v_1y$ is a common factor of Q_1 and Q_2. For this reason, following [Bruce and Nogueira (1998)], we associate to each tangent direction $\mathbf{v}$ a line in $\mathbb{R}P^2$ which consists of quadratic forms having $(v_2x - v_1y)$ as a factor. This line is the tangent line to the conic Γ at $(v_2x - v_1y)^2$.

Proposition 7.14 ([Bruce and Nogueira (1998)]). (i) *Suppose that p is not an inflection point. The direction of projection yields a cross-cap unless the line it determines passes through one of the points of intersection of Γ with the pencil (Q_1, Q_2).*

(ii) *At an inflection point, where we have an associated single quadratic form $Q = Q_1 = Q_2$, most directions of projection yield an S_k or B_k singularity, and we have a direction of projection of type H_k provided the pencil it determines contains Q. If Q is inside (resp. outside) the conic Γ there are 0 (resp. 2) such projections.*

Proof. (i) We prove the statement for the case p a hyperbolic or parabolic point; the argument for the elliptic points is similar. We take $(Q_1, Q_2) = (x^2, y^2)$ at a hyperbolic point so that $j^2P_{\mathbf{v}}(x, y) = (v_2x - v_1y, x^2, y^2)$. If $v_2 \neq 0$, we can make the change of coordinates $X = v_2x - v_1y$, $Y = y$ and obtain a map-germ $\mathcal{A}^{(2)}$-equivalent to $(X, \frac{2v_1}{v_2^2}XY, Y^2)$. We get a cross-cap if and only if $v_1 \neq 0$. When $v_1 = 0$, $j^2P_{\mathbf{v}}$ is $\mathcal{A}^{(2)}$-equivalent to $(X, 0, Y^2)$. If $v_2 = 0$, then $j^2P_{\mathbf{v}}$ is $\mathcal{A}^{(2)}$-equivalent $(Y, X^2, 0)$. Therefore, the singularity is a cross-cap unless $v_1 = 0$ or $v_2 = 0$. Observe that the degenerate singularities of $P_{\mathbf{v}}$ are S_k, B_k or C_k.

At a parabolic point and with $(Q_1, Q_2) = (x^2, xy)$, we get $j^2P_{\mathbf{v}}(x, y) = (v_2x - v_1y, x^2, xy)$. If $v_2 = 0$, the singularity is a cross-cap. If $v_2 \neq 0$, the same change of coordinates as above reduces the 2-jet to $(X, 2v_1XY + v_1^2Y^2, XY + v_1Y^2)$ which is a cross-cap unless $v_1 = 0$. If $v_1 = 0$, $j^2P_{\mathbf{v}}(x, y)$ is $\mathcal{A}^{(2)}$-equivalent to $(X, XY, 0)$, which leads to a projection P-generic surface to H_2, H_3 or $P_3(c)$ singularity.

As before, the point $(A : B : C) \in \mathbb{R}P^2$ represents the quadratic form $Ax^2 + 2Bxy + cy^2$. The conic Γ has equation $AC - B^2 = 0$ and any point on it represents perfect squared $(ax + by)^2$, so has the form $(a^2 : ab : b^2)$. It follows that the tangent line to Γ at $(a^2 : ab : b^2)$ has equation $Ca^2 - 2Bab + Ab^2 = 0$.

Now the pencil determined by $\mathbf{v}$ consists of quadratic forms having $(v_2x - v_1y)$ as a factor, so is the line tangent to Γ at $(v_2^2, -v_1v_2, v_1^2)$. Therefore it has equation $Cv_2^2 + 2Bv_1v_2 + Av_1^2 = 0$. The pencil determined by (Q_1, Q_2) intersects Γ at $(1 : 0 : 0)$ and $(0 : 0 : 1)$. The point $(1 : 0 : 0)$ (resp.

$(0:0:1)$) is on the line determined by $\mathbf{v}$ if and only if $v_2 = 0$ (resp. $v_1 = 0$) which are precisely the conditions for the projection to have a singularity more degenerate than a cross-cap.

(ii) At a non-degenerate inflection point, we take $(Q_1, Q_2) = (x^2 \pm y^2, 0)$, so that $j^2 P_{\mathbf{v}}(x, y) = (v_2 x - v_1 y, x^2 \pm y^2, 0)$. Following the same calculations as in (i) above, if $v_2 \neq 0$, we can reduce the $j^2 P_{\mathbf{v}}$ to $(X, 2v_1 XY + (v_1^2 \pm v_2^2)Y^2, 0)$. Completing the square in the second component, we can reduce further the 2-jet to $(X, Y^2, 0)$ when $v_1^2 \pm v_2^2 \neq 0$ or to $(X, XY, 0)$ when $v_1^2 \pm v_2^2 = 0$ (but $v_2 \neq 0$). This means that most directions of projections yield singularities of type S_k and B_k, and there are 2 directions (when the inflection point is of real type) or 0 direction (when the inflection point is of imaginary type) yielding singularities of H_k-type. □

Asymptotic directions can also be described via the singularities of projections to hyperplanes.

Proposition 7.15 ([Mond (1982); Bruce and Nogueira (1998)]). *A tangent direction $\mathbf{v}$ at p on M is an asymptotic direction if and only if the projection in the direction $\mathbf{v}$ yields a singularity more degenerate than a cross-cap.*

Proof. We take the surface as before in Monge form. If p is not an inflection point, a tangent direction $\mathbf{v} = (v_1, v_2, 0, 0)$ at p is an asymptotic direction if and only if there exists b_1, b_2 such that $b_1 Q_1 + b_2 Q_2$ is degenerate at p and $\mathbf{v}$ is in the kernel of its Hessian at p, that is, $(v_2 x - v_1 y)^2 = b_1 Q_1 + b_2 Q_2$. But this means that the pencil generated by Q_1 and Q_2 meets the conic Γ at $(v_2 x - v_1 y)^2$, which by Proposition 7.14 is equivalent to the projection along $\mathbf{v}$ having a singularity more degenerate than a cross-cap.

If p is an inflection point then every tangent direction at p is an asymptotic at p, and the projection along these directions is more degenerate than a cross-cap. □

Proposition 7.16. *Let M be a projection P-generic surface in $\mathbb{R}^4$. The $\mathcal{A}_e$-codimension 2 singularities of the family P occur on curves on the surface and the $\mathcal{A}_e$-codimension 3 singularities are special points on these curves.*

Proof. This follows from the assumption that P is a versal unfolding, hence it is transversal to the $\mathcal{A}$-orbits in jet space. □

Definition 7.8. We call S_2-curve (resp. B_2-curve and H_2-curve) the closure of the set of points p on M for which there exists a projection $P_{\mathbf{v}}$

having an S_2 (resp. B_2, H_2)-singularity at p.

Proposition 7.17. *Let M be a projection P-generic surface in $\mathbb{R}^4$. Then the H_2-curve coincides with the set $\Delta = 0$ and the B_2-curve coincides with the closure of the set of point where the height functions has an A_3-singularity.*

Proof. The statement on the H_2-curve follows from the proof of Proposition 7.14.

Again, from Proposition 7.14, we can take the point p to be a hyperbolic point. We work with the surface in Monge form $\phi(x, y) = (x, y, f_1(x, y), f_2(x, y))$ with $j^3\phi(x, y)$ given by

$$(x, y, x^2+a_{30}x^3+a_{31}x^2y+a_{32}xy^2+a_{33}y^3, y^2+b_{30}x^3+b_{31}x^2y+b_{32}xy^2+b_{33}y^3).$$

Consider the projection along the asymptotic direction $\mathbf{v} = (0, 1, 0, 0)$ so that $P_{\mathbf{v}}(x, y) = (x, f_1(x, y), f_2(x, y))$. Then $j^3P(x, y)$ is $\mathcal{A}^{(3)}$-equivalent to $(x, a_{31}x^2y + a_{33}y^3, y^2)$. The singularity is of type B_k, $k \geq 1$, if and only if $a_{31} \neq 0$ and of type $B_{\geq 2}$ if further more $a_{33} = 0$.

The binormal direction associated to $\mathbf{v}$ is $\mathbf{w} = (0, 0, 1, 0)$ and the height function along $\mathbf{w}$ is given by $h_{\mathbf{w}} = x^2 + a_{30}x^3 + a_{31}x^2y + a_{32}xy^2 + a_{33}y^3$. It has an $A_{\geq 3}$ singularity if and only if $a_{33} = 0$. Thus the B_2-curve (of $P_{\mathbf{v}}$ with $\mathbf{v}$ an asymptotic direction) is also the locus of points where the height function $h_{\mathbf{w}}$ (along the binormal direction $\mathbf{w}$ associated to $\mathbf{v}$) has an $A_{\geq 3}$-singularity. □

Remark 7.7. (1). The results in Proposition 7.17 follow in fact from the duality in [Bruce and Nogueira (1998)] between certain strata of the bifurcation set of the family of height functions with that of the family of projections.

(2) The A_4-singularities $h_{\mathbf{w}}$ are in general not related to the B_3-singularities of $P_{\mathbf{v}}$.

(3) The C_3-singularity of $P_{\mathbf{v}}$ occurs at a point of intersection of the S_2-curve and B_2-curve. The S_2-curve is in general not related to the singularities of the height functions.

(4) For a projection generic surface, the robust curves captured by the singularities of $P_{\mathbf{v}}$ are as in Figure 7.7. It is shown in [Bruce and Tari (2002)] that the B_2-curve and the Δ-set (i.e., the parabolic curve) meet tangentially. The point of tangency is a $P_3(c)$-singularity of a $P_{\mathbf{v}}$. The two asymptotic direction fields in the hyperbolic region can be coloured, so we have two different coloured B_2-curves. The B_2-curve changes colour at a

$P_3(c)$-singularity. The $P_3(c)$-singularity is also a point where the asymptotic configurations have a folded singularity, i.e., they are as in Figure 7.2.

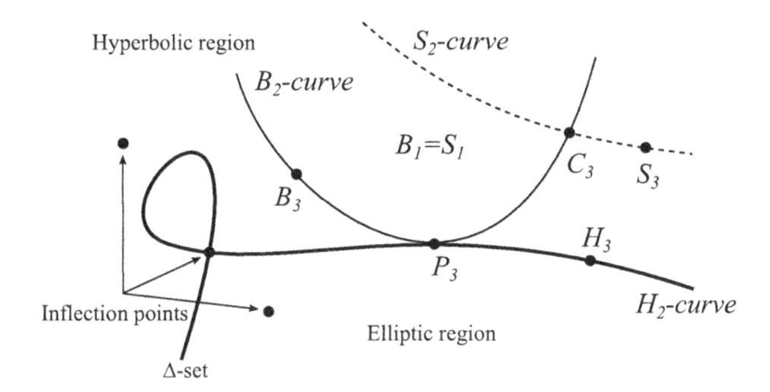

Fig. 7.7 Robust curves on the surface depicted by its contact with lines.

7.7.1 *The geometry of the projections*

The surface $P_{\mathbf{v}}(M)$ lives in $\mathbf{v}^{\perp} = T_{\mathbf{v}}S^3$, and we are interested in its geometry as a surface in a 3-dimensional Euclidean space. For this reason, we consider its contact with planes in $T_{\mathbf{v}}S^3$ which is measured by the height functions on $P_{\mathbf{v}}(M)$ in $T_{\mathbf{v}}S^3$. These planes can be parametrised by unit vectors $\mathbf{w}$ in $T_{\mathbf{v}}S^3$, i.e., $\mathbf{w}\cdot\mathbf{v} = 0$ and $\mathbf{w}\cdot\mathbf{w} = 1$. We denote by

$$\mathcal{D} = \{(\mathbf{v},\mathbf{w}) \in S^3 \times S^3 \,:\, \langle \mathbf{v},\mathbf{w}\rangle = 0\}.$$

Given $(\mathbf{v},\mathbf{w}) \in \mathcal{D}$, the height function on the projected surface $P_{\mathbf{v}}(M)$ along the vector $\mathbf{w}$ is given by

$$H_{(\mathbf{v},\mathbf{w})}(u) = \langle P_{\mathbf{v}}(u),\mathbf{w}\rangle = \langle \mathbf{x}(u) - \langle \mathbf{x}(u),\mathbf{v}\rangle\mathbf{v},\mathbf{w}\rangle = \langle \mathbf{x}(u),\mathbf{w}\rangle.$$

This is precisely the height function on M along the direction $\mathbf{w}$. In particular, it follows that

Proposition 7.18. *The height function $H_{(\mathbf{v},\mathbf{w})}$ on $P_{\mathbf{v}}(M)$ along the direction $\mathbf{w}$ has the same singularity type at $P_{\mathbf{v}}(p)$ as the height function $H_{\mathbf{w}}$ on M along $\mathbf{w}$ at p.*

The family $H : U\times\mathcal{D} \to \mathbb{R}$ has parameters in $\mathcal{D}$ which is a 5-dimensional manifold. However, it is trivial along the parameter $\mathbf{v}$. Thus, the generic singularities that can appear in $H_{(\mathbf{v},\mathbf{w})}$ are those of $\mathcal{R}_e$-codimension ≤ 3.

For $\mathbf{v}$ fixed, $\mathbf{w}$ varies in a 2-dimensional sphere, so for a generic M and for most directions $\mathbf{v}$, the height function on $P_{\mathbf{v}}(M)$ has singularities of type $A_1^{\pm}$, A_2 and $A_3^{\pm}$, and these are versally unfolded by varying $\mathbf{w}$. (Recall from Chapter 6 that the closure of the A_2-singularities form the parabolic set on $P_{\mathbf{v}}(M)$, the A_3-singularities are its cusps of Gauss.) For isolated directions $\mathbf{v}$, we expect the following singularities: A_4, $D_4^{\pm}$ and an A_3-singularity which is not versally unfolded by the family $H_{\mathbf{v}}$. We denote the latter by NVA_3.

Definition 7.9. We call the pre-image on M by $P_{\mathbf{v}}$ of the parabolic set of $P_{\mathbf{v}}(M)$ as a surface in the 3-space $T_{\boldsymbol{v}v}S^3$ the $\mathbf{v}$-pre-parabolic set and denote it by $\mathbf{v}$-PPS.

We consider how the geometric data of $P_{\mathbf{v}}(M)$ at $P_{\mathbf{v}}(p)$ can provide geometric information about M at p. (More details on this can be found in [Nuño-Ballesteros and Tari (2007); Oset Sinha and Tari (2010)].)

We start with the case when $\mathbf{v}$ is not a tangent direction at p. We write $\mathbf{v} = \mathbf{v}_T + \mathbf{v}_N$ where $\mathbf{v}_T$ is the orthogonal projection of $\mathbf{v}$ to the tangent space T_pM and $\mathbf{v}_N$ is its orthogonal projection to the normal space N_pM. Since $\mathbf{v}_N \neq 0$, the surface $P_{\mathbf{v}}(M)$ is a smooth surface at $P_{\mathbf{v}}(p)$.

Proposition 7.19. *The height function $H_{(\mathbf{v},\mathbf{w})}$ on $P_{\mathbf{v}}(M)$ is singular at $P_{\mathbf{v}}(p)$ if and only if $\mathbf{w} \in N_pM$. For a generic surface, the singularity of $H_{(\mathbf{v},\mathbf{w})}$ at $P_{\mathbf{v}}(p)$ is of type*

- A_2: *if p is a hyperbolic or parabolic point, $\mathbf{w} = \mathbf{v}_N^{\perp}$ and is a binormal direction.*
- A_3: *$\mathbf{w} = \mathbf{v}_N^{\perp}$ is a binormal direction, p is on the B_2-curve and $\mathbf{v}$ is away from a circle of directions C in the sphere $\mathbf{w}^{\perp} \in \mathcal{D}$. Then the $\mathbf{v}$-PPS is a regular curve.*
- NVA_3 : *$\mathbf{w} = \mathbf{v}_N^{\perp}$ is a binormal direction, p is on the B_2-curve and $\mathbf{v} \in C$. For generic $\mathbf{v} \in C$ the singularity of the $\mathbf{v}$-PPS is an A_1. For isolated directions in C the singularity becomes an A_2, and for special points on the B_2-curve it becomes an A_3-singularity.*
- A_4 *$\mathbf{w} = \mathbf{v}_N^{\perp}$ is a binormal direction, p is an A_4-point on the B_2-curve.*
- D_4: *$\mathbf{w} = \mathbf{v}_N^{\perp}$ is a binormal direction, p is an inflection point.*

Proof. The proof can be found in [Oset Sinha and Tari (2010)]; see also [Nuño-Ballesteros and Tari (2007)]. □

Suppose now that $\mathbf{v} \in S^3$ is a tangent direction at $p \in M$. Then $P_{\mathbf{v}}(M)$ is a singular surface $P_{\mathbf{v}}(p)$. The map-germ $P_{\mathbf{v}}$ is of corank 1 and these can be written, in some coordinate system, in the form

$$\psi(x,y) = (x, p(x,y), q(x,y))$$

with $p, q \in \mathcal{M}^2(x,y)$. We denote by $Q_1(x,y) = j^2p(x,y) = p_{20}x^2 + p_{21}xy + p_{22}y^2$ and $Q_2(x,y) = j^2q(x,y) = q_{20}x^2 + q_{21}xy + q_{22}y^2$. We can consider the action of $GL(2,\mathbb{R}) \times GL(2,\mathbb{R})$ on (Q_1, Q_2) as in §7.2.1 to define an affine property of the singular surface S, image of ψ, at its singular point.

Definition 7.10. The singular point of the surface S is called hyperbolic (resp. elliptic, parabolic) or (a generic) inflection point if the $GL(2,\mathbb{R}) \times GL(2,\mathbb{R})$-class of (Q_1, Q_2) can be represented by (x^2, y^2) (resp. $(x^2, x^2 - y^2)$, (x^2, xy)) or $(x^2 \pm y^2, 0)$, as in Table 7.1.

The surface $S = P_{\mathbf{v}}(M)$ has a cross-cap singularity if and only if $\mathbf{v} \in T_pM$ is not an asymptotic direction at p. It is shown in [Bruce and West (1998); West (1995)] that a parametrisation of a cross-cap can be taken, by a suitable choice of a coordinate system in the source and affine changes of coordinates in the target, in the form

$$\psi(x,y) = (x, xy + f_1(y), y^2 + ax^2 + f_2(x,y)), \tag{7.14}$$

where $f_1 \in \mathcal{M}^4(y)$ and $f_2 \in \mathcal{M}^3(x,y)$. The following is also shown in [West (1995)]. When $a < 0$, the height function along any normal direction at the cross-cap point has an A_1-singularity. Such cross-caps are labelled *hyperbolic cross-caps* as the surface has negative Gaussian curvature at all its regular points (Figure 7.8, left). When $a > 0$, there are two normal directions $(0, \pm 2\sqrt{a}, 1)$ at the cross-cap point along which the height function has a singularity more degenerate than A_1 (i.e., of type $A_{\geq 2}$). Such a cross-cap is labelled *elliptic cross-cap* (Figure 7.8, right). The singularity of the height function along the degenerate normal direction is precisely of type A_2 if and only if $j^3 f_2(\mp\frac{1}{\sqrt{a}}, 1) \neq 0$. When $a = 0$, there is a unique normal direction at the cross-cap point where the height function has a singularity more degenerate than A_1. The singularity of its corresponding height function is of type A_2 if and only if $\frac{\partial^3 f_2}{\partial x^3}(0,0) \neq 0$. Such a cross-cap is labelled *parabolic cross-cap* (Figure 7.8, centre).

Theorem 7.10. *A cross-cap is hyperbolic (resp. elliptic, parabolic) if and only if its singular point is elliptic (resp. hyperbolic, parabolic) as in* Definition 7.10.

Fig. 7.8 Hyperbolic cross-cap (left) and elliptic cross-cap (right) separated by a parabolic cross-cap (centre).

Proof. The pair of quadratic forms associated to ψ in (7.14) is $(xy, y^2 + ax^2)$. This is $GL(2,\mathbb{R}) \times GL(2,\mathbb{R})$-equivalent to $(xy, x^2 - y^2)$, (x^2, y^2) or (x^2, xy) in Table 7.1 if and only if $a < 0$, $a > 0$ or $a = 0$, and the result follows from the discussion above. $\square$

The surface $S = P_{\mathbf{v}}(M)$ has a cross-cap singularity when $\mathbf{v} \in T_pM$ is not an asymptotic direction at p. We can say more about $P_{\mathbf{v}}(M)$ and M.

Theorem 7.11. *Suppose that $\mathbf{v} \in T_pM$ but is not an asymptotic direction (in particular, p is not an inflection point).*

(i) *The $\mathbf{v}$-PPS has a Morse singularity if $p \notin \Delta$-set. Furthermore, the singularity is of type A_1^- (i.e. $P_{\mathbf{v}}(M)$ is an elliptic cross-cap) if p is a hyperbolic point and of type A_1^+ (i.e. $P_{\mathbf{v}}(M)$ is a hyperbolic cross-cap) if p is an elliptic point.*

(ii) *At points on Δ-set that are not on the B_2-curve, the $\mathbf{v}$-PPS has an A_2-singularity (i.e. $P_{\mathbf{v}}(M)$ is a parabolic cross-cap). At the point of tangency of the B_2-curve with Δ-set, the singularity of the $\mathbf{v}$-PPS is generically of type A_3.*

(iii) *The tangent directions to the $\mathbf{v}$-PPS are along the asymptotic directions to M at p.*

Proof. We take M in Monge form $\phi(x,y) = (x, y, f_1(x,y), f_2(x,y))$ at the origin p. Suppose that p is a hyperbolic point, so we can take $(Q_1, Q_2) = (x^2, y^2)$. The asymptotic directions at p are $(1,0,0,0)$ and $(0,1,0,0)$. We consider a tangent vector $\mathbf{v} = (\alpha, \beta, 0, 0)$ which is not an asymptotic direction, that is, $\alpha\beta \neq 0$.

Calculating the Gaussian curvature of $P_{\mathbf{v}}(M)$ away from the singular point, one can get the $\mathbf{v}$-PPS as the zero set of the function

$$\tilde{K}_{\mathbf{v}} = \det(\mathbf{v}, \phi_x, \phi_y, \phi_{xx}) \det(\mathbf{v}, \phi_x, \phi_y, \phi_{yy}) - \det(\mathbf{v}, \phi_x, \phi_y, \phi_{xy})^2.$$

We have $j^2\tilde{K}_{\mathbf{v}}(x,y) = -16\alpha\beta xy$, which has an A_1^--singularity at the origin. It is clear that its tangent directions coincide with the asymptotic directions of M at the origin.

Analogously, if p is an elliptic point, we take $(Q_1, Q_2) = (xy, x^2 - y^2)$ and $\mathbf{v} = (\alpha, \beta, 0, 0)$, with $\alpha^2 + \beta^2 = 1$. (Recall that there are no asymptotic directions at an elliptic point.) Then $j^2\tilde{K}_{\mathbf{v}}(x,y) = -4(x^2 + y^2)$ and this has an A_1^+-singularity at the origin, that is the $\mathbf{v}$-PPS is locally an isolated point.

Suppose now that $p \in \Delta$. We take $(Q_1, Q_2) = (x^2, xy)$ and let $\mathbf{v} = (\alpha, \beta, 0, 0)$ be a tangent vector with $\alpha \neq 0$ ($\alpha = 0$ gives the unique asymptotic direction at p). We have $j^2\tilde{K}_{\mathbf{v}}(x,y) = -4\alpha^2x^2$, so the $\mathbf{v}$-PPS has an A_k-singularity, with $k \geq 2$. Note that we have one tangent direction to the $\mathbf{v}$-PPS which is exactly the unique asymptotic direction at the origin. The coefficients of y^3 in $j^3\tilde{K}_{\mathbf{v}}(x,y)$ is $12a_{33}\alpha^2$. Hence, for a generic point on the Δ-set the $\mathbf{v}$-PPS has an A_2-singularity. The points where $a_{33} = 0$ correspond to points where the height function along the unique binormal direction $(0,0,1,0)$ has an $A_{\geq 3}$-singularity. That is p is also on the B_2-curve, so p is the point of tangency of Δ-set and the B_2-curve. For a generic surface, $\tilde{K}_{\mathbf{v}}$ has an A_3-singularity at such a point. □

Definition 7.11. We call an *elliptic cross-cap* where the height function has an A_i-singularity along one degenerate direction and an A_j-singularity along the other degenerate direction an *elliptic cross-cap of type* A_iA_j or an A_iA_j*-elliptic cross-cap*. Likewise, we label an A_k*-parabolic cross-cap* one where the height function has a degenerate singularity (of type A_k) along the unique degenerate normal direction.

Proposition 7.20. *Suppose that* $\mathbf{v} \in T_pM$ *but is not an asymptotic direction at* p.

(i) *If* p *is a hyperbolic point, then* $P_{\mathbf{v}}(M)$ *is a surface with an elliptic cross-cap of type* A_2A_2 *if* p *is not on the* B_2*-curve. If it is, the elliptic cross-cap becomes of type* A_2A_3 *and at isolated points on this curve it can be of type* A_2A_4 *or* A_3A_3.

(ii) *If* p *is a parabolic point, then* $P_{\mathbf{v}}(M)$ *is in general an* A_2*-parabolic cross-cap and becomes an* A_3*-parabolic cross-cap if* p *is the point of tangency of the* B_2*-curve with the* Δ*-set.*

Proof. The type of the cross-cap is determined by the singularities of the height function $H_{(\mathbf{v},\mathbf{w}_i)}$ on $P_{\mathbf{v}}(M)$ at $P_{\mathbf{v}}(p)$ along the binormal directions $\mathbf{w}_i$, $i = 1, 2$. The result follows from Proposition 7.18 that these are the same as the singularities of the height function $H_{\mathbf{w}_i}$ on M at p and from Proposition 7.17.

In (i), the A_2A_4 cross-cap occurs at special points on the B_2-curve where the height function has an A_4-singularity and these are distinct in general from the B_3 and C_3-points. The A_3A_3 cross-cap occurs at the point of intersection of two B_2-curves associated to the two binormal directions. $\square$

When the direction of projection is asymptotic at p, $P_{\mathbf{v}}$ has a singularity more degenerate than a cross-cap. The $\mathbf{v}$-PPS has also singularities more degenerate than Morse. For a projection P-generic surface, these are as in Table 7.4.

Table 7.4 The singularities of $P_{\mathbf{v}}$ and of the $\mathbf{v}$-PPS.

$P_{\mathbf{v}}$	$B_1^{\pm}$	B_2	B_3	S_2	S_3	C_3	H_2	H_3	P_3
$\mathbf{v}$-PPS	$D_4^{\mp}$	D_5	D_5	E_7	J_{10}	$X_{1,0}$	D_5	D_5	J_{10}

Recall that the parabolic curve on a surface in 3-space is the locus of degenerate singularities of the height function. From Proposition 7.18, the degenerate singularities of the height function on $P_{\mathbf{v}}(M)$ are along the binormal directions of the surface M. Thus, at a hyperbolic point where are two distinct binormal directions, we can split the $\mathbf{v}$-PPS into two components with each component corresponding to one of the binormal direction. We denote $\mathcal{L}_1$ and $\mathcal{L}_2$ these components. Analysing in Table 7.5 the singularities of $\mathcal{L}_1$ and $\mathcal{L}_2$ give a better understanding why the singularities of the $\mathbf{v}$-PPS (which is that of $\mathcal{L}_1 \cup \mathcal{L}_2$) is highly degenerate (see Table 7.4). In Table 7.5, "tg" is for tangency and "$\pitchfork$" is for transversality between the components $\mathcal{L}_1$ and $\mathcal{L}_2$.

7.8 Contact with planes

We discuss briefly in this section orthogonal projections of a surface to a plane. Some of their aspects are studied in [Nogueira (1998)].

Consider the orthogonal projection from $\mathbb{R}^4$ to a 2-dimensional vector space π (which we refer to as a plane). Its kernel is the plane $\pi^\perp$, the orthogonal complement of π. The set of all planes in $\mathbb{R}^4$ is the Grassmanian $\mathrm{Gr}(2,4)$, which can be used to parametrise all the orthogonal projections from $\mathbb{R}^4$ to planes. Clearly, $\mathrm{Gr}(2,4)$ can be used to parametrise either the planes of projections or their orthogonal complements.

Let $\{\mathbf{u}, \mathbf{v}\}$ be an orthonormal basis of $\pi^\perp$ and denote by $\Pi_{(\mathbf{u},\mathbf{v})}$ the

Table 7.5 The generic structure of the $\mathbf{v}$-PPS and of its two components.

S	B_1	B_2	B_3
$\mathcal{L}_1$	$A_1^{\pm}$ 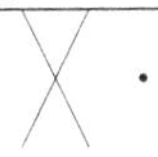	A_2 	A_2 
$\mathcal{L}_2$	A_0 (⫛) 	A_0 (⫛) 	A_0 (⫛) 
$\mathbf{v}$-PPS	$D_4^{\pm}$ 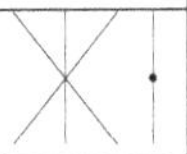	D_5 	D_5 

S	S_2	S_3	C_3
$\mathcal{L}_1$	A_2 	$A_3^{\pm}$ 	$D_4^{\pm}$ 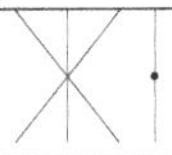
$\mathcal{L}_2$	A_0 (tg) 	A_0 (tg) 	A_0 (tg) 
$\mathbf{v}$-PPS	E_7 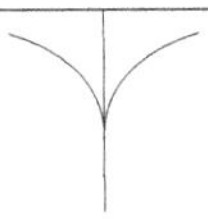	J_{10} 	$X_{1,0}$ 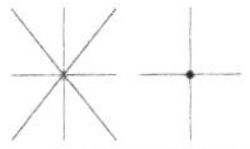

orthogonal projection from $\mathbb{R}^4$ to π. Then,

$$\Pi_{(\mathbf{u},\mathbf{v})}(p) = p - \langle p, \mathbf{u}\rangle\mathbf{u} - \langle p, \mathbf{v}\rangle\mathbf{v}. \tag{7.15}$$

Now $(\mathbf{u}, \mathbf{v})$ is in $\mathcal{D} = \{(\mathbf{u}, \mathbf{v}) \in S^3 \times S^3 : \langle \mathbf{u}, \mathbf{v}\rangle = 0\}$ which is a 5-dimensional manifold. However, the choice of an orthonormal basis in $\pi^{\perp}$ is determined up-to rotations, so we have an action of SO(2) on each plane. This gives the quotient space $\mathcal{D}/\mathrm{SO}(2)$ which is a 4-dimensional manifold and can be identified locally with $\mathrm{Gr}(2, 4)$.

Given a surface M in $\mathbb{R}^4$, we still denote by $\Pi_{(\mathbf{u},\mathbf{v})}$ the restriction of $\Pi_{(\mathbf{u},\mathbf{v})}$ to M. This map can be considered locally at a point $p \in M$ as a map-germ $\Pi_{(\mathbf{u},\mathbf{v})} : \mathbb{R}^2, 0 \to \mathbb{R}^2, 0$. Since $\mathcal{D}/\mathrm{SO}(2)$ is a 4-dimensional manifold, using the transversality theorem, we expect the map $\Pi_{(\mathbf{u},\mathbf{v})}$ to have only simple singularities of $\mathcal{A}_e$-codimension ≤ 4 or non-simple singularities with the $\mathcal{A}_e$-codimension of the stratum ≤ 4.

The corank 1 singularities of $\mathcal{A}_e$-codimension ≤ 4 are given in Table 7.6. They are taken from the classification in [Rieger (1987)].

The $\mathcal{A}$-simple germs of corank 2 are classified in [Rieger and Ruas (1991)]. There are two such families:

$$I_{2,2}^{l,m} : (x^2 + y^{2l+1}, y^2 + x^{2m+1}),\ l \geq m \geq 1$$
$$II_{2,2}^{l} : \ (x^2 - y^2 + x^{2l+1}, xy), \quad l \geq 1.$$

Table 7.6 Corank 1 singularities of projections of surfaces in $\mathbb{R}^4$ to 2-planes.

Name	Normal form	$\mathcal{A}_e$-codimension
1 : Immersion	(x, y)	0
2 : Fold	(x, y^2)	0
3 : Cusp	$(x, xy + y^3)$	0
4_k	$(x, y^3 \pm x^k y), k \geq 2$	$2 \leq k \leq 5$
5	$(x, xy + y^4)$	1
6	$(x, xy + y^5 \pm y^7)$	2
7	$(x, xy + y^5)$	3
8	$(x, xy + y^6 \pm y^8 + ay^9)$	$5(4^{(*)})$
9	$(x, xy + y^6 + y^9)$	4
10	$(x, xy + y^7 \pm y^9 + ay^{10} + by^{11})$	$6(4^{(*)})$
11_{2k+1}	$(x, xy^2 + y^4 + y^{2k+1}), k \geq 2$	$k = 2, 3, 4$
12	$(x, xy^2 + y^5 + y^6)$	3
13	$(x, xy^2 + y^5 \pm y^9)$	4
15	$(x, xy^2 + y^6 + y^7 + ay^4)$	$5(4^{(*)})$
16	$(x, x^2y + y^4 \pm y^5)$	3
17	$(x, x^2y + y^4)$	4
18	$(x, x^2y + xy^3 + ay^5 + y^6 + by^7)$	$6(4^{(*)})$

$(*)$ The codimension is that of the stratum.

The classification of corank 2 non-simple singularities with $\mathcal{A}_e$-codimension of the stratum ≤ 4 so far is not complete. They are $\mathcal{K}$-equivalent to $(x^2 + y^3, xy)$ when the $\mathcal{A}_e$-codimension of the stratum is 3 and to $(x^2 + y^4, xy)$ when this codimension is 4.

An orthogonal projection of M to a plane π is singular at $p \in M$ if and only if $\dim(T_pM \cap \pi^\perp) \geq 1$. If the dimension is 1, the singularity is of corank 1, otherwise it is of corank 2.

Suppose that $\dim(T_pM \cap \pi^\perp) = 1$ and let $\mathbf{u}$ be a unit vector in $T_pM \cap \pi^\perp$. Let $\mathbf{v}$ be a unit vector in $\pi^\perp$ orthogonal to $\mathbf{u}$. (Then $\mathbf{v}$ belongs to the 2-sphere $\{(\mathbf{u}, \mathbf{v}) \in S^3 \times S^3 : \langle \mathbf{u}, \mathbf{v} \rangle = 0\}$ with the poles $\pm \mathbf{u}^\perp$ removed, where $\mathbf{u}^\perp$ is orthogonal to $\mathbf{u}$ in T_pM.) To simplify notation, we take M in Monge form at the origin and suppose, without loss of generality, that $\mathbf{u} = (1, 0, 0, 0)$. Then $\mathbf{v} = (0, v_2, v_3, v_4)$ with either v_3 or v_4 not zero. It follows from (7.15) that

$$\begin{aligned}
\Pi_{(\mathbf{u},\mathbf{v})}(x, y) &= (x, y, f_1(x, y)), f_2(x, y)) - (x, 0, 0, 0) \\
&\quad -(v_2 y + v_3 f_1(x, y) + v_4 f_2(x, y))(0, v_2, v_3, v_4) \\
&= (0, y - (v_2 y + v_3 f_1(x, y) + v_4 f_2(x, y))v_2, \\
&\quad f_1(x, y) - (v_2 y + v_3 f_1(x, y) + v_4 f_2(x, y))v_3, \\
&\quad f_2(x, y) - (v_2 y + v_3 f_1(x, y) + v_4 f_2(x, y))v_4).
\end{aligned}$$

This is $\mathcal{A}$-equivalent to

$$((1 - v_2^2)y - (v_3 f_1(x, y) + v_4 f_2(x, y))v_2, v_4 f_1(x, y) - v_3 f_2(x, y)).$$

Observe that $(1 - v_2^2) \neq 0$ so at the 2-jet level we have

$$j^2\Pi_{(\mathbf{u},\mathbf{v})}(x,y) \simeq_{\mathcal{A}^{(2)}} (y, v_4Q_1(x,y) - v_3Q_2(x,y)),$$

with the second component tracing the pencil (Q_1, Q_2) in $\mathbb{RP}^2$. As we fixed $\mathbf{u}$, we cannot take (Q_1, Q_2) as one of the normal forms in Table 7.1. We have

$$j^2\Pi_{(\mathbf{u},\mathbf{v})} \simeq_{\mathcal{A}^{(2)}} (y, (v_4l_1 - v_3l_2)x^2 + 2(v_4m_1 - v_3m_2)xy + (v_4n_1 - v_3n_2)y^2),$$

where l_i, m_i, n_i, $i = 1, 2$ denote the coefficients of the second fundamental form at the origin.

If $v_4l_1 - v_3l_2 \neq 0$, the singularity is a fold, and if $v_4l_1 - v_3l_2 = 0$ but $v_4m_1 - v_3m_2 \neq 0$, then $j^2\Pi_{(\mathbf{u},\mathbf{v})} \simeq_{\mathcal{A}^{(2)}} (y, xy)$. This 2-jet leads to the singularities of type $2, 3$ and $5, \cdots, 10$ in Table 7.6. These singularities are characterised by the property that the singular set $\Sigma_{(\mathbf{u},\mathbf{v})}$ of $\Pi_{(\mathbf{u},\mathbf{v})}$ in M is a smooth curve at p.

Now $v_4l_1 - v_3l_2 = v_4m_1 - v_3m_2 = 0$ if and only if $l_1m_2 - l_2m_1 = 0$, equivalently, $\mathbf{u}$ is an asymptotic direction at p (see Theorem 7.3). The condition $v_4l_1 - v_3l_2 = v_4m_1 - v_3m_2 = 0$ also means that the direction $\mathbf{w} = (0, 0, -v_4, v_3)$ is the binormal direction associated to $\mathbf{u}$. Observe that the vector $\mathbf{w}$ is in $N_pM \cap \pi$. We conclude that at an elliptic point only the singularities of type $2, 3$ and $5, \cdots, 10$ can occur. At a non-elliptic point, we get these singularities too but also those with a 2-jet equivalent to $(y, 0)$. The latter singularities occur when the plane of projections is in the pencil of planes in the 3-space $\mathbf{u}^\perp$ that contain the binormal direction associated to the asymptotic direction $\mathbf{u}$. At a hyperbolic point there are two such pencils associated to the two distinct asymptotic directions and at a parabolic point which is not an inflection point there is only one such pencil. At an inflection point, there is a unique binormal direction but any tangent direction is asymptotic. Thus any plane of projection that contains the binormal direction gives a singularity with a singular critical set.

We consider now the corank 2 singularities of $\Pi_{(\mathbf{u},\mathbf{v})}$ and take M in Monge form as above. The vectors $\mathbf{u}$ and $\mathbf{v}$ generate T_pM, so we can take them to be $\mathbf{u} = (1, 0, 0, 0)$ and $\mathbf{v} = (0, 1, 0, 0)$. Then $\Pi_{(\mathbf{u},\mathbf{v})}(x, y) = (f_1(x, y), f_2(x, y))$. At the 2-jet level, we have $j^2\Pi_{(\mathbf{u},\mathbf{v})}(x, y) = (Q_1(x, y), Q_2(x, y))$ and we have the classification in the 2-jet space given by that of pairs of quadratic forms in Theorem 7.1 (see also Table 7.2). Of course that classification is related to the partition of the surface into elliptic, hyperbolic, parabolic and inflection points.

7.9 Contact with hyperspheres

We give in this section geometric characterisations of the generic singularities of the distance-squared functions on the surface. We refer to [Montaldi (1983)] for the study of the contact of the surface with k-spheres, $k = 1, 2, 3$.

The family of distance squared functions $D : U \times \mathbb{R}^4 \to \mathbb{R}$ on M is given by

$$D(u, a) = ||\mathbf{x}(u) - a||^2,$$

and the extended family of distance squared functions $\tilde{D} : U \times \mathbb{R}^4 \times \mathbb{R} \to \mathbb{R}$ is defined by

$$\tilde{D}(x, a, r) = ||\mathbf{x}(u) - a||^2 - r^2.$$

The functions D_a and $\tilde{D}_{a,r}$ have a singularity at u if and only if the direction $a - \mathbf{x}(u)$ is normal to M at $p = \mathbf{x}(u)$.

The catastrophe set C_D of the family D coincides with the normal bundle

$$NM = \{(u, a) \in U \times \mathbb{R}^4 |\, a = \mathbf{x}(u) + \lambda \mathbf{v},\ \mathbf{v} \in (N_pM)_1,\ \lambda \in \mathbb{R}\}$$

and its bifurcation set B_D is the focal set of M. It follows that the catastrophe map of the family D coincides with the map $G^e : NM \to \mathbb{R}^4$ defined by $G^e(u, a) = a$, which we call the *normal exponential map* of M.

Theorem 7.12. *There is an open and dense set $\mathcal{O}_D$ in $Imm(U, \mathbb{R}^4)$ such that for any $\mathbf{x} \in \mathcal{O}_D$, the surface $M = \mathbf{x}(U)$ has the following properties:*

i) *The distance squared function D_a has only singularities of $\mathcal{R}$-type A_k, $1 \le k \le 5$, D_4 and D_5 at any point p in M. Similarly, the function $\tilde{D}_{a,r}$ has only singularities of $\mathcal{K}$-type A_k, $1 \le k \le 5$, D_4 and D_5 at p.*
ii) *The singularities of D_a are $\mathcal{R}^+$-versally unfolded by the family D, and the singularities of $\tilde{D}_{a,r}$ are $\mathcal{K}$-versally unfolded by the family $\tilde{D}$.*
iii) *The normal exponential map G^e is stable as a Lagrangian map.*

Proof. The proofs of (i) and (ii) follow from Theorem 4.8. Since the catastrophe map of a Morse family of functions can be identified with the Lagrangian map (cf. §5.4), G^e is a Lagrangian map. The proof of (iii) follows from Theorem 5.4. □

Definition 7.12. An immersion $\mathbf{x} : M \to \mathbb{R}^4$ is called *distance squared function generic* if it belongs to the set $\mathcal{O}_D$ in Theorem 7.12.

Definition 7.13. The points of B_D are called *focal points* of M. If $p \in M$ is a degenerate singularity of D_a, with $a \in B_D$, we say that the hypersphere centred at a and tangent to M at $p = \mathbf{x}(u)$ is an *osculating hypersphere* of M at p. In such case, the kernel of the Hessian of D_a at u, $Hess(D_a)(u)$, is non-zero and the directions in this kernel are called *spherical contact directions* of M at p.

We prove next that the spherical contact directions are principal directions associated to normal fields pointing towards the focal points of the surface.

Lemma 7.2. *For* $p = \mathbf{x}(u) \in M$, $a = p + \lambda\mathbf{v}$, *with* $\mathbf{v} \in (N_pM)_1$,

$$Hess(D_a(u)) = 2\big(Id - \lambda Hess(h_{\mathbf{v}})(u)\big).$$

Proof. The proof follows from the definitions of D_a and $h_{\mathbf{v}}$. □

Lemma 7.3. *Let a be a focal centre of M at $p = \mathbf{x}(u)$ and let $\mathbf{w} \in T_pM$ be a spherical contact direction associated to the distance squared function D_a, with $a = p + \lambda\mathbf{v}$. Then $\mathbf{w}$ is an eigenvector of the shape operator $W_p^{\mathbf{v}}$ in the normal direction $\mathbf{v}$ at p with eigenvalue $1/\lambda$.*

Proof. The matrices of $W_p^{\mathbf{v}}$ and $Hess(h_{\mathbf{v}})(u)$ coincide at p. Then $\mathbf{w}$ is a spherical contact direction associated to the distance squared function d_a, with $a = p\lambda\mathbf{v}$ if and only if $\det Hess(d_a)(u) = 0$. It follows from Lemma 7.2 that $W_p^{\mathbf{v}}(\mathbf{w}) = 1/\lambda\mathbf{w}$. Clearly, $1/\lambda$ is a $\mathbf{v}$-principal curvature at p. □

Definition 7.14. The *strong principal directions* at $p \in M$ are the spherical contact directions corresponding to singularities at p of type $A_k, k \geq 3$, of the distance squared functions.

The following result on the number of strong principal directions at the points of a surface generically immersed in 4-space was proved by Montaldi in [Montaldi (1983)].

Theorem 7.13. *For a distance squared function generic immersion of a surface in $\mathbb{R}^4$ there are at least 1 and at most 5 strong principal directions at each point.*

Definition 7.15. We define the *rib of order k* of M as the set of points of $\mathbb{R}^4$ that determine distance squared functions with (corank 1) singularities of type $A_k, k \geq 3$.

Definition 7.16. A point $p \in M$ which is a singular point of type A_k, $k \geq 4$ for some distance-squared function D_a is said to be a *ridge point of order* k. We call *ridge of order* k the set of all k-order ridge points for a given $k \geq 4$.

Observe that the ridge of order k is the projection of the (regular) curve $S_{1,1,1}(G^e)$ in the normal bundle through the canonical projection $\pi_N : NM \to M$.

Proposition 7.21. *On a distance squared generic immersed surface, the set of* 4*-order ridge points is either empty or is a smooth curve called ridge. The ridge points of order* 5*, if they exist, are isolated points on the ridge.*

Proof. The ridge of order 4 is the projection of the (regular) curve $S_{1,1,1}(G)$ in the normal bundle of M by the canonical projection π_M : $NM \to M$. For a generically immersed surface, the kernel of this projection is transversal to this curve at every point. Therefore, its image in M is a regular curve too.

The ridge points of order 5 are the images of singular points of type A_4 of G, which are generically isolated in $S_{1,1,1}(G)$. □

Another characterisation of 5-order ridge points is given in [Romero Fuster and Sanabria Codesal (2002)].

We consider now the focal hyperspheres whose centres define distance-squared functions with corank 2 singularities on M. In this case, $\text{Hess}(D_a)$ vanishes at the given point.

Definition 7.17. An *umbilical focus* is a point $a \in \mathbb{R}^4$ for which the distance squared function D_a has a corank 2 singularity. We refer to the focal 3-spheres centred at umbilical foci as *umbilical focal hyperspheres.*

Theorem 7.14 ([Montaldi (1983)]). *A point* $p = \mathbf{x}(u) \in M$ *is a semi-umbilic and not an inflection point if and only if it is a singularity of corank* 2 *of some distance squared function on* M.

Proof. Let

$$\alpha(p) = \begin{bmatrix} a_3 & b_3 & c_3 \\ a_4 & b_4 & c_4 \end{bmatrix},$$

be the matrix of the second fundamental form of $\mathbf{x}$ at p.

The curvature ellipse is given by

$$\eta(\theta) = \mathbf{H} + \cos(2\theta)\mathbf{B} + \sin(2\theta)\mathbf{C}.$$

Therefore, $p = \mathbf{x}(u)$ is a semiumbilic point if and only if rank $(\mathbf{B}, \mathbf{C}) = 1$.

For $\mathbf{v} = v_3 e_3 + v_4 e_4 \in N_p M$, we have

$$\text{Hess}(h_{\mathbf{v}})(u) = \begin{bmatrix} \sum_{i=3}^4 a_i v_i & \sum_{i=3}^4 b_i v_i \\ \sum_{i=3}^4 b_i v_i & \sum_{i=3}^4 c_i v_i \end{bmatrix}.$$

Then, if we denote $\lambda = \frac{1}{\|a-p\|}$, we get

$$\text{Hess}(D_a)(u) = \begin{bmatrix} 1 - \frac{1}{\lambda} \sum_{i=3}^4 a_i v_i & \frac{1}{\lambda} \sum_{i=3}^4 b_i v_i \\ \frac{1}{\lambda} \sum_{i=3}^4 b_i v_i & 1 - \frac{1}{\lambda} \sum_{i=3}^4 c_i v_i \end{bmatrix},$$

so the rank of $\text{Hess}(D_a)(u)$ vanishes if and only if

$$\lambda - \sum_{i=3}^4 a_i v_i = \lambda - \sum_{i=3}^4 c_i v_i = \sum_{i=3}^4 b_i v_i = 0.$$

This is equivalent to

$$\sum_{i=3}^{4} a_i v_i = \sum_{i=3}^{4} c_i v_i \quad , \text{ and } \quad \sum_{i=3}^{4} b_i v_i = 0,$$

which imply that $\mathbf{v}$ is orthogonal to both $\mathbf{B}$ and $\mathbf{C}$ in $N_p M$. Equivalently, rank $(\mathbf{B}, \mathbf{C}) = 1$, that is, p is a semiumbilic point.

Conversely, if $p = \mathbf{x}(u)$ is a semiumbilic point and $\mathbf{v}$ is a unit normal vector orthogonal to the curvature segment at p in $N_p M$, then the distance squared function D_a, with $a = p + 1/\langle \mathbf{H}, \mathbf{v} \rangle \mathbf{v}$, has a corank 2 singularity at u. □

Observe that $|\langle \mathbf{H}, \mathbf{v} \rangle|$ is the distance of p to the affine line determined by the curvature segment. When this distance vanishes, the point p belongs to the curvature segment, so p is an inflection point and the focal hypersphere becomes the (unique) osculating hyperplane.

Proposition 7.22 ([Montaldi (1983)]). *The semiumbilics of a surface generically immersed in 4-space form regular curves.*

Corollary 7.4. *If p is a semiumbilic point of M with umbilical focus at $a \in \mathbb{R}^4$, then p is a $\mathbf{v}$-umbilic point for the unit normal direction $\mathbf{v} = (a-p)/\|a-p\|$ and the corresponding $\mathbf{v}$-curvature is $\langle \mathbf{H}, \mathbf{v} \rangle$.*

Proof. The result follows from Theorem 7.14 and Lemma 7.3. □

7.10 Notes

Montaldi studied in [Montaldi (1983)] the extrinsic geometry of surfaces in $\mathbb{R}^4$ based on their generic contacts with 2-spheres and hyperspheres. The contact of the surface with hyperspheres is studied in [Romero Fuster and Sanabria Codesal (2002)]. Mond applied in [Mond (1982)] his classification of simple singularities of map-germs $\mathbb{R}^2, 0 \to \mathbb{R}^3, 0$ to study generic central projections of surfaces in $\mathbb{R}^4$ to hyperplanes.

We touched briefly on the asymptotic curves. There are various other pairs of geometrical foliations on the surface of interest. These can be found, for example, in [Garcia, Mello and Sotomayor (2005); Garcia and Sotomayor (2000); Gutierrez, Guadalupe, Tribuzy and Guíñez (1997); Gutierrez, Guadalupe, Tribuzy and Guíñez (2001); Gutierrez and Guíñez (2003); Little (1969); Mello (2003); Ramírez-Galarza and Sánchez-Bringas (1995); Tari (2009)].

For a generic immersion of a surface in $\mathbb{R}^4$, the height function has isolated double points, which correspond to bi-tangencies of the immersion. Bi-tangency properties of generic immersions of surfaces in $\mathbb{R}^4$ are studied by Dreibelbis in [Dreibelbis (2001, 2004, 2006, 2007)].

Conformal properties of surfaces in $\mathbb{R}^4$ are studied in [Romero Fuster and Sanabria Codesal (2004, 2008, 2013)].

Chapter 8

Surfaces in the Euclidean 5-space

With the aim of illustrating how the singularity techniques can be applied to analyse the extrinsic geometry of surfaces in higher codimensions we discuss in this chapter the geometrical properties associated to the contacts of surfaces with hyperplanes and hyperspheres in $\mathbb{R}^5$. In this, as well as in higher codimensional cases, the curvature ellipse at each point of the surface determines a proper subspace of the normal space and both, its dimension and relative positions with respect to the considered point are relevant in the description of the second order geometry of the surface at this point. Moreover, differently from the case of surfaces in $\mathbb{R}^4$, the directions leading to degenerate singularities of height functions at each point are not finite. These directions determine a cone in the normal space at the considered point. Attending to the behaviour of this cone we can distinguish among different types of points. The corresponding properties are described in §8.1. The generic singularities of the height functions on surfaces in $\mathbb{R}^5$ are analysed in §8.2. Analogously to the case of surfaces in $\mathbb{R}^4$, this setting leads naturally to the introduction of binormal and asymptotic directions at each point of the surface. Since the number of normal directions leading to degenerate singularities of height functions at each point is not finite, we use the concept of binormal direction for those leading to singularities of height functions on the surface with codimension higher than one, which only occur in a finite number of normal directions at every point. The corresponding contact directions are the asymptotic directions at the point. Alternative characterisations of asymptotic directions in terms of normal sections and of the geometrical behaviour of orthogonal projections of the surface into 3- and 4-spaces are also given. We see in Proposition 8.9 that the number of asymptotic directions at each point is at least one and at most five. The possible local configurations, described in Proposition 8.10, were determined in [Romero Fuster, Ruas and Tari (2008)].

Another relevant subset in this context is the flat ridges curve. Here we shall understand by flat ridge points as those at which there is some height function having a singularity of type $A_{k\geq 4}$ or worse.

We use the stereographical projection to transport these results on contacts of surfaces with hyperplanes in 5-space into results on contacts of surfaces with hyperspheres in 4-space.

We include in 8.3 a description of the generic behaviour of orthogonal projections onto hyperplanes, 3-spaces and planes. Finally, in §8.3 we analyse the generic singularities of distance squared functions on surfaces in $\mathbb{R}^5$. This setting leads to the introduction of geometrical concepts such as rib and ridge points, as well as umbilical foci and umbilical curvature.

The basic references for this chapter are [Mochida (1993); Mochida, Romero Fuster and Ruas (2003); Romero Fuster, Ruas and Tari (2008); Costa, Moraes and Romero Fuster (2009)].

8.1 The second order geometry of surfaces in $\mathbb{R}^5$

Let M be a closed surface in $\mathbb{R}^5$, and $\mathbf{x} : U \to \mathbb{R}^5$ a local parametrisation of M. Let $\{\mathbf{e}_1, \mathbf{e}_2, \mathbf{e}_3, \mathbf{e}_4, \mathbf{e}_5\}$ be a positively oriented orthonormal frame in $\mathbb{R}^5$ such that at any $u \in U$, $\{\mathbf{e}_1(u), \mathbf{e}_2(u)\}$ is a basis for the tangent plane T_pM and $\{\mathbf{e}_3(u), \mathbf{e}_4(u), \mathbf{e}_5(u)\}$ is a basis for the normal plane N_pM at $p = \mathbf{x}(u)$. By using the notation of Chapter 7, we can represent the matrix of the second fundamental form associated to the embedding as

$$\alpha_p = \begin{bmatrix} a_1 & b_1 & c_1 \\ a_2 & b_2 & c_2 \\ a_3 & b_3 & c_3 \end{bmatrix}$$

where $a_i = \langle \mathbf{x}_{u_1u_1}, \mathbf{e}_{i+2} \rangle$,$b_i = \langle \mathbf{x}_{u_1u_2}, \mathbf{e}_{i+2} \rangle$ and $c_i = \langle \mathbf{x}_{u_2u_2}, \mathbf{e}_{i+2} \rangle$, $i = 1, 2, 3$. We can naturally extend the concept of curvature ellipse to surfaces immersed with higher codimension:

Definition 8.1. The *curvature ellipse* at a point p of the surface M is the image $\eta : S^1 \to N_pM$, obtained by assigning to each tangent direction θ the curvature vector of the normal section γ_θ.

We can apply the same arguments to those applied to surfaces in 4-space to obtain the following expression for the curvature ellipse,

$$\eta(\theta) = \mathbf{H} + \cos(2\theta)\mathbf{B} + \sin(2\theta)\mathbf{C}, \tag{8.1}$$

where

$$\begin{aligned}\mathbf{H} &= \tfrac{1}{2}(a_1 + c_1)\mathbf{e_3} + \tfrac{1}{2}(a_2 + c_2)\mathbf{e_4} + \tfrac{1}{2}(a_3 + c_3)\mathbf{e_5},\\ \mathbf{B} &= \tfrac{1}{2}(a_1 - c_1)\mathbf{e_3} + \tfrac{1}{2}(a_2 - c_2)\mathbf{e_4} + \tfrac{1}{2}(a_3 - c_3)\mathbf{e_5},\\ \mathbf{C} &= b_1\mathbf{e_3} + b_2\mathbf{e_4} + b_3\mathbf{e_5}.\end{aligned}$$

Definition 8.2. The normal field

$$\mathbf{H} = \frac{1}{2}(a_1 + c_1)\mathbf{e_3} + \frac{1}{2}(a_2 + c_2)\mathbf{e_4} + \frac{1}{2}(a_3 + c_3)\mathbf{e_5}$$

evaluated at a point p is the *mean curvature* vector of M at p.

As in the previous chapter, we consider (affine) invariant properties of the surface under the action of $\mathrm{GL}(2,\mathbb{R}) \times \mathrm{GL}(3,\mathbb{R})$ on $T_pM \times N_pM$, where $\mathrm{GL}(j,\mathbb{R})$, $j = 2,3$ denotes the general linear group. This action can be viewed as a change of basis in T_pM and N_pM.

Let $\{\mathbf{f}_1, \mathbf{f}_2\} = \{\mathbf{x}_{u_1}, \mathbf{x}_{u_2}\}$ be the basis of T_pM given by the parametrisation $\mathbf{x} : U \to \mathbb{R}^5$ of the surface . One can complete it at each point p on M to obtain a basis $\{\mathbf{f}_1, \mathbf{f}_2, \mathbf{f}_3, \mathbf{f}_4, \mathbf{f}_5\}$ of $\mathbb{R}^5$ varying smoothly with p, with $\{\mathbf{f}_3, \mathbf{f}_4, \mathbf{f}_5\}$ a basis of N_pM (which is also not necessarily orthonormal).

The analysis of the effects of the change of basis on the coefficients of the second fundamental form and on the curvature ellipse follows similarly to Theorems 7.1, 7.2 and 7.3. We state the corresponding results for surfaces in $\mathbb{R}^5$ for completeness.

Let

$$\begin{aligned}\mathbf{f}_1 &= \alpha_1\mathbf{e}_1 + \beta_1\mathbf{e}_2,\\ \mathbf{f}_2 &= \alpha_2\mathbf{e}_1 + \beta_2\mathbf{e}_2.\end{aligned}$$

Then we have the following.

Proposition 8.1. *Denote by a_i, b_i, c_i, $i = 1,2,3$ the coefficients of the second fundamental form with respect to the basis $\{\mathbf{e}_1, \mathbf{e}_2, \mathbf{e}_3, \mathbf{e}_4, \mathbf{e}_5\}$ and by l_i, m_i, n_i, $i = 1,2,3$ its coefficients with respect to the basis $\{\mathbf{f}_1, \mathbf{f}_2, \mathbf{e}_3, \mathbf{e}_4, \mathbf{e}_5\}$. Then*

$$\begin{pmatrix} l_i \\ m_i \\ n_i \end{pmatrix} = \mathbf{\Lambda} \begin{pmatrix} a_i \\ b_i \\ c_i \end{pmatrix} \quad i = 1,2,3,$$

with

$$\mathbf{\Lambda} = \begin{pmatrix} \alpha_1^2 & 2\alpha_1\beta_1 & \beta_1^2 \\ \alpha_1\alpha_2 & \alpha_1\beta_2 + \alpha_2\beta_1 & \beta_1\beta_2 \\ \alpha_2^2 & 2\alpha_2\beta_2 & \beta_2^2 \end{pmatrix} \quad \textit{and} \quad \det\mathbf{\Lambda} = (\det\Omega)^3.$$

We write

$$\mathbf{f}_3 = \alpha_1\mathbf{e}_3 + \beta_1\mathbf{e}_4 + \gamma_1\mathbf{e}_5,$$
$$\mathbf{f}_4 = \alpha_2\mathbf{e}_3 + \beta_2\mathbf{e}_4 + \gamma_2\mathbf{e}_5,$$
$$\mathbf{f}_5 = \alpha_3\mathbf{e}_3 + \beta_3\mathbf{e}_4 + \gamma_3\mathbf{e}_5$$

and denote by

$$a_i = \langle \mathbf{x}_{u_1u_1}, \mathbf{e}_{i+2}\rangle, \quad b_i = \langle \mathbf{x}_{u_1u_2}, \mathbf{e}_{i+2}\rangle, \quad c_i = \langle \mathbf{x}_{u_2u_2}, \mathbf{e}_{i+3}, \rangle, i = 1, 2, 3$$

and

$$l_i = \langle \mathbf{x}_{u_1u_1}, \mathbf{f}_{i+2}\rangle, \quad m_i = \langle \mathbf{x}_{u_1u_2}, \mathbf{f}_{i+2}\rangle, \quad n_i = \langle \mathbf{x}_{u_2u_2}, \mathbf{f}_{i+2}\rangle, i = 1, 2, 3$$

the coefficients of the second fundamental form with respect to the basis $\{\mathbf{e}_3, \mathbf{e}_4\}$ and $\{\mathbf{f}_3, \mathbf{f}_4, \mathbf{f}_5\}$ respectively. Also denote by

$$E_n = \langle \mathbf{f}_3, \mathbf{f}_3\rangle,\ F_n = \langle \mathbf{f}_3, \mathbf{f}_4\rangle\ G_n = \langle \mathbf{f}_3, \mathbf{f}_5\rangle,$$
$$H_n = \langle \mathbf{f}_4, \mathbf{f}_4\rangle,\ I_n = \langle \mathbf{f}_4, \mathbf{f}_5\rangle,\ J_n = \langle \mathbf{f}_5, \mathbf{f}_5\rangle.$$

Proposition 8.2. *Denote by a_i, b_i, c_i, $i = 1, 2, 3$ the coefficients of the second fundamental form with respect to the basis $\{\mathbf{e}_1, \mathbf{e}_2, \mathbf{e}_3, \mathbf{e}_4, \mathbf{e}_5\}$ and by l_i, m_i, n_i, $i = 1, 2, 3$ its coefficients with respect to the basis $\{\mathbf{e}_1, \mathbf{e}_2, \mathbf{f}_3, \mathbf{f}_4, \mathbf{f}_5\}$. Then*

$$\begin{pmatrix} a_1 & b_1 & c_1 \\ a_2 & b_2 & c_2 \\ a_3 & b_3 & c_3 \end{pmatrix} = \begin{pmatrix} \alpha_1 & \alpha_2 & \alpha_3 \\ \beta_1 & \beta_2 & \beta_3 \\ \gamma_1 & \gamma_2 & \gamma_3 \end{pmatrix} \begin{pmatrix} E_n & F_n & G_n \\ F_n & H_n & I_n \\ G_n & I_n & J_n \end{pmatrix}^{-1} \begin{pmatrix} l_1 & m_1 & n_1 \\ l_2 & m_2 & n_2 \\ l_3 & m_3 & n_3 \end{pmatrix}.$$

Definition 8.3. We define the subsets

$$M_i = \{p \in M |\ \mathrm{rank}\ \alpha_p = i\}, i \leq 3.$$

The property that rank $\alpha_p = i$ is invariant by the action of $\mathrm{GL}(2, \mathbb{R}) \times \mathrm{GL}(3, \mathbb{R})$ on $T_pM \times N_pM$.

Given $p \in M$, the second fundamental form induces a linear map, $A_p : N_pM \to \mathcal{Q}_2$, from the normal 3-space of M at p to the space $\mathcal{Q}_2$ of quadratic forms on T_pM. It is defined by $A_p(\mathbf{v}) = \mathrm{II}_p^{\mathbf{v}}$, where $\mathrm{II}_p^{\mathbf{v}} : T_pM \to \mathbb{R}$ is the second fundamental form at p with respect to any normal direction $\mathbf{v} \in N_pM$. With respect to the basis $B(\mathbf{x}) = \{\mathbf{x}_{u_1}, \mathbf{x}_{u_2}\}$ of T_pM, we have $\mathbf{w} = w_1\mathbf{x}_{u_1} + w_2\mathbf{x}_{u_1}$, and we can write

$$\mathcal{Q}_2 = \{aw_1^2 + 2bw_1w_2 + cw_2^2 \mid a, b, c \in \mathbb{R}\}.$$

If $\mathbf{v} \in N_pM$ is represented by its coordinates (v_1, v_2, v_3) with respect to the basis $\{\mathbf{e}_3, \mathbf{e}_4, \mathbf{e}_5\}$, then

$$A_p(v_3, v_4, v_5) = \sum_{i=1}^{3} v_i(l_iw_1^2 + 2m_iw_1w_2 + n_iw_2^2).$$

We can identify $\mathcal{Q}_2$ with the space of real symmetric matrices

$$\mathcal{Q}_2 = \left\{ \begin{pmatrix} a & b \\ b & c \end{pmatrix} \,\middle|\, a, b, c \in \mathbb{R} \right\}.$$

Then we have the canonical basis

$$\left\{ \begin{pmatrix} 1 & 0 \\ 0 & 0 \end{pmatrix}, \begin{pmatrix} 0 & 1 \\ 1 & 0 \end{pmatrix}, \begin{pmatrix} 0 & 0 \\ 0 & 1 \end{pmatrix} \right\}$$

of $\mathcal{Q}_2$. We remark that the matrix α_p^T is the representation matrix of the linear mapping A_p with respect to the basis $\{\mathbf{e}_3, \mathbf{e}_4, \mathbf{e}_5\}$ of N_pM and the above basis of $\mathcal{Q}_2$.

Let C be the cone of *degenerate quadratic forms* in $\mathcal{Q}_2$, which is defined by $ac - b^2 = 0$ in the above representation and denote by C_p the subset $A_p^{-1}(C) \subset N_pM$. The following properties are an immediate consequence of the definition of A_p and do not depend on the choices of basis on T_pM and N_pM.

1) If $p \in M_3$, the linear map A_p has maximal rank, so C_p is a cone in N_pM.
2) If $p \in M_2$, the image of A_p is a plane through the origin in $\mathcal{Q}_2$. Now, according to the relative position of the image of A_p with respect to the cone C, we have the following three possible cases for the set C_p.

   (2a) *Hyperbolic type* (denoted by M_2^h): $ImA_p \cap C$ consists of two lines and then C_p is the union of two planes intersecting along the line $\ker \alpha(p)$.
   (2b) *Elliptic type* (denoted by M_2^e): $ImA_p \cap C = \{0\}$ and thus C_p is the line $= \ker \alpha(p)$.
   (2c) *Parabolic type* (denoted by M_2^p): ImA_p is tangent to C along a line, in which case C_p is a plane containing the line $\ker \alpha(p)$.

3) If $p \in M_1$, the image of A_p is a line through the origin in $\mathcal{Q}_2$ and C_p can be either a plane coinciding with $KerA_p$, or the whole N_pM.

Proposition 8.3 ([Mochida, Romero Fuster and Ruas (2003)]). *Given a closed surface M, there exists a residual set $\mathcal{O}$ in Emb$(M, \mathbb{R}^5)$ such that for any $\mathbf{x} \in \mathcal{O}_H$, we have that $M = M_3 \cup M_2$. Moreover,*

a) *M_3 is an open subset of M.*
b) *M_2 is a regularly embedded curve.*

Proof. We show first that generically the rank of the second fundamental form at any point is at least 2. Let $\mathbf{x} : U \to \mathbb{R}^5$ a local parametrisation of M, and $\{\mathbf{x}_{u_1}, \mathbf{x}_{u_1}, \mathbf{e}_3, \mathbf{e}_4, \mathbf{e}_5\}$ a frame in U. With respect to these coordinates, for each $p \in U$, the second fundamental form at p is represented by the 3×3 matrix

$$\alpha_p = \begin{pmatrix} l_1 & m_1 & n_1 \\ l_2 & m_2 & n_2 \\ l_3 & m_3 & n_3 \end{pmatrix}$$

where

$$l_i = \langle \mathbf{x}_{u_1u_1}, \mathbf{e}_{i+2} \rangle, \quad m_i = \langle \mathbf{x}_{u_1u_2}, \mathbf{e}_{i+2} \rangle, \quad n_i = \langle \mathbf{x}_{u_2u_2}, \mathbf{e}_{i+2} \rangle, \; i = 1, 2, 3.$$

Let $\mathcal{M}(3)$ be the set of real 3×3 matrices and $\mathcal{M}^i(3) = \{\alpha \in \mathcal{M}(3) \, |\text{rank}\, \alpha \leq i, \, i = 0, 1, 2, 3\}$. Then $\mathcal{M}^i(3)$ is a singular variety of codimension $(3 - i)^2$, whose singular set is $\mathcal{M}^{i-1}(3)$ ([Arbarello, Cornalba, Griffiths and Harris (1985)]). The decomposition $\mathcal{M}(3) = \cup_{i=0}^3 \mathcal{S}_i$, where $\mathcal{S}_i = (\mathcal{M}^i(3) \setminus \mathcal{M}^{i-1}(3))$ is a stratification of $\mathcal{M}(3)$.

We consider now the following diagram

$$U \stackrel{j^2\mathbf{x}}{\longrightarrow} J^2(U, \mathbb{R}^5) \stackrel{\Pi^*}{\longrightarrow} \mathcal{M}(3),$$

where $\Pi^*(j^2\mathbf{x}(p)) = \alpha_p$.

The mapping $\Pi^* : J^2(U, \mathbb{R}^5) \to \mathcal{M}(3)$ is a submersion, as we can take the variables l_i, m_i and n_i as coordinates in the jet space $J^2(U, \mathbb{R}^5)$. Then, the stratification $(\mathcal{S}_i)$ in $\mathcal{M}(3)$ pulls back to a stratification $(\mathcal{T}_i)$ in $J^2(U, \mathbb{R}^5)$, such that $\text{cod}\, \mathcal{T}_i = \text{cod}\, \mathcal{S}_i$.

Then, it follows from Thom's Transversality Theorem that for a generic embedding $\mathbf{x} : U \to \mathbb{R}^5$, the map $j^2\mathbf{x} : U \to J^2(U, \mathbb{R}^5)$ does not intersect the strata $\mathcal{T}_i$, $i \leq 1$, since they have codimension greater than or equal to 4. Hence, for a generic embedding $\mathbf{x}$, $\text{rank}\, \alpha_p \geq 2$, $\forall p \in U$.

Let $\Delta(p) = det(\alpha_p)$, with $p = \mathbf{x}(u) \in M$. We have that $M - M_3 = \Delta^{-1}(0)$ and since Δ is a continuous function on M, it follows that M_3 must be an open region in M.

We can consider a local representation of M in its Monge form at a point $p \in M$,

$$\begin{aligned} \phi : \mathbb{R}^2, 0 &\longrightarrow \mathbb{R}^5 \\ (x, y) &\longmapsto (x, y, f_1(x, y), f_2(x, y), f_3(x, y)). \end{aligned}$$

In these coordinates

$\Delta(p) = f_{1xx}f_{2xy}f_{3yy} - f_{1xy}f_{2xx}f_{3yy} - f_{1xx}f_{2yy}f_{3xy} + f_{1yy}f_{2xx}f_{3xy} + f_{1xy}f_{2yy}f_{3xx} - f_{1yy}f_{2xy}f_{3xx}$.

Now, it follows from this expression that, under appropriate transversality conditions on the 3-jet of ϕ, the set $\Delta = 0$ represents a curve possibly with isolated singular points determined by the vanishing of the derivatives of the function Δ. Since the orthogonality property of the frame $\{\mathbf{e}_1, \mathbf{e}_2, \mathbf{e}_3, \mathbf{e}_4, \mathbf{e}_5\}$ is irrelevant for our study, we can take $\{\mathbf{e}_3, \mathbf{e}_4, \mathbf{e}_5\}$ such that $\mathbf{e}_5$ generates $\mathrm{Ker}(A_p)$.

If $p \in M_2^h$, we choose $\{\mathbf{e}_3, \mathbf{e}_4\}$ as the two degenerate directions in N_pM. Furthermore, we can also make linear changes of coordinates in source and target, such that the two degenerate directions correspond to the quadratic forms x^2 and y^2 in C. Thus f can be locally written as

$$\phi(x, y) = (x, y, x^2 + R_1(x, y), y^2 + R_2(x, y), R_3(x, y)),$$

where $R_i \in m^3$, i.e., all the derivatives of the R_i vanish up to order $3, i = 1, 2, 3$.

If $p \in M_2^e$, then $Im(A_p) \cap C = \{0\}$ and we take $\mathbf{e}_5$ as the generator of $\ker A_p = A_p^{-1}(Im(A_p) \cap C)$. With additional changes of coordinates, f can be written as

$$f(x, y) = (x, y, x^2 - y^2 + R_1(x, y), xy + R_2(x, y), R_3(x, y)).$$

If $p \in M_2^p$ then analogously, f can be written as

$$f(x, y) = (x, y, x^2 + R_1(x, y), xy + R_2(x, y), R_3(x, y)).$$

In each of the above cases it is a simple (but tedious) calculation to verify that under generic conditions on the 3-jet of ϕ at $(0, 0)$, the point p is a regular point of $\Delta^{-1}(0)$. □

Definition 8.4. We denote by E_p and $\mathcal{A}ff_p$ the vector subspace and the affine subspace determined by the curvature ellipse respectively in N_pM. That is, $\mathcal{A}ff_p$ is the affine plane spanned by the vectors $\mathbf{C} - \mathbf{H}$ and $\mathbf{B} - \mathbf{H}$ and E_p is the parallel plane through the origin (spanned by $\mathbf{C}$ and $\mathbf{B}$). The orthogonal complement of E_p in N_pM will be denoted by $E_p^{\perp}$.

The points of M are characterised as follows in terms of the relative positions of these subspaces:

1) If $p \in M_3$, then $\mathcal{A}ff_p$ is a plane that does not contain the origin of N_pM.

2) If $p \in M_2$, then $\mathcal{A}ff_p$ is either a plane through the origin of N_pM (and thus coincides with E_p), or a line that does not contain the origin of N_pM.
3) If $p \in M_1$, then $\mathcal{A}ff_p$ is a line through the origin p of N_pM, that is $\mathcal{A}ff_p = E_p$.

We thus have that the M_2 points are either semiumbilics, or points at which $\mathcal{A}ff_p$ passes through the origin p of N_pM. In the last case, it is not difficult to see that, similarly to the case of surfaces in $\mathbb{R}^4$, a point p is of type M_2^e, M_2^h or M_2^p according to the origin p of N_pM lies inside, outside, or on the curvature ellipse at p. The semiumbilic points can be considered as points of type M_2^h.

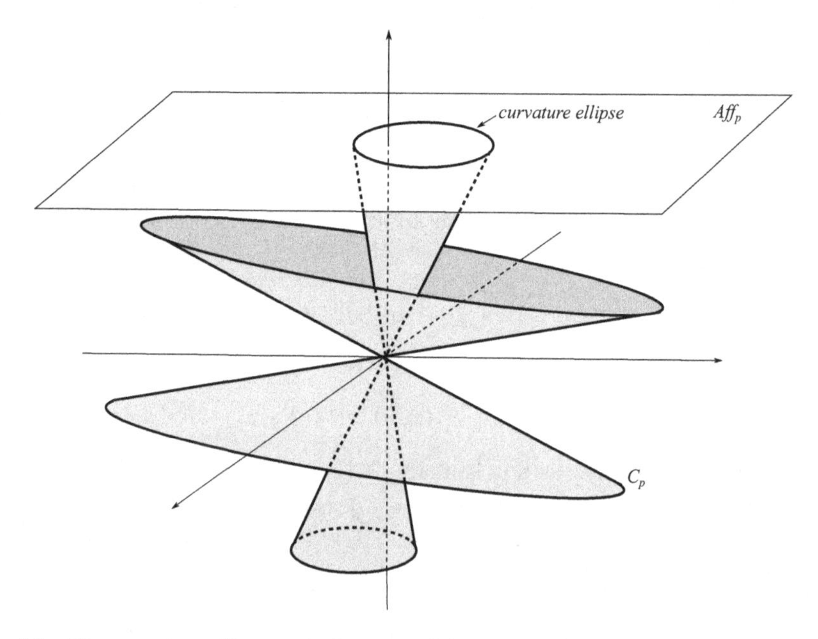

Fig. 8.1 The curvature ellipse and the cone of degenerate directions at a point on a surface in $\mathbb{R}^5$.

Proposition 8.4. *The cone whose basis is the curvature ellipse is perpendicular to the cone C_p of degenerate directions at p.*

Proof. Given any normal direction $\mathbf{v}$, the $\mathbf{v}$-principal curvatures at p are the extremal values of the projection of the normal curvature vectors at p, that is the extremal values of the projection of the curvature ellipse on the vector line determined by $\mathbf{v}$ in N_pM. Then it follows that $\mathbf{v}$ is perpendicular to some direction on the cone whose basis is the curvature ellipse if and only if one of these $\mathbf{v}$-principal curvatures vanishes, which implies that $\mathbf{v}$ lies in the cone of degenerate directions at p. $\square$

8.2 Contacts with hyperplanes

Consider the family of height functions,

$$\begin{aligned} H : M \times S^4 &\to \quad \mathbb{R} \\ (p, \mathbf{v}) &\mapsto h_{\mathbf{v}}(p). \end{aligned}$$

We observe that, analogously to the case of surfaces in $\mathbb{R}^4$ if we denote by $\bar{H}$ the height functions family on the canal hypersurface CM, it follows that $p \in M$ is a (degenerate) singular point of $h_{\mathbf{v}}$ if and only if $(p, \mathbf{v}) \in CM$ is a (degenerate) singular point (of the same $\mathcal{K}$-type) of $\bar{h}_{\mathbf{v}}$ and, as a consequence of Theorem 4.4, we can state:

Theorem 8.1. *There is an open and dense set $\mathcal{O}_H$ in Imm$(U, \mathbb{R}^5)$ such that for any $\mathbf{x} \in \mathcal{O}_H$, the surface $M = \mathbf{x}(U)$ has the following properties:*

i) *For any $\mathbf{v} \in S^4$, the height function $h_{\mathbf{v}}$ along the normal direction $\mathbf{v}$ at a point $p \in M$ (and hence $\bar{h}_{\mathbf{v}}$ at $(p, \mathbf{v}) \in CM$) has only local singularities of $\mathcal{K}$- type $A_1, A_2, A_3, A_4, A_5, D_4$ and D_5.*
ii) *The singularities of $h_{\mathbf{v}}$ (resp. $\bar{h}_{\mathbf{v}}$) are $\mathcal{R}$-versally unfolded by the family H (resp. $\bar{H}$).*
iii) *The generalised Gauss map $G : CM \to S^4$ is stable as a Lagrangian map.*

Definition 8.5. A surface in $\mathbb{R}^5$ is called (*locally*) *height function generic* if any of its local parametrisations belongs to the set $\mathcal{O}_H$ in Theorem 8.1.

Definition 8.6. Given $\mathbf{v} \in C_p \subset N_pM$, the point p is a non-stable singularity of $h_{\mathbf{v}}$. We say that $\mathbf{v}$ is a *degenerate* normal direction at p. In such case, $\ker(\mathcal{H}(h_{\mathbf{v}})(p)) \neq \{0\}$ and any direction $\mathbf{u} \in \ker \mathcal{H}(h_{\mathbf{v}})(p)$ is called *flat contact direction associated to* $\mathbf{v}$. Here, $\mathcal{H}(h_{\mathbf{v}})(p)$ is the Hessian matrix of $h_{\mathbf{v}}$ at p.

Definition 8.7. A degenerate direction $\mathbf{v} \in C_p$ for which $h_{\mathbf{v}}$ has a singularity A_3 or worse (i.e., the $\mathcal{K}$-codimension of $h_{\mathbf{v}}$ is at least 2) is said to be a *binormal direction* at p. The tangent hyperplane orthogonal to the binormal direction has higher order contact with M at p and we call it *osculating hyperplane* at p. The corresponding contact directions are called *asymptotic directions* on M. In the special case of the corank 2 singularities $D_4^{\pm}$ and D_5 we refer to them as *umbilic binormals.*

We say that a point $p \in M$ is *2-regular* if for some coordinate chart at p, we have that $\operatorname{rank}\{\mathbf{x}_{u_1}, \mathbf{x}_{u_2}, \mathbf{x}_{u_1u_1}, \mathbf{x}_{u_1u_2}, \mathbf{x}_{u_2u_2}\} = 5$. If p fails to be 2-regular, we say that it is *2-singular.*

We have the following geometrical characterisation for the corank 2 singularities of the height functions on M:

Proposition 8.5. *For a local parametrisation $\mathbf{x} : M \to \mathbb{R}^5$, the following conditions are equivalent:*

a) $p \in M_2$.
b) $p \in M$ *is a singularity of corank* 2 *for some height function on* M.
c) p *is* 2*-singular.*

Proof. This follows easily by taking the embedding in Monge form and observing that the matrix of the second fundamental form in a normal direction $\mathbf{v}$ at p coincides with the Hessian matrix of the height function $h_{\mathbf{v}}$ at p. □

Consequently the generic singularities of the height functions at points of type M_3 are of type $A_k, k = 2, 3, 4, 5$. At a point of type M_2, there is some normal direction defining a height function with a singularity of type $D_4^{\pm}$ or D_5.

The binormal and asymptotic directions at the M_3 points can be characterised in terms of normal sections as follows: let $\mathbf{v}$ be a degenerate direction at $p \in M_3$ and let θ be a tangent direction in $\ker(\mathcal{H}(h_{\mathbf{v}})(p))$. Denote by γ_θ the normal section of the surface M in the tangent direction θ, that is, γ_θ is the curve obtained by the intersection of M with the 4-space $V_\theta = N_pM \oplus \langle \theta \rangle$.

Proposition 8.6. *The direction $\theta \in T_pM$ is an asymptotic direction at $p \in M$ corresponding to the binormal $\mathbf{v}$ if and only if $\mathbf{v}$ is the binormal direction at q of the curve γ_θ in the 4-space V_θ.*

Proof. Observe that the restriction of the family of height functions H to (some parametrisation of) the curve γ_θ gives the family of height functions on this curve. Now, the result follows from the fact that a normal vector

$\mathbf{v}$ is the binormal of a curve in 4-space if and only if the height function in the direction $\mathbf{v}$ over this curve has a singularity of type A_k, $k \geq 3$. □

We can also characterise the asymptotic directions on the M_3 points in terms of the geometry of the orthogonal projections of the surface into 3-spaces as follows ([Romero Fuster, Ruas and Tari (2008)]): Given $\mathbf{v} \in N_qM$ denote by $M_\mathbf{v}$ the surface patch obtained by projecting M orthogonally to the 3-space $T_qM \oplus \langle \mathbf{v} \rangle$ (considered as an affine space through q).

Proposition 8.7. *Given a point* $q \in M_3$, *we have*

(1) *A direction* $\mathbf{v} \in N_qM$ *is degenerate if and only if* q *is a parabolic point of* $M_\mathbf{v}$. *In this case, the unique principal asymptotic direction of* $M_\mathbf{v}$ *at* q *coincides with the contact direction associated to* $\mathbf{v}$.

(2) *A direction* $\theta \in T_qM$ *is asymptotic if and only if there exists* $\mathbf{v} \in N_qM$ *such that* q *is a cusp of Gauss of* $M_\mathbf{v}$ *and* θ *is its unique asymptotic direction there.*

Proof. Both assertions follow from observing that in appropriate coordinates, the matrix of the Gauss map of the surface patch $M_\mathbf{v}$ at p coincides with the Hessian matrix of the height function $h_\mathbf{v}$ at this point. □

Let $\mathcal{K}_c : CM \to \mathbb{R}$ be the Gaussian curvature function on CM. The parabolic set of the hypersurface CM is the singular set of the Gauss map $G : CM \to S^4$, given by $\mathcal{K}_c^{-1}(0)$. This is a stratified subset that can be decomposed in terms of the Thom-Boardman symbols of the singularities of the generalised Gauss map G as $\mathcal{K}_c^{-1}(0) = S_1(G) \cup S_2(G)$, where

$$S_1(G) = S_{1,0}(G) \cup S_{1,1,0}(G) \cup S_{1,1,1,0}(G) \cup S_{1,1,1,1,0}(G)$$

and

$$S_2(G) = S_{2,0}(G) \cup S_{2,1,0}(G).$$

We recall that

$S_{1,1}(G) = \{(p, \mathbf{v}) \in S_1(G) \,|\, \mathrm{Ker}\, d(G) \subset T_{(p,\mathbf{v})} S_1(G)\}$.

$S_{1,1,1}(G) = \{(p, \mathbf{v}) \in S_{1,1}(G) \,|\, \mathrm{Ker}\, d(G) \subset T_{(p,\mathbf{v})} S_{1,1}(G)\}$.

$S_{2,1}(G) = \{(p, \mathbf{v}) \in S_2(G) \,|\, \mathrm{dim} Ker\ (d(G) \cap T_{(p,\mathbf{v})} S_2(G)) = 1\}$.

Then $S_{1,0}(G) = S_1(G) - S_{1,1}(G)$, $S_{1,1,0}(G) = S_{1,1}(G) - S_{1,1,1}(G)$ and so on. Moreover, from the properties of Thom-Boardman symbols ([Golubitsky and Guillemin (1973)]) we get that for a generic immersion

a) $S_1(G)$ is a 3-dimensional submanifold of CM.

b) $S_{1,1}(G)$ is a regular surface.

c) $S_{1,1,1}(G)$ and $S_2(G)$ are regular curves in CM.

d) $S_{1,1,1,1}(G)$ and $S_{2,1}(G)$ are made of isolated points.

Taking into account the relation between the singularities of the generalised Gauss map G, as a catastrophe map of the height functions family, and the singularities of the height functions on the surface M (see Remark 3.2), we observe:

a) $(p, \mathbf{v}) \in S_{1,0}(G)$ if and only if p is a singularity of type A_2 of $h_{\mathbf{v}}$.
b) $(p, \mathbf{v}) \in S_{1,1,0}(G)$ if and only if p is a singularity of type A_3 of $h_{\mathbf{v}}$.
c) $(p, \mathbf{v}) \in S_{1,1,1,0}(G)$ if and only if p is a singularity of type A_4 of $h_{\mathbf{v}}$.
d) $(p, \mathbf{v}) \in S_{1,1,1,1,0}(G)$ if and only if p is a singularity of type A_5 of $h_{\mathbf{v}}$.
e) $(p, \mathbf{v}) \in S_{2,0}(G)$ if and only if p is a singularity of type $D_k^{\pm}, k = 4, 5$ of $h_{\mathbf{v}}$.

Then we can characterise the flat contact directions and the asymptotic directions as follows:

1) $(p, \mathbf{v}) \in S_1(G)$ if and only if $\mathbf{v}$ is a degenerate direction at p.
2) $(p, \mathbf{v}) \in S_{1,1}(G)$ if and only if $\mathbf{v}$ is an asymptotic direction at p.

Denote by $\xi : CM \to M$ the natural projection and by $\xi_1 : S_1(G) \to M$ and $\xi_2 : S_{1,1}(G) \to M$ its restrictions to the 3-manifold $S_1(G)$ and to the surface $S_{1,1}(G)$ respectively. Observe that there is a unique principal asymptotic direction (in the sense that it has zero principal curvature) at each point of $S_1(G)$. The following result is a consequence of the above considerations relating the singularities of the generalised Gauss map G on CM and the singularities of height functions on M.

Proposition 8.8. *The flat contact directions at $p \in M$ are the projections of the principal asymptotic directions of CM at the points $(p, \mathbf{v}) \in \xi^{-1}(p)$ through the derivative $d\xi_1(p, \mathbf{v}) : T_{(p,\mathbf{v})}S_1(G) \to T_pM$.*
The (unique) principal asymptotic direction at any point $(p, \mathbf{v}) \in S_{1,1}(G)$ is projected through $d\xi_2$ onto an asymptotic direction at $p \in M$.

We can study now the possible numbers and the distribution of asymptotic directions over the points of a generic surface M.

Proposition 8.9 ([Mochida, Romero Fuster and Ruas (2003)]). *There are at least one and at most 5 asymptotic directions at each point of M_3.*

Proof. Let M be given in the Monge form

$$\phi(x,y) = (x, y, f_1(x,y), f_2(x,y), f_3(x,y))$$

in a neighbourhood of $p = (0,0) \in \mathbb{R}^2$, with $f_i(x,y) = Q_i(x,y) + K_i(x,y) + R_i(x,y)$, where Q_i are quadratic forms, K_i are cubic forms and $R_i \in \mathcal{M}_2^4$, $i = 1,2,3$. Since $p \in M_3$, the 3 quadratic forms $Q_1(x,y), Q_2(x,y)$ and $Q_3(x,y)$ are linearly independent and without loss of generality we can take local coordinates at p in such a way that $Q_3(x,y) = -(x^2+y^2)$ and $K_3(x,y) = 0$. To simplify the notation we denote as $Q_i(\mathbf{w},\mathbf{z}), i = 1,2,3$ the bilinear form $d^2 f_i(x,y)(\mathbf{w},\mathbf{z})$, where $\mathbf{w},\mathbf{z} \in T_qM$ and $q = \phi(x,y)$ varies in a small enough neighbourhood of p in $\mathbb{R}^2$. Analogously $K_i(\mathbf{w}^3), i = 1,2,3$ denotes the cubic form associated to $\mathbf{x}$ at the point q acting on a vector $\mathbf{w} \in T_qM$.

Let $\mathbf{v} \in N_qM$ be a solution of the equation $A_q(\mathbf{v}) = 0$. Then $\mathcal{H}(h_{\mathbf{v}}(q))$ is a degenerate quadratic form and so there is $\mathbf{w} \in T_qM$ such that $\mathcal{H}(h_{\mathbf{v}}(q))(\mathbf{w},\mathbf{z}) = 0$, for any $\mathbf{z} \in T_qM$. By writing $\mathbf{v} = v_1\mathbf{e}_1 + v_2\mathbf{e}_2 + v_3\mathbf{e}_3 \in N_qM$ in terms of the normal frame $\{\mathbf{e}_1, \mathbf{e}_2, \mathbf{e}_3\}$ we have $v_1Q_1(\mathbf{w},\mathbf{z}) + v_2Q_2(\mathbf{w},\mathbf{z}) - v_3\langle\mathbf{w},\mathbf{z}\rangle = 0$. This expression must be true in particular for the vector $\mathbf{w}$ and a vector $\mathbf{z} \in T_qM$ orthogonal to $\mathbf{w}$, so we have the equations,

$$v_1Q_1(\mathbf{w},\mathbf{w}) + v_2Q_2(\mathbf{w},\mathbf{w}) - v_3(w_1^2 + w_2^2) = 0 \quad (1)$$

$$v_1Q_1(\mathbf{w},\mathbf{z}) + v_2Q_2(\mathbf{w},\mathbf{z}) = 0 \quad (2)$$

On the other hand, q is a singular point of cusp type or worse if the vector $\mathbf{w}$ satisfies $v_1K_1(\mathbf{w}^3) + v_2K_2(\mathbf{w}^3) + v_3K_3(\mathbf{w}^3) = 0$ (see [Mochida, Romero Fuster and Ruas (2003)]). And since in the chosen local coordinates $K_3(w_1,w_2) = 0$, this gives

$$v_1K_1(\mathbf{w}^3) + v_2K_2(\mathbf{w}^3) = 0 \quad (3)$$

Given v_1 and v_2 we can obtain v_3 from (1). On the other hand, eliminating v_1 and v_2 in (2) and (3) gives

$$K_1(\mathbf{w}^3)Q_2(\mathbf{w},\mathbf{z}) - K_2(\mathbf{w}^3)Q_1(\mathbf{w},\mathbf{z}) = 0.$$

Since the coordinates of the vector $\mathbf{z}$ (orthogonal to $\mathbf{w}$ in T_qM) can be given as a linear combination of those of $\mathbf{w}$, we obtain that for each q in a neighbourhood of $p = (0,0)$, the above equation is defined by a quintic form in two variables. This gives the differential equation for the asymptotic curves on M. We observe that this equation cannot be identically zero on regular points of order 2 of a generic surface. □

A different proof of the above result appears in [Romero Fuster, Ruas and Tari (2008)]. The integral lines of the fields of asymptotic directions are the *asymptotic curves* on M. They are the solutions of a quintic equation of the form:

$$A_0 dy^5 + A_1 dx dy^4 + A_2 dx^2 dy^3 + A_3 dx^3 dy^2 + A_4 dx^4 dy + A_5 dx^5 = 0$$

whose coefficients A_i $(i = 0, ..., 5)$ depend on those of the second fundamental form, $(l_1, m_1, n_1), (l_2, m_2, n_2), (l_3, m_3, n_3)$ and their first order partial derivatives, and are given by

$$A_0 = [\tfrac{\partial \mathbf{n}}{\partial y}, \mathbf{m}, \mathbf{n}]$$

$$A_1 = [\tfrac{\partial \mathbf{n}}{\partial x}, \mathbf{m}, \mathbf{n}] + 2[\tfrac{\partial \mathbf{m}}{\partial y}, \mathbf{m}, \mathbf{n}] + [\tfrac{\partial \mathbf{n}}{\partial y}, \mathbf{l}, \mathbf{n}]$$

$$A_2 = [\tfrac{\partial \mathbf{n}}{\partial x}, \mathbf{l}, \mathbf{n}] + 2[\tfrac{\partial \mathbf{m}}{\partial x}, \mathbf{m}, \mathbf{n}] + [\tfrac{\partial \mathbf{l}}{\partial y}, \mathbf{m}, \mathbf{n}] + 2[\tfrac{\partial \mathbf{m}}{\partial y}, \mathbf{l}, \mathbf{n}] + [\tfrac{\partial \mathbf{n}}{\partial y}, \mathbf{l}, \mathbf{m}]$$

$$A_3 = [\tfrac{\partial \mathbf{l}}{\partial x}, \mathbf{m}, \mathbf{n}] + 2[\tfrac{\partial \mathbf{m}}{\partial x}, \mathbf{l}, \mathbf{n}] + [\tfrac{\partial \mathbf{n}}{\partial x}, \mathbf{l}, \mathbf{m}] + 2[\tfrac{\partial \mathbf{m}}{\partial y}, \mathbf{l}, \mathbf{m}] + [\tfrac{\partial \mathbf{l}}{\partial y}, \mathbf{l}, \mathbf{n}]$$

$$A_4 = [\tfrac{\partial \mathbf{l}}{\partial x}, \mathbf{l}, \mathbf{n}] + 2[\tfrac{\partial \mathbf{m}}{\partial x}, \mathbf{l}, \mathbf{m}] + [\tfrac{\partial \mathbf{l}}{\partial y}, \mathbf{l}, \mathbf{m}]$$

$$A_5 = [\tfrac{\partial \mathbf{l}}{\partial x}, \mathbf{l}, \mathbf{m}]$$

where $\mathbf{l} = (l_1, l_2, l_3)$, $\mathbf{m} = (m_1, m_2, m_3)$, $\mathbf{n} = (n_1, n_2, n_3)$.

Definition 8.8. The singular set of $\xi_2 : S_{1,1}(G) \to M$ is known as the *criminant set* and its image by ξ_2, denoted by Δ, is the *discriminant set.*

The discriminant set is a (non necessarily connected) curve separating M into regions made of points with a constant number of asymptotic directions. The possible local configurations of the asymptotic curves on a surface generically immersed in $\mathbb{R}^5$ are described in the following proposition whose proof can be found in [Romero Fuster, Ruas and Tari (2008)].

Proposition 8.10.

(1) *The local configurations of the asymptotic curves of a height function generic surface in* $\mathbb{R}^5$ *are modeled by super-imposing in each quadrant in* Figure 8.2 ① *and* ② *one figure from the left column with one from the right column.*

(2) *Let* $q \in M_3$ *be a point on the discriminant* Δ *of the asymptotic IDE and* $\mathbf{u}$ *the double asymptotic direction there. Then* q *is a folded-singularity of the asymptotic IDE at* $(q, \mathbf{u})$ *if and only if* q *is an* A_4*-singularity of the height function along* $\mathbf{u}^*$ (Figure 8.2 ① (a or b)+(g, h or i)).

(3) *The discriminant* Δ *intersects transversally the* M_2*-curve at* M_2^p *and* D_5*-points (it may also intersect it at other points). The* D_5*-points are generically not folded singularities, so the configuration of the asymptotic curves at such points is as in* Figure 8.2 ① (a or b)+(f). *An* M_2^p*-point is (at the appropriate direction) a folded singularity of the IDE of the asymptotic curves and the configurations there are as in Figure* 8.2 ① (a or b)+(g, h or i).

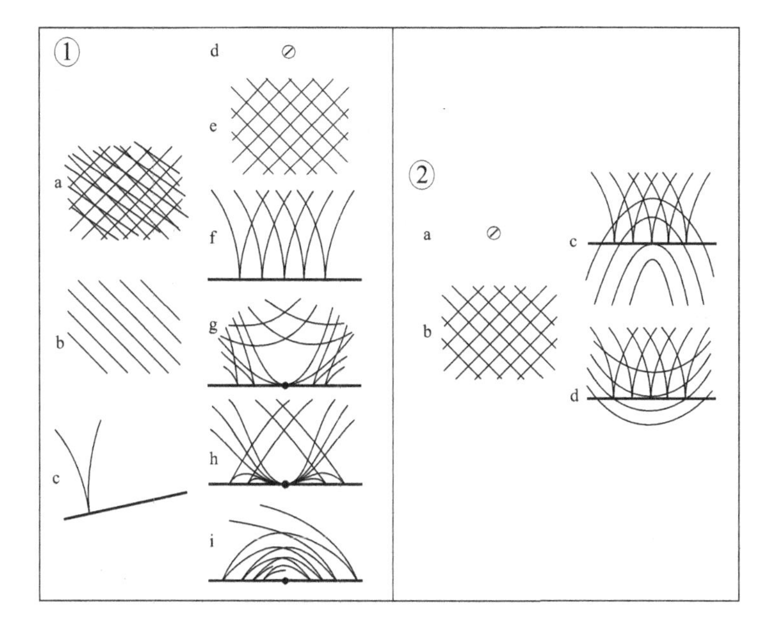

Fig. 8.2 Configurations of the asymptotic curves on surfaces in $\mathbb{R}^5$.

Definition 8.9. The *flat ridge* $\mathcal{FR}$ of M is the set of points for which there exists a height function (in some binormal direction), having a singularity of type $A_k,\ k \geq 4$.

It is easy to see that

$$\mathcal{FR} = \xi(S_{1,1,1}(G)).$$

Proposition 8.11. *Let* $q \in M_3$ *and* $\mathbf{v} \in N_qM$ *a binormal direction. Let* θ *be its corresponding asymptotic direction and* γ_θ *the corresponding normal section of* M*. Then*

(1) $q = \gamma_\theta(0)$ *is a flat ridge point of* M *if and only if* q *is a flattening of* γ_θ *(as a curve in the 4-space* V_θ*).*

(2) $q = \gamma_\theta(0)$ *is a higher order flat ridge point of* M *if and only if* q *is a degenerate flattening of* γ_θ*.*

Proof. A point q is a singularity of type A_k of the height function $h_{\mathbf{v}}$ on M if and only if it is a singularity of type A_k of $h_{\mathbf{v}}|_{\gamma_\theta}$. Therefore it is a flattening (resp. degenerate flattening) of γ_θ if and only if it is a flat ridge point (resp. higher order flat ridge point) of M. □

Let $P \in S^4$ be the north pole, it follows from Theorem 4.11 that the stereographic projection $\psi : S^4 - \{P\} \to \mathbb{R}^4$ takes the singularities of type A_k and D_k of height functions on a surface $M \subset S^4$ (considered as a surface in $\mathbb{R}^5$) into the singularities of the same type for distance squared functions on $\psi(M) \subset \mathbb{R}^4$. Moreover, the differential of ψ takes contact directions of M with hyperplanes to contact directions of $\psi(M)$ with hyperspheres. Therefore, we have the following:

Corollary 8.1. *The stereographic projection* ψ *maps diffeomorphically:*

1) *The asymptotic curves of* M *into the strong principal lines of* $\psi(M)$*.*
2) *The flat ridges of* M *to the ridges of* $\psi(M)$*.*
3) *The* M_2 *points of* M *to the semiumbilic points of* $\psi(M)$*.*

As a consequence of this, together with the previous results regarding the generic contacts of surfaces with hyperplanes in $\mathbb{R}^5$, we get the following results concerning generic contacts of surfaces with hyperspheres in $\mathbb{R}^4$ as announced in §7.9. We point out that from Theorem 4.11 and Remark 4.4 we get that a surface immersed in S^4 is generic from the viewpoint of its contacts with hyperplanes of $\mathbb{R}^5$ if and only if the surface $\psi(M)$ is generic from the viewpoint of its contacts with hyperspheres in $\mathbb{R}^4$. We thus have that the assertion 2) of the following Corollary provides an alternative proof of Theorem 7.14.

Corollary 8.2. *Any distance squared function generic immersion of a surface in* $\mathbb{R}^4$ *satisfies the following properties:*

1) *There are at least one and at most five strong principal lines through any point of* M*. The generic local qualitative behaviour of the strong principal foliations of* M *corresponds to the one described above for the asymptotic curves of surfaces generically immersed in* $\mathbb{R}^5$ *(see Figure 8.2).*

2) *The semiumbilic points of M form a smooth (non necessarily connected) curve.*
3) *The discriminant curve Υ of the strong principal directions IDE of a surface M immersed in $\mathbb{R}^4$ is the image of the discriminant Δ of a surface immersed in $S^4 \subset \mathbb{R}^5$ by the stereographic projection ς. Moreover,*
   a) *The intersection of the ridge curve with Υ are folded singularities of the strong principal directions IDE.*
   b) *The semiumbilic points of type D_5 lie at transversal intersections of Υ with the curve of semiumbilics, but are not folded singularity of the strong principal directions IDE.*
   c) *If q is a non semiumbilic point of Υ, then q is a folded singularity of the strong principal directions IDE $\Leftrightarrow$ q is a ridge point of type A_4.*

8.3 Orthogonal projections onto hyperplanes, 3-spaces and planes

We describe now how the asymptotic directions can be characterised in terms of singularities of projections onto hyperplanes, 3-spaces and planes. We analyse the different possibilities for these three cases.

8.3.1 *Contact with lines*

If TS^4 denotes the tangent bundle of the 4-sphere S^4, the family of projections to 4-planes is given by

$$\begin{aligned} P : M \times S^4 &\to \quad TS^4 \\ (q, \mathbf{v}) &\to (q, P_{\mathbf{v}}(q)) \end{aligned}$$

where $P_{\mathbf{v}}(q) = q - \langle q, \mathbf{v}\rangle \mathbf{v}$. Observe that for a given $\mathbf{v} \in S^4$, the map $P_{\mathbf{v}}$ can be considered locally as a germ of a smooth map $\mathbb{R}^2, 0 \to \mathbb{R}^4, 0$. As in the case of surfaces in 4-space, it follows from Theorem 4.12 that there is an open and dense subset $\mathcal{O}_P$ in $\mathrm{Imm}(U, \mathbb{R}^5)$ for which the family P is a generic family of mappings. So the singularities of $P_{\mathbf{v}}$ that occur in an irremovable way in the family P are those of $\mathcal{A}_e$-codimension ≤ 4, and these are versally unfolded by the family P. For a generic surface, the singularities of $P_{\mathbf{v}}$ are simple and are given in Table 4.3.

Definition 8.10. A surface in $\mathbb{R}^5$ is called *projection P-generic* if any of its local parametrisations belongs to the set $\mathcal{O}_P$.

The bifurcation set of the family of projections P is the set of parameters $\mathbf{v} \in S^4$ where $P_{\mathbf{v}}$ has a non-stable singularity at some point $q \in M$, i.e. has a singularity of $\mathcal{A}_e$-codimension ≥ 1. The following result provides a geometric characterisation of the singularities of projections of surfaces immersed in $\mathbb{R}^5$ into hyperplanes.

Proposition 8.12 ([Romero Fuster, Ruas and Tari (2008)]). *For a projection P-generic surface parametrised by* $\mathbf{x} : M \to \mathbb{R}^5$, *we have*

1) *Given a point* $p \in M_3$, *a direction* $\mathbf{u} \in T_pM$ *is asymptotic if and only if the orthogonal projection of* M *to the* 4*-space* $\mathbf{u}^{\perp}$ *has a singularity of type* I_2 *or worse.*
2) *For* $p \in M_2$ *there are at most* 3 *and at least* 1 *directions* $\mathbf{u} \in T_pM$, *where* $p_{\mathbf{u}}$ *has a singularity of type* I_2. *Moreover,*

   2a) *If* $p \in M_2^h$, *there are two asymptotic directions where* $p_{\mathbf{u}}$ *has a singularity of type* II_2.
   2b) *If* $p \in M_2^e$, *there are no asymptotic directions where* $p_{\mathbf{u}}$ *has a singularity of type* II_2 *or more degenerate.*
   2c) *If* $p \in M_2^p$, *there is one asymptotic direction where* $p_{\mathbf{u}}$ *has a singularity of type* VII_1 *at* p.

3) *There may be a curve in* M_3 *where projecting along one of the asymptotic directions yields a singularity of type* I_3 *and isolated points of this curve where the singularity is of type* I_4.
4) *There may be isolated points where projecting along one of the asymptotic directions yields a singularity of type* II_3. *There may also be isolated* M_2^h*-points where projecting along one of the directions yields a singularity of type* $\mathrm{III}_{2,3}$. *The above points are in general distinct from the* D_5*-points.*

The proof follows the standard procedure of making successive changes of coordinates in order to reduce the appropriate jet of $P_{\mathbf{v}}$ to a normal form.

8.3.2 *Contact with planes*

An orthogonal projection from $\mathbb{R}^5$ to a 3-dimensional subspace is determined by its kernel, so we can parametrise all these projections by the Grassmanian $G(2,5)$ of 2-planes in $\mathbb{R}^5$. If $\mathbf{w_1}, \mathbf{w_2}$ are two linearly independent vectors in $\mathbb{R}^5$, we denote by $\{\mathbf{w_1}, \mathbf{w_2}\}$ the plane they generate and by $\pi_{(\mathbf{w_1},\mathbf{w_2})}$ the orthogonal projection from $\mathbb{R}^5$ to the orthogonal complement of $\langle \mathbf{w_1}, \mathbf{w_2} \rangle$. The restriction of $\pi_{(\mathbf{w_1},\mathbf{w_2})}$ to M, $\pi_{(\mathbf{w_1},\mathbf{w_2})}|_M$, can be considered locally at a point $q \in M$ as a map-germ $\pi_{(\mathbf{w_1},\mathbf{w_2})}|_M : \mathbb{R}^2, 0 \to \mathbb{R}^3, 0$. We distinguish between the two following cases:

1) The plane $\{\mathbf{w_1}, \mathbf{w_2}\}$ is transverse to T_qM, so the projection $\pi_{(\mathbf{w_1},\mathbf{w_2})}|_M$ is locally an immersion.
2) The projection $\pi_{(\mathbf{w_1},\mathbf{w_2})}|_M$ is singular.

In the first case, given any $\mathbf{v} \in N_qM$, let $M_\mathbf{v}$ be the surface patch obtained by projecting M orthogonally to the 3-space $T_qM \oplus \langle \mathbf{v} \rangle$ (considered as an affine space through q). Now, we can take the Gauss map on $M_\mathbf{v}$ considered as a surface in the Euclidean 3-space $T_qM \oplus \langle \mathbf{v} \rangle$. The local stable singularities of this map are folds and cusps. We have the following result characterising the degenerate normal directions and the asymptotic directions at a point $q \in M$ in terms of the geometry of the surface patch $M_\mathbf{v}$ at the point q.

Proposition 8.13 ([Romero Fuster, Ruas and Tari (2008)]). (1) *Suppose that $q \in M_3$ and let $\mathbf{v} \in N_qM$.*

(a) *The direction $\mathbf{v}$ is degenerate if and only if q is a parabolic point of $M_\mathbf{v}$. In this case, the unique principal asymptotic direction of $M_\mathbf{v}$ at q coincides with the contact direction associated to $\mathbf{v}$.*
(b) *A direction $\mathbf{u} \in T_pM$ is asymptotic if and only if there exists $\mathbf{v} \in N_pM$ such that q is a cusp of Gauss of $M_\mathbf{v}$ and $\mathbf{u}$ is its unique asymptotic direction there.*

(2) *Suppose that $q \in M_2$. Then there are two distinct directions $\mathbf{v} \in N_qM$ if $q \in M_2^h$, none if $q \in M_2^e$, and a unique direction if $q \in M_2^p$, where q is a cusp of Gauss of $M_\mathbf{v}$ and $\mathbf{v}^*$ is the unique asymptotic direction of $M_\mathbf{v}$ at q. In addition, there is a unique direction $\bar{\mathbf{v}} \in N_qM$ where $M_{\bar{\mathbf{v}}}$ has a flat umbilic at q. The asymptotic directions of M at q associated to $\bar{\mathbf{v}}$ are the tangent directions to the separatrices of the asymptotic curves of $M_{\bar{\mathbf{v}}}$ at q (see* Figure 8.3).

We deal now with the case when $\pi_{(\mathbf{w_1},\mathbf{w_2})}|_M$ is singular. This means that the kernel of the projection $\pi_{(\mathbf{w_1},\mathbf{w_2})}$ contains a tangent direction at q. When $\{\mathbf{w_1}, \mathbf{w_2}\} = T_qM$, the map-germ $\pi_{(\mathbf{w_1},\mathbf{w_2})}|_M$ has rank zero at the origin and does not identify the asymptotic directions. We shall assume that $\{\mathbf{w_1}, \mathbf{w_2}\}$ is distinct from T_qM. Then $\pi_{(\mathbf{w_1},\mathbf{w_2})}|_M$ has rank 1 at the origin.

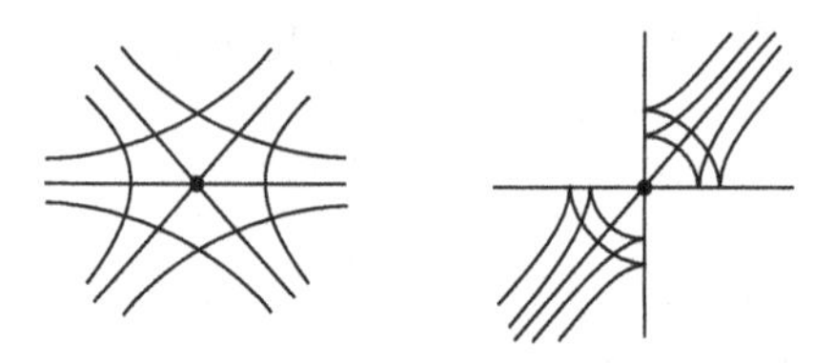

Fig. 8.3 Configurations of the asymptotic curves on surfaces in $\mathbb{R}^3$ at a flat umbilic point (elliptic left and hyperbolic right). There are three separatrices at an elliptic umbilic and one separatrix at a hyperbolic umbilic ([Bruce and Tari (1995)]).

The map $\pi_{(\mathbf{w_1},\mathbf{w_2})}$ can be considered locally as a corank 1 smooth map-germ $\mathbb{R}^2,0 \to \mathbb{R}^3,0$. Since $\pi_{(\mathbf{w_1},\mathbf{w_2})}$ is a 6-parameter family, we expect the map $\pi_{(\mathbf{w_1},\mathbf{w_2})}$ to have only simple singularities of $\mathcal{A}_e$-codimension ≤ 6 or non-simple singularities with the $\mathcal{A}_e$-codimension of the stratum ≤ 6 (as $\dim G(2,5) = 6$). The $\mathcal{A}$-simple singularities of map-germs $\mathbb{R}^2,0 \to \mathbb{R}^3,0$ of $\mathcal{A}_e$-codimension ≤ 6 are given in Table 4.2. The complete classification of non-simple singularities of $\mathcal{A}_e$-codimension of the stratum ≤ 6 is not known so far.

We now state the following result for a generic immersion of a surface in $\mathbb{R}^5$ whose detailed proof can be found in [Romero Fuster, Ruas and Tari (2008)].

Proposition 8.14. *Let* $\mathbf{u} \in T_qM$ *and* $\mathbf{v}$ *in the unit sphere* $S^2 \subset N_qM$.

(1) *The projection* $\pi_{(\mathbf{u},\mathbf{v})}|_M$ *has a cross-cap singularity for almost all* $\mathbf{v} \in S^2$.

(2) *On a circle of directions* $\mathbf{v}$ *in* S^2 *minus a point,* $\pi_{(\mathbf{u},\mathbf{v})}|_M$ *has a singularity with 2-jet* $\mathcal{A}$*-equivalent to* $(x, y^2, 0)$.

(3) *There is a unique direction* $\mathbf{v} \in S^2$ *where* $\pi_{(\mathbf{u},\mathbf{v})}|_M$ *has a singularity of type* H_k *provided* $\mathbf{u}$ *is not an asymptotic direction. If* $\mathbf{u}$ *is asymptotic, then the singularity becomes non-simple with 2-jet* $\mathcal{A}$*-equivalent to* $(x, xy, 0)$.

8.3.3 *Contact with 3-spaces*

An orthogonal projection from $\mathbb{R}^5$ to a 2-dimensional subspace is also determined by its kernel and hence we can parametrise all these projections by the Grassmanian $G(3,5)$ of 3-planes in $\mathbb{R}^5$. However, $G(3,5)$ can be identified with $G(2,5)$, so the projections can also be parametrised by $\{\mathbf{w_1}, \mathbf{w_2}\} \in G(2,5)$, where $\{\mathbf{w_1}, \mathbf{w_2}\}$ is the orthogonal complement of the kernel of the projection. We denote the associated projection by $\Pi_{(\mathbf{w_1},\mathbf{w_2})}$. The restriction $\Pi_{(\mathbf{w_1},\mathbf{w_2})}|_M$ of $\Pi_{(\mathbf{w_1},\mathbf{w_2})}$ to M can be considered, locally at a point $q \in M$, as a map-germ

$$\Pi_{(\mathbf{w_1},\mathbf{w_2})}|_M : \mathbb{R}^2, 0 \to \mathbb{R}^2, 0.$$

As in the previous section, we expect that simple singularities of $\mathcal{A}_e$-codimension ≤ 6 or non-simple singularities with the $\mathcal{A}_e$-codimension of the stratum ≤ 6 to occur for generic surfaces. The list of corank 1 singularities of map-germs $\mathbb{R}^2, 0 \to \mathbb{R}^2, 0$, of $\mathcal{A}_e$-codimension ≤ 6 is given by Rieger ([Rieger (1987)]). The corank 1 simple singularities are given in Table 7.6. The $\mathcal{A}$-simple germs of corank 2 are classified in [Rieger and Ruas (1991)]. The complete classification of non-simple singularities from the plane to the plane of $\mathcal{A}_e$-codimension of the stratum ≤ 6 is not known so far.

We start with the corank 1 singularities. Let $\mathbf{u} \in T_qM$, $\mathbf{u}^\perp$ an orthogonal vector to $\mathbf{u}$ in T_qM and take $\mathbf{v} = (v_1, v_2, v_3) \in N_qM$. We consider the projection $\Pi_{(\mathbf{u}^\perp,\mathbf{v})}|_M$. We take M in Monge form at the origin and suppose, without loss of generality, that the intersection of the kernel of $\Pi_{(\mathbf{u}^\perp,\mathbf{v})}$ with T_qM occurs along $\mathbf{u} = (1, 0)$. Then we have

$$\Pi_{(\mathbf{u}^\perp,\mathbf{v})}|_M(x,y) = (y, v_1(x^2+f_1(x,y))+v_2(xy+f_2(x,y))+v_3(cy^2+f_3(x,y))),$$

where c is equal to 0 or 1 according to q being an M_3 or an M_2 point. Observe that the $\mathcal{A}$-type of the singularities of the above map-germ is independent of c. Therefore, the corank 1 singularities of the projections to planes do not distinguish between the M_3 and M_2 points.

If $v_1 \neq 0$, then $\Pi_{(\mathbf{u}^\perp,\mathbf{v})}|_M$ is $\mathcal{A}$-equivalent to a fold map-germ.

If $v_1 = 0$ and $v_2 \neq 0$, then $j^2\Pi_{(\mathbf{u}^\perp,\mathbf{v})}|_M \sim_{\mathcal{A}} (y, xy)$. The $\mathcal{A}$-singularities of $\Pi_{(\mathbf{u}^\perp,\mathbf{v})}|_M$ are given by the normal forms 5, 6 and 7 in Table 7.6. Non-simple singularities such that the codimension of the stratum is $\mathcal{A}_e$-codimension ≤ 6 may also occur.

If $v_1 = v_2 = 0$, $\Pi_{(\mathbf{u}^\perp,\mathbf{v})}|_M(x,y) = (y, f_3(x,y))$, and the singularities are of type 4_k unless $f_{3xxx}(0,0) = 0$. In this case, the singularities are of type 11_{2k+1} (Table 7.6) or more degenerate. The condition $f_{3xxx}(0,0) = 0$ is precisely the condition for $\mathbf{u} = (1, 0)$ to be an asymptotic direction at the origin. Therefore we can characterise asymptotic directions in terms of corank 1 singularities of projections to 2-planes.

Proposition 8.15. *Let* $\mathbf{u} \in T_qM$ *and* $\mathbf{v}$ *in the unit sphere* $S^2 \subset N_qM$.

(1) *The projection* $\Pi_{(\mathbf{u}^\perp,\mathbf{v})}|_M$ *has a fold singularity for almost all* $\mathbf{v} \in S^2$.

(2) *On a circle of directions* $\mathbf{v}$ *in* S^2 *minus a point* $\Pi_{(\mathbf{u}^\perp,\mathbf{v})}|_M$ *has a singularity with 2-jet* $\mathcal{A}$*-equivalent to* (x, xy) *(equivalently, it is not a fold and has a smooth critical set).*
(3) *There is a unique direction* $\mathbf{v} \in S^2$ *where* $\Pi_{(\mathbf{u}^\perp,\mathbf{v})}|_M$ *has a singularity of type* 4_k *provided* $\mathbf{u}$ *is not an asymptotic direction. If* $\mathbf{u}$ *is asymptotic, then the singularity is* $\mathcal{A}$*-equivalent to* 11_{2k+1} *or is more degenerate.*

We analyse now the corank 2 singularities of the projection. Let $\{\mathbf{w_1}, \mathbf{w_2}\}$ be a plane in N_qM and denote by $M_{(\mathbf{w_1},\mathbf{w_2})}$ the surface patch obtained by projecting M orthogonally to the 4-space $T_qM \oplus \{\mathbf{w_1}, \mathbf{w_2}\}$ (considered as an affine space through the point q). The map-germ $\Pi_{(\mathbf{w_1},\mathbf{w_2})}|_M$ has then a corank 2 singularity at the origin, and this singularity can be characterised as follows in terms of the geometry of $M_{(\mathbf{w_1},\mathbf{w_2})}$.

Proposition 8.16 ([Romero Fuster, Ruas and Tari (2008)]). *The assertions hold for a generic immersed surface* M *in* $\mathbb{R}^5$.

(1) *The 2-jet of the projection* $\Pi_{(\mathbf{w_1},\mathbf{w_2})}|_M$ *is* $\mathcal{A}$*-equivalent to* (x^2, y^2), $(x^2 - y^2, xy)$ *or* (x^2, xy) *if and only if* q *is, respectively, a hyperbolic, elliptic or parabolic point of* $M_{(\mathbf{w_1},\mathbf{w_2})}$.
(2) *The 2-jet of the projection* $\Pi_{(\mathbf{w_1},\mathbf{w_2})}|_M$ *is* $\mathcal{A}$*-equivalent to* $(x^2 + y^2, 0)$, $(x^2 - y^2, 0)$, *or* $(x^2, 0)$ *if and only if* q *is, respectively, an inflection point of real type, of imaginary type or of flat type of* $M_{(\mathbf{w_1},\mathbf{w_2})}$.
Moreover, if $q \in M_3$ *then* $\Pi_{(\mathbf{w_1},\mathbf{w_2})}|_M$ *satisfies (1) for every plane* $\{\mathbf{w_1}, \mathbf{w_2}\} \subset N_qM$. *The point* $q \in M_2$ *if and only if there exists a direction* $\mathbf{w}_2 \in N_qM$ *such that* q *is an inflection point of* $M_{(\mathbf{w_1},\mathbf{w_2})}$, *for any* $\mathbf{w}_1 \in N_qM$.

The proof of these assertions follows directly from the classification of points of surfaces in $\mathbb{R}^4$ described in Chapter 7 and the $\mathcal{A}$-classification of map-germs from the plane to the plane ([Rieger (1987); Rieger and Ruas (1991)]).

8.4 Contacts with hyperspheres

Consider the family of distance squared functions on $M = \mathbf{x}(\mathbb{R}^2) \subset \mathbb{R}^5$,

$$\begin{aligned} D : \mathbb{R}^2 \times \mathbb{R}^4 &\to \mathbb{R} \\ (u, a) &\longmapsto \|\mathbf{x}(u) - a\|^2 = d_a(u) \end{aligned}$$

The catastrophe manifold and map are respectively given by the normal bundle of M in $\mathbb{R}^5$, $NM = \{(u,a) \in U \times \mathbb{R}^5 | \, a = \mathbf{x}(u) + \lambda \mathbf{v}, \, \mathbf{v} \in (N_pM)_1\}$ and the *normal exponential map*,

$$\begin{aligned} G^e : NM &\to \mathbb{R}^5 \\ (u,a) &\longmapsto a. \end{aligned}$$

Then we have the following description of the generic singularities of the distance squared functions as a consequence of Theorem 4.4.

Theorem 8.2. *There is an open and dense set $\mathcal{O}_D$ of proper immersions $\mathbf{x} : U \to \mathbb{R}^5$ such that for any $\mathbf{x} \in \mathcal{O}_D$, the surface $M = \mathbf{x}(U)$ has the following properties:*

(i) *Given any point $p \in M$, the distance squared function d_a from a point $a \in \mathbb{R}^5$ such that $a - p \in N_pM$ has only singularities of $\mathcal{K}$- type $A_1, A_2, A_3, A_4, A_5, A_6, D_4, D_5$ and D_6.*
(ii) *The singularities of d_a are $\mathcal{R}$-versally unfolded by the family D.*
(iii) *The normal exponential map G^e is stable as a Lagrangian map.*

Definition 8.11. A surface in $\mathbb{R}^5$ is called *distance squared function generic* if any of its local parametrisations belongs $\mathcal{O}_D$ in Theorem 8.2.

Definition 8.12. A *focal centre* at $p \in M$ is a point a at which D_a has a degenerate singularity. The directions lying in the kernel of the corresponding Hessian quadratic form are said to be *spherical contact directions* at p. A focal centre a is said to be an *umbilical focus* at p if D_a has a singularity of corank 2 at p. The corresponding focal hypersphere is called *umbilical focal hypersphere*.

The set of the focal centres of the points of M is said to be the *focal set* of M. We denote it by $\mathcal{F}$.

We observe that the proofs of Lemmas 7.2 and 7.3 can be easily adapted to surfaces in $\mathbb{R}^5$, so we can state the following:

Proposition 8.17. *For a point $a = p + \mu\mathbf{v}$ lying in the normal plane of M at p (i. e. $\mathbf{v} \in N_pM$), we have that $\mathcal{H}(d_a)(u) = 2(I - \mu\mathcal{H}(h_\mathbf{v})(u))$ and moreover, if $a = p + \mu\mathbf{v}$ is a focal centre of M at p, the spherical contact directions associated to the distance squared functions from focal centres at p are eigenvectors of the shape operators $S_\mathbf{v}$ in the normal direction $\mathbf{v}$ at p with eigenvalues $1/\mu$.*

The *focal set* of M clearly coincides with the bifurcation set of the family D, given by

$$\mathcal{F} = \{a \in \mathbb{R}^5 : \exists p = \mathbf{x}(u) \in M | d_a \text{ has a degenerate singularity at } u\}.$$

The focal set $\mathcal{F}$ of a distance squared generic embedding is an algebraic variety of dimension 4 in $\mathbb{R}^5$.

Definition 8.13. The *rib set* of M is the singular set of the focal set $\mathcal{F}$.

As in the case of height functions, we can relate the singularity types of the normal exponential map N_e, as a catastrophe map of the family of distance squared functions, with those of the distance squared functions on the surface M (see Remark 3.2):

a) $(u, a) \in S_{1,0}(N_e)$ if and only if u is a singularity of type A_2 of d_a.
b) $(u, a) \in S_{1,1,0}(N_e)$ if and only if u is a singularity of type A_3 of d_a.
c) $(u, a) \in S_{1,1,1,0}(N_e)$ if and only if u is a singularity of type A_4 of d_a.
d) $(u, a) \in S_{1,1,1,1,0}(N_e)$ if and only if u is a singularity of type A_5 of d_a.
e) $(u, a) \in S_2(N_e)$ if and only if u is a singularity of type $D_k^{\pm}, K = 4, 5$ of d_a.

So the focal set $\mathcal{F}$ is a stratified set that can be decomposed as $S_{1,1}(N_e) \cup S_2(N_e)$, where $S_{1,1}(N_e)$ can be decomposed in turn as a union $S_{1,1,0}(N_e) \cup S_{1,1,1,0}(N_e) \cup S_{1,1,1,1,0}(N_e) \cup S_{1,1,1,1,1,0}(N_e)$ and $S_2(N_e) = S_{2,0}(N_e) \cup S_{2,1,0}(N_e)$. These are respectively stratified 3- and 2-dimensional subsets of $\mathcal{F}$ in $\mathbb{R}^5$.

Definition 8.14. The *ridge curves* on M are the connected components of the image of $S_{1,1,1,1}(N_e)$ by the natural projection $\pi_N : NM \to M$.

The ridge curves are made of singularities of type $A_{j\geq 5}$ of the distance squared functions on M. For a distance squared function generic surface, the ridge points of type A_6 are isolated points on curves of points of type A_5. On the other hand, the projection of $S_{2,1,0}(N_e)$ by π_N on M is a curve made of singularities of type D_5 of distance squared functions on M with isolated points of type $D_6^{\pm}$.

Given an immersion $\mathbf{x}$ of M in $\mathbb{R}^5$ for $p = \mathbf{x}(0) \in M$ and $\mathbf{v} = a - p \in N_pM$ we have

$$\mathcal{H}(d_a)(0,0) = -2 \begin{bmatrix} \langle \mathbf{x}_{u_1u_1}(0,0), \mathbf{v}\rangle - 1 & \langle \mathbf{x}_{u_1u_2}(0,0), \mathbf{v}\rangle \\ \langle \mathbf{x}_{u_1u_2}(0,0), \mathbf{v}\rangle & \langle \mathbf{x}_{u_2u_2}(0,0), \mathbf{v}\rangle - 1 \end{bmatrix}.$$

So the focal set of M at p is given by

$$\mathcal{F}_p = \left\{ a \in \mathbb{R}^5 | a = p + \mathbf{v}, \text{ with } \mathbf{v} \in \mathcal{W}_p \right\},$$

where $\mathcal{W}_p$ is the subset of N_pM made of vectors $\mathbf{v} \in N_pM$ satisfying,

$$(\langle \mathbf{x}_{u_1u_1}(0,0), \mathbf{v}\rangle - 1)(\langle \mathbf{x}_{u_2u_2}(0,0), \mathbf{v}\rangle - 1) - \langle \mathbf{x}_{u_1u_2}(0,0), \mathbf{v}\rangle^2 = 0.$$

A straightforward calculation shows that

$$\mathcal{W}_p = \left\{ \mathbf{v} \in N_pM | (\langle \mathbf{H}, \mathbf{v}\rangle - 1)^2 = \langle \mathbf{B}, \mathbf{v}\rangle^2 + \langle \mathbf{C}, \mathbf{v}\rangle^2 \right\},$$

where $\mathbf{H}, \mathbf{B}$ and $\mathbf{C}$ are the vectors that determine the curvature ellipse of M at p. We can express the focal set at p, in terms of these vectors:

$$\mathcal{F}_p = \left\{ a = p + \mathbf{v} \in N_pM; \ (\langle \mathbf{H}, \mathbf{v}\rangle - 1)^2 = \langle \mathbf{B}, \mathbf{v}\rangle^2 + \langle \mathbf{C}, \mathbf{v}\rangle^2 \right\}.$$

The analysis of the above expression for $\mathcal{W}_p$ at the different points of a surface generically immersed in $\mathbb{R}^5$ leads to the following result.

Proposition 8.18 ([Costa, Moraes and Romero Fuster (2009)]). *Given a surface M immersed in $\mathbb{R}^5$, a point $p \in M$, let $\mathbf{H}^\perp$ be the projection of the mean curvature vector $\mathbf{H}$ on the orthogonal complement $E_p^\perp$ of E_p in N_pM.*

(1) *If $p \in M_3$ then $\mathcal{F}_p$ is a cone with vertex at $\frac{\mathbf{H}^\perp}{\|\mathbf{H}^\perp\|^2}$.*
(2) *For $p \in M_2$ and $\dim(E_p) = 2$ all focal points at p have corank 1 and we may have:*
   (a) *If $p \in M_2^e$ then $\mathcal{W}_p = \varnothing$.*
   (b) *If $p \in M_2^h$ then $\mathcal{W}_p$ is a hyperbola in E_p.*
   (c) *If $p \in M_2^p$ then $\mathcal{W}_p$ is a parabola in E_p.*

We can now analyse the distribution of the umbilical foci of M on its different types of points.

Theorem 8.3.

(a) *Given a point $p \in M_3$, there exists a unique umbilical focus at p, given by*

$$a_p = p + \frac{1}{d(p, E_p)} v_p,$$

with v_p = unit normal vector orthogonal to the plane E_p, such that $\langle v_p, \mathbf{H}\rangle > 0$.

(b) *If $p \in M_2$ is not semiumbilic, the umbilical focus lies at infinity and the umbilical focal hypersphere becomes an osculating hyperplane.*
(c) *If $p \in M_2$ is semiumbilic then there is a line of umbilical foci at p contained in the vector plane orthogonal to the affine line E_p.*

Proof. With analogous argument to that of Theorem 7.14 we can see that if a point $a = p + \lambda \mathbf{v}$ is an umbilical focus then the (unit) vector $\mathbf{v}$ must be orthogonal to the vectors $\mathbf{B}$ and $\mathbf{C}$, i.e. $\mathbf{v} \in E_p^{\perp}$.

Then, provided E_p is a plane, we get from the condition that rank Hess $D_a(p) = 0$ that $\lambda = \frac{1}{\langle \mathbf{H}, \mathbf{v} \rangle}$. And the statement a) follows from observing that $\langle \mathbf{H}, \mathbf{v} \rangle = \pm d(p, E_p)$, with sign $+$ or $-$ according $\langle \mathbf{v_p}, \mathbf{H} \rangle > 0$ or < 0. For statement b) we just observe that $\mathbf{H}, \mathbf{B}, \mathbf{C}$ are linearly dependent and $d(p, E_p) = 0$. In this case, there is no proper umbilical focal hypersphere, but the height function in the direction $\mathbf{v_p}$ has a corank 2 singularity at p and hence, the hyperplane with orthogonal direction $\mathbf{v_p}$ is an osculating hyperplane (with corank 2 contact) that can be considered as a degenerate umbilical focal hypersphere.

In the case that p is a semiumbilic point, $E_p^{\perp}$ is a plane and the condition rank Hess $D_a(p) = 0$ is fulfilled by a whole line of points in $E_p^{\perp}$. □

Definition 8.15. Given $M\mathbb{R}^5$ and $p \in M$, we define the *umbilical curvature* of M at p as

$$\kappa_u(p) = d(p, E_p).$$

Provided M has no semiumbilic points, we can give a sign to the function $\kappa_u(p)$ by defining

$$\kappa(u) = \langle \mathbf{H}, \frac{\mathbf{B} \times \mathbf{C}}{\|\mathbf{B} \times \mathbf{C}\|} \rangle,$$

where $\mathbf{B}, \mathbf{C}, \mathbf{H}$ are the 3 vectors that determine the curvature ellipse at the point p.

Corollary 8.3.

(a) *If $p \in M_3$, then $\kappa_u(p)$ is the curvature of the unique umbilic focal hypersphere of M at p.*
(b) *If $p \in M_2$ is a non semiumbilic point, then $\kappa_u(p) = 0$.*
(c) *If $p \in M_2$ is a semiumbilic point, $\kappa_u(p)$ is the maximum curvature among those of all the tangent hyperspheres centred at umbilical foci of M at p.*
(d) *If p is 2-regular then $\kappa_u(p) \neq 0$.*

Proof. The proofs of all these assertions follow easily from the definition of κ_u. □

8.5 Notes

Relative mean curvature foliations: The study of the mean directional configurations carried out by Mello ([Mello (2003)]) on surfaces in $\mathbb{R}^4$ has been generalised to surfaces in $\mathbb{R}^n, n \geq 5$ in [Gonçalves, Martínez Alfaro, Montesinos Amilibia and Romero Fuster (2007)].

We point out first that surfaces immersed in codimension higher than 2 do not admit mean curvature directions in the way defined for surfaces in $\mathbb{R}^4$. Nevertheless it is possible to introduce certain foliations on surfaces immersed in $\mathbb{R}^n$, that in the particular case of surfaces in 4-space coincide with the mean curvature foliations. The procedure is based on the fact that, from a qualitative viewpoint, all the principal configurations on S arise from normal vector fields parallel to the subspace determined by the curvature ellipse at every point. In fact, any normal vector $v \in N_pS$ can be decomposed into a sum $v^\top + v^\perp$, with $v^\top$ and $v^\perp$ respectively parallel and orthogonal to the plane determined by the curvature ellipse. Since the shape operator associated to $v^\perp$ is a multiple of the identity ([Moraes and Romero Fuster (2005)]), then the eigenvectors of the shape operator W_v and $W_{v^\top}$ must coincide. This idea lead in [Gonçalves, Martínez Alfaro, Montesinos Amilibia and Romero Fuster (2007)] to the definition of the *relative mean curvature directions* at a point p of a surface immersed in $\mathbb{R}^n$ with $n > 4$ as those inducing normal sections with curvature vector parallel to $H(p)^\top$. This gives rise to two orthogonal foliations whose critical points are the semiumbilics and the pseudo-umbilics (with inflection points and minimal points considered as non generic particular cases). Interesting global consequences of this setting are the following facts ([Gonçalves, Martínez Alfaro, Montesinos Amilibia and Romero Fuster (2007)]):

a) The mean curvature vector of a 2-sphere generically immersed in $\mathbb{R}^n, n \geq 5$ becomes orthogonal to the normal subspace determined by the curvature ellipse in at least 4 points.
b) Closed oriented surfaces with non-vanishing Euler number immersed into $\mathbb{R}^n, n \geq 5$ always have either semiumbilic, pseudo-umbilic, inflection, or minimal points.
c) Closed oriented 2-regular surfaces with non-vanishing Euler number in

$\mathbb{R}^n, n \geq 5$, always have pseudo-umbilic points.

Immersions in higher codimension: The curvature locus at a point p of an m-manifold M immersed in $\mathbb{R}^n$ is a projection of the Veronese submanifold of dimension $m-1$ into the normal space of the submanifold at p. The possible shapes and topological types of this projection for 3-manifolds immersed in $\mathbb{R}^n$ are analysed in [Binotto (2008)] (see also [Binotto, Costa and Romero Fuster (2015)]). Figure 8.4 illustrates the curvature locus of the immersion $f : \mathbb{R}^3 \to \mathbb{R}^6; f(x_1, x_2, x_3) = (x_1, x_2, x_3, x_1^2 + x_2^2 - x_3^3, x_1x_3, x_1x_2)$ at a point near the origin.

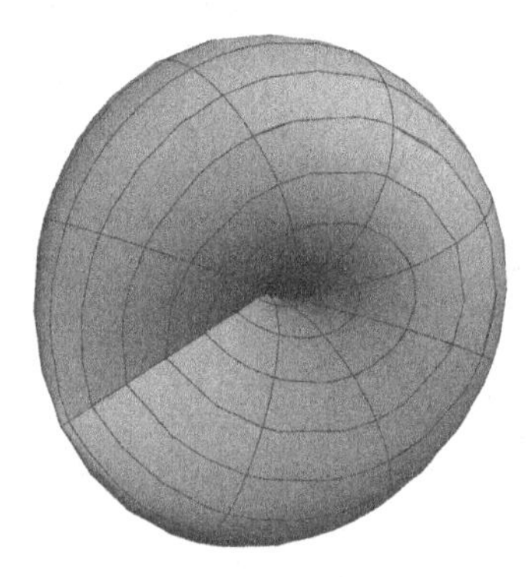

Fig. 8.4 Curvature locus at a generic point of a 3-manifold in $\mathbb{R}^6$.

The generic behaviour of the ridges and flat ridges of 3-manifolds in $\mathbb{R}^n, n \geq 4$ has been analysed in [Nabarro and Romero Fuster (2009)] showing that they can be seen as the image of a stable map from a convenient surface (non necessarily connected) into the 3-manifold. On the other hand, Dreibelbis ([Dreibelbis (2012)]) generalises to n-manifolds in $\mathbb{R}^{2n}$ the notions of asymptotic directions and parabolic and inflection points of surfaces in $\mathbb{R}^4$, analysing in detail the case of 3-manifolds in $\mathbb{R}^6$. An interesting fact, described in this work is that the parabolic subset of a 3-manifold M generically immersed in $\mathbb{R}^6$ is a surface with normal crossings and possible isolated cross-caps. Moreover, the cross-caps (resp. triple points) are the points of M for which the origin of the normal space is a cross-cap (resp.a triple) point of the curvature locus.

kth-regular immersions: The concept of kth-regular immersion of a submanifold in Euclidean space was introduced by Pohl ([Pohl (1962)]) and Feldman ([Feldman (1965)]) in terms of maps between osculating bundles. In the case of curves, the k-regularity condition means that the first k derivatives are linearly independent. An interesting question arises when we consider the problem from a global viewpoint: *Under what conditions can we ensure the existence of* 2*-regular immersions from a given class of closed submanifold?* For surfaces in $\mathbb{R}^4$, the 2-singular points coincide with the inflection points and local convexity is an obstruction for the 2-regularity of closed surfaces with non-vanishing Euler characteristic (see Chapter 7). In the case of surfaces in $\mathbb{R}^5$, a classically known example of 2-regular immersion of S^2 into $\mathbb{R}^5$ is given by the Veronese surface. Examples of 2-regular immersions of closed orientable surfaces with non-zero genus are not known so far. The existence of 2-regular embeddings of orientable closed surfaces in S^4 is investigated in [Romero Fuster (2004)]. Some obstructions to the 2-regularity of surfaces in $\mathbb{R}^5$ are given in terms of the umbilical curvature ([Romero Fuster (2007)]).

Chapter 9

Spacelike surfaces in the Minkowski space-time

The Lorentz-Minkowski space, also called Minkowski space-time, provides a mathematical setting in which Einstein's theory of special relativity is most conveniently formulated. In this setting, the three ordinary dimensions of space are combined with a single dimension of time to form a four-dimensional manifold for representing space-time. The geometry on this space is defined by the Poincaré's group or the group of Lorentz transformations. The geometrical properties of submanifolds of Minkowski space can be studied, in a similar way to those of Euclidean space, through the analysis of their contact with model submanifolds invariant by the Lorentz transformations group. The situation is richer in this case than in the Euclidean one due to the different possibilities for the type of the submanifolds and of the models. The submanifolds can be spacelike, timelike or lightlike and the models can be taken to be the lightlike hyperplanes or hyperspheres.

The geometrical properties associated to the contact of spacelike submanifolds with spacelike models do not differ much from those of submanifolds of Euclidean space. A new interesting and geometrically rich situation arises when considering the contact of submanifolds with lightlike hyperplanes or with lightcones. We call the geometric properties obtained from such contact *Lightlike Geometry* and devote this chapter to introducing it. We point out that Lightlike Geometry can be seen as a generalisation of *Horospherical Geometry*, which is concerned with the study of geometric properties derived from the contact of submanifolds with hyperhorospheres in hyperbolic space.

We shall restrict in this chapter to spacelike surfaces in 4-dimensional Minkowski space-time (surfaces in Euclidean 3-space and surfaces in hyperbolic space being special cases). However, the results are valid for any codimension 2 spacelike submanifold of a higher dimensional Minkowski space-time.

We highlight the following important features of Lightlike Geometry of surfaces in Minkowski 4-space.

(i) The study of contact of spacelike surfaces with lightlike hyperplanes in Minkowski 4-space has interesting similarities to that of surfaces in Euclidean 3-space with planes. For instance, as we shall see in Chapter 10, some global results in Euclidean Geometry, such as the Gauss-Bonnet Theorem, can be formulated in a natural way in Lightlike Geometry.
(ii) Lightlike 3-manifolds, whose study is of special interest in Cosmology, present a challenge for the classical differential geometry techniques. However, they can be viewed (at least locally) as line bundles with lightlike fibres over spacelike surfaces and their contact with lightlike hyperplanes provides a powerful tool for studying their geometry. In particular, it allows us to describe their generic singularities as well as to introduce properties, such as flatness, which are invariant under Lorentz transformations (see [Izumiya and Romero Fuster (2007)]).

The structure of this chapter is as follows. In the first three sections we discuss the basic tools, leading to the concepts of Lightcone Gauss map, normalised lightcone Gauss map and their corresponding Gauss-Kronecker curvatures on a spacelike surfaces. In §9.5 we define the family of lightcone height functions and study their properties. This allows us to describe in §9.6 the normalised lightcone Gauss map as a Lagrangian map whose generating family is the family of lightcone height functions on the considered surface. As a consequence, we define the lightcone pedal in §9.7 and study its properties from the Legendrian viewpoint. In §9.8 we study some special cases of spacelike surfaces. These are surfaces in spacelike hyperplanes (i.e. in Euclidean 3-space), and spacelike surfaces contained in either timelike hyperplanes (i.e. in Minkowski 3-space), de Sitter 3-space or the hyperbolic 3-space. In §9.9 we study the contact of spacelike surfaces with lightcones and its link with lightlike 3-dimensional submanifolds. In §9.10 we introduce several Legendrian dualities between pseudo-spheres generalising the well known classical projective (and spherical) duality. Finally in §9.11 we use these dualities to study spacelike surfaces in the lightcone and prove an analogous Theorema Egregium of Gauss for these surfaces, showing that the mean curvature (as defined in §9.11) is an intrinsic invariant and is equal to the section curvature.

9.1 Minkowski space-time

We give a brief introduction to Minkowski space-time and refer to [O'Neill (1983)] for more details.

The 4-dimensional Minkowski space-time $(\mathbb{R}^4, \langle , \rangle_1)$, denoted by $\mathbb{R}^4_1$, is the vector space $\mathbb{R}^4$ endowed with the pseudo scalar product

$$\langle \mathbf{u}, \mathbf{v} \rangle_1 = -u_0v_0 + u_1v_1 + u_2v_2 + u_3v_3,$$

with $\mathbf{u} = (u_0, u_1, u_2, u_3)$ and $\mathbf{v} = (v_0, v_1, v_2, v_3)$ any two vectors in $\mathbb{R}^4$. A non-zero vector $\mathbf{u}$ is said to be

$$\begin{array}{lll} \textit{spacelike} & \text{if} & \langle \mathbf{u}, \mathbf{u} \rangle_1 > 0, \\ \textit{timelike} & \text{if} & \langle \mathbf{u}, \mathbf{u} \rangle_1 < 0, \\ \textit{lightlike} & \text{if} & \langle \mathbf{u}, \mathbf{u} \rangle_1 = 0. \end{array}$$

Two vectors $\mathbf{u}$, $\mathbf{v}$ are said to be *pseudo-orthogonal* if $\langle \mathbf{u}, \mathbf{v} \rangle_1 = 0$.

The *norm* of a vector $\mathbf{u}$ is defined to be $\|\mathbf{u}\|_1 = \sqrt{|\langle \mathbf{u}, \mathbf{u} \rangle_1|}$.

A linear plane (i.e., a 2-dimensional vector subspace of $\mathbb{R}^4_1$) is said to be *spacelike* if all its non-zero vectors are spacelike. It is *timelike* if it contains a timelike and a spacelike vector. It is *lightlike* if it contains a single lightlike direction.

Any linear hyperplane (i.e., a 3-dimensional vector subspace of $\mathbb{R}^4_1$) can be defined as the set of vectors $\mathbf{u} \in \mathbb{R}^4_1$ which are pseudo-orthogonal to a given non-zero vector $\mathbf{v}$. The hyperplane is said to be *spacelike* (resp. *timelike*, *lightlike*) if $\mathbf{v}$ is timelike (resp. spacelike, lightlike).

We choose a system of coordinates $(O, \{\mathbf{e}_0, \mathbf{e}_1, \mathbf{e}_2, \mathbf{e}_3\})$ in $\mathbb{R}^4_1$, where O is referred to as the origin and where $\{\mathbf{e}_0, \mathbf{e}_1, \mathbf{e}_2, \mathbf{e}_3\}$ denotes the canonical basis of the vector space $\mathbb{R}^4$. We can now view $\mathbb{R}^4_1$, in the usual way, as a set of points by identifying a point p with the vector Op.

For any non-zero vector $\mathbf{v}$ and $c \in \mathbb{R}$, the hyperplane with the pseudo-normal $\mathbf{v}$ is the set of points

$$HP(\mathbf{v}, c) = \{p \in \mathbb{R}^4_1 \,|\, \langle p, \mathbf{v} \rangle_1 = c\}.$$

We say that $HP(\mathbf{v}, c)$ is *spacelike* (resp. *timelike*, *lightlike*) if $\mathbf{v}$ is time-like (resp. spacelike, lightlike).

A hypersphere of centre q and radius $\sqrt{|r|}$ is the set of points

$$\{p \in \mathbb{R}^4_1 \,|\, \langle p - q, p - q \rangle_1 = r\}.$$

There are three types of hyperspheres determined by $r < 0$, $r = 0$, $r > 0$. The hyperspheres of centre q are the translations to the centre q of the following hyperspheres of centre the origin.

The *hyperbolic* 3*-space*:

$$H^3_+(-1) = \{p \in \mathbb{R}^4_1 \,|\, \langle p, p\rangle_1 = -1,\ p_0 > 0\}.$$

The *de Sitter* 3*-space*:

$$S^3_1 = \{p \in \mathbb{R}^4_1 \,|\, \langle p, p\rangle_1 = 1\}.$$

The *open lightcone*:

$$LC^* = \{p \in \mathbb{R}^4_1 \,|\, p \neq O,\ \langle p, p\rangle_1 = 0\}.$$

We sometimes view $H^3_+(-1)$, S^3_1 and LC^* as sets of vectors and denote their elements by $\mathbf{u}$.

Since $\mathbb{R}^4_1$ is time-orientable ([O'Neill (1983)]), we choose $\mathbf{e}_0$ as *the future direction.* We say that a timelike vector $\mathbf{u}$ is *future directed* (resp. *past directed*) if $\langle \mathbf{u}, \mathbf{e}_0\rangle_1 < 0$ (resp. $\langle \mathbf{u}, \mathbf{e}_0\rangle_1 > 0$)

The Minkowski wedge product of three vectors $\mathbf{u}_i = (u^i_0, u^i_1, u^i_2, u^i_3)$, $i = 1, 2, 3$, is defined as the vector

$$\mathbf{u}_1 \wedge \mathbf{u}_2 \wedge \mathbf{u}_3 = \begin{vmatrix} -\mathbf{e}_0 & \mathbf{e}_1 & \mathbf{e}_2 & \mathbf{e}_3 \\ u^1_0 & u^1_1 & u^1_2 & u^1_3 \\ u^2_0 & u^2_1 & u^2_2 & u^2_3 \\ u^3_0 & u^3_1 & u^3_2 & u^3_3 \end{vmatrix}.$$

Clearly, $\langle \mathbf{u}, \mathbf{u}_1 \wedge \mathbf{u}_2 \wedge \mathbf{u}_3\rangle_1 = \det(\mathbf{u}, \mathbf{u}_1, \mathbf{u}_2, \mathbf{u}_3)$. Therefore, $\mathbf{u}_1 \wedge \mathbf{u}_2 \wedge \mathbf{u}_3$ is pseudo orthogonal to the three vectors $\mathbf{u}_i, i = 1, 2, 3$.

9.1.1 *The hyperbolic space and the Poincaré ball model*

The hyperbolic space $H^3_+(-1)$ with the induced metric provides the model for the Gauss-Bolyai-Lobachevski non-Euclidean geometry. Another model for the hyperbolic space is the *Poincaré ball model* which is constructed as follows. Let $D = \{(x_1, x_2, x_3) \in \mathbb{R}^3 \,|\, x_1^2 + x_2^2 + x_3^2 < 1\}$ be the open ball in $\mathbb{R}^3$. We can consider $\mathbb{R}^3$ as a subset of $\mathbb{R}^4_1$ by identifying it with the hyperplane $x_0 = 0$. The Poincaré ball model of the hyperbolic space is the set D endowed with the hyperbolic metric

$$ds^2 = \frac{4(dx_1^2 + dx_2^2 + dx_3^2)}{1 - x_1^2 - x_2^2 - x_3^2}.$$

Let $x = (x_0, x_1, x_2, x_3)$ be a point in $H^3_+(-1)$ and consider the line joining x and the fixed point $(-1, 0, 0, 0)$. This line intersects the unit disc

D at a unique point whose coordinates are $(\frac{x_1}{x_0+1}, \frac{x_2}{x_0+1}, \frac{x_3}{x_0+1})$. The map $\Phi : H^3_+(-1) \to D$, given by

$$\Phi(x_0, x_1, x_2, x_3) = \left(\frac{x_1}{x_0+1}, \frac{x_2}{x_0+1}, \frac{x_3}{x_0+1} \right)$$

is an isometry. The hyperbolic metric ds^2 on D is conformally equivalent to the Euclidean space.

The hyperbolic space $H^3_+(-1)$ can be considered as a homogeneous space. Let $\mathrm{SO}_0(1,3)$ be the identity component of the matrix group

$$\mathrm{SO}(1,3) = \{ g \in GL(4,\mathbb{R}) \mid g \cdot I_{1,3} \cdot g^t = I_{1,3} \},$$

where

$$I_{1,3} = \left(\begin{array}{c|ccc} -1 & 0 & 0 & 0 \\ \hline 0 & 1 & 0 & 0 \\ 0 & 0 & 1 & 0 \\ 0 & 0 & 0 & 1 \end{array} \right) \in GL(4,\mathbb{R}).$$

The group $\mathrm{SO}_0(1,3)$ acts transitively on $H^3_+(-1)$ and the isotropic group at $p = (1,0,0,0)$ is $\mathrm{SO}(3)$, which is naturally embedded in $\mathrm{SO}_0(1,3)$. Moreover, the action induces isometries on $H^3_+(-1)$. Therefore, we have an isometry between $\mathrm{SO}_0(1,3)/\mathrm{SO}(3)$ and $H^3_+(-1)$. By the uniqueness of the model of the hyperbolic space, any model of the hyperbolic space is isometric to the homogeneous space $\mathrm{SO}_0(1,3)/\mathrm{SO}(3)$. Therefore, $H^3_+(-1)$ is isometric to the Poincaré disc model of the hyperbolic 3-space.

9.2 The lightcone Gauss maps

Consider the orientation of $\mathbb{R}^4_1$ given by the volume form $\mathbf{e}_0^* \wedge \mathbf{e}_1^* \wedge \mathbf{e}_2^* \wedge \mathbf{e}_3^*$, where $\{\mathbf{e}_0^*, \mathbf{e}_1^*, \mathbf{e}_2^*, \mathbf{e}_3^*\}$ is the dual basis of the canonical basis of $\mathbb{R}^4_1$. The Minkowski space $\mathbb{R}^4_1$ has also a *time-orientation* given by the choice of $\mathbf{e}_0$ as a future timelike vector field.

As in the previous chapters, we are interested in the local geometric properties of smooth surfaces in $\mathbb{R}^4_1$. For this reason we consider a surface patch M which is the image of an embedding $\mathbf{x} : U \to \mathbb{R}^4_1$, where U is an open set of $\mathbb{R}^2$ (so $M = \mathbf{x}(U)$).

We say that M is a *spacelike surface* if its tangent plane T_pM is spacelike at all points $p \in M$. Then the pseudo-normal space N_pM of M at p is a timelike plane. We denote by $N(M)$ the pseudo-normal bundle over M. Since this is a trivial bundle, there are two well defined transverse lightlike smooth direction fields on M. These direction fields can be defined by two smooth vector fields $\mathbb{L}_i$, $i = 1, 2$ on M. Of course, the two vector fields are not unique.

Lemma 9.1. *For any given smooth lightlike vector field* $\mathbb{L}$ *on* M*, there is a unique unit timelike normal vector field* $\mathbf{u}$ *on* M *and a unique unit spacelike normal vector field* $\mathbf{v}$ *on* M *such that* $\mathbb{L} = \mathbf{u} + \mathbf{v}$. *Observe that* $\mathbf{u}$ *and* $\mathbf{v}$ *are necessarily pseudo-orthogonal on* M.

Proof. Since the normal bundle $N(M)$ is trivial we can work on N_pM at any given point p on M. Choose a local coordinate system in N_pM, identify N_pM with $\mathbb{R}^2_1$ and represent $\mathbb{L}(p)$ by a vector $\mathbf{w} = (a, a)$, with $a \neq 0$. We are seeking two vectors $\mathbf{u} = (u_0, u_1)$ and $\mathbf{v} = (v_0, v_1)$ with the following properties

$$\begin{aligned} -u_0^2 + u_1^2 &= -1, \\ -v_0^2 + v_1^2 &= 1, \\ u_0 + v_0 &= a, \\ u_1 + v_1 &= a. \end{aligned}$$

Straightforward calculations show that

$$\mathbf{u} = (\frac{a^2+1}{2a}, \frac{a^2-1}{2a}), \ \mathbf{v} = (\frac{a^2-1}{2a}, \frac{a^2+1}{2a}).$$

□

Remark 9.1. The result in Lemma 9.1 shows that any normal lightlike vector field on M can be obtained in the following way. Suppose that M is parametrised by $\mathbf{x} : U \to \mathbb{R}^4_1$ and choose a unit timelike normal vector field $\mathbf{n}^T$ on M (this can always be done as $N(M)$ is a trivial bundle). Then

$$\mathbf{n}^S(u) = \frac{\mathbf{n}^T(u) \wedge \mathbf{x}_{u_1}(u) \wedge \mathbf{x}_{u_2}(u)}{\|\mathbf{n}^T(u) \wedge \mathbf{x}_{u_1}(u) \wedge \mathbf{x}_{u_2}(u)\|_1} \tag{9.1}$$

is a unit spacelike normal vector field which is pseudo orthogonal to $\mathbf{n}^T(u)$. The vector fields $\mathbf{n}^T + \mathbf{n}^S$ and $\mathbf{n}^T - \mathbf{n}^S$ are lightlike normal vector fields on M and determine at each point $p \in M$ the two lightlike directions in N_pM. Any smooth lightlike vector field $\mathbb{L}$ on M coincides with $\mathbf{n}^T + \mathbf{n}^S$ or $\mathbf{n}^T - \mathbf{n}^S$ for some chosen $\mathbf{n}^T$ (take $\mathbf{n}^T$ to be $\mathbf{u}$ in Lemma 9.1 so $\mathbf{v}$ in Lemma 9.1 coincides with $\mathbf{n}^S$ or $-\mathbf{n}^S$).

Since M is spacelike, $\mathbf{e}_0$ is a future directed timelike vector field along M which is transverse to M at all points. We can write any $\mathbf{v} \in T_p\mathbb{R}^4_1|_M$ in the form $\mathbf{v} = \mathbf{v}_1 + \mathbf{v}_2$, where $\mathbf{v}_1 \in T_pM$ and $\mathbf{v}_2 \in N_pM$. If $\mathbf{v}$ is timelike, then $\mathbf{v}_2$ is timelike. Let $\pi_{N_pM} : T_p\mathbb{R}^4_1|_M \to N_pM$ be the canonical projection, where p is any point on M. Then $\pi_{N_pM}(\mathbf{e}_0)$ is a future directed timelike normal vector of M at the point p. Therefore, we can always define a future directed unit timelike normal vector field along M, i.e., globally. It follows that $\mathbf{n}^T$ can be defined on a closed spacelike surface in $\mathbb{R}^4_1$ (i.e., in a global sense) even when the surface is not orientable. However, $\mathbf{n}^S$ depends on the orientation of the surface M and can be defined only for orientable surfaces in $\mathbb{R}^4_1$.

Remark 9.2. There is nothing special about choosing a future directed normal frame. We could choose $\mathbf{n}^T$ to be a past unit normal vector field and construct a *past directed normal frame* $(\mathbf{n}^T, \mathbf{n}^S)$ along M. The results in this chapter are the same for past directed normal frames. (We choose a future directed frame because we like the future!)

Definition 9.1. Let $\mathbf{n}^T$ be a future unit normal timelike vector field on M and let $\mathbf{n}^S$ be its associated unit normal spacelike vector field on M as in (9.1). We call $(\mathbf{n}^T, \mathbf{n}^S)$ a *future directed normal frame* along M and set $\mathbb{L}^+ = \mathbf{n}^T + \mathbf{n}^S$ and $\mathbb{L}^- = \mathbf{n}^T - \mathbf{n}^S$ the two lightlike normal vector fields defined by $\mathbf{n}^T$. (Of course $\mathbb{L}^+$ and $\mathbb{L}^-$ are uniquely determined by $\mathbf{n}^T$ and depend on the choice of $\mathbf{n}^T$.) The vector fields $\mathbb{L}^\pm$ determine the two lightlike normal direction fields on M. In rest of the chapter, $\mathbb{L}$ indicates one of the lightlike vector fields $\mathbb{L}^+$ or $\mathbb{L}^-$.

Remark 9.3. In Chapter 7 and for a surface M in the Euclidean space $\mathbb{R}^4$, we associated to a unit normal vector field ν on M a ν-shape operator. For a spacelike surface in $\mathbb{R}^4_1$, and following Definition 9.1, we have two distinguished normal direction fields on M parallel to $\mathbb{L} = \mathbb{L}^\pm$. The vector fields $\mathbb{L}$ are lightlike, so have zero length at each point. However, we can still define an $\mathbb{L}$-shape operator, and the rest of the chapter is devoted to deriving geometric properties of M from $\mathbb{L}$.

The vector field $\mathbb{L}$ is a map $M \to LC^*$ which we call the *lightcone Gauss map*. We consider its derivative map at a point $p \in M$ which is linear map

$$d\mathbb{L}_p : T_pM \to T_p\mathbb{R}^4_1 = T_pM \oplus N_pM.$$

We have the canonical projections

$$\pi_1 : T_pM \oplus N_pM \to T_pM \quad \text{and} \quad \pi_2 : T_pM \oplus N_pM \to N_pM.$$

We define the $\mathbb{L}$-*shape operator* of M at $p \in M$, and denote it by $W(\mathbb{L})_p$, as the linear map $T_pM \to T_pM$, given by

$$W(\mathbb{L})_p = -\pi_1 \circ d\mathbb{L}_p.$$

We also consider the shape operators $W(\mathbf{n}^T)_p = -\pi_1 \circ d(\mathbf{n}^T)_p$ and $W(\mathbf{n}^S)_p = -\pi_1 \circ d(\mathbf{n}^S)_p$ which we call, respectively, the $\mathbf{n}^T$-*shape operator* and the $\mathbf{n}^S$-*shape operator* of M at p. We have, by definition

$$W(\mathbb{L}^{\pm})_p = W(\mathbf{n}^T)_p \pm W(\mathbf{n}^S)_p.$$

We call the eigenvalues of $W(\mathbb{L})_p$ the $\mathbb{L}$-*lightcone principal curvatures* and denote them by κ_1 and κ_2. (We shall use the notation $\kappa_1^{\pm}$ and $\kappa_2^{\pm}$ when needed.)

A point p on M is said to be an $\mathbb{L}$-*umbilic point* if $\kappa_1(p) = \kappa_2(p)$, equivalently, $W(\mathbb{L})p$ is a multiple of the identity map 1_{T_pM}.

We also define the $\mathbf{n}^T$-*principal curvature* $\kappa_i(\mathbf{n}^T)$, $i = 1, 2$, at p (resp. $\mathbf{n}^S$-*principal curvature* $\kappa_i(\mathbf{n}^S)$, $i = 1, 2$) as the eigenvalues of $W(\mathbf{n}^T)_p$ (resp. $W(\mathbf{n}^S)_p$).

Proposition 9.1. *With notation as above and for* $i = 1, 2$,

$$\kappa_i^{\pm} = \kappa_i(\mathbf{n}^T) \pm \kappa_i(\mathbf{n}^S).$$

Proof. This follows from the fact that $W(\mathbb{L}^{\pm})_p = W(\mathbf{n}^T)_p \pm W(\mathbf{n}^S)_p$.□

Definition 9.2. The $\mathbb{L}$-*lightcone Gauss-Kronecker curvature* of M at p is defined to be

$$K_\ell(p) = \det(W(\mathbb{L})_p).$$

The $\mathbb{L}$-*lightcone mean curvature* of M at p is defined to be

$$H_\ell(p) = \frac{1}{2}\mathrm{Trace}(W(\mathbb{L})_p).$$

A point p on M is said to be an $\mathbb{L}$-*parabolic point* if $K_\ell(p) = 0$.

Since M is a spacelike surface, we have a Riemannian metric (the *first fundamental form*) on M defined by

$$ds^2 = Edu_1^2 + 2Fdu_1du_2 + Gdu_2^2 = g_{11}du_1^2 + 2g_{12}du_1du_2 + g_{22}du_2^2,$$

where

$$\begin{aligned} E = g_{11} &= \langle \mathbf{x}_{u_1}, \mathbf{x}_{u_1} \rangle_1, \\ F = g_{12} = g_{21} &= \langle \mathbf{x}_{u_1}, \mathbf{x}_{u_2} \rangle_1, \\ G = g_{22} &= \langle \mathbf{x}_{u_2}, \mathbf{x}_{u_2} \rangle_1. \end{aligned}$$

The second fundamental form associated to the normal vector field $\mathbb{L}$, which we refer to as the *lightcone second fundamental form*, is given by

$\mathrm{II}_p(\mathbf{u}) = \langle W(\mathbb{L})_p(\mathbf{u}), \mathbf{u}\rangle$, with $\mathbf{u} \in T_pM$. The coefficients of the lightcone second fundamental form are given by

$$\begin{aligned} l = h_{11} &= -\langle \mathbb{L}_{u_1}, \mathbf{x}_{u_1}\rangle_1 = \langle \mathbb{L}, \mathbf{x}_{u_1u_1}\rangle_1, \\ m = h_{12} = h_{21} &= -\langle \mathbb{L}_{u_1}, \mathbf{x}_{u_2}\rangle_1 = -\langle \mathbb{L}_{u_2}, \mathbf{x}_{u_1}\rangle_1 = \langle \mathbb{L}, \mathbf{x}_{u_1u_2}\rangle_1, \\ n = h_{22} &= -\langle \mathbb{L}_{u_2}, \mathbf{x}_{u_2}\rangle_1 = \langle \mathbb{L}, \mathbf{x}_{u_2u_2}\rangle_1. \end{aligned}$$

We shall use the notation $l^\pm, m^\pm, n^\pm$ (or, $h_{11}^\pm, h_{12}^\pm, h_{22}^\pm$) when necessary. We denote by

$$A_p = \begin{pmatrix} a_{11} & a_{12} \\ a_{21} & a_{22} \end{pmatrix}$$

the matrix of the shape operator $W(\mathbb{L})_p$ with respect to the basis $\{\mathbf{x}_{u_1}, \mathbf{x}_{u_2}\}$ of T_pM at $p = \mathbf{x}(u)$.

Proposition 9.2 (The lightcone Weingarten formula). *We have the following formulae:*

$$\begin{aligned} \mathbb{L}_{u_1} &= -(a_{11}\mathbf{x}_{u_1} + a_{21}\mathbf{x}_{u_2}) \pm \langle \mathbf{n}^S, \mathbf{n}^T_{u_1}\rangle_1 \mathbb{L}, \\ \mathbb{L}_{u_2} &= -(a_{12}\mathbf{x}_{u_1} + a_{22}\mathbf{x}_{u_2}) \pm \langle \mathbf{n}^S, \mathbf{n}^T_{u_2}\rangle_1 \mathbb{L} \end{aligned}$$

and the following expression for the matrix of the shape operator $W(\mathbb{L})_p$

$$A_p = \frac{1}{EG - F^2}\begin{pmatrix} G & -F \\ -F & E \end{pmatrix}\begin{pmatrix} l & m \\ m & n \end{pmatrix} = \begin{pmatrix} g^{11} & g^{12} \\ g^{12} & g^{22} \end{pmatrix}\begin{pmatrix} h_{11} & h_{12} \\ h_{12} & h_{22} \end{pmatrix}, \tag{9.2}$$

where

$$\begin{pmatrix} g^{11} & g^{12} \\ g^{12} & g^{22} \end{pmatrix} = \begin{pmatrix} g_{11} & g_{12} \\ g_{12} & g_{22} \end{pmatrix}^{-1} = \frac{1}{g_{11}g_{22} - g_{12}^2}\begin{pmatrix} g_{22} & -g_{12} \\ -g_{12} & g_{11} \end{pmatrix}.$$

Proof. The vectors $\mathbf{x}_{u_1}, \mathbf{x}_{u_2}, \mathbf{n}^T, \mathbf{n}^S$ form a basis of $\mathbb{R}^4_1$, so there exist real numbers λ, μ such that

$$\mathbb{L}_{u_1} = -(a_{11}\mathbf{x}_{u_1} + a_{21}\mathbf{x}_{u_2}) + \lambda\mathbf{n}^T + \mu\mathbf{n}^S.$$

Since $\mathbb{L} = \mathbf{n}^T \pm \mathbf{n}^S$, we have $\langle \mathbb{L}_{u_1}, \mathbf{n}^T\rangle_1 = -\lambda$. On the other hand $\mathbb{L}_{u_1} = \mathbf{n}^T_{u_1} \pm \mathbf{n}^S_{u_1}$, so $\langle \mathbb{L}_{u_1}, \mathbf{n}^T\rangle_1 = \langle \mathbf{n}^T_{u_1} \pm \mathbf{n}^S_{u_1}, \mathbf{n}^T\rangle_1$. But $\langle \mathbf{n}^T, \mathbf{n}^T\rangle_1 = -1$ implies $\langle \mathbf{n}^T_{u_1}, \mathbf{n}^T\rangle_1 = 0$, and $\langle \mathbf{n}^T, \mathbf{n}^S\rangle_1 = 0$ implies $\langle \mathbf{n}^T, \mathbf{n}^S_{u_1}\rangle_1 = -\langle \mathbf{n}^T_{u_1}, \mathbf{n}^S\rangle_1$. Therefore,

$$\lambda = -\langle \mathbf{n}^T_{u_1} \pm \mathbf{n}^S_{u_1}, \mathbf{n}^T\rangle_1 = \pm\langle \mathbf{n}^T_{u_1}, \mathbf{n}^S\rangle_1.$$

Similarly,

$$\mu = \langle \mathbb{L}_{u_1}, \mathbf{n}^S\rangle_1 = \langle \mathbf{n}^T_{u_1} \pm \mathbf{n}^S_{u_1}, \mathbf{n}^S\rangle_1 = \langle \mathbf{n}^T_{u_1}, \mathbf{n}^S\rangle_1.$$

It follows that

$$\begin{aligned}\mathbb{L}_{u_1} &= -(a_{11}\mathbf{x}_{u_1} + a_{21}\mathbf{x}_{u_2}) \pm \langle \mathbf{n}^T_{u_1}, \mathbf{n}^S\rangle_1 \mathbf{n}^T + \langle \mathbf{n}^T_{u_1}, \mathbf{n}^S\rangle_1 \mathbf{n}^S \\ &= -(a_{11}\mathbf{x}_{u_1} + a_{21}\mathbf{x}_{u_2}) \pm \langle \mathbf{n}^T_{u_1}, \mathbf{n}^S\rangle_1 (\mathbf{n}^T \pm \mathbf{n}^S) \\ &= -(a_{11}\mathbf{x}_{u_1} + a_{21}\mathbf{x}_{u_2}) \pm \langle \mathbf{n}^T_{u_1}, \mathbf{n}^S\rangle_1 \mathbb{L}.\end{aligned}$$

The formula for $\mathbb{L}_{u_2}$ follows similarly. It follows now from the expressions of $\mathbb{L}_{u_1}$ and $\mathbb{L}_{u_2}$ that

$$\begin{aligned}h_{11} &= -\langle \mathbb{L}_{u_1}, \mathbf{x}_{u_1}\rangle_1 = a_{11}g_{11} + a_{21}g_{12}, \\ h_{12} &= -\langle \mathbb{L}_{u_1}, \mathbf{x}_{u_2}\rangle_1 = a_{11}g_{12} + a_{21}g_{22}, \\ h_{12} &= -\langle \mathbb{L}_{u_2}, \mathbf{x}_{u_1}\rangle_1 = a_{12}g_{11} + a_{22}g_{12}, \\ h_{22} &= -\langle \mathbb{L}_{u_2}, \mathbf{x}_{u_2}\rangle_1 = a_{12}g_{12} + a_{22}g_{22}.\end{aligned}$$

Writting the above equations in matrix form

$$\begin{pmatrix} h_{11} & h_{12} \\ h_{12} & h_{22}\end{pmatrix} = \begin{pmatrix} a_{11} & a_{21} \\ a_{12} & a_{22}\end{pmatrix}\begin{pmatrix} g_{11} & g_{12} \\ g_{12} & g_{22}\end{pmatrix}$$

gives

$$A_p^t = \begin{pmatrix} a_{11} & a_{21} \\ a_{12} & a_{22}\end{pmatrix} = \begin{pmatrix} h_{11} & h_{12} \\ h_{12} & h_{22}\end{pmatrix}\begin{pmatrix} g_{11} & g_{12} \\ g_{12} & g_{22}\end{pmatrix}^{-1}$$

so that

$$A_p = \begin{pmatrix} g_{11} & g_{12} \\ g_{12} & g_{22}\end{pmatrix}^{-1}\begin{pmatrix} h_{11} & h_{12} \\ h_{12} & h_{22}\end{pmatrix} = \frac{1}{g_{11}g_{22} - g_{12}^2}\begin{pmatrix} g_{22} & -g_{12} \\ -g_{12} & g_{11}\end{pmatrix}\begin{pmatrix} h_{11} & h_{12} \\ h_{12} & h_{22}\end{pmatrix}.$$

□

Corollary 9.1. *It follows from the Weingarten formulae that:*

(i) *The shape operators $W(\mathbb{L})_p$, $W(\mathbf{n}^T)_p$ and $W(\mathbf{n}^S)_p$ are self-adjoint operators on T_pM.*
(ii) *The $\mathbb{L}$-lightcone Gauss-Kronecker curvature is given by*

$$K_\ell = \frac{ln - m^2}{EG - F^2} = \frac{h_{11}h_{22} - h_{12}^2}{g_{11}g_{22} - g_{12}^2}.$$

(iii) *The $\mathbb{L}$-lightcone mean curvature is given by*

$$H_\ell = \frac{lG - 2mF + nE}{2(EG - F^2)} = \frac{h_{11}g_{22} - 2h_{12}g_{12} + h_{22}g_{11}}{2(g_{11}g_{22} - g_{12}^2)}.$$

9.3 The normalised lightcone Gauss map

Define the *lightcone unit 2-sphere* as the set

$$\begin{aligned} S^2_+ &= \{\mathbf{u} = (u_0, u_1, u_2, u_3) \in LC^* \,|\, u_0 = 1\} \\ &= \{\mathbf{u} = (1, u_1, u_2, u_3) \in LC^* \mid u_1^2 + u_2^2 + u_3^2 = 1\}. \end{aligned}$$

If $\mathbf{u}$ is a lightlike vector, then $u_0 \neq 0$. Denote

$$\widetilde{\mathbf{u}} = \left(1, \frac{u_1}{u_0}, \frac{u_2}{u_0}, \frac{u_3}{u_0}\right)$$

the unique vector in S^2_+ parallel to $\mathbf{u}$.

We define the projection $\pi^L_S : LC^* \to S^2_+$ by $\pi^L_S(\mathbf{u}) = \widetilde{\mathbf{u}}$.

Given a spacelike surface M in $\mathbb{R}^4_1$, the normal lightlike vector field $\mathbb{L}$ depends on the choice of the unit timelike normal vector field $\mathbf{n}^T$ (see Definition 9.1). However, the normal lightlike vector field $\widetilde{\mathbb{L}}$ does not depend on the choice of $\mathbf{n}^T$. Indeed, given any two unit timelike normal vector fields $\mathbf{n}^T_1$ and $\mathbf{n}^T_2$, we have $\pi^L_S(\mathbf{n}^T_1 \pm \mathbf{n}^S_1) = \pi^L_S(\mathbf{n}^T_2 \pm \mathbf{n}^S_2)$. We can thus introduce the following map which is independent of the choice of $\mathbf{n}^T$.

Definition 9.3.

Let M be a spacelike surface patch parametrised by $\mathbf{x} : U \to \mathbb{R}^4_1$. The *normalised lightcone Gauss map* of M is the map $\widetilde{\mathbb{L}} : U \to S^2_+$ given by

$$\widetilde{\mathbb{L}}(u) = \pi^L_S(\mathbb{L}(u)).$$

We define the *normalised lightcone shape operator* of M at p as the linear map $W(\widetilde{\mathbb{L}})_p = -\pi_1 \circ d\widetilde{\mathbb{L}}_p$ from the tangent plane T_pM to itself.

The eigenvalues $\widetilde{\kappa}_1(p)$ and $\widetilde{\kappa}_2(p)$ of $W(\widetilde{\mathbb{L}})_p$ are called the *normalised lightcone principal curvatures.*

The *normalised lightcone Gauss-Kronecker curvature* and *normalised lightcone mean curvature* of M are defined, respectively, to be

$$\widetilde{K}_\ell(p) = \det(W(\widetilde{\mathbb{L}})_p) \quad \text{and} \quad \widetilde{H}_\ell(p) = \frac{1}{2}\mathrm{Trace}(W(\widetilde{\mathbb{L}})_p).$$

A point p is said to be a *lightcone parabolic point* if $\widetilde{K}_\ell(p) = 0$.

Let $\widetilde{A}_p$ denote the matrix of the normalised lightcone shape operator with respect to the basis $\{\mathbf{x}_{u_1}, \mathbf{x}_{u_2}\}$ of T_pM at $p = \mathbf{x}(u)$. We write

$$\mathbb{L}(u) = (\ell_0(u), \ell_1(u), \ell_2(u), \ell_3(u)).$$

Proposition 9.3. *We have, at* $p = \mathbf{x}(u)$,

$$\widetilde{A}_p = \frac{1}{\ell_0(u)} A_p.$$

Proof. We have, by definition, $\ell_0\widetilde{\mathbb{L}} = \mathbb{L}$, so $\ell_0\widetilde{\mathbb{L}}_{u_i} = \mathbb{L}_{u_i} - (\ell_0)_{u_i}\widetilde{\mathbb{L}}$ for $i = 1, 2$. Since $\widetilde{\mathbb{L}}(u) \in N_pM$, we have $\pi_1 \circ \widetilde{\mathbb{L}}_{u_i} = \frac{1}{\ell_0}\mathbb{L}_{u_i}, i = 1, 2,$. □

Corollary 9.2. *It follows from* Proposition 9.3 *that, for* $i = 1, 2$,

$$\widetilde{\kappa}_i = \frac{1}{\ell_0}\kappa_i, \quad \widetilde{K}_\ell = \left(\frac{1}{\ell_0}\right)^2 K_\ell, \quad \widetilde{H}_\ell = \frac{1}{\ell_0}H_\ell.$$

The following also follows from Proposition 9.3.

Proposition 9.4. *For any future directed normal frame* $(\mathbf{n}^T, \mathbf{n}^S)$ *defining* $\mathbb{L}$, $\widetilde{K}_\ell(p) = 0$ *if and only if* $K_\ell(p) = 0$, *and* $\widetilde{H}_\ell(p) = 0$ *if and only if* $H_\ell(p) = 0$.

Remark 9.4. Observe that the eigenvectors of $W(\mathbb{L})_p$ coincide with those of $W(\widetilde{\mathbb{L}})_p$. Therefore, the $\mathbb{L}$-principal directions and $\mathbb{L}$-umbilic points are concepts that depend only on $\widetilde{\mathbb{L}}$.

9.4 Marginally trapped surfaces

We define the *mean curvature vector* of M at p as the vector

$$\mathbf{H}(p) = \mathrm{Trace}(W(\mathbf{n}^T)_p)\mathbf{n}^T(p) + \mathrm{Trace}(W(\mathbf{n}^S)_p)\mathbf{n}^S(p).$$

The mean curvature vector is independent of the choice of $(\mathbf{n}^T, \mathbf{n}^S)$.

The concept of trapped surfaces in space-time, introduced by Penrose in [Penrose (1965)], plays an important role in cosmology and general relativity. (In particular, it plays a principal role in the proofs of the theorems of space-time singularities, gravitational collapse, the cosmic censorship and Penrose inequality.)

A spacelike surface in a space-time is said to be *marginally trapped* if its mean curvature vector is lightlike at each point.

Proposition 9.5. *The following assertions are equivalent:*

(i) $H_\ell^+(p) = 0$ *or* $H_\ell^-(p) = 0$.
(ii) *The mean curvature vector* $\mathbf{H}(p)$ *is isotropic (i.e., it is lightlike or zero).*
(iii) $\widetilde{H}_\ell^+(p) = 0$ *or* $\widetilde{H}_\ell^-(p) = 0$.

Proof. We have $W(\mathbb{L})_p = W(\mathbf{n}^T)_p \pm W(\mathbf{n}^S)_p$, so

$$\mathrm{Trace}(W(\mathbb{L})_p) = \mathrm{Trace}(W(\mathbf{n}^T)_p) \pm \mathrm{Trace}(W(\mathbf{n}^S)_p).$$

On the other hand,

$$\langle \mathbf{H}(p), \mathbf{H}(p)\rangle_1 = -\mathrm{Trace}(W(\mathbf{n}^T)_p)^2 + \mathrm{Trace}(W(\mathbf{n}^S)_p)^2.$$

Thus, $\langle \mathbf{H}(p), \mathbf{H}(p)\rangle_1 = 0$ if and only if $\mathrm{Trace}(W(\mathbf{n}^T)_p) = \pm\mathrm{Trace}(W(\mathbf{n}^S)_p)$, that is, if and only if condition (i) holds. Since $\widetilde{H}_\ell^\pm(p) = (1/\ell_0(p)) = H_\ell^\pm(p)$, the conditions (i) and (iii) are equivalent. □

It follows from Proposition 9.5 that a spacelike surface M in $\mathbb{R}^4_1$ is marginally trapped if and only if $H_\ell^+ \equiv 0$ or $H_\ell^- \equiv 0$ for any future directed normal frame $(\mathbf{n}^T, \mathbf{n}^S)$. Surfaces satisfying $H_\ell^+ \equiv 0$ and $H_\ell^- \equiv 0$ are called *minimal surfaces*. However, this class of spacelike surfaces contains the class of maximal spacelike surfaces in $\mathbb{R}^4_1$ ([Cheng and Yau (1973); Kobayashi (1983)]), where a spacelike surface in $\mathbb{R}^3_1$ is *maximal* if its mean curvature vanishes identically. For this reason, we say that M is *strongly marginally trapped* if $H_\ell^+ \equiv 0$ and $H_\ell^- \equiv 0$.

9.5 The family of lightcone height functions

The family of lightcone height functions $\mathcal{H} : \mathbb{R}^4_1 \times S^2_+ \to \mathbb{R}$ is defined by

$$\mathcal{H}(p, \mathbf{v}) = \langle p, \mathbf{v}\rangle_1.$$

The *family of lightcone height functions* $H : U \times S^2_+ \to \mathbb{R}$ on M parametrised by $\mathbf{x} : U \to \mathbb{R}^4_1$ is the restriction of the family $\mathcal{H}$ to M. Thus,

$$H(u, \mathbf{v}) = \langle \mathbf{x}(u), \mathbf{v}\rangle_1.$$

For $\mathbf{v}$ fixed, we write $h_{\mathbf{v}}(u) = H(u, \mathbf{v})$ for the lightcone height function on M along $\mathbf{v}$.

We have the following characterisation of the lightlike parabolic points and lightlike flat points in terms of the lightcone height function.

Proposition 9.6. *Let $p_0 = \mathbf{x}(u_0)$ be a point on M and $\mathbf{v}_0$ a vector in S^2_+.*

(i) *The function $h_{\mathbf{v}_0}$ is singular at u_0 if and only if $\mathbf{v}_0 = \widetilde{\mathbb{L}}(u_0)$.*
(ii) *Suppose that $\mathbf{v}_0 = \widetilde{\mathbb{L}}(u_0)$. Then,*
   (a) *The point p_0 is a lightcone parabolic point if and only if $\det \mathrm{Hess}(h_{\mathbf{v}_0})(u_0) = 0$.*
   (b) *The point p_0 is a lightcone flat point if and only if $\mathrm{rank}\, \mathrm{Hess}(h_{\mathbf{v}_0})(u_0) = 0$.*

Proof. (i) Since $\{\mathbf{x}_{u_1}(u), \mathbf{x}_{u_2}(u), \mathbf{n}^T(u), \mathbf{n}^S(u)\}$ is a basis of $T_p\mathbb{R}^4_1$ for all $p = \mathbf{x}(u)$ on M, there exist real numbers $\xi_1, \xi_2, \lambda_1, \lambda_2$ such that

$$\mathbf{v}_0 = \lambda_1 \mathbf{x}_{u_1}(u_0) + \lambda_2 \mathbf{x}_{u_2}(u_0) + \mu_1 \mathbf{n}^T(u_0) + \mu_2 \mathbf{n}^S(u_0).$$

We have

$$\langle \frac{\partial h_{\mathbf{v}_0}}{\partial u_1}, \mathbf{v}_0 \rangle_1 = \langle \mathbf{x}_{u_1}, \mathbf{v}_0 \rangle_1 = \lambda_1 E + \lambda_2 F,$$
$$\langle \frac{\partial h_{\mathbf{v}_0}}{\partial u_2}, \mathbf{v}_0 \rangle_1 = \langle \mathbf{x}_{u_2}, \mathbf{v}_0 \rangle_1 = \lambda_1 F + \lambda_2 G.$$

Therefore $\partial h_{\mathbf{v}_0}/\partial u_1(u_0) = \partial h_{\mathbf{v}_0}/\partial u_2(u_0) = 0$ if and only if $\lambda_1 = \lambda_2 = 0$ ($EG - F^2 > 0$ as M is spacelike). Thus, when $h_{\mathbf{v}_0}$ is singular at u_0, we have $\mathbf{v}_0 = \mu_1 \mathbf{n}^T(u_0) + \mu_2 \mathbf{n}^S(u_0)$. Because $\mathbf{v}_0$ lightlike, $-\mu_1^2 + \mu_2^2 = 0$. This means that $\mathbf{v}_0 = \mu_1 \mathbb{L}(u_0)$, and as $\mathbf{v}_0$ is in S^2_+, we have $\mathbf{v}_0 = \widetilde{\mathbb{L}}(u_0)$.

(ii)-(a) We have

$$\begin{aligned}\mathrm{Hess}(h_{\mathbf{v}_0})(u_0) &= \begin{pmatrix} \langle \mathbf{x}_{u_1u_1}(u_0), \widetilde{\mathbb{L}}^{\pm}(u_0)\rangle_1 & \langle \mathbf{x}_{u_1u_2}(u_0), \widetilde{\mathbb{L}}^{\pm}(u_0)\rangle_1 \\ \langle \mathbf{x}_{u_1u_2}(u_0), \widetilde{\mathbb{L}}^{\pm}(u_0)\rangle_1 & \langle \mathbf{x}_{u_2u_2}(u_0), \widetilde{\mathbb{L}}^{\pm}(u_0)\rangle_1 \end{pmatrix} \\ &= \frac{1}{\ell_0(u_0)} \begin{pmatrix} h_{11}(u_0) & h_{12}(u_0) \\ h_{12}(u_0) & h_{22}(u_0) \end{pmatrix}.\end{aligned}$$

It follows from Propositions 9.2 and 9.3 that

$$\mathrm{Hess}(h_{\mathbf{v}_0})(u_0) = \widetilde{A}_p(u_0) \begin{pmatrix} g_{11}(u_0) & g_{12}(u_0) \\ g_{12}(u_0) & g_{22}(u_0) \end{pmatrix},$$

so

$$\widetilde{K}_\ell(u_0) = \det(\widetilde{A}_p(u_0)) = \frac{\det(\mathrm{Hess}(h_{\mathbf{v}_0})(u_0))}{g_{11}g_{22} - g_{12}^2}.$$

(ii)-(b) The point p is a lightcone umbilic point if and only

$$\widetilde{A}_p = \kappa I,$$

where I is the identity matrix and κ one of the equal normalised lightcone principal curvatures at p. Equivalently,

$$\mathrm{Hess}(h_{\mathbf{v}_0}) = \kappa \begin{pmatrix} g_{11} & g_{12} \\ g_{12} & g_{22} \end{pmatrix}.$$

The point p is a lightcone flat point if and only if $\kappa = 0$, if and only if $\operatorname{rank} \mathrm{Hess}(h_{\mathbf{v}_0})(u_0) = 0$. □

Corollary 9.3. *The following assertions are equivalent at any $p = \mathbf{x}(u)$ on M.*

(i) *The point p is a $\mathbb{L}$-parabolic point for any future directed normal frame $(\mathbf{n}^T, \mathbf{n}^S)$.*
(ii) *The point p is a singular point of the lightcone Gauss map $\widetilde{\mathbb{L}}$.*
(iii) $\widetilde{K}_\ell(p) = 0$.
(iv) $\det \mathrm{Hess}(h_{\mathbf{v}})(u) = 0$ *for* $\mathbf{v} = \widetilde{\mathbb{L}}(u)$.

Corollary 9.4. *The following assertions are equivalent at any* $p = \mathbf{x}(u)$ *on* M.

(i) *The point* p *is a* $\mathbb{L}$*-flat point for any future directed normal frame* $(\mathbf{n}^T, \mathbf{n}^S)$.
(ii) *There exists a future directed normal frame* $(\mathbf{n}^T, \mathbf{n}^S)$ *such that* p *is a* $\mathbb{L}$*-flat point.*
(iii) *The point* p *is a lightcone flat point with respect to* $\widetilde{\mathbb{L}}$.

We have the following characterisation of lightcone flatness of spacelike surfaces.

Proposition 9.7. *The following assertions are equivalent.*

(i) M *is totally* $\mathbb{L}$*-flat.*
(ii) *The normalised lightcone Gauss map* $\widetilde{\mathbb{L}}$ *is a constant map.*
(iii) *There exists a lightlike vector* $\mathbf{v}$ *and a real number* c *such that* M *is contained in the lightlike hyperplane* $HP(\mathbf{v}, c)$.

Proof. Suppose that M is totally $\mathbb{L}$-flat. This means that the matrix A_p in Proposition 9.2 is identically zero for any future directed frame $(\mathbf{n}^T, \mathbf{n}^S)$. Therefore, by Proposition 9.2, have

$$\mathbb{L}_{u_i}(u) = \pm\langle \mathbf{n}^S(u), \mathbf{n}^T_{u_i}(u)\rangle_1 \mathbb{L}(u)$$

at all points $u \in U$ and for $i = 1, 2$. Then

$$\widetilde{\mathbb{L}}_{u_i}(u) = (-\frac{(\ell_0)_{u_i}(u)}{\ell_0(u)^2} \pm \langle \mathbf{n}^S(u), \mathbf{n}^T_{u_i}(u)\rangle_1)\mathbb{L}(u), i = 1, 2,$$

and this implies that $\widetilde{\mathbb{L}}_{u_i}(u)$ is the zero vector for $i = 1, 2$, as being a spacelike vector it cannot be a non-zero multiple of a lightlike vector. It follows that $\widetilde{\mathbb{L}}$ is a constant map, so (i) implies (ii). For the converse if $\widetilde{\mathbb{L}}$ is constant then $\widetilde{A}_p$ is identically zero and so is A_p by Proposition 9.3, so all points are $\mathbb{L}$-flat umbilic points by Proposition 9.6.

Suppose now that the normalised lightcone Gauss map $\widetilde{\mathbb{L}}(u) = \mathbf{v}$ is constant. We consider a function $F : U \to \mathbb{R}$ defined by $F(u) = \langle \mathbf{x}(u), \mathbf{v}\rangle_1$. By definition, we have

$$\frac{\partial F}{\partial u_i}(u) = \langle \mathbf{x}_{u_i}(u), \mathbf{v}\rangle_1 = \langle \mathbf{x}_{u_i}(u), \widetilde{\mathbb{L}}(u)\rangle_1 = 0,$$

for $i = 1, 2$. Therefore, $F(u) = \langle \mathbf{x}(u), \mathbf{v}\rangle_1 = c$ is constant. Since $\mathbf{v}$ is lightlike, M is a subset of the lightlike hyperplane $HP(\mathbf{v}, c)$, (ii) implies (iii).

Suppose that M is a subset of a lightlike hyperplane $H(\mathbf{v}, c)$. For any point $p \in M$, the tangent space of $HP(\mathbf{v}, c)$ can be identified with $HP(\mathbf{v}, 0)$. Since $M \subset HP(\mathbf{v}, c)$, $T_pM \subset HP(\mathbf{v}, 0)$, so that $N_p(M) \cap HP(\mathbf{v}, 0)$ is the line generated by $\mathbf{v}$. For any future directed timelike unit normal vector field $\mathbf{n}^T$ along M, there exists a lightlike vector $\overline{\mathbf{v}}$ such that $\overline{\mathbf{v}}$ is parallel to $\mathbf{v}$ and $\overline{\mathbf{v}} - \mathbf{n}^T$ is a spacelike unit normal vector field along M. We write $\mathbf{n}^S = \overline{\mathbf{v}} - \mathbf{n}^T$, so that we have a future directed normal frame $(\mathbf{n}^T, \mathbf{n}^S)$ along M with

$$\widetilde{\mathbb{L}}(u) = \pi_S^L(\mathbf{n}^T(u) + \mathbf{n}^S(u)) = \pi_S^L(\overline{\mathbf{v}}).$$

This means that the corresponding lightcone Gauss map $\widetilde{\mathbb{L}}$ is constant. This completes the proof. □

Corollary 9.5. *The following assertions are equivalent.*
(i) *M is totally $\mathbb{L}^+$-flat and $\mathbb{L}^-$-flat.*
(ii) *M is a part of a spacelike plane.*

Proof. By Proposition 9.7, condition (i) is equivalent to the condition that there exist two linearly independent lightlike vectors $\mathbf{v}^{\pm}$ and real numbers $c^{\pm}$ such that $M \subset HP(\mathbf{v}^+, c^+) \cap HP(\mathbf{v}^-, c^-)$. (The set $HP(\mathbf{v}^+, c^+) \cap HP(\mathbf{v}^-, c^-)$ is a spacelike plane.) □

9.6 The Lagrangian viewpoint

The space $\mathbb{R}_0^3 = \{(x_0, x_1, x_2, x_3) \in \mathbb{R}_1^4 \,|\, x_0 = 0\}$ can be identified with the Euclidean 3-space. We consider the canonical projection $\pi : \mathbb{R}_1^4 \to \mathbb{R}_0^3$ defined by

$$\pi(x_0, x_1, x_2, x_3) = (0, x_1, x_2, x_3).$$

Lemma 9.2. *Let M be a spacelike surface patch parametrised by $\mathbf{x} : U \to \mathbb{R}_1^4$. Then the direction of the vector field $\pi \circ \widetilde{\mathbb{L}}$ is transversal to $\pi(M)$ in $\mathbb{R}_0^3$, that is,*

$$\mathbb{R}.\{\pi \circ \widetilde{\mathbb{L}}(u)\} + d(\pi \circ \mathbf{x})_u(T_uU) = T_p\mathbb{R}_0^3$$

at any $p = \mathbf{x}(u) \in M$.

Proof. Since $\widetilde{\mathbb{L}}(u)$ is lightlike and $\mathrm{Ker}(d\pi_p)$ is a timelike one-dimensional subspace of $\mathbb{R}_1^4$, $\widetilde{\mathbb{L}}(u) \notin \mathrm{Ker}(d\pi_p)$. The fact that $d\mathbf{x}_u(T_uU)$ is spacelike implies that

$$\mathbb{R}.\{\widetilde{\mathbb{L}}(u), \mathrm{Ker}(d\pi_p)\} + d\mathbf{x}_u(T_uU) = T_p\mathbb{R}_1^4 \tag{9.3}$$

at any point $p = \mathbf{x}(u) \in M$.

Suppose that there exists a point $u \in U$ such that the direction of the vector field $\pi \circ \widetilde{\mathbb{L}}^{\pm}(u)$ is not transversal to $\pi \circ \mathbf{x}(U)$ in $\mathbb{R}^3_0$ at $p = \mathbf{x}(u)$. Since $\pi \circ \mathbf{x}(U)$ is a smooth surface in $\mathbb{R}^3_0$, our assumption implies that $\pi \circ \widetilde{\mathbb{L}}(u) \in d(\pi \circ \mathbf{x})_u(T_uU)$. This means that

$$\widetilde{\mathbb{L}}(u) \in d\mathbf{x}_u(T_uU) + \mathrm{Ker}(d\pi_p).$$

But this implies in turn that the dimension of $\mathbb{R}.\{\widetilde{\mathbb{L}}(u), \mathrm{Ker}(d\pi_p)\} + d\mathbf{x}_u(T_uU)$ is at most 3, which contradicts equality (9.3). $\square$

Proposition 9.8. *Let M be a spacelike surface patch in $\mathbb{R}^4_1$. Then family of lightcone height functions H is a Morse family of functions.*

Proof. Let $\mathbf{v} = (1, v_1, v_2, v_3)$ be in S^2_+ and consider the chart of S^2_+ with $v_1 > 0$ (the result follows similarly for the other charts). Then $v_1 = \sqrt{1 - v_2^2 - v_3^2}$ and

$$H(u, \mathbf{v}) = -x_0(u) + x_1(u)\sqrt{1 - v_2^2 - v_3^2} + x_2(u)v_2 + x_3(u)v_3,$$

where $\mathbf{x}(u) = (x_0(u), x_1(u), x_2(u), x_3(u))$. We show that the mapping

$$\Delta H = (\frac{\partial H}{\partial u_1}, \frac{\partial H}{\partial u_2}) = (\langle \mathbf{x}_{u_1}, \mathbf{v}\rangle_1, \langle \mathbf{x}_{u_2}, \mathbf{v}\rangle_1)$$

is not singular at any point in $C_H = (\Delta H)^{-1}(0)$. The Jacobian matrix of ΔH is given by

$$\begin{pmatrix} \langle \mathbf{x}_{u_1u_1}, \mathbf{v}\rangle_1 & \langle \mathbf{x}_{u_1u_2}, \mathbf{v}\rangle_1 & -x_{1u_1}\dfrac{v_2}{v_1} + x_{2u_1} & -x_{1u_1}\dfrac{v_3}{v_1} + x_{3u_1} \\ \langle \mathbf{x}_{u_2u_1}, \mathbf{v}\rangle_1 & \langle \mathbf{x}_{u_2u_2}, \mathbf{v}\rangle_1 & -x_{1u_2}\dfrac{v_2}{v_1} + x_{2u_2} & -x_{1u_2}\dfrac{v_3}{v_1} + x_{3u_2} \end{pmatrix}.$$

Consider the 2×2-matrix A given by the last two columns of the Jacobian matrix of ΔH, that is,

$$A = \begin{pmatrix} -x_{1u_1}\dfrac{v_2}{v_1} + x_{2u_1} & -x_{1u_1}\dfrac{v_3}{v_1} + x_{3u_1} \\ -x_{1u_2}\dfrac{v_2}{v_1} + x_{2u_2} & -x_{1u_2}\dfrac{v_3}{v_1} + x_{3u_2} \end{pmatrix},$$

and let $\mathbf{a}_i = \begin{pmatrix} x_{iu_1} \\ x_{iu_2} \end{pmatrix}$ for $i = 1, 2, 3$. Then,

$$A = \left(-\mathbf{a}_1\frac{v_2}{v_1} + \mathbf{a}_2, -\mathbf{a}_1\frac{v_3}{v_1} + \mathbf{a}_3\right)$$

and

$$\det A = \det(\mathbf{a}_2, \mathbf{a}_3) - \frac{v_2}{v_1}\det(\mathbf{a}_1, \mathbf{a}_3) + \frac{v_3}{v_1}\det(\mathbf{a}_1, \mathbf{a}_2).$$

Since $\mathbf{x}$ is a spacelike embedding, $\overline{\mathbf{x}} = \pi \circ \mathbf{x} : U \to \mathbb{R}^3_0$ is an immersion, so that $\overline{\mathbf{x}}_{u_1} \times \overline{\mathbf{x}}_{u_2} \neq 0$. (Here "$\times$" denotes the vector product in $\mathbb{R}^3_0$, identified with the Euclidean 3-space.) Then,

$$\det A = \langle (\frac{v_1}{v_1}, \frac{v_2}{v_1}, \frac{v_3}{v_1}), \overline{\mathbf{x}}_{u_1} \times \overline{\mathbf{x}}_{u_2} \rangle_1 = \frac{1}{v_1} \langle \mathbf{v}, \overline{\mathbf{x}}_{u_1} \times \overline{\mathbf{x}}_{u_2} \rangle_1.$$

Since $(u, \mathbf{v}) \in C_H = (\Delta H)^{-1}(0)$, by Proposition 9.6 we may take $\mathbf{v} = \widetilde{\mathbb{L}}(u)$. By Lemma 9.2, the direction of $\pi(\mathbf{v}) = \pi \circ \widetilde{\mathbb{L}}(u)$ is transverse to the tangent space of $\pi(M)$ at $\pi \circ \mathbf{x}(u)$. Therefore, $\det A \neq 0$ and so ΔH has maximal rank. □

We have shown in Chapter 5 how to construct a germ of a Lagrangian immersion from a Morse family of functions. We use below that construction and the result in Proposition 9.8 to define a germ of a Lagrangian immersion whose generating family is the family of lightcone height functions H on M.

Consider the local chart $U_1 = \{v = (1, v_1, v_2, v_3) \in S^2_+ \mid v_1 > 0\}$ of S^2_+ (we proceed similarly for the other local charts). By Proposition 9.6, we have

$$C_H = \{(u, \widetilde{\mathbb{L}}(u)) \mid u \in U\}.$$

Since $T^*S^2_+|U_1$ is a trivial bundle, we can define the map $L(H) : C_H \to T^*S^2_+|U_1$ by

$$L(H)(u, \widetilde{\mathbb{L}}^{\pm}(u)) = (\widetilde{\mathbb{L}}^{\pm}(u), x_2(u) \mp x_1(u)\frac{\ell_2^{\pm}(u)}{\ell_1^{\pm}(u)}, x_3(u) \mp x_1(u)\frac{\ell_3^{\pm}(u)}{\ell_1^{\pm}(u)}),$$

where $\mathbb{L}(u) = \mathbb{L}^{\pm}(u) = (\ell_0^{\pm}(u), \ell_1^{\pm}(u), \ell_2^{\pm}(u), \ell_3^{\pm}(u)) \in LC^*$. The following result follows from Proposition 9.8.

Corollary 9.6. *The map-germ $L(H)$ is a Lagrangian immersion whose generating family is the family of the lightcone height functions H on M.*

Consider the lightcone height function $\mathfrak{h}_{\mathbf{v}} : \mathbb{R}^4_1 \to \mathbb{R}$, given by $\mathfrak{h}_{\mathbf{v}}(p) = \mathcal{H}(p, \mathbf{v}) = \langle p, \mathbf{v} \rangle_1$. Clearly, $\mathfrak{h}_{\mathbf{v}}$ is a submersion. Let $\mathbf{v}_0 = \widetilde{\mathbb{L}}(u_0)$. By Proposition 9.6, we have for $i = 1, 2$,

$$\frac{\partial \mathfrak{h}_{\mathbf{v}_0} \circ \mathbf{x}}{\partial u_i}(u_0) = \frac{\partial H}{\partial u_i}(u_0, \mathbf{v}_0) = 0.$$

This means that the lightlike hyperplane $\mathfrak{h}^{-1}_{\mathbf{v}_0}(c) = HP(\mathbf{v}_0, c)$ is tangent to M at $p_0 = \mathbf{x}(u_0)$, where $c = \langle \mathbf{x}(u_0), \mathbf{v}_0 \rangle_1$. We say that $HP(\mathbf{v}_0, c)$ is a *lightlike tangent hyperplane* of M with pseudo-normal direction $\mathbf{v}_0$ and denote it by $THP(\mathbf{x}, u_0)$.

Let ϵ be a sufficiently small positive real number. For any $t \in (c-\epsilon, c+\epsilon)$, we have a lightlike hyperplane $HP(\mathbf{v}_0, t) = \mathfrak{h}_{\mathbf{v}_0}^{-1}(t)$. Clearly,

$$\mathcal{F}_{h_{v_0}} = \{HP(\mathbf{v}_0, t) \mid t \in (c-\epsilon, c+\epsilon)\}$$

is a family of parallel lightlike hyperplanes around p_0.

Let M and $\bar{M}$ be two germs spacelike surfaces parametrised by $\mathbf{x} : (U, u) \to (\mathbb{R}^4_1, p)$ and $\bar{\mathbf{x}} : (\bar{U}, \bar{u}) \to (\mathbb{R}^4_1, \bar{p})$, respectively. Set $h_{1,\mathbf{v}}(u) = h_{\mathbf{v}} \circ \mathbf{x}(u)$ and $h_{2,\bar{\mathbf{v}}}(u) = h_{\bar{\mathbf{v}}} \circ \bar{\mathbf{x}}(u)$, where $\mathbf{v} = \widetilde{\mathbb{L}}(u)$ and $\bar{\mathbf{v}} = \widetilde{\mathbb{L}}(\bar{u})$.

We have the following result.

Theorem 9.1. *Let M and $\bar{M}$ as above and suppose that $h_{1,\mathbf{v}}$ and $h_{2,\bar{\mathbf{v}}}$ satisfy the Milnor condition. Then the following assertions are equivalent.*

(i) $K(M, \mathcal{F}_{h_{1,\mathbf{v}}}; \mathbf{x}(u)) = K(\bar{M}, \mathcal{F}_{h_{2,\bar{\mathbf{v}}}}; \bar{\mathbf{x}}(\bar{u}))$
(ii) $h_{1,\mathbf{v}_1}$ *and* $h_{2,\mathbf{v}_2}$ *are* $\mathcal{R}^+$*-equivalent.*
(iii) H_1 *and* H_2 *are* P-$\mathcal{R}^+$*- equivalent.*
(iv) $L(H_1)$ *and* $L(H_2)$ *are Lagrangian equivalent.*
(v) *The rank and signature of* $\mathrm{Hess}(h_{1,\mathbf{v}})(u)$ *and* $\mathrm{Hess}(h_{2,\bar{\mathbf{v}}})(\bar{u})$ *are equal, and there is an isomorphism* $\gamma : \mathcal{R}_2(h_{1,\mathbf{v}}) \to \mathcal{R}_2(h_{2,\bar{\mathbf{v}}})$ *such that* $\gamma(\overline{h_{1,\mathbf{v}}}) = \overline{h_{2,\bar{\mathbf{v}}}}$.

Proof. The statements (i) and (ii) are equivalent by Proposition 4.1. Since the germs $L(H_1)$ and $L(H_2)$ are Lagrangian stable, the germs H_1 and H_2 are $\mathcal{R}^+$-versal unfoldings of $h_{1,\mathbf{v}}$ and $h_{2,\bar{\mathbf{v}}}$, respectively. Then (ii) is equivalent to (iii) by the uniqueness of $\mathcal{R}^+$-versal unfoldings.

By Theorem 5.4, (iii) is equivalent to (iv) and it also follows from that theorem that $h_{1,\mathbf{v}}$ and $h_{2,\bar{\mathbf{v}}}$ satisfy the Milnor condition. Then we can apply Proposition 5.1 to show that (ii) is equivalent to (v). □

Corollary 9.7. *Let* $\mathrm{Emb}_s\,(U, \mathbb{R}^4_1)$ *be the set of spacelike embeddings* $U \to \mathbb{R}^4_1$ *endowed with the Whitney* C^∞*-topology. Then there exists a residual subset* $\mathcal{O} \subset \mathrm{Emb}_s\,(U, \mathbb{R}^4_1)$ *such that for any* $\mathbf{x} \in \mathcal{O}$*, the following properties hold.*

(i) *The lightcone parabolic set is a regular curve.*
(ii) *The lightcone parabolic curve consists of fold singularities of the normalised lightcone Gauss map except possibly at isolated points where it has a cusp singularity.*

Proof. We remark that $\mathrm{Emb}_s\,(U, \mathbb{R}^4_1)$ is an open subset of $C^\infty(U, \mathbb{R}^4_1)$. The proof is similar to those of Theorems 4.7 and 6.1. We use the lightcone height function $H : U \times S^2_+ \to \mathbb{R}$ here instead of the height function in Theorem 4.7. For a generic spacelike embedding $\mathbf{x} : U \to \mathbb{R}^4_1$, the $\mathcal{R}^+$-singularities of $h_{\mathbf{v}}$ are type A_k, $k = 1, 2, 3$ and H is an $\mathcal{R}^+$-versal unfolding of $h_{\mathbf{v}}$ at any point $(u, \mathbf{v}) \in U \times S^2_+$ (Theorems 3.13 and 4.4). In this case the catastrophe map-germ π_{C_H} is $\mathcal{A}$-equivalent to the fold singularity or the cusp singularity by the calculation of §3.9.2. Moreover, the singular set of the catastrophe map is a regular curve. Since the catastrophe map of H is identified with the normalised lightcone Gauss map $\widetilde{\mathbb{L}}$, the singular set of the catastrophe map is the lightcone parabolic curve. This completes the proof. □

Remark 9.5. Let $\mathcal{O}$ be the set given in Corollary 9.7 endowed with the Whitney C^∞-topology. For any $\mathbf{x} \in \mathcal{O}$, we can show that the corresponding Lagrangian submanifold $L(H)$ is stable under the perturbation of $\mathbf{x}$, and it follows from this that the set $\mathcal{O}$ is open. The full details of the proof of this observation requires the use of more tools from singularity theory and is omitted.

9.7 The lightcone pedal and the extended lightcone height function: the Legendrian viewpoint

We shall associate to the surface M a mapping (or its image) $M \to LC^*$ whose singularities correspond to the those of the lightcone Gauss map of M.

Definition 9.4. We define the *lightcone pedal* of M as the mapping $\mathbb{LP}_M : U \to LC^*$ given by

$$\mathbb{LP}_M(u) = \langle \mathbf{x}(u), \widetilde{\mathbb{L}}(u) \rangle_1 \widetilde{\mathbb{L}}(u).$$

We call the image of $\mathbb{LP}_M$ the *lightcone pedal surface* (associated to M).

We call the family of functions $\bar{\mathcal{H}} : \mathbb{R}^4_1 \times LC^* \to \mathbb{R}$ given by

$$\bar{\mathcal{H}}(p, \mathbf{v}) = \bar{\mathfrak{h}}_{\mathbf{v}}(p) = \langle p, \mathbf{v} \rangle_1 - v_0,$$

where $\mathbf{v} = (v_0, v_1, v_2, v_3)$, the *extended family of lightcone height functions*. Given a spacelike surface patch M parametrised by $\mathbf{x} : U \to \mathbb{R}^4_1$, we call $\bar{H}$, restriction of $\bar{\mathcal{H}}$ to M, the extended family of lightcone height functions on M. The family $\bar{H} : U \times LC^* \to \mathbb{R}$ is given by

$$\bar{H}(u, \mathbf{v}) = \bar{h}_{\mathbf{v}}(u) = \langle \mathbf{x}(u), \mathbf{v} \rangle_1 - v_0.$$

Since $\partial\bar{H}/\partial u_i = \partial H/\partial u_i$ for $i = 1, 2$ and $\mathrm{Hess}(\bar{h}_{\mathbf{v}}) = \mathrm{Hess}(h_{\mathbf{v}})$, we have the following result as a corollary of Proposition 9.6.

Proposition 9.9. *Let M be a spacelike surface patch parametrised by $\mathbf{x} : U \to \mathbb{R}^4_1$ and $p_0 = \mathbf{x}(u_0)$. Then,*

(i) $\bar{H}(u_0, \mathbf{v}) = \dfrac{\partial\bar{H}}{\partial u_1}(u_0, \mathbf{v}) = \dfrac{\partial\bar{H}}{\partial u_2}(u_0, \mathbf{v}) = 0$ *if and only if* $\mathbf{v} = \langle \mathbf{x}(u_0), \widetilde{\mathbb{L}}(u_0)\rangle_1 \widetilde{\mathbb{L}}(u_0)$.

(ii) *Suppose that* $\mathbf{v}_0 = \langle \mathbf{x}(u_0), \widetilde{\mathbb{L}}(u_0)\rangle_1 \widetilde{\mathbb{L}}(u_0)$. *Then*

(a) *p_0 is a lightcone parabolic point if and only if* $\det \mathrm{Hess}(\bar{h}_{\mathbf{v}_0})(u_0) = 0$.

(b) *p_0 is a lightcone flat point if and only if* $\mathrm{rank}\, \mathrm{Hess}(\bar{h}_{\mathbf{v}_0})(u_0) = 0$.

Proposition 9.9(i) implies that the discriminant set of $\bar{H}$ is the lightcone pedal surface, that is, $D_{\bar{H}} = \mathbb{LP}_M(U)$. Moreover, the set of singular points of $\mathbb{LP}_M$ is the lightcone parabolic set of M.

Proposition 9.10. *The extended family of lightcone height functions $\bar{H}$ is a Morse family of hypersurfaces.*

Proof. Consider the canonical diffeomorphism $\Phi : S^2_+ \times (\mathbb{R} \setminus \{0\}) \to LC^*$ given by

$$\Phi(\mathbf{v}, r) = r\mathbf{v}.$$

and denote by $\widetilde{H} = \bar{H} \circ \Phi : U \times (S^2_+ \times (\mathbb{R} \setminus \{0\})) \to \mathbb{R}$ the composite map given by

$$\widetilde{H}(u, \mathbf{v}, r) = \langle \mathbf{x}(u), \mathbf{v}\rangle_1 - r.$$

We only need to show that $\widetilde{H}$ is a Morse family of hypersurfaces.

Let $(\mathbf{v}, r) = ((1, v_1, v_2, v_3), r) \in S^2_+ \times (\mathbb{R} \setminus \{0\})$, and assume that $v_1 > 0$, so that $v_1 = \sqrt{1 - v_2^2 - v_3^2}$ (the proof follows similarly for the other charts of S^2_+). It follows that

$$\widetilde{H}(u, \mathbf{v}, r) = -x_0(u) + x_1(u)\sqrt{1 - v_2^2 - v_3^2} + x_2(u)v_2 + x_3(u)v_3 - r,$$

where $\mathbf{x}(u) = (x_0(u), x_1(u), x_2, x_3(u))$. We show that the mapping

$$\Delta^*\widetilde{H} = (\widetilde{H}, \frac{\partial\widetilde{H}}{\partial u_1}, \frac{\partial\widetilde{H}}{\partial u_2}) = (\widetilde{H}, \frac{\partial\widetilde{H}}{\partial u_1}, \frac{\partial\widetilde{H}}{\partial u_2})$$

is not singular at any point in $\Sigma_{\widetilde{H}} = (\Delta^* \widetilde{H})^{-1}(0)$. The Jacobian matrix of $\Delta^* \widetilde{H}$ is given by

$$\begin{pmatrix} \langle \mathbf{x}_{u_1}, \mathbf{v} \rangle_1 & \langle \mathbf{x}_{u_2}, \mathbf{v} \rangle_1 & -1 & -x_1 \frac{v_2}{v_1} + x_2 & -x_1 \frac{v_3}{v_1} + x_3 \\ \langle \mathbf{x}_{u_1 u_1}, \mathbf{v} \rangle_1 & \langle \mathbf{x}_{u_1 u_2}, \mathbf{v} \rangle_1 & 0 & -x_{1u_1} \frac{v_2}{v_1} + x_{2u_1} & -x_{1u_1} \frac{v_3}{v_1} + x_{3u_1} \\ \langle \mathbf{x}_{u_2 u_1}, \mathbf{v} \rangle_1 & \langle \mathbf{x}_{u_2 u_2}, \mathbf{v} \rangle_1 & 0 & -x_{1u_2} \frac{v_2}{v_1} + x_{2u_2} & -x_{1u_2} \frac{v_3}{v_1} + x_{3u_2} \end{pmatrix}.$$

Consider the matrix

$$A = \begin{pmatrix} -x_{1u_1} \frac{v_2}{v_1} + x_{2u_1} & -x_{1u_1} \frac{v_3}{v_1} + x_{3u_1} \\ -x_{1u_2} \frac{v_2}{v_1} + x_{2u_2} & -x_{1u_2} \frac{v_3}{v_1} + x_{3u_3} \end{pmatrix}$$

which is the same matrix in the proof of Proposition 9.8, so $\det A \neq 0$ at any point of $\Sigma_{\widetilde{H}} = (\Delta^* \widetilde{H})^{-1}(0)$. Therefore rank $\Delta^* \widetilde{H} = 3$ on $\Sigma_{\widetilde{H}}$. □

We define now a Legendrian immersion whose generating family is the family of extended lightcone height functions on M.

Consider the projective cotangent bundle $\pi : PT^*(LC^*) \to LC^*$ with the canonical contact structure (observe that $PT^*(LC^*) \cong LC^* \times P(\mathbb{R}^4)^*$ is a trivial bundle). Given any point $\mathbf{v} = (v_0, v_1, v_2, v_3) \in LC^*$, we have $v_0 = \pm\sqrt{v_1^2 + v_2^2 + v_3^2}$, so we can adopt the coordinates (v_1, v_2, v_3) for the manifold LC^*.

We define the map $\mathscr{L}(\bar{H}) : \Sigma_{\bar{H}} \to PT^*(LC^*)$ as follows. Given $(u, \mathbf{v}) \in \Sigma_{\bar{H}}$, with $\mathbf{v} = (v_0, v_1, v_2, v_3) = \mathbb{LP}_M(u)$ and $\mathbf{x}(u) = (x_0(u), x_1(u), x_2(u), x_3(u))$, we set

$$\mathscr{L}(\bar{H})(u, \mathbf{v}) = (\mathbf{v}, r, z),$$

with

$$z = [v_0 x_1(u) \mp v_1 x_0(u) : v_0 x_2(u) \mp v_2 x_0(u) : v_0 x_3(u) \mp v_3 x_0(u)]$$

for $\mathbb{L} = \mathbb{L}^{\pm}$. An immediate consequence of the definition of $\mathscr{L}(\bar{H})$ is the following result.

Proposition 9.11. *The map $\mathscr{L}(\bar{H})$ is a Legendrian immersion and the extended family of lightcone height functions $\bar{H}$ on M is a generating family of $\mathscr{L}(\bar{H})$ at any point of the domain U of the parametrisation of M.*

Observe that the corresponding wavefront of the Legendrian immersion $\mathscr{L}(\bar{H})$ is the lightcone pedal surface $\mathbb{LP}_M(U)$. Therefore, the singularities of the lightcone pedal are Legendrian singularities.

Recall that $THP(\mathbf{x}, u)$ denotes the tangent lightlike hyperplane to M at $\mathbf{x}(u)$.

Lemma 9.3. *Let M be a spacelike surface patch parametrised by $\mathbf{x} : U \to \mathbb{R}^4_1$. For any two points $u, \bar{u} \in U$, $\mathbb{LP}_M(u) = \mathbb{LP}_M(\bar{u})$ if and only if $THP(\mathbf{x}, u) = THP(\mathbf{x}, \bar{u})$.*

Let $\mathbb{LP}_M : (U, u) \to (LC^*, \mathbf{v})$ and $\mathbb{LP}_{\bar{M}} : (U, \bar{u}) \to (LC^*, \bar{\mathbf{v}})$ be two germs of lightlike pedal mappings of spacelike surface patches parametrised by $\mathbf{x} : (U, u) \to \mathbb{R}^4_1$ and $\bar{\mathbf{x}} : (\bar{U}, \bar{u}) \to \mathbb{R}^4_1$, respectively. If the regular sets of $\mathbb{LP}_M$ and $\mathbb{LP}_{\bar{M}}$ are dense in their domain, then by Theorem 5.10, $\mathbb{LP}_M$ and $\mathbb{LP}_{\bar{M}}$ are $\mathcal{A}$-equivalent if and only if their corresponding germs of Legendrian immersions $\mathscr{L}_{\bar{H}_1} : (U, u) \to PT^*(LC^*)$ and $\mathscr{L}_{\bar{H}_2} : (\bar{U}, \bar{u}) \to PT^*(LC^*)$ are Legendrian equivalent. Equivalently, by Theorem 5.11, the two generating families $\bar{H}_1$ and $\bar{H}_2$ are P-$\mathcal{K}$-equivalent.

On the other hand, if we denote $\bar{h}_{1,\mathbf{v}}(u) = \bar{H}_1(u, \mathbf{v})$, and $\bar{h}_{2,\bar{\mathbf{v}}}(u) = \bar{H}_2(u, \bar{\mathbf{v}})$, then by Theorem 4.1, $K(M, THP(\mathbf{x}, u), \mathbf{v}) = K(\bar{M}, THP(\bar{\mathbf{x}}, \bar{u}), \bar{\mathbf{v}})$ if and only if $\bar{h}_{1,\mathbf{v}}$ and $\bar{h}_{2,\bar{\mathbf{v}}}$ are $\mathcal{K}$-equivalent.

Let $Q_r(\mathbf{x}, u)$ (resp. $Q_r(\bar{\mathbf{x}}, \bar{u})$) denote the local ring of the function germ $\bar{h}_{1,\mathbf{v}}$ (resp. $\bar{h}_{2,\bar{\mathbf{v}}}$) with $\mathbf{v} = \mathbb{LP}_M(u)$ (resp. $\bar{\mathbf{v}} = \mathbb{LP}_{\bar{M}}(\bar{u})$) (see Chapter 5 for the explicit expression of $Q_r(\mathbf{x}, u)$).

Theorem 9.2. *Let $\mathbf{x} : (U, u) \to \mathbb{R}^4_1$ and $\bar{\mathbf{x}} : (U, \bar{u}) \to \mathbb{R}^4_1$ be parametrisations of spacelike surface patches such that the germs of the Legendrian germs $\pi \circ \mathscr{L}(\bar{H}_1) : (U, u) \to (LC^*, \mathbf{v})$ and $\pi \circ \mathscr{L}(\bar{H}_2) : (U, \bar{u}) \to (LC^*, \bar{\mathbf{v}})$ are Legendrian stable. Then the following statements are equivalent.*

(i) *The lightcone pedal germs $\mathbb{LP}_M$ and $\mathbb{LP}_{\bar{M}}$ are $\mathcal{A}$-equivalent.*
(ii) *$\bar{H}_1$ and $\bar{H}_2$ are P-$\mathcal{K}$-equivalent.*
(iii) *$\bar{h}_{1,\mathbf{v}}$ and $\bar{h}_{2,\bar{\mathbf{v}}}$ are $\mathcal{K}$-equivalent.*
(iv) *$K(M, THP(\mathbf{x}, u), \mathbf{v}) = K(\bar{M}, THP(\bar{\mathbf{x}}, \bar{u}), \bar{\mathbf{v}})$.*
(v) *$Q_4(\mathbf{x}, u)$ and $Q_4(\bar{\mathbf{x}}, \bar{u})$ are isomorphic as $\mathbb{R}$-algebras.*

Proof. Assertion (iii) is equivalent to (iv) by the arguments in the proof of Lemma 9.3. The equivalence between the other assertions follows from Theorem 5.11 and Proposition 5.2. □

Corollary 9.8. *Let M be a surface patch parametrised by $\mathbf{x} \in \mathcal{O}$ with $\mathcal{O}$ as in* Corollary 9.7. *Then the following hold.*

(i) *The lightcone pedal surface is a cuspidal edge at each point of the lightlike parabolic curve except possibly at isolated points. At such points it is a swallowtail surface.*

(ii) *A lightcone parabolic point is a fold of the normalised lightcone Gauss map $\widetilde{\mathbb{L}}$ if and only if it is a cuspidal edge point of the lightcone pedal map $\mathbb{LP}_M$.*

(iii) *A lightcone parabolic point is a cusp of the normalised lightcone Gauss map $\widetilde{\mathbb{L}}$ if and only if it is a swallowtail point of the lightcone pedal map $\mathbb{LP}_M$.*

Proof. Corollary 9.7, the family of lightcone height function germ H at any point $(u, \mathbf{v}) \in U \times S^2_+$ is the $\mathcal{R}^+$-versal unfolding of $h_{\mathbf{v}}$. Therefore, it is P-$\mathcal{R}^+$-equivalent to one of the following germs:

$$(A_1)\ \ F(q, x_1, x_2) = \pm q^2,$$
$$(A_2)\ \ F(q, x_1, x_2) = q^3 + x_1 q,$$
$$(A_3)\ \ F(q, x_1, x_2) = \pm q^4 + x_1 q + x_2 q^2.$$

For each F, we have $\overline{F}(q, x_1, x_2, y) = F(q, x_1, x_2) - y$. On the other hand, $\overline{H}$ is the family of extended lightcone height functions such that $\mathscr{L}(\overline{H})(\Sigma_*(\overline{H}) = \mathfrak{L}_H(C_H)$ is a germ of a graph-like Legendrian submanifold at any point and its corresponding Lagrangian map-germ is $\widetilde{\pi} \circ \mathfrak{L}_H|_{C_H} = \widetilde{\mathbb{L}}$. Thus, if H is P-$\mathcal{R}^+$-equivalent to F, then $\overline{H}$ is P-$\mathcal{K}$-equivalent to $\overline{F}$. If H is of A_1-type, then $\widetilde{\pi} \circ \mathfrak{L}_H|_{C_H}$ is Lagrangian non-singular. If H is of A_2-type, then $\widetilde{\pi} \circ \mathfrak{L}_H|_{C_H}$ is fold map-germ. In this case, the discriminant $D_{\overline{F}}$ is a cuspidal edge. Moreover, the discriminant $D_{\overline{H}}$ is the image of the lightcone pedal map $\mathbb{LP}_M$. This means that we have assertion (ii). If H is of A_3-type, then we have assertion (iii). The set of singular points of $\widetilde{\mathbb{L}}$ is the lightcone parabolic curve, which is equal to the set of singular points of $\mathbb{LP}_M$. This completes the proof. □

9.8 Special cases of spacelike surfaces

The Euclidean space $\mathbb{R}^3_0$ and any spacelike hyperplanes in $\mathbb{R}^4_1$ are Riemannian manifolds so any surface embedded in these spaces is a spacelike surface. We illustrate below some properties concerning the lightcone Gauss map of surfaces in these spaces. We also deal with the case of spacelike surfaces contained in de Sitter 3-space and treat with more details the case of surfaces in the hyperbolic 3-space.

9.8.1 *Surfaces in Euclidean 3-space*

Let M be a surface patch contained in Euclidean space $\mathbb{R}^3_0 \subset \mathbb{R}^4_1$ and suppose that it is parametrised by $\mathbf{x} : U \to \mathbb{R}^3_0$. Let S^2 denote the unit sphere in $\mathbb{R}^3_0$.

Here, we take $\mathbf{n}^T = \mathbf{e}_0$, so that

$$\mathbf{n}^S(u) = \frac{\mathbf{e}_0 \wedge \mathbf{x}_{u_1}(u) \wedge \mathbf{x}_{u_2}(u)}{\|\mathbf{e}_0 \wedge \mathbf{x}_{u_1}(u) \wedge \mathbf{x}_{u_2}(u)\|_1} \in S^2$$

is the Euclidean unit normal of $M \subset \mathbb{R}^3_0$ at $p = \mathbf{x}(u)$.

In this case the lightcone height functions measure the contact of M with planes in $\mathbb{R}^3_0$, and hence coincides with the usual family of height functions in Euclidean 3-space.

The lightcone Gauss map, given by

$$\mathbb{L}^{\pm}(u) = \mathbf{e}_0 \pm \mathbf{n}^S(u),$$

coincides with the classical Gauss map on surfaces in Euclidean 3-space. The normalised lightcone shape operator is given by

$$W(\widetilde{\mathbb{L}}^{\pm})_p = -d\mathbb{L}^{\pm}_p = \mp d\mathbf{n}^S_p,$$

and coincides with the Weingarten map of M considered as a surface in Euclidean space $\mathbb{R}^3_0$. It follows that $\widetilde{K}^{\pm}_{\ell}(u) = \pm K(u)$ and $\bar{H}^{\pm}_{\ell}(u) = \pm H(u)$, where K and H are, respectively, the Gauss curvature and mean curvature of M as a surface in Euclidean space $\mathbb{R}^3_0$. Consequently, in $\mathbb{R}^3_0$, the lightcone flat spacelike surfaces are developable surfaces and the marginally trapped surfaces are minimal surfaces.

In general, when $\mathbf{n}^T(u) = \mathbf{v}$ is a constant timelike unit vector, the spacelike surface M is contained in a spacelike hyperplane $HP(\mathbf{v}, c)$. Since $HP(\mathbf{v}, c)$ is isometric to $\mathbb{R}^3_0$, the results for surfaces in $\mathbb{R}^3_0$ hold for surfaces in $HP(\mathbf{v}, c)$.

9.8.2 *Spacelike surfaces in de Sitter 3-space*

Let M be a spacelike surface in de Sitter 3-space parametrised by $\mathbf{x} : U \to S^3_1$. Then $\mathbf{x}(u)$ is a unit spacelike vector pseudo-orthogonal to M at $\mathbf{x}(u)$, so we can take $\mathbf{n}^S(u) = \mathbf{x}(u)$ for all $u \in U$ and we get

$$\mathbf{n}^T(u) = \frac{\mathbf{x}_{u_1}(u) \wedge \mathbf{x}_{u_2}(u) \wedge \mathbf{x}(u)}{\|\mathbf{x}_{u_1}(u) \wedge \mathbf{x}_{u_2}(u) \wedge \mathbf{x}(u)\|_1} \in H^3(-1).$$

Given a lightlike vector $\mathbf{v}$, the set $HP(\mathbf{v}, c) \cap S^3_1$ is called a *de Sitter horosphere.* Thus, the lighcone height functions measure the contact of M with de Sitter horospheres.

The lightcone Gauss map is given by

$$\mathbb{L}^{\pm}(u) = \mathbf{x}(u) \pm \mathbf{n}^T(u).$$

The following result follows from Proposition 9.7.

Proposition 9.12. *Let M be a spacelike surface patch in de Sitter 3-space. Then one of the lightcone Gauss maps ($\mathbb{L}^+$ or $\mathbb{L}^-$) is a constant vector if and only if M is part of a de Sitter horosphere.*

We refer to [Kasedou (2009)] for further results on contact of surfaces with de Sitter horospheres.

9.8.3 *Spacelike surfaces in Minkowski 3-space*

The 3-dimensional Minkowski space $\mathbb{R}^3_1$ can be identified isometrically with the set

$$\{(x_0, x_1, x_2, x_3) \in \mathbb{R}^4_1 \,|\, x_3 = 0\}.$$

Let M be a spacelike surface patch in $\mathbb{R}^3_1 \subset \mathbb{R}^4_1$, parametrised by $\mathbf{x} : U \to \mathbb{R}^3_1$ and denote by $H^2(-1)$ the pseudo sphere $\{\mathbf{u} \in \mathbb{R}^3_1 : \langle \mathbf{u}, \mathbf{u} \rangle = -1\}$. Here we set $\mathbf{n}^S(u) = \mathbf{e}_2$, so that

$$\mathbf{n}^T(u) = \frac{\mathbf{x}_{u_1}(u) \wedge \mathbf{x}_{u_2}(u) \wedge \mathbf{e}_2}{\|\mathbf{x}_{u_1}(u) \wedge \mathbf{x}_{u_2}(u) \wedge \mathbf{e}_2\|_1} \in H^2(-1)$$

is the timelike unit normal vector of M as a surface in $\mathbb{R}^3_1$ at $p = \mathbf{x}(u)$.
The lightcone Gauss map is given by

$$\mathbb{L}^{\pm}(u) = \mathbf{n}^T(u) \pm \mathbf{e}_3,$$

and the normalised lightcone shape operator is given by

$$W(\widetilde{\mathbb{L}}^{\pm})_p = -d\mathbf{n}^T(u).$$

It is the spacelike shape operator of M as a surface in Minkowski 3-space. We thus have $\widetilde{K}^{\pm}_{\ell}(u) = K_S(u)$ and $\bar{H}^{\pm}_{\ell}(u) = H_S(u)$, where K_S (resp. H_S) is the spacelike Gauss-Kronecker curvature (resp. mean curvature) defined in [Cheng and Yau (1973); Kobayashi (1983)]. It follows that the lightcone flat spacelike surfaces in $\mathbb{R}^3_1 \subset \mathbb{R}^4_1$ are spacelike developable surfaces and the marginally trapped surface are maximal surfaces.

9.8.4 *Surfaces in hyperbolic 3-space*

The hyperbolic 3-space $H^3_+(-1) \subset \mathbb{R}^4_1$, with the induced metric, is a Riemannian 3-manifold. Therefore, any embedded surface in $H^3_+(-1)$ is spacelike. The intersection of $H^3_+(-1)$ with a spacelike (resp. timelike, lightlike) hyperplane is called a *sphere* (resp. an *equidistant surface*, a *horosphere*). An equidistant surface is said to be a *plane* if it is given by the intersection of $H^3_+(-1)$ with a timelike hyperplane through the origin. The study of contact of a surface M in $H^3_+(-1)$ with lightlike hyperplanes is the study of the contact of M with the horospheres of $H^3_+(-1)$. We refer to the geometric properties of M derived from this contact as the *horospherical properties* of M.

Let M be a surface patch in the hyperbolic 3-space parametrised by $\mathbf{x} : U \to H^3_+(-1)$. Here we take $\mathbf{n}^T(u) = \mathbf{x}(u)$ so that

$$\mathbf{n}^S(u) = \frac{\mathbf{x}(u) \wedge \mathbf{x}_{u_1}(u) \wedge \mathbf{x}_{u_2}(u)}{\|\mathbf{x}(u) \wedge \mathbf{x}u_1(u) \wedge \mathbf{x}_{u_2}(u)\|_1} \in S^3_1.$$

We define *the de Sitter Gauss map* $\mathbb{E} : U \to S^3_1$, as the map given by

$$\mathbb{E}(u) = \mathbf{n}^S(u).$$

The lightcone Gauss map $\mathbb{L}^{\pm} : U \to LC^*$ is given by

$$\mathbb{L}^{\pm}(u) = \mathbf{x}(u) \pm \mathbb{E}(u).$$

Remark 9.6. The normalised lightcone Gauss map is the same as the hyperbolic Gauss map defined [Epstein (1986); Bryant (1987); Kobayashi (1986)] in the Poincaré disk model of the hyperbolic space.

Remark 9.7. The intersection of lightlike hyperplanes with $H^3_+(-1)$ are the horospheres. Therefore, the family of lightcone height functions on $M \subset H^3_+(-1)$ measures the contact of M with horospheres.

Proposition 9.13. *Let M be a surface patch in the hyperbolic 3-space parametrised by $\mathbf{x} : U \to H^3_+(-1)$. If one of the hyperbolic Gauss maps $\widetilde{\mathbb{L}}^+$ or $\widetilde{\mathbb{L}}^-$ is constant, then the surface M is contained in a horosphere.*

Proof. It follows from Proposition 9.7 that $\widetilde{\mathbb{L}}^+$ or $\widetilde{\mathbb{L}}^-$ is a constant lightlike vector $\mathbf{v}$ if and only if $M \subset H(\mathbf{v}, c) \cap H^3_+(-1)$, equivalently, M is a subset of a horosphere. □

For a general spacelike surface M in $\mathbb{R}^4_1$, the image of the differential map $d\mathbb{L}_p$ is not a subset of T_pM. In order to obtain a shape operator, we considered the composite of $d\mathbb{L}_p$ with the canonical projection $\pi_1 : T_p\mathbb{R}^4_1 \to T_pM$, i.e., $W(\mathbb{L})_p = -\pi_1 \circ d\mathbb{L}_p$. However, for surfaces in hyperbolic space the situation is rather different as shown by the following result, where $D_{\mathbf{v}}$ denotes the covariant derivative with respect to the tangent vector $\mathbf{v} \in T_pM$.

Lemma 9.4. *Let M be a surface patch in $H^3_+(-1)$. For any $p \in M$ and $\mathbf{v} \in T_pM$, we have $D_{\mathbf{v}}\mathbb{E} \in T_pM$ and $D_{\mathbf{v}}\mathbb{L} \in T_pM$.*

Proof. Let $\mathbf{x} : U \to H^3_+(-1)$ be a parametrisation of M. Since $\{\mathbf{x}_{u_1}, \mathbf{x}_{u_2}, \mathbf{x}, \mathbb{E}\}$ is a basis of $\mathbb{R}^4_1$, we have at $p = \mathbf{x}(u)$,

$$D_{\mathbf{v}}\mathbb{E} = \lambda_1 \mathbf{x}_{u_1} + \lambda_2 \mathbf{x}_{u_2} + \mu_1 \mathbf{x} + \mu_2 \mathbb{E}$$

for some real numbers $\lambda_1, \lambda_2, \mu_1, \mu_2$.

It follows from the fact that $\langle \mathbb{E}, \mathbb{E} \rangle_1 = 1$ that $\langle D_{\mathbf{v}}\mathbb{E}, \mathbb{E} \rangle_1 = 0$, so $\mu_2 = 0$.

From the identity $\langle \mathbb{E}, \mathbf{x} \rangle_1 = 0$ we get $\langle D_{\mathbf{v}}\mathbb{E}, \mathbf{x} \rangle_1 = -\langle \mathbb{E}, D_{\mathbf{v}}\mathbf{x} \rangle$. But $\mathbb{E}$ is a normal vector and $D_{\mathbf{v}}\mathbf{x}$ is a tangent vector to M, so $\langle \mathbb{E}, D_{\mathbf{v}}\mathbf{x} \rangle = 0$, which implies $\langle D_{\mathbf{v}}\mathbb{E}, \mathbf{x} \rangle_1 = 0$. Therefore $\mu_1 = 0$. Consequently,

$$D_{\mathbf{v}}\mathbb{E} = \lambda_1 \mathbf{x}_{u_1} + \lambda_2 \mathbf{x}_{u_2}$$

is a vector in T_pM. As $D_{\mathbf{v}}\mathbf{x}$ is in T_pM, $D_{\mathbf{v}}\mathbb{L}^{\pm} = D_{\mathbf{v}}\mathbf{x} \pm D_{\mathbf{v}}\mathbb{E}$ is also a vector in T_pM. □

As a consequence of Lemma 9.4, the differential maps $d\mathbb{E}_p$ and $d\mathbb{L}^{\pm}_p = 1_{T_pM} \pm d\mathbb{E}(u)$ are linear transformations $T_pM \to T_pM$, at $p = \mathbf{x}(u)$, and they are both self-adjoint operators.

Remark 9.8. Let $W(\mathbb{E})_p = -d\mathbb{E}_p$ denote the shape operator associated to the de Sitter Gauss map and let H_d be the de Sitter mean curvature, so $H_d(u) = \frac{1}{2}\mathrm{Trace}(W(\mathbb{E})_p)$, with $p = \mathbf{x}(u)$. Then

$$H^{\pm}_{\ell}(u) = \pm H_d(u) - 1.$$

It follows that surfaces in hyperbolic space with $H_d \equiv \pm 1$ correspond to surfaces with $H^{\pm}_{\ell} \equiv 0$ which are marginally trapped in $\mathbb{R}^4_1$.

Proposition 9.14. *Suppose that M is a totally $\mathbb{L}$-umbilic surface patch in hyperbolic 3-space parametrised by $\mathbf{x} : U \to H^3_+(-1)$. Then the lightcone principal curvatures κ_1 and κ_2 are equal to a constant κ, and we have the following classification of the surface M.*

(i) *Suppose that $\kappa \neq 0$.*

  (a) *If $\kappa \neq -1$ and $|\kappa + 1| < 1$, then M is a part of an equidistant surface.*

(b) *If* $\kappa \neq -1$ *and* $|\kappa + 1| > 1$, *then* M *is a part of a sphere.*
(c) *If* $\kappa = -1$, *then* M *is a part of a plane.*

(ii) *If* $\kappa = 0$, *then* M *is a part of a horosphere.*

Proof. We deal with the case for $\mathbb{L}^+$, the case $\mathbb{L}^-$ follows similarly. By definition, we have $-\mathbb{L}^+_{u_i} = \kappa \mathbf{x}_{u_i}$ for $i = 1, 2$. Therefore, for $i, j = 1, 2$,

$$-\mathbb{L}^+_{u_i u_j} = \kappa_{u_j} \mathbf{x}_{u_i} + \kappa \mathbf{x}_{u_i u_j}.$$

Since $-\mathbb{L}^+_{u_i u_j} = -\mathbb{L}^+_{u_j u_i}$ and $\kappa \mathbf{x}_{u_i u_j} = \kappa \mathbf{x}_{u_j u_i}$, we have

$$\kappa_{u_j} \mathbf{x}_{u_i} - \kappa_{u_i} \mathbf{x}_{u_j} = 0.$$

As $\mathbf{x}_{u_1}, \mathbf{x}_{u_2}$ are linearly independent, $\kappa_{u_i} = 0, i = 1, 2$, and hence κ is constant.

We have $d\mathbb{L}^+ = 1_{T_pM} + d\mathbb{E}$, so $-\mathbb{L}^+_{u_i} = \kappa \mathbf{x}_{u_i}$ is equivalent to $-\mathbb{E}_{u_i} = \kappa_d \mathbf{x}_{u_i}$, with $\kappa = -1 + \kappa_d$.

Suppose now that $\kappa \neq 0$. If $\kappa \neq -1$, then $\kappa_d \neq 0$, so $\mathbf{x}_{u_i} = (1/\kappa_d)\mathbb{E}_{u_i}$. Therefore, there exists a constant vector $\mathbf{a}$ such that $\mathbf{x} = \mathbf{a} - (1/\kappa_d)\mathbb{E}$. But since $\langle \mathbf{x} - \mathbf{a}, \mathbf{x} - \mathbf{a} \rangle_1 = (1/\kappa_d)^2$,

$$\langle \mathbf{a}, \mathbf{x} \rangle_1 = -\frac{1}{2}\left(\frac{1}{\kappa_d^2} + 1 - \langle \mathbf{a}, \mathbf{a} \rangle_1\right) = -\frac{1}{2}\left(\frac{1}{\kappa_d^2} + 1 + 1 - \frac{1}{\kappa_d^2}\right) = -1,$$

which means that $M =\subset HP(\mathbf{a}, -1) \cap H^3_+(-1)$. If $|\kappa_d| < 1$, $\mathbf{a} = \mathbf{x} + (1/\kappa_d)\mathbb{E}$ is spacelike and we have assertion (i)-(a), and in case $|\kappa_d| > 1$ we get assertion (i)-(b).

If $\kappa = -1$, then we have $\mathbb{E}_{u_i} = 0$, so $\mathbb{E}$ is a constant vector $\mathbf{a}$. Since $\mathbf{a}$ is a spacelike vector and we have $\langle \mathbf{x}, \mathbf{a} \rangle_1 = 0$, $M \subset HP(\boldsymbol{a}, 0) \cap H^3_+(-1)$. This means that M is a part of a plane.

Finally, we assume that $\kappa = 0$. In this case, we have $\mathbb{L}^+_{u_i} = 0$, so that $\mathbb{L}^+$ is a constant lightlike vector $\boldsymbol{a}$. This means that the $\mathbb{L}^+$-lightcone Gauss map is constant, and assertion (ii) follows from Proposition 9.13. □

9.9 Lorentzian distance squared functions

The *family of Lorentzian distance squared functions* in $\mathbb{R}^4_1$ is the map $\mathcal{G} : \mathbb{R}^4_1 \times \mathbb{R}^4_1 \to \mathbb{R}$ given by

$$\mathcal{G}(p, a) = \langle p - a, p - a \rangle_1.$$

The fibres of the map $\mathcal{G}$ are the pseudo spheres in $\mathbb{R}^4_1$ of centre a.

Let M be a spacelike surface patch parametrised by $\mathbf{x} : U \to \mathbb{R}^4_1$. The *family of Lorentzian distance squared functions* $G : U \times \mathbb{R}^4_1 \to \mathbb{R}$ on M is the restriction of $\mathcal{G}$ to M and is given by

$$G(u, a) = \langle \mathbf{x}(u) - a, \mathbf{x}(u) - a \rangle_1.$$

For a fixed $a_0 \in \mathbb{R}^4_1$, we write $g_{a_0}(u) = G(u, a_0)$.

Proposition 9.15. *Let M be a spacelike surface patch parametrised by $\mathbf{x} : U \to \mathbb{R}^4_1$. Suppose that $a_0 \neq p_0 = \mathbf{x}(u_0)$. Then*

(i) *u_0 is a singular point of g_{a_0} with $g_{a_0}(u_0) = 0$ if and only if*

$$a_0 = \mathbf{x}(u_0) - \lambda \widetilde{\mathbb{L}}(u_0)$$

for real non-zero scalar λ.

(ii) *u_0 is a degenerate singular point of g_{a_0} with $g_{a_0}(u_0) = 0$ if and only if*

$$a_0 = \mathbf{x}(u_0) + \frac{1}{\widetilde{\kappa}_i(u_0)} \widetilde{\mathbb{L}}(u_0),$$

for $i = 1$ or $i = 2$.

Proof. (i) The condition $g_{a_0}(u_0) = 0$ means that $p_0 - a_0 \in LC^*$. We observe that $dg_{a_0}(u_0) = \langle d\mathbf{x}(u_0), \mathbf{x}(u_0) - a_0 \rangle_1 = 0$ if and only if $p_0 - a_0 \in N_pM$. Hence $g_{a_0}(u_0) = dg_{a_0}(u_0) = 0$ if and only if $p_0 - a_0 \in N_{p_0}M \cap LC^*$. This is equivalent to $p_0 - a_0 = \lambda \widetilde{\mathbb{L}}(u_0)$ for some real non-zero scalar λ.

(ii) We have, for $i = 1, 2$,

$$\frac{\partial g}{\partial u_i} = 2\langle \mathbf{x}_{u_i}, \mathbf{x} - a_0 \rangle_1 \quad \text{and} \quad \frac{\partial^2 g}{\partial u_i \partial u_j} = 2(\langle \mathbf{x}_{u_i u_j}, \mathbf{x} - a_0 \rangle_1 + \langle \mathbf{x}_{u_i}, \mathbf{x}_{u_j} \rangle_1).$$

When $p_0 - a_0 = \lambda \widetilde{\mathbb{L}}(u_0)$, $\langle \mathbf{x}_{u_i u_j}(u_0), \mathbf{x}(u_0) - a_0 \rangle_1 = \lambda \langle \mathbf{x}_{u_i u_j}(u_0), \widetilde{\mathbb{L}}(u_0) \rangle_1$ so

$$\text{Hess}(g)(u_0) = 2\lambda \frac{1}{\ell_0} \begin{pmatrix} h_{11} & h_{11} \\ h_{12} & h_{22} \end{pmatrix} + 2 \begin{pmatrix} g_{11} & g_{12} \\ g_{12} & g_{22} \end{pmatrix}$$

where all the entries of the right hand side of the above equality are evaluated at u_0 and ℓ_0 is the first coordinate of $\mathbb{L}$. Therefore,

$$\text{Hess}(g)(u_0) \begin{pmatrix} g_{11} & g_{12} \\ g_{12} & g_{22} \end{pmatrix}^{-1} = 2(\lambda \widetilde{A}_{p_0} + I)$$

with I denoting the 2×2 identity matrix. Thus, $\det(\text{Hess}(g_{a_0}))(u_0) = 0$ if and only if $-1/\lambda$ is an eigenvalue of $\widetilde{A}_{p_0}$, that is, $\lambda = -1/\widetilde{\kappa}_1(u_0)$ or $\lambda = -1/\widetilde{\kappa}_2(u_0)$. □

It can be shown by straightforward calculations that G is a Morse family in a neighbourhood of each point $(u,a) \in G^{-1}(0)$, so we have the Legendrian immersion $\mathscr{L}(G) : \Sigma_G \to PT^*(\mathbb{R}^4_1)$, with

$$\mathscr{L}(G)(u,a) = (a, [(\mathbf{x}_0(u) - a_0) : (a_1 - \mathbf{x}_1(u)) : (a_2 - \mathbf{x}_2(u)) : (a_3 - \mathbf{x}_3(u))]),$$

where $a = (a_0, a_1, a_2, a_3)$ and $\mathbf{x}(u) = (\mathbf{x}_0(u), \mathbf{x}_1(u), \mathbf{x}_2(u), \mathbf{x}_3(u))$.

By Proposition 9.15, we have

$$\Sigma_G = (\Delta_* G)^{-1}(0) = \{(u,a) \in U \times \mathbb{R}^4_1 \mid a = \mathbf{x}(u) + \lambda \widetilde{\mathbb{L}}(u), \lambda \in \mathbb{R} \setminus \{0\}\}.$$

This means that G is a generating family of the Legendrian immersion $\mathscr{L}(G)$ whose discriminant set is given by

$$D_G = \{a = \mathbf{x}(u) + \lambda \widetilde{\mathbb{L}}(u), u \in U, \lambda \in \mathbb{R} \setminus \{0\}\}. \tag{9.4}$$

Theorem 9.3. *There exists a residual subset $\mathcal{O} \subset \mathrm{Emb}_s(U, \mathbb{R}^4_1)$ such that for any $\mathbf{x} \in \mathcal{O}$, the germ of the Legendrian immersion $\mathscr{L}(G)$ at each point is Legendrian stable.*

Proof. This is a consequence of Theorems 4.3, 5.11 and 5.12. □

9.9.1 *Lightlike hypersurfaces*

Definition 9.5. A hypersurface S in $\mathbb{R}^4_1$ is said to be *lightlike* if its tangent space at all of its points is a lightlike hyperplane.

Let M be a spacelike surface patch in $\mathbb{R}^4$ parametrised by $\mathbf{x} : U \to \mathbb{R}^4_1$. We can construct a lightlike hypersurface S from M, as the image of the map $\mathbb{LH}_M : U \times \mathbb{R} \to \mathbb{R}^4_1$ with

$$\mathbb{LH}_M(u,\lambda) = \mathbf{x}(u) + \lambda \widetilde{\mathbb{L}}(u).$$

The hypersurface $\mathbb{LH}_M(U \times \mathbb{R})$ is a ruled hypersurface with M its base surface and $\widetilde{\mathbb{L}}$ its ruling. In fact, any lightlike hypersurface is, at least locally, a ruled hypersurface with base surface some spacelike surface patch and ruling one of the lightlike normal directions of the base surface ([Kossowski (1989)]). It is clear that the lightlike hypersurface $\mathbb{LH}_M(U \times (\mathbb{R} \setminus \{0\}))$ is the discriminant D_G of the family of Lorentzian distance squared functions G on M; see (9.4). In particular, $\mathbb{LH}_M(U \times (\mathbb{R} \setminus \{0\}))$ is the wavefront of the Legendrian immersion $\mathscr{L}(G)$. Since $\mathbb{LH}_M$ is non-singular at $(u,0) \in U \times \mathbb{R}$, $\mathbb{LH}_M(U \times \mathbb{R})$ is a wavefront set. As a consequence (see Proposition 9.15), a point (u,λ) is a singular point of $\mathbb{LH}_M(U \times \mathbb{R})$ if and only if

$$a = \mathbf{x}(u) + \frac{1}{\widetilde{\kappa}_i(u)} \widetilde{\mathbb{L}}(u)$$

for $i = 1$ or $i = 2$.

We define

$$\mathbb{F}_M = \bigcup_{i=1}^{2} \left\{ \mathbf{x}(u) + \frac{1}{\widetilde{\kappa}_i(u)} \widetilde{\mathbb{L}}(u) \mid \kappa_i(u) \neq 0, u \in U \right\},$$

and call it the *lightlike focal set* of M.

From Theorem 9.3 and from the classification of stable Legendrian mappings (Theorem 5.12), we obtain the following classification of germs of generic lightlike hypersurface.

Corollary 9.9. *For* $\mathbf{x} \in \mathcal{O}$, *with* $\mathcal{O}$ *as in* Theorem 9.3, *the germ at any point* (u, λ) *of* $\mathbb{LH}_M$ *of the parametrisation of the lightlike hypersurface determined by* $M = \mathbf{x}(U)$ *is* $\mathcal{A}$*-equivalent to one of the following map-germs* $(\mathbb{R}^3, 0) \to (\mathbb{R}^4, 0)$:

(A_1) $f(x, y, z) = (x, y, z, 0)$,
(A_2) $f(x, y, z) = (3x^2, 2x^3, y, z)$,
(A_3) $f(x, y, z) = (4x^3 + 2xy, 3x^4 + yx^2, y, z)$,
(A_4) $f(x, y, z) = (5x^4 + 3zx^2 + 2xz, 4x^5 + 2yx^3 + zx^2, x, y)$,
(D_4^+) $f(x, y, z) = (2(x^2 + y^2) + xyz, 3x^2 + yz, 3y^2 + xz, z)$,
(D_4^-) $f(x, y, z) = (2(x^3 - xy^2) + (x^2 + y^2)z, y^2 - 3x^2 - 2xz, xy - yz, z)$.

Proof. By Theorems 5.11 and 9.3, the Lorentzian distance squared function G is a $\mathcal{K}$-versal deformation of g_{a_0} at each point $(u, a_0) \in U \times \mathbb{R}$. Therefore, we can apply the classification of Legendrian stable map-germs (Theorem 5.12). Then the generating family is P-$\mathcal{K}$-equivalent to one for the following normal forms:

(A_k) $F(q_1, q_2, \mathbf{x}) = x^{k+1} \pm q_2^2 + x_1 + x_2 q_1 + \cdots + x_k q_1^{k-1}$, $1 \leq k \leq 4$,
(D_4^+) $F(q_1, q_2, \mathbf{x}) = q_1^3 + q_2^3 + x_1 + x_2 q_1 + x_3 q_2 + x_4 q_1 q_2$,
(D_4^-) $F(q_1, q_2, \mathbf{x}) = q_1^3 - q_1 q_2^2 + x_1 + x_2 q_1 + x_3 q_2 + x_4 (q_1^2 + q_2^2)$

with $\mathbf{x} = (x_1, x_2, x_3, x_4)$. We consider the D_4^+-case and take F as above (the other cases follow by similar calculations). Then $\Sigma_*(F)$ is given by

$$\{(q_1, q_2, 2(q_1^2 + q_2^2) + x_4 q_1 q_2, -3q_1^2 - x_4 q_2, -3q_2^2 - x_4 q_1, x_4) \mid (q_1, q_2, x_4) \in \mathbb{R}^3\}.$$

If we change the parameters into $(q_1, q_2, x_4) = (x, y, z)$ and apply the linear transformation $\Phi(X, Y, Z, W) = (X, -Y, -Z, W)$, the corresponding Legendrian map-germ is

$$f(x, y, z) = (2(x^2 + y^2) + xyz, 3x^2 + yz, 3y^2 + xz, z).$$

□

9.9.2 *Contact of spacelike surfaces with lightcones*

We have the following characterisation of spacelike surfaces contained in a given lightcone $LC^*(a) = \{p \in \mathbb{R}^4_1 : \langle p - a, p - a\rangle_1 = 0\}$.

Proposition 9.16. *Let M be a spacelike surface patch without lightcone parabolic points and suppose that it is parametrised by $\mathbf{x} : U \to \mathbb{R}^4_1$. Then M is contained in a lightcone $LC^*(a_0)$ if and only if the lightlike hypersurface $\mathbb{LH}_M(U \times \mathbb{R})$ is contained in $LC^*(a_0)$ and a_0 is an isolated singular value of the map $\mathbb{LH}_M$.*

Proof. In the first place, we remark that $\widetilde{K}_\ell(u) \neq 0$ if and only if the three vectors $\widetilde{\mathbb{L}}(u), \widetilde{\mathbb{L}}_{u_1}(u), \widetilde{\mathbb{L}}_{u_2}(u)$ are linearly independent.

By definition, M is contained in a lightcone $LC^*(a_0)$ if and only if $g_{a_0}(u) = 0$ for all $u \in U$, where g_{a_0} is the Lorentzian distance squared function on M. It follows from Proposition 9.15 that there exists a smooth function $\mu : U \to \mathbb{R} \setminus \{0\}$ such that

$$\mathbf{x}(u) = a_0 + \mu(u)\widetilde{\mathbb{L}}(u).$$

We have then

$$\mathbb{LH}_M(u, \lambda) = a_0 + (\lambda + \mu(u))\widetilde{\mathbb{L}}(u),$$

which implies that $\mathbb{LH}_M(U \times \mathbb{R})$ is contained in $LC^*(a_0)$. Moreover, it follows that

$$\frac{\partial \mathbb{LH}_M}{\partial \lambda}(u, \lambda) = \widetilde{\mathbb{L}}(u),$$
$$\frac{\partial \mathbb{LH}_M}{\partial u_i}(u, \lambda) = \mu_{u_i}(u)\widetilde{\mathbb{L}}(u) + (\lambda + \mu(u))\widetilde{\mathbb{L}}_{u_i}(u), i = 1, 2,$$

from which we get

$$(\frac{\partial \mathbb{LH}_M}{\partial \mu} \wedge \frac{\partial \mathbb{LH}_M}{\partial u_1} \wedge \frac{\partial \mathbb{LH}_M}{\partial u_2})(u, \lambda) = (\lambda + \mu(u))^2 \widetilde{\mathbb{L}}(u) \wedge \widetilde{\mathbb{L}}_{u_1}(u) \wedge \widetilde{\mathbb{L}}_{u_2}(u).$$

Under the assumption $\widetilde{K}_\ell(u) \neq 0$, $\widetilde{\mathbb{L}}(u) \wedge \widetilde{\mathbb{L}}_{u_1}(u) \wedge \widetilde{\mathbb{L}}_{u_2}(u) \neq 0$, so that $(\frac{\partial \mathbb{LH}_M}{\partial \mu} \wedge \frac{\partial \mathbb{LH}_M}{\partial u_1} \wedge \frac{\partial \mathbb{LH}_M}{\partial u_2})(u, \lambda) = 0$ if and only if $\lambda + \mu(u) = 0$. This means that a_0 is an isolated singularity value of $\mathbb{LH}_M$. Since $M \subset \mathbb{LH}_M(U \times \mathbb{R})$, the converse assertion holds. □

By Proposition 9.16, the lightlike hypersurface has the most degenerate singular point if and only if the spacelike surface M is a subset of a lightcone. In this case, the lightlike hypersurface itself is a part of a lightcone. Therefore, we can consider lightcones as model lightlike hypersurfaces and study the contact of spacelike surfaces with lightcones.

From Proposition 9.15, the lightcone $LC^*(a_0)$ is tangent to the surface M at $p_0 = \mathbf{x}(u_0)$ if and only if u_0 is a singularity of g_{a_0} and $g_{a_0}(u_0) = 0$. We call $LC^*(a_0)$ a tangent lightcone to M at p_0.

Let M and $\bar{M}$ be two germs spacelike surfaces parametrised by $\mathbf{x} : (U, u) \to (\mathbb{R}^4_1, p)$ and $\bar{\mathbf{x}} : (\bar{U}, \bar{u}) \to (\mathbb{R}^4_1, \bar{p})$, respectively. Let $\mathbb{LH}_M : (U, u) \to (\mathbb{R}^4_1, p)$ and $\mathbb{LH}_{\bar{M}} : (\bar{U}, \bar{u}) \to (\mathbb{R}^4_1, \bar{p})$, be parametrisations of the germs of the lightlike hypersurfaces associated to M and $\bar{M}$ respectively.

If the regular sets of $\mathbb{LH}_M$ and $\mathbb{LH}_{\bar{M}}$ are dense in their respective domaines, then by Theorem 5.10, $\mathbb{LH}_M$ and $\mathbb{LH}_{\bar{M}}$ are $\mathcal{A}$-equivalent if and only if the germs of their corresponding Legendrian lift are Legendrian equivalent. Equivalently, by Theorem 5.11, the generating families G_1 and G_2 of the Legendrian lifts are P-$\mathcal{K}$-equivalent. Here $G_1 : (U \times \mathbb{R}^4_1, (u, a)) \to \mathbb{R}$ (resp. $G_2 : (\bar{U} \times \mathbb{R}^4_1, (\bar{u}, \bar{a})) \to \mathbb{R}$) denotes the germ of the family of the Lorentzian distance squared function on M (resp. $\bar{M}$).

Let $g_{1,a}(u) = G_1(u, a)$, and $g_{2,\bar{a}}(u) = G_2(u, \bar{a})$. By Proposition 5.2, $K(M, LC^*(a), p) = K(\bar{M}, LC^*(\bar{a}), \bar{p})$ if and only if $g_{1,a}$ and $g_{2,\bar{a}}$ are $\mathcal{K}$-equivalent, so we can apply the results in Proposition 5.2.

We denote by $Q_r(\mathbf{x}, u)$ (resp. $Q_r(\bar{\mathbf{x}}, \bar{u})$) the local ring of the germ of function $g_{1,a}$ (resp. $g_{2,\bar{a}}$), which is defined by

$$Q_r(\mathbf{x}, u) = \frac{\mathcal{E}_2}{g^*_{1,a}(\mathcal{M}_1)\mathcal{E}_2 + \mathcal{M}_2^{r+1}}.$$

Theorem 9.4. *Let M and $\bar{M}$ be two germs of spacelike surfaces parametrised by $\mathbf{x} : (U, u) \to (\mathbb{R}^4_1, p)$ and $\bar{\mathbf{x}} : (\bar{U}, \bar{u}) \to (\mathbb{R}^4_1, \bar{p})$, respectively. Suppose that their corresponding Legendrian lifts are Legendrian stable. Then the following statements are equivalent.*

(i) *The germs $\mathbb{LH}_M$ and $\mathbb{LH}_{\bar{M}}$ are $\mathcal{A}$-equivalent.*
(ii) *G_1 and G_2 are P-$\mathcal{K}$-equivalent.*
(iii) *$g_{1,a}$ and $g_{2,\bar{a}}$ are $\mathcal{K}$-equivalent.*
(iv) *$K(M, LC^*(a), p) = K(\bar{M}, LC^*(\bar{a}), \bar{p})$*
(v) *$Q_5(\mathbf{x}, u)$ and $Q_5(\bar{\mathbf{x}}, \bar{u})$ are isomorphic as $\mathbb{R}$-algebras.*

Proof. The discussion proceeding the proposition shows that (iii) and (iv) are equivalent. The equivalence of the other statements follow by Proposition 5.2. □

9.10 Legendrian dualities between pseudo-spheres

Given a Legendrian double fibration $\pi_1 : E \to B_1$ and $\pi_2 : E \to B_2$ and L a Legendrian submanifold of E, we say that the projections $\pi_1(L)$ and $\pi_2(L)$ are *Legendrian dual* to each other.

The Legendrian duality is a generalisation of the classical projective duality and the spherical duality [Shcherbak (1986); Nagai (2012)].

We present here a theorem on Legendrian dualities for pseudo-spheres in Lorentz-Minkowski space. It plays a fundamental role in the study of the extrinsic differential geometry of submanifolds in these pseudo-spheres from the view point of singularity theory ([Izumiya (2009); Chen and Izumiya (2009); Izumiya and Tari (2008, 2010b,a); Izumiya and Yıldırım (2011); Izumiya and Saji (2010)]).

Although the duality theorem holds for pseudo-spheres in general dimensions, we shall state it here for the pseudo-spheres in $\mathbb{R}^4_1$.

Let $\mathbf{v} = (v_0, v_1, v_2, v_3)$ and $\mathbf{w} = (w_0, w_1, w_2, w_3)$ be two non-zero vectors in $\mathbb{R}^4_1$. The 1-forms $\langle d\mathbf{v}, \mathbf{w}\rangle_1$ and $\langle \mathbf{v}, d\mathbf{w}\rangle_1$ are defined as follows:

$$\begin{aligned}\langle d\mathbf{v}, \mathbf{w}\rangle_1 &= -w_0 dv_0 + w_1 dv_1 + w_2 dv_2 + w_3 dv_3,\\ \langle \mathbf{v}, d\mathbf{w}\rangle_1 &= -v_0 dw_0 + v_1 dw_1 + v_2 dw_2 + v_3 dw_3.\end{aligned}$$

Consider the following double fibrations and 1-forms:

(1) In $H^3(-1) \times S^3_1 \supset \Delta_1 = \{(\mathbf{v}, \mathbf{w}) \mid \langle \mathbf{v}, \mathbf{w}\rangle_1 = 0\ \}$,

(a) $\pi_{11} : \Delta_1 \to H^3(-1)$, $\pi_{12} : \Delta_1 \to S^3_1$,
(b) $\theta_{11} = \langle d\mathbf{v}, \mathbf{w}\rangle_1 | \Delta_1$, $\theta_{12} = \langle \mathbf{v}, d\mathbf{w}\rangle_1 | \Delta_1$.

(2) In $H^3(-1) \times LC^* \supset \Delta_2^{\pm} = \{(\mathbf{v}, \mathbf{w}) \mid \langle \mathbf{v}, \mathbf{w}\rangle_1 = \pm 1\ \}$,

(a) $\pi_{21}^{\pm} : \Delta_2^{\pm} \to H^n(-1)$, $\pi_{22}^{\pm} : \Delta_2^{\pm} \to LC^*$,
(b) $\theta_{21}^{\pm} = \langle d\mathbf{v}, \mathbf{w}\rangle_1 | \Delta_2^{\pm}$, $\theta_{22}^{\pm} = \langle \mathbf{v}, d\mathbf{w}\rangle_1 | \Delta_2^{\pm}$.

(3) In $LC^* \times S^3_1 \supset \Delta_3^{\pm} = \{(\mathbf{v}, \mathbf{w}) \mid \langle \mathbf{v}, \mathbf{w}\rangle_1 = \pm 1\ \}$,

(a) $\pi_{31}^{\pm} : \Delta_3^{\pm} \to LC^*$, $\pi_{32}^{\pm} : \Delta_3^{\pm} \to S^n_1$,
(b) $\theta_{31}^{\pm} = \langle d\mathbf{v}, \mathbf{w}\rangle_1 | \Delta_3^{\pm}$, $\theta_{32}^{\pm} = \langle \mathbf{v}, d\mathbf{w}\rangle_1 | \Delta_3^{\pm}$.

(4) In $LC^* \times LC^* \supset \Delta_4^{\pm} = \{(\mathbf{v}, \mathbf{w}) \mid \langle \mathbf{v}, \mathbf{w}\rangle_1 = \pm 2\ \}$,

(a) $\pi_{41}^{\pm} : \Delta_4^{\pm} \to LC^*$, $\pi_{42}^{\pm} : \Delta_4^{\pm} \to LC^*$,
(b) $\theta_{41}^{\pm} = \langle d\mathbf{v}, \mathbf{w}\rangle_1 | \Delta_4^{\pm}$, $\theta_{42}^{\pm} = \langle \mathbf{v}, d\mathbf{w}\rangle_1 | \Delta_4^{\pm}$.

If we consider a surface $\mathbf{x} : U \to H^3(-1)$, then we have the de Sitter Gauss map $\mathbb{E} : U \to S^3_1$. It follows that $\langle \mathbf{x}(u), \mathbb{E}(u)\rangle_1 = 0$. Moreover, we have $\mathbb{L}^{\pm}(u) = \mathbf{x}(u) \pm \mathbb{E}(u)$, so that $\langle \mathbf{x}(u), \mathbf{x}(u) - \mathbb{E}(u)\rangle_1 = -1$ and $\langle \mathbf{x}(u) + \mathbb{E}(u), \mathbb{E}(u)\rangle_1 = 1$. We also have $\langle \mathbf{x}(u) + \mathbb{E}(u), \mathbf{x}(u) - \mathbb{E}(u)\rangle_1 = -2$. Thereofore, if we start with a suface in the hyperbolic space, $\Delta_1, \Delta_2^-, \Delta_3^+$ and Δ_4^- are natural sets to consider. However, there is no reason to exclude the other cases from the mathematical view point.

Remark 9.9. Observe that ${\theta_{11}}^{-1}(0)$ and ${\theta_{12}}^{-1}(0)$ define the same field of tangent hyperplanes over Δ_1, denoted by K_1. Also ${\theta_{i1}^{\pm}}^{-1}(0)$ and ${\theta_{i2}^{\pm}}^{-1}(0)$ define the same field of tangent hyperplanes over $\Delta_i^{\pm}$, $i = 2, 3, 4$, denoted by $K_i^{\pm}$.

We have the following duality theorem.

Theorem 9.5 ([Izumiya (2009)]). *The contact manifolds (Δ_1, K_1) and $(\Delta_i^{\pm}, K_i^{\pm})$, $i = 2, 3, 4$, are contact manifolds such that π_{1j} and $\pi_{ij}^{\pm}$, $j = 1, 2$, are Legendrian fibrations. Moreover, these contact manifolds are contact diffeomorphic to each other.*

Proof. It follows from the definition of Δ_1 and $\Delta_i^{\pm}$, $i = 2, 3, 4$, that these sets are smooth submanifolds of $\mathbb{R}^4_1 \times \mathbb{R}^4_1$ and that π_{1j} and $\pi_{ij}^{\pm}$, $i = 2, 3, 4; j = 1, 2$, are smooth fibrations.

It is shown in [Izumiya (2009)] that (Δ_1, K_1) is a contact manifold. We give here an outline of the proof. The restriction of the pseudo-scalar product $\langle , \rangle_1$ to $H^3(-1)$ is a Riemannian metric. Let $\pi : S(TH^3(-1)) \to H^3(-1)$ be the unit tangent sphere bundle of $H^3(-1)$.

For $\mathbf{v} \in H^3(-1)$, a vector $\mathbf{w} \in \mathbb{R}^4_1$ is in $T_{\mathbf{v}}H^3(-1)$ if and only if $\langle \mathbf{v}, \mathbf{w}\rangle = 0$. Therefore, $\mathbf{w} \in S(T_{\mathbf{v}}H^3(-1))$ if and only if $\langle \mathbf{w}, \mathbf{w}\rangle_1 = 1$ and $\langle \mathbf{v}, \mathbf{w}\rangle_1 = 0$. The last two conditions are equivalent to $(\mathbf{v}, \mathbf{w}) \in \Delta_1$. This means that we can identify canonically $S(TH^3(-1))$ with Δ_1. The canonical contact structure on $S(TH^3(-1))$ is given by the 1-form $\theta(V) = \langle d\pi(V), \tau(V)\rangle_1$, where $\tau : TS(TH^3(-1)) \to S(TH^3(-1))$ is the tangent bundle of $S(TH^3(-1))$ (see [Blair (1976); Cecil (1980)]). This form can be represented by $\theta_{11} = \langle d\mathbf{v}, \mathbf{w}\rangle_1|\Delta_1$ in the above identification. Thus, (Δ_1, K_1) is a contact manifold.

For $\Delta_i^{\pm}$, $i = 2, 3, 4$, we define the smooth mappings $\Psi_{1i}^{\pm} : \Delta_1 \to \Delta_i^{\pm}$ by

$$\begin{aligned} \Psi_{12}^{\pm}(\mathbf{v}, \mathbf{w}) &= (\mathbf{v}, \mp\mathbf{v} + \mathbf{w}), \\ \Psi_{13}^{\pm}(\mathbf{v}, \mathbf{w}) &= (\mathbf{v} \pm \mathbf{w}, \mathbf{w}), \\ \Psi_{14}^{\pm}(\mathbf{v}, \mathbf{w}) &= (\mathbf{v} \pm \mathbf{w}, \mp\mathbf{v} + \mathbf{w}). \end{aligned}$$

The above mappings have smooth inverses, so they are diffeomorphisms.

Consider, for example, the map $\Psi^{\pm}_{12}$. We have

$$\begin{aligned}(\Psi^{\pm}_{12})^*\theta^{\pm}_{21} &= \langle d\mathbf{v}, \mp\mathbf{v} + \mathbf{w}\rangle_1|\Delta_1 \\ &= \langle d\mathbf{v}, \mp\mathbf{v}\rangle_1|\Delta_1 + \langle d\mathbf{v}, \mathbf{w}\rangle_1|\Delta_1 \\ &= \langle d\mathbf{v}, \mathbf{w}\rangle_1|\Delta_1 = \theta_{11}.\end{aligned}$$

This means that $(\Delta^{\pm}_2, K^{\pm}_2)$ is a contact manifold and $\Psi^{\pm}_{12}$ is a contactomorphism. Similar argument shows that $(\Delta^{\pm}_i, K^{\pm}_i)$, $i = 3, 4$, are contact manifolds and $\Psi^{\pm}_{1i}$, $i = 3, 4$, are contactomorphisms. □

For a surface $\mathbf{x} : U \to H^3(-1)$, we have an embedding $\mathscr{L}_1 : U \to \Delta_1$ defined by $\mathscr{L}_1(u) = (\mathbf{x}(u), \mathbb{E}(u))$. By the definition of the contact structure K_1, $\mathscr{L}(U)$ is a Legendrian submanifold such that the image of the de Sitter Gauss map $\mathbb{E}(U)$ is the wavefront set of $\mathscr{L}_1(U)$ with respect to the Legendrian fibration $\pi_{12} : \Delta_1 \to S^3_1$. Moreover, $\mathbb{L}^-(U)$ is the wavefront set of a certain Legendrian submanifold in Δ^-_2. Even if we start with a spacelike surface in other pseudo spheres, we can apply the theory of Legendrian singularities to obtain geometric invariants.

We can also consider the contactomorphisms $\Psi^{\pm}_{ij} : \Delta^{\pm}_i \to \Delta^{\pm}_j$ for the other pairs (i, j) by $\Psi^{\pm}_{ij} = \Psi^{\pm}_{i1} \circ \Psi^{\pm}_{1j}$, where $\Psi^{\pm}_{i1} = (\Psi^{\pm}_{1i})^{-1}$. All these contactomorphisms give rise to the following commutative diagram

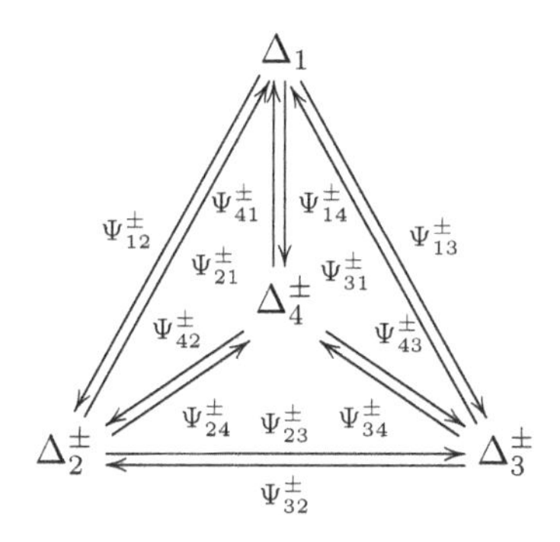

The above commutative diagram has a similar structure to a religious picture in Buddhism called "mandala". For this reason, it is called *mandala of Legendrian dualities* in [Chen and Izumiya (2009)].

9.11 Spacelike surfaces in the lightcone

The induced metric on the lightcone is degenerate, so one cannot apply directly the results on differential geometry of Riemannian and Lorentzian surfaces to surfaces in the lightcone. We construct here the basic tools for the study of the extrinsic differential geometry of spacelike surfaces in the lightcone LC^* as an application of the mandala of Legendrian dualities.

Let M be a spacelike surface patch in the lightcone parametrised by $\mathbf{x} : U \to LC^*$. We shall show the existence and uniqueness of the lightcone normal vector to M as a consequence of Theorem 9.5.

Consider the double Legendrian fibration $\pi_{41} : \Delta_4 \to LC^*$, $\pi_{42} : \Delta_4 \to LC^*$ and let $\mathbf{v} \in LC^*$, where we set $\Delta_4 = \Delta_4^-, \pi_{41} = \pi_{41}^-, \pi_{42} = \pi_{42}^-$. The fibre of $\pi_{41}^{-1}(\mathbf{v})$ can be identified with

$$\{\mathbf{w} \in LC^* \mid \langle \mathbf{v}, \mathbf{w} \rangle_1 = -2\},$$

which is the intersection of LC^* with a lightlike hyperplane, so it is a two dimensional spacelike submanifold. For any $p = \mathbf{x}(u) \in M$, the normal space N_pM is a timelike plane, so there exists two lightlike lines on N_pM. One of the lines is generated by $p = \mathbf{x}(u)$. We remark that a lightlike plane consists of lightlike vectors and spacelike vectors only. Moreover, all lightlike vectors are linearly dependent. Therefore, if another lightlike line is generated by $\mathbf{w}$, then we have $\langle \mathbf{w}, \mathbf{x}(u) \rangle_1 = c \neq 0$. If necessary, we consider $\widetilde{\mathbf{w}} = -c\mathbf{w}/2$. Then we have $\langle \widetilde{\mathbf{w}}, \mathbf{x}(u) \rangle_1 = -2$. Therefore, the intersection of $\pi_{41}^{-1}(\mathbf{v})$ with the normal plane of M (a timelike plane) in $\mathbb{R}^4_1$ consists of only one point at each point on M. Since $\pi_{41} : \Delta_4 \to LC^*$ is a Legendrian fibration, there is a Legendrian submanifold parametrised by $\mathcal{L}_4 : U \to \Delta_4$ such that $\pi_{41} \circ \mathcal{L}_4(u) = \mathbf{x}(u)$. It follows that we have a smooth map $\mathbf{x}^\ell : U \to LC^*$ such that $\mathcal{L}_4(u) = (\mathbf{x}(u), \mathbf{x}^\ell(u))$, i.e., $\pi_{42} \circ \mathcal{L}_4 = \mathbf{x}^\ell$. Since $\mathcal{L}_4$ is a Legendrian embedding, we have $\langle d\mathbf{x}(u), \mathbf{x}^\ell(u) \rangle_1 = 0$, so $\mathbf{x}^\ell(u)$ belongs to the normal plane of M at $\mathbf{x}(u)$.

Given another Legendrian embedding $\mathcal{L}_4^1(u) = (\mathbf{x}(u), \mathbf{x}_1^\ell(u))$, we have that $\mathbf{x}^\ell(u)$ and $\mathbf{x}_1^\ell(u)$ are parallel. However, the relation $\langle \mathbf{x}(u), \mathbf{x}^\ell(u) \rangle_1 = \langle \mathbf{x}(u), \mathbf{x}_1^\ell(u) \rangle_1 = -2$ holds, so $\mathbf{x}^\ell(u) = \mathbf{x}_1^\ell(u)$. This means that $\mathcal{L}_4$ is the unique Legendrian lift of $\mathbf{x}$.

Definition 9.6. We call the vector $\mathbf{x}^\ell(u) = \pi_{42} \circ \mathcal{L}_4(u)$ the *lightcone normal vector* to M at $p = \mathbf{x}(u)$. The map $\mathbf{x}^\ell : U \to LC^*$ (or, its image) is called the *lightcone dual* of M. The map $\widetilde{\mathbf{x}}^\ell : U \to S^2_+$, with $\widetilde{\mathbf{x}}^\ell(u) = \pi^L_S(\mathbf{x}^\ell(u))$, is called the *lightcone Gauss map* of M.

We can construct directly the lightcone normal vector $\mathbf{x}^\ell(u)$ as follows. Let $\pi : \mathbb{R}^4_1 \to \mathbb{R}^3_0$ be the canonical projection given by $\pi(x_0, x_1, x_2, x_3) = (x_1, x_2, x_3)$ and denote $\mathbf{r}(u) = \pi \circ \mathbf{x}(u)$. Since $\pi | LC^* : LC^* \to \mathbb{R}^3_0$ is an embedding, $\mathbf{r} : U \to \mathbb{R}^3_0$ is an embedding too. Therefore, we have the ordinary Euclidean unit normal $N(u)$ of $\mathbf{r}(U) = \pi(M)$ and the Euclidean Gauss map $N : U \to S^2 \subset \mathbb{R}^3_0$. We can now define a transversal vector field $\mathbf{r}^\ell(u)$ to M along $\pi(M)$ in $\mathbb{R}^3_0$ by

$$\mathbf{r}^\ell(u) = \frac{\mathbf{r}(u) - 2(\mathbf{r}(u) \cdot N(u))N(u)}{(\mathbf{r}(u) \cdot N(u))^2},$$

where "$\cdot$" is the usual Euclidean scalar product. It follows by the uniqueness of the vector $\mathbf{x}^\ell$ that

$$\mathbf{x}^\ell(u) = (\frac{\|\mathbf{r}(u)\|}{(\mathbf{r}(u) \cdot N(u))^2}, \mathbf{r}^\ell(u)).$$

We can study the extrinsic differential geometry of M using the normal vector field $\mathbf{x}^\ell$. Here too, as in the case of surfaces in the hyperbolic space, the differential of $\mathbf{x}^\ell$ at each point is a linear transformation of T_pM. Indeed,

Lemma 9.5. *For any $p = \mathbf{x}(u) \in M$ and $\mathbf{v} \in T_pM$, we have $D_\mathbf{v}\mathbf{x}^\ell(u) \in T_pM$.*

Proof. The proof is similar to that of Lemma 9.4. We have

$$D_\mathbf{v}\mathbf{x}^\ell = \lambda_1\mathbf{x}_{u_1} + \lambda_2\mathbf{x}_{u_2} + \mu_1\mathbf{x} + \mu_2\mathbf{x}^\ell$$

for some real numbers $\lambda_1, \lambda_2, \mu_1, \mu_2$.

Since $\langle \mathbf{x}^\ell, \mathbf{x}\rangle_1 = -2$, $D_\mathbf{v}\langle \mathbf{x}^\ell, \mathbf{x}\rangle_1 = 0$, so that $\langle D_\mathbf{v}\mathbf{x}^\ell, \mathbf{x}\rangle_1 = -\langle \mathbf{x}^\ell, D_\mathbf{v}\mathbf{x}\rangle_1$. As $D_\mathbf{v}\mathbf{x}$ is a tangent vector and $\mathbf{x}^\ell$ is a normal vector we get $\langle D_\mathbf{v}\mathbf{x}^\ell, \mathbf{x}\rangle_1 = \langle \mathbf{x}^\ell, D_\mathbf{v}\mathbf{x}\rangle_1 = 0$, so $\mu_2 = 0$.

Similarly, $\langle \mathbf{x}^\ell, \mathbf{x}^\ell\rangle_1 = 0$ so $\langle D_\mathbf{v}\mathbf{x}^\ell, \mathbf{x}^\ell\rangle_1 = 0$, and from that we get $\mu_1 = 0$. (We have used in both calculations above the fact that the vectors $\mathbf{x}_{u_i}$, $i = 1, 2$, are pseudo orthogonal to $\mathbf{x}$ and $\mathbf{x}^\ell$.) It follows then that $D_\mathbf{v}\mathbf{x}^\ell = \lambda_1\mathbf{x}_{u_1} + \lambda_2\mathbf{x}_{u_2} \in T_pM$. □

We define the lightcone shape operator $W_p^\ell = -d\mathbf{x}^\ell(u) : T_pM \to T_pM$. Following the same steps in the previous sections, we can define its associated principal curvature and the Gauss-Kronecker and mean curvatures. We also obtain a similar Weingarten formula as well as the formulae for the Gauss-Kronecker and mean curvatures.

Taking into account the fact that the defining equation for Δ_4 is $\langle \mathbf{v}, \mathbf{w}\rangle_1 = -2$, we define the family of height functions $H : U \times LC^* \to \mathbb{R}$ on M by

$$H(u, \mathbf{v}) = h_\mathbf{v}(u) = \langle \mathbf{x}(u), \mathbf{v}\rangle_1 + 2$$

and call it the *family of lightcone height functions* on M.

One can obtain similar results to those in the previous sections about the family H and derive accordingly information about the extrinsic geometry of M in LC^*.

Remark 9.10. The study of surfaces in hyperbolic space in §9.8.4 can also be carried out using the duality Theorem 9.5. For any regular surface patch parametrised by $\mathbf{x} : U \to H^3(-1)$, we have the lightcone Gauss map $\mathbb{L} : U \to LC^*$. By definition, we have a Legendrian embedding $\mathcal{L}_2 : U \to \Delta_2$ given by $\mathcal{L}_2(u) = (\mathbf{x}(u), \mathbb{L}(u))$. Since $\pi_{22} : \Delta_2 \to LC^*$ is a Legendrian fibration, $\mathbb{L} = \pi_{22} \circ \mathcal{L}_2$ is a Legendrian map. All the results in §9.8.4 (and in [Izumiya, Pei and Sano (2003)]) can be interpreted using this construction.

9.11.1 *The Lightcone Theorema Egregium*

The duality Theorem 9.5 can be used to obtain other curvatures of spacelike surfaces in the lightcone. Let M be a spacelike surface patch in the lightcone parametrised by $\mathbf{x} : U \to LC^*$. The map $\Phi_{41} : \Delta_4 \to \Delta_1$, given by

$$\Phi_{41}(\mathbf{v}, \mathbf{w}) = \left(\frac{\mathbf{v} + \mathbf{w}}{2}, \frac{\mathbf{v} - \mathbf{w}}{2} \right)$$

is a diffeomorphism and

$$(\Phi_{41})^* \theta_{12} = \langle d\frac{\mathbf{v} + \mathbf{w}}{2}, \frac{\mathbf{v} - \mathbf{w}}{2} \rangle_1 |_{\Delta_4} = -\frac{1}{4} \langle d\mathbf{v}, \mathbf{w} \rangle_1 + \frac{1}{4} \langle d\mathbf{w}, \mathbf{v} \rangle_1 |_{\Delta_4} = \frac{1}{2} \theta_{42}.$$

Therefore, Φ_{41} is a contactomorphism. It follows that the map $\mathscr{L}_1 : U \to \Delta_1$, given by

$$\mathscr{L}_1(u) = \Phi_{41} \circ \mathscr{L}_4(u),$$

is a Legendrian embedding, where $\mathscr{L}_4$ is as in §9.11. If we write $\mathscr{L}_1(u) = (\mathbf{x}^h(u), \mathbf{x}^d(u))$, then

$$\mathbf{x}^h(u) = \frac{\mathbf{x}(u) + \mathbf{x}^\ell(u)}{2}, \quad \mathbf{x}^d(u) = \frac{\mathbf{x}(u) - \mathbf{x}^\ell(u)}{2}.$$

Definition 9.7. We call $\mathbf{x}^h(u)$ (resp. $\mathbf{x}^d(u)$) the *hyperbolic normal vector* (resp. *de Sitter normal vector*) of M at $p = \mathbf{x}(u)$.

As a consequence of Lemma 9.5, we have the *hyperbolic shape operator*

$$W_p^h = -d\mathbf{x}^h(u) : T_pM \to T_pM$$

and the *de Sitter shape operator*

$$W_p^d = -d\mathbf{x}^d(u) : T_pM \to T_pM$$

at $p = \mathbf{x}(u)$. From these, we define the following curvatures of M at p:

$$\begin{aligned}
\textit{Hyperbolic Gauss-Kronecker curvature}: K_h(u) &= \det(W_p^h)\\
\textit{de Sitter Gauss-Kronecker curvature}: K_d(u) &= \det(W_p^d)\\
\textit{Hyperbolic mean curvature}: H_h(u) &= \tfrac{1}{2}\mathrm{Trace}(W_p^h)\\
\textit{de Sitter mean curvature}: H_d(u) &= \tfrac{1}{2}\mathrm{Trace}(W_p^d).
\end{aligned}$$

Let l, m, n denote the coefficients of the second fundamental form associated to $\mathbf{x}^\ell$, i.e., $l = -\langle \mathbf{x}^\ell_{u_1}, \mathbf{x}^\ell_{u_1}\rangle_1, m = -\langle \mathbf{x}^\ell_{u_1}, \mathbf{x}^\ell_{u_2}\rangle_1, n = -\langle \mathbf{x}^\ell_{u_2}, \mathbf{x}^\ell_{u_2}\rangle_1$.

Proposition 9.17. *We have the following formulae:*

$$K_h = \frac{(E-l)(G-n)-(F-m)^2}{4(EG-F^2)},$$

$$K_d = \frac{(E+l)(G+n)-(F+m)^2}{4(EG-F^2)}.$$

Proof. We denote by

$$A_p^\ell = \begin{pmatrix} a^\ell_{11} & a^\ell_{12}\\ a^\ell_{21} & a^\ell_{22}\end{pmatrix}$$

the matrix of the shape operator $W_p^\ell = -d\mathbf{x}^\ell(u)$ with respect to the basis $\{\mathbf{x}_{u_1}, \mathbf{x}_{u_2}\}$ of T_pM. Following similar arguments in the proof of Proposition 9.2, we get a similar formula to (9.2), namely,

$$\begin{pmatrix} E & F\\ F & G\end{pmatrix} A_p^\ell = \begin{pmatrix} l & m\\ m & n\end{pmatrix}.$$

Since $\mathbf{x}^h = (\mathbf{x}+\mathbf{x}^\ell)/2$, we have

$$\begin{aligned}
\mathbf{x}^h_{u_1} &= \tfrac{1}{2}\left((1-a^\ell_{11})\mathbf{x}_{u_1} - a^\ell_{21}\mathbf{x}_{u_2}\right),\\
\mathbf{x}^h_{u_2} &= \tfrac{1}{2}\left(-a^\ell_{12}\mathbf{x}_{u_1} + (1-a^\ell_{22})\mathbf{x}_{u_2}\right)
\end{aligned}$$

so the matrix of $d\mathbf{x}^h$ with respect to the basis $\{\mathbf{x}_{u_1}, \mathbf{x}_{u_2}\}$ of T_pM is

$$\begin{aligned}
\tfrac{1}{2}\begin{pmatrix} 1-a^\ell_{11} & -a^\ell_{12}\\ -a^\ell_{21} & 1-a^\ell_{22}\end{pmatrix} &= \tfrac{1}{2}\left(\begin{pmatrix} 1 & 0\\ 0 & 1\end{pmatrix} - \begin{pmatrix} a^\ell_{11} & a^\ell_{12}\\ a^\ell_{21} & a^\ell_{22}\end{pmatrix}\right)\\
&= \tfrac{1}{2}\begin{pmatrix} E & F\\ F & G\end{pmatrix}^{-1}\left(\begin{pmatrix} E & F\\ F & G\end{pmatrix} - \begin{pmatrix} E & F\\ F & G\end{pmatrix}\begin{pmatrix} a^\ell_{11} & a^\ell_{12}\\ a^\ell_{21} & a^\ell_{22}\end{pmatrix}\right)\\
&= \tfrac{1}{2}\begin{pmatrix} E & F\\ F & G\end{pmatrix}^{-1}\left(\begin{pmatrix} E & F\\ F & G\end{pmatrix} - \begin{pmatrix} l & m\\ m & n\end{pmatrix}\right)\\
&= \tfrac{1}{2}\begin{pmatrix} E & F\\ F & G\end{pmatrix}^{-1}\begin{pmatrix} E-l & F-m\\ F-m & G-n\end{pmatrix}.
\end{aligned}$$

The curvature K_h is the determinant of the matrix in the left hand side of the equality above, and the result follows by taking the determinant of the right hand side of that equality.

The formula for K_d follows the same steps as above with an appropriate change of signs as $\mathbf{x}^d = (\mathbf{x} - \mathbf{x}^\ell)/2$. □

We denote, as usual, by Γ^k_{ij} the Christoffel symbols of M (which is a Riemannian manifold), where

$$\Gamma^k_{ij} = \frac{1}{2}\sum_{m=1}^{2} g^{km}\left\{\frac{\partial g_{jm}}{\partial u_i} + \frac{\partial g_{im}}{\partial u_j} - \frac{\partial g_{ij}}{\partial u_m}\right\}.$$

Above, (g_{ij}) is the matrix of the first fundamental form and (g^{km}) is its inverse matrix. Using the notation E, F, G for the coefficients of the first fundamental form, the Christoffel symbols are given by the following six functions:

$$\Gamma^1_{11} = \frac{GE_{u_1} - 2FF_{u_1} + FE_{u_2}}{2(EG - F^2)},\ \Gamma^2_{11} = \frac{2EF_{u_1} - EE_{u_2} - FE_{u_1}}{2(EG - F^2)},$$
$$\Gamma^1_{12} = \Gamma^1_{21} = \frac{GE_{u_2} - FG_{u_1}}{2(EG - F^2)},\ \Gamma^2_{12} = \Gamma^2_{21} = \frac{EG_{u_1} - FE_{u_2}}{2(EG - F^2)},$$
$$\Gamma^1_{22} = \frac{2GF_{u_2} - GG_{u_1} - FG_{u_2}}{2(EG - F^2)},\ \Gamma^2_{22} = \frac{EG_{u_2} - 2FF_{u_2} + FG_{u_1}}{2(EG - F^2)}.$$

Proposition 9.18. *Let M be a spacelike surface patch in the lightcone parametrised by $\mathbf{x} : U \to LC^*$ and let l, m, n denote the coefficients of the second fundamental form associated to $\mathbf{x}^\ell$. Then the lightcone Gauss equations are given by*

$$\begin{aligned}
\mathbf{x}_{u_1u_1} &= \Gamma^1_{11}\mathbf{x}_{u_1} + \Gamma^2_{11}\mathbf{x}_{u_2} - \tfrac{1}{2}(l\mathbf{x} + E\mathbf{x}^\ell),\\
\mathbf{x}_{u_1u_2} &= \Gamma^1_{12}\mathbf{x}_{u_1} + \Gamma^2_{12}\mathbf{x}_{u_2} - \tfrac{1}{2}(m\mathbf{x} + F\mathbf{x}^\ell),\\
\mathbf{x}_{u_2u_2} &= \Gamma^1_{22}\mathbf{x}_{u_1} + \Gamma^2_{22}\mathbf{x}_{u_2} - \tfrac{1}{2}(n\mathbf{x} + G\mathbf{x}^\ell).
\end{aligned}$$

Proof. Since $\{\mathbf{x}_{u_1}, \mathbf{x}_{u_2}, \mathbf{x}, \mathbf{x}^\ell\}$ is a basis of $\mathbb{R}^4_1$, we can write

$$\mathbf{x}_{u_iu_j} = \Lambda^1_{ij}\mathbf{x}_{u_1} + \Lambda^2_{ij}\mathbf{x}_{u_2} + \lambda_{ij}\mathbf{x} + \mu_{ij}\mathbf{x}^\ell,$$

for some scalars Λ^k_{ij}, λ_{ij} and μ_{ij}. From $\langle\mathbf{x}, \mathbf{x}\rangle_1 = 0$ we get $\langle\mathbf{x}_{u_1}, \mathbf{x}\rangle_1 = 0$ and differentiating again we get $\langle\mathbf{x}_{u_1u_1}, \mathbf{x}\rangle_1 = -\langle\mathbf{x}_{u_1}, \mathbf{x}_{u_1}\rangle_1 = -E$. But $\langle\mathbf{x}_{u_1u_1}, \mathbf{x}\rangle_1 = \mu_{11}\langle\mathbf{x}^\ell, \mathbf{x}\rangle_1 = -2\mu_{11}$, so $\mu_{11} = -(1/2)E$.

By definition $\langle\mathbf{x}_{u_1u_1}, \mathbf{x}^\ell\rangle_1 = l$, hence $-2\lambda_{11} = l$, so that $\lambda_{11} = -(1/2)l$. It follows that

$$\mathbf{x}_{u_1u_1} = \Lambda^1_{11}\mathbf{x}_{u_1} + \Lambda^2_{11}\mathbf{x}_{u_2} - \frac{1}{2}(l\mathbf{x} + E\mathbf{x}^\ell).$$

The expressions for $\mathbf{x}_{u_1u_2}$ and $\mathbf{x}_{u_2u_2}$ follow similarly. Moreover, we have

$$\frac{1}{2}E_{u_1} = \langle \mathbf{x}_{u_1u_1}, \mathbf{x}_{u_1}\rangle_1 = \Lambda^1_{11}E + \Lambda^2_{11}F.$$

By the similar arguments to the above, we have

$$\begin{aligned} \frac{1}{2}E_{u_2} &= \Lambda^1_{12}E + \Lambda^2_{12}F, \ \frac{1}{2}G_{u_1} = \Lambda^1_{12}F + \Lambda^2_{12}G, \\ F_{u_2} - \frac{1}{2}G_{u_1} &= \Lambda^1_{22}E + \Lambda^2_{22}F, \ \frac{1}{2}G_{u_2} = \Lambda^1_{22}F + \Lambda^2_{22}G. \end{aligned}$$

It follows that

$$\begin{pmatrix} E & F \\ F & G \end{pmatrix} \begin{pmatrix} \Lambda^1_{11} & \Lambda^1_{12} & \Lambda^1_{22} \\ \Lambda^2_{11} & \Lambda^2_{12} & \Lambda^2_{22} \end{pmatrix} = \frac{1}{2} \begin{pmatrix} E_{u_1} & E_{u_2} & 2F_{u_2} - G_{u_1} \\ 2F_{u_1} - E_{u_2} & G_{u_1} & G_{u_2} \end{pmatrix},$$

which gives $\Lambda^k_{ij} = \Gamma^k_{ij}$. □

Now from $\mathbf{x}^h = (\mathbf{x}+\mathbf{x}^\ell)/2$ and $\mathbf{x}^d = (\mathbf{x}-\mathbf{x}^\ell)/2$ we get $\mathbf{x} = \mathbf{x}^h + \mathbf{x}^d$ and $\mathbf{x}^\ell = \mathbf{x}^h - \mathbf{x}^d$. Substituting these expressions in Propostion 9.18 yields the following corollary.

Corollary 9.10. *Let M be a spacelike surface patch in the lightcone parametrised by $\mathbf{x} : U \to LC^*$. Then,*

$$\begin{aligned} \mathbf{x}_{u_1u_1} &= \Gamma^1_{11}\mathbf{x}_{u_1} + \Gamma^2_{11}\mathbf{x}_{u_2} - \tfrac{1}{2}(l+E)\mathbf{x}^h - \tfrac{1}{2}(l-E)\mathbf{x}^d, \\ \mathbf{x}_{u_1u_2} &= \Gamma^1_{12}\mathbf{x}_{u_1} + \Gamma^2_{12}\mathbf{x}_{u_2} - \tfrac{1}{2}(m+F)\mathbf{x}^h - \tfrac{1}{2}(m-F)\mathbf{x}^d, \\ \mathbf{x}_{u_2u_2} &= \Gamma^1_{22}\mathbf{x}_{u_1} + \Gamma^2_{22}\mathbf{x}_{u_2} - \tfrac{1}{2}(n+G)\mathbf{x}^h - \tfrac{1}{2}(n-G)\mathbf{x}^d. \end{aligned}$$

Let K_s denote the sectional curvature of M. It is known that

$$\begin{aligned} K_s = {} & \frac{E(E_{u_2}G_{u_2} - 2F_{u_1}G_{u_2} + G^2_{u_1})}{4(EG-F^2)^2} \\ & + \frac{F(E_{u_1}G_{u_2} - E_{u_2}G_{u_1} - 2E_{u_2}F_{u_2} - 2F_{u_1}G_{u_1} + 4F_{u_1}F_{u_2})}{4(EG-F^2)^2} \\ & + \frac{G(E_{u_1}G_{u_1} - 2E_{u_1}F_{u_2} + E^2_{u_2})}{4(EG-F^2)^2} - \frac{E_{u_2u_2} - 2F_{u_1u_2} + G_{u_1u_1}}{2(EG-F^2)}. \end{aligned}$$

It follows that K_s depends only on the first fundamental form, which means that is an intrinsic property of the surface M.

Proposition 9.19. *Let M be a spacelike surface in the lightcone. Then,*

$$K_h - K_d = K_s.$$

Proof. We have $(\langle \mathbf{x}_{u_1u_1}, \mathbf{x}_{u_2}\rangle_1)_{u_2} = \langle \mathbf{x}_{u_1u_1u_2}, \mathbf{x}_{u_2}\rangle_1 + \langle \mathbf{x}_{u_1u_1}, \mathbf{x}_{u_2u_2}\rangle_1$. By the relations $\Lambda^k_{ij} = \Gamma^k_{ij}$ in the proof of Proposition 9.18, we have

$$\begin{aligned}\langle \mathbf{x}_{u_1u_1u_2}, \mathbf{x}_{u_2}\rangle_1 &= \left(F_{u_1} - \frac{1}{2}E_{u_2}\right)_{u_2} - \frac{1}{2}E_{u_1}\Gamma^1_{22} - \left(F_{u_1} - \frac{1}{2}E_{u_2}\right)\Gamma^2_{22} \\ &\quad + \frac{1}{4}(\ell + E)(n+G) - \frac{1}{4}(\ell - E)(n-G).\end{aligned}$$

We also have

$$\begin{aligned}\langle \mathbf{x}_{u_1u_2u_1}, \mathbf{x}_{u_2}\rangle_1 &= \frac{1}{2}G_{u_1u_1} - \frac{1}{2}E_{u_2}\Gamma^1_{12} - \frac{1}{2}G_{u_1}\Gamma^2_{12} \\ &\quad + \frac{1}{4}(m+F)^2 - \frac{1}{4}(m-F)^2.\end{aligned}$$

Since $\mathbf{x}_{u_1u_1u_2} = \mathbf{x}_{u_1u_2u_1}$, we have

$$\begin{aligned}\left(F_{u_1} - \frac{1}{2}E_{u_2}\right)_{u_2} &- \frac{1}{2}E_{u_1}\Gamma^1_{22} - \left(F_{u_1} - \frac{1}{2}E_{u_2}\right)\Gamma^2_{22} \\ &- \frac{1}{2}G_{u_1u_1} + \frac{1}{2}E_{u_2}\Gamma^1_{12} + \frac{1}{2}G_{u_1}\Gamma^2_{12} \\ &= -\frac{1}{4}(\ell + E)(n+G) + \frac{1}{4}(\ell - E)(n-G) \\ &\quad + \frac{1}{4}(m+F)^2 - \frac{1}{4}(m-F)^2.\end{aligned}$$

If we substitute the Christoffel symbols Γ^k_{ij} by their expressions in terms of the coefficients of the first fundamental form in the above equality, we get the left hand side of the equality equal to $(EG - F^2)K_s$. Therefore we have

$$\begin{aligned}K_s &= \frac{((\ell - E)(n-G) - (m-F)^2 - (\ell + E)(n+G) + (m+F)^2)}{4(EG - F^2)} \\ &= K_h - K_d.\end{aligned}$$

□

The classical "Theorema Egregium of Gauss" asserts that the Gaussian curvature of a surface in Euclidean space is equal to its sectional curvature. For a spacelike surface in the lightcone, we have the following result which asserts that the lightcone mean curvature is equal to the sectional curvature of M. In particular, the lightcone mean curvature is an intrinsic property of M.

Theorem 9.6 (The Lightcone Theorema Egregium). *For a spacelike surface M in the lightcone LC^*, we have*

$$-K_s = K_d - K_h = H_\ell = H_h - H_d.$$

Proof. Let $p = \mathbf{x}(u)$ be a point on M. We denote the eigenvalues of W_p^ℓ by $\kappa_1^\ell(u)$ and $\kappa_2^\ell(u)$, those of W_p^h by $\kappa_1^h(u)$ and $\kappa_2^h(u)$, and those of W_p^d by $\kappa_1^d(u)$ and $\kappa_2^d(u)$.

From the equality $W_p^\ell = W_p^h - W_p^d$, we get $\kappa_i^\ell(u) = \kappa_i^h(u) - \kappa_i^d(u)$, for $i = 1, 2$. It follows that $H_\ell(u) = H_h(u) - H_d(u)$.

On the other hand, we have

$$\mathbf{x}^h(u) = \frac{\mathbf{x}(u) + \mathbf{x}^\ell(u)}{2}, \quad \mathbf{x}^d(u) = \frac{\mathbf{x}(u) - \mathbf{x}^\ell(u)}{2},$$

so that

$$W_p^h = \frac{1}{2}(-1_{T_pM} + W_p^\ell), \quad W_p^d = \frac{1}{2}(-1_{T_pM} - W_p^\ell).$$

Therefore, we have for $i = 1, 2$,

$$\kappa_i^h(u) = \frac{1}{2}(-1 + \kappa_i^\ell(u)), \quad \kappa_i^d(u) = \frac{1}{2}(-1 - \kappa_i^\ell(u)),$$

and from these equalities we get

$$K_h(u) = \kappa_1^h(u)\kappa_2^h(u) = \frac{1}{4}(1 - 2H_\ell(u) + K_\ell(u)),$$
$$K_d(u) = \kappa_1^d(u)\kappa_2^d(u) = \frac{1}{4}(1 + 2H_\ell(u) + K_\ell(u)).$$

It follows that

$$K_d(u) - K_h(u) = H_\ell(u).$$

The relation $-K_s(u) = K_d(u) - K_h(u)$ is the result in Proposition 9.19. □

9.12 Notes

The lightlike geometry in the hyperbolic space is called Horospherical Geometry and describes the contact of submanifolds in Hyperbolic space with hyperhorospheres. The basic study of the horospherical geometry for hypersurfaces in Hyperbolic space with general codimension is carried out in [Izumiya, Pei and Sano (2003)]. A detailed study of the contact of surfaces in Hyperbolic 4-space with hyperhorospheres is carried out in [Izumiya, Pei and Romero-Fuster (2006)] via the singularities of the lightcone height functions. That led to the introduction of the concepts of osculating hyperhorospheres, horobinormals, horoasymptotic directions and horospherical points and to certain conditions ensuring their existence.

Kasedou [Kasedou (2010a,b)] studied spacelike submanifolds immersed in codimension higher than one in de Sitter space. The results are analogous to those of submanifolds immersed in Hyperbolic space studied in [Izumiya, Pei, Romero Fuster and Takahashi (2005)]. Kasedou introduced the notion of horospherical hypersurface and spacelike canal hypersurface by using timelike unit normal vector fields, and showed that the horospherical hypersurface of a spacelike submanifold is the wavefront set of horospherical height functions. The singularities of the lightcone Gauss map and the lightcone Gauss image of hypersurfaces in de Sitter space are investigated in [Kasedou (2009)]

The study of the lightlike geometry of codimension two submanifolds of a Lorentz-Minkowski space ([Izumiya and Romero Fuster (2007)] for the general dimension case) is generalised in [Izumiya and Kasedou (2014)] to higher codimension submanifolds. This is done by introducing the notion of codimension two spacelike canal submanifold and applying the results presented in this chapter or [Izumiya and Romero Fuster (2007)].

The behaviour of lightlike hypersurfaces in Lorentz-Minkowski space is investigated in [Izumiya, Romero Fuster and Saji (2009)] where a concept of flatness for lightlike hypersurfaces is introduced. The flat lightlike hypersurfaces are characterised as envelopes of certain families of lightlike hyperplanes. Their generic singularities are the suspended cuspidal edge, the suspended swallowtail, the suspended cuspidal cross-cap and the A_4-type hypersurface singularity. It is interesting to notice that the $D_4^{\pm}$-type singularities do not appear generically as singularities of flat lightlike hypersurfaces, whereas the suspended cuspidal cross-cap does not appear as a generic singularity in the general case of lightlike hypersurfaces (Corollary 9.9).

Surfaces in the Minkowski 3-space are also studied using the singularitry theory approach exposed in this book. The induced pseudo-metric on any closed surface in $\mathbb{R}^3_1$ degenerates at some point on the surface. We call the locus of such points *the locus of degeneracy of the metric* (LD). In general, the LD is a smooth curve and separates locally the surface into a spacelike and a timelike region. Very interesting problems arrise when trying to understand how the Riemannian geometry on one side and the Lorentzian geometry on the other side meet on the LD (and in general, on surfaces with degenerate metrics, see for example, [Genin, Khesin and Tabachnikov (2007); Kossowski (1987); Pelletier (1995); Tabachnikov (1997); Remizov (2009); Steller (2006)].) Here, the contact theory can be very usefull. Contact with lines and planes does not depend on the metric, so the results in Chapter 6 apply. When considering the contact with lines (i.e., projections of the surface to planes), we can restrict to projecting along the lightlike directions. These can be parametrised by a circle on the lightcone. Varying the lightlike direction of projections gives a 1-parameter family of contour generators and apparent contours. It is shown in [Izumiya and Tari (2013)], amongst other things, that the families of contour generators and apparent contours are solutions of certain first order ordinary differential equations. The caustic of a surface in $\mathbb{R}^3_1$ can also be defined without the use of the metric. In [Tari (2012)] is studied the behaviour of the induced metric on the caustic, including at the LD of the surface. (For caustics of plane curves, see [Saloom and Tari (2012)].) The lines of principal curvatures and the way they extend across the LD are studied in [Izumiya and Tari (2010c)].

Chapter 10

Global viewpoint

In the previous chapters we used singularity theory techniques to study a submanifold M of Euclidean space and of Minkowski space-time locally at a given point on M. In this chapter, we illustrate how those techniques can be used to obtain *global* results on M (which we shall assume to be closed, i.e., compact without boundary). The approaches we adopt here are the following.

The first uses a stratification of the parameters space of the family of functions and mappings defined on M. The parameters space can be stratified according to the singularity types of the different functions associated to each parameter. In fact, these stratifications can be regarded as pull-backs of convenient stratifications on the space of smooth functions (or mappings) on M [Gibson, Wirthmüller, du Plessis and Looijenga (1976)]. The multi-transversality conditions imposed by the genericity theorems on these families provide all the information we need on the behaviour of strata in the parameter spaces. Then, appropriate global topological considerations on the induced strata lead to the desired results. This allows us to obtain, for instance, relations between different geometric invariants associated to M.

The second approach is a topological one. We compute the Euler characteristic of a surface (or a given geometric set such as the wavefront) either in terms of the total curvatures defined in the previous chapters or in terms of the number of certain stable singularity types of the members of the families of functions and maps on the surface.

The third approach gravitates around the Poincaré-Hopf formula. We have on the surface (or part of it) direction fields defined by the contact directions associated to the families of functions and mappings defined on M. The integral curves of these fields define foliations whose singular points are of interest. (For example, the umbilic points on a surface in $\mathbb{R}^3$ are the singular points of the integral curves of the principal directions, which are the contact directions of the distance squared functions on the surface.) One can use topological arguments, such as the Poincaré-Hopf formula, to obtain lower bounds for the number of such points.

We outline several applications of the above approaches on surfaces M in $\mathbb{R}^3$ and $\mathbb{R}^4$ (§10.1). We also consider the case of spacelike surfaces in Minkowski space-time $\mathbb{R}^4_1$ (§10.2). We comment in §10.3 on some other global results on submanifolds of Euclidean and Minkowski spaces.

We emphasise that our aim in this chapter is to give some applications of singularity theory to the study of global properties of submanifolds, related mainly to the work of the authors on the subject.

10.1 Submanifolds of Euclidean space

10.1.1 *Surfaces in* $\mathbb{R}^3$

Definition 10.1. The *convex hull* $\mathcal{H}(S)$ of a subset S in $\mathbb{R}^n$ is the intersection of all the convex subsets of $\mathbb{R}^n$ containing S, that is the minimal convex subset of $\mathbb{R}^n$ that contains S.

The convex hull $\mathcal{H}(M)$ of a surface in $\mathbb{R}^3$ is homeomorphic to a closed 3-disc and its boundary $H(M)$ is a C^1-surface which is C^1-diffeomorphic to the standard 2-sphere S^2 (see for example [Romero Fuster (1981)]).

Definition 10.2. A closed surface M in $\mathbb{R}^3$ is said to be *convex* if it coincides with the boundary $H(M)$ of its convex hull. Points on the surface M that lie on $H(M)$ are called *external points*. Other points on M are called *internal points*.

It is a well established geometrical property that a surface M is convex if and only if its Gaussian curvature is non-negative.

A point $p \in M$ is an external point if and only if it is an absolute minimum of some height function $h_{\mathbf{v}}$, $\mathbf{v} \in S^2$, on M. Then $\mathbf{v}$ is, up to a sign, the unit normal vector to M at p.

Definition 10.3. A *stratification* of a subset S of a manifold M is a locally finite partition $\mathcal{S}$ of S into locally closed submanifolds of M, called *strata*. The pair $(S, \mathcal{S})$ is said to be a *stratified subset* of M.

We can view a finite plane graph as a stratified set. The vertices and edges being respectively the 0- and 1-dimensional strata. Moreover, if this graph lies on a surface, we can view the whole surface as a 2-dimensional stratified set whose 2-dimensional strata are the connected components of the complement of the graph.

Any embedding f of a surface M in $\mathbb{R}^3$ induces a stratification of the 2-sphere. This stratification is constructed according to the type of the absolute minima of the family of height functions on the embedded surface M. Provided that the embedding f is height functions generic, the multi-transversality conditions on the height functions family imply that we only have the following possibilities for the absolute minima of the height functions on the surface M (a detailed definition for the general case of hypersurfaces in $\mathbb{R}^n$ can be found in [Romero Fuster (1983)]):

(1) *Morse strata*: The absolute minimum of $h_{\mathbf{v}}$ is attained at a unique point of type A_1. We say that $\mathbf{v} \in S^2$ is of type A_1. All points of type A_1 form an open region in S^2 whose complement is a graph, called the *Maxwell subset* associated to the embedding f.
(2) *Conflict strata*: There are $k = 2$ or $k = 3$ absolute minima of $h_{\mathbf{v}}$, each one of type A_1. Then $h_{\mathbf{v}}$ lies in a codimension $k - 1$ stratum of the Maxwell subset in S^2.
(3) *Bifurcation strata*: The absolute minimum of $h_{\mathbf{v}}$ is attained at a unique point of type A_3. The function $h_{\mathbf{v}}$ lies in a codimension 2 stratum of the Maxwell subset of f in S^2.

We call the above the *Maxwell stratification of S^2 associated to the embedding f*. It is shown in [Romero Fuster (1983)] that the strata of the Maxwell stratification is a Whitney regular stratification satisfying the frontier condition. Moreover, we have the following relation

$$\chi(S^2 - \mathcal{M}) + \sum_{j=0}^{3}(-1)^j(\chi(B_j) + \chi(M_j) + \chi(C_j)) = 2 \qquad (10.1)$$

where $\mathcal{M}$ is the union of strata of codimension ≥ 1 in the Maxwell stratification and B_j, M_j and C_j are, respectively, the union of bifurcation, mixed and conflict strata of codimension j, and $\chi(X)$ denotes the Euler characteristic of the set X.

From the characterisation of the cusp points of the Gauss map in terms of height functions, it follows that the extremal points of the Maxwell graph (bifurcation strata of type A_2) correspond to external cusps of the Gauss map (i.e., cusps of the Gauss map lying on the boundary of the convex hull of the surface). The other vertices of the graph (conflict strata of type $A_1A_1A_1$) are of degree 3 (i.e., they are the end points of exactly 3 edges) and correspond to the (isolated) tri-tangent support planes of the surface. The edges of the graph (conflict strata of type A_1A_1) correspond to the normal directions to the 1-parameter family of support bi-tangent planes of the surface.

Applying equality (10.1) to surfaces in $\mathbb{R}^3$ leads to the following result.

Corollary 10.1 ([Romero Fuster (1988)]). *Given a height function generic surface M in $\mathbb{R}^3$, the numbers C of external cusps of Gauss and T of tri-tangent support planes of M satisfy*

$$C - T = 4 - 2\chi(M \cap H(M)).$$

It follows from Corollary 10.1 that the existence of support tri-tangent planes implies the existence of external cusps of the Gauss map.

The above considerations can be applied to the canal surface of a closed space curve γ. When the radius of the circle of intersection of the canal surface with the normal plane to the curve is small enough, the canal surface is a torus embedded in $\mathbb{R}^3$. One can show that the Maxwell graph of the family of height functions on γ coincides with the Maxwell graph of the canal surface. Comparing the singularities of the height function on the curve and on its canal surface leads to the following result.

Theorem 10.1 ([Romero Fuster (1988)]). *Given a closed curve γ generically immersed in $\mathbb{R}^3$, denote respectively by C, T and ρ the numbers of external torsion zero points, support tri-tangent planes and connected components of $\gamma - \gamma \cap H(\gamma)$. Then the following relation holds:*

$$C - T = 4 - 2\rho.$$

A consequence of the above theorem is the following result for *convex curves*, i.e. closed curves lying on the boundary of their convex hull. This is known as the 4-*vertex theorem for closed curves* in $\mathbb{R}^3$.

Corollary 10.2. *Any convex closed curve generically immersed in $\mathbb{R}^3$ has at least 4 torsion zero points.*

An extension of 4-vertex theorem for closed curves in $\mathbb{R}^3$ to (non necessarily generic) convex closed curves in $\mathbb{R}^3$ with no zero curvature points is given in [Sedykh (1992)]. Other generalisations to curves with isolated points of zero curvature and to singular curves are obtained in [Romero Fuster and Sedykh (1995); Costa and Romero Fuster (1997); Romero Fuster and Sedykh (1997)].

10.1.2 *Wavefronts*

For a Legendrian fibration $\pi : E \to N$, where $\dim N = 3$, we consider a Legendrian immersion $i : L \to E$. The wavefront $W(L)$ of a generic Legendrian immersion has singularitites of type cuspidal edges (A_2), swallowtails (A_3) and points of transversal self-intersection ($A_1A_1, A_1A_2, A_1A_1A_1$); see [Arnol'd (1990)]. (In Theorem 10.2 and Corollary 10.3 bellow, generic means that the wavefront has only the above local and multi-local singularities.) If L is a closed surface, then the numbers of swallowtails and triple points are finite and satisfy the following relation.

Theorem 10.2 ([Izumiya and Marar (1993)]). *Let $i : L \to E$ be a generic Legendrian immersion of a closed surface. Then,*

$$\chi(W(L)) = \chi(L) + T(L) + \frac{1}{2}S(L),$$

where $T(L)$ is the number of triple points on $W(L)$ and $S(L)$ is the number of swallowtail points.

Proof. We define the following sets:

$$D^2(L) = \overline{\{x \in L \mid \sharp(\pi|_L)^{-1}(x) \geq 2\}},$$
$$D^3(L) = \{x \in D^2(L) \mid \sharp(\pi|_L)^{-1}(\pi(x)) = 3\},$$
$$D^2(L,(2)) = \{x \in D^2(L) \mid \sharp(\pi|_L)^{-1}(\pi(x)) = 1\},$$

where $\overline{X}$ is the topological closure of X. Since $W(L)$ is a generic wavefront, $D^2(L)$ is a union of circles and curves on L with self-intersection, $D^3(L)$ is the inverse image of triple points, and $D^2(L,(2))$ is the set of swallowtails of $\pi|_L$. It follows that these are immersed submanifolds of L with $\dim D^2(L) = 1$ and $\dim D^3(L) = \dim D^2(L,(2)) = 0$.

We consider the following equation

$$\chi(W(L)) = \alpha\chi(L) + \beta\chi(D^2(L)) + \gamma\chi(D^2(L,(2)) + \delta\chi(D^3(L)), \quad (10.2)$$

in the unknown variables $\alpha, \beta, \gamma, \delta$. We solve the equation by a purely combinatorial method. We consider a triangulation T_L of the stratified set $W(L)$ as follows. We start to triangulate $W(L)$ by including the image of $D^2(L,(2))$ and the image of $D^3(L)$ among the vertices of K_L.

After this, we build up the one-skeleton $K_L^{(1)}$ of K_L so that the image of $D^2(L)$ is a sub-complex of $K_L^{(1)}$. We complete our procedure by constructing the two-skeleton $K_L^{(2)}$. Since $\pi|_{D^2(L)}, \pi|_{D^2(L,(2))}, \pi|_{D^3(L)}$ are proper finite to one maps, we can pull back K_L to obtain a triangulation of $L, D^2(L), D^2(L,(2))$ and $D^3(L)$. Let C_j^X be the number of j-cells in X, where $X = W(L), L, D^2(L), D^2(L,(2))$ or $D^3(L)$. Then equation (10.2) can be written as

$$\sum_j (-1)^j C_j^{W(L)} = \alpha \sum_j (-1)^j C\ L_j + \beta \sum_j (-1)^j C_j^{D^2(L)} \\ +\gamma \sum_j (-1)^j C_j^{D^2(L,(2))} + \delta \sum_j (-1)^j C_j^{D^3(L)},$$

where $C_j^X = 0$ for $i > \dim X$. Therefore, if we can find real numbers α, β, γ and δ such that

$$C_j^{W(L)} = \alpha C_j^L + \beta C_j^{D^2(L)} + \gamma C_j^{D^2(L,(2))} + \delta C_j^{D^3(L)},$$

for any j, then we have solutions of equation (10.2). We deal with the case $j = 0$. We remark that $\pi|_L$ is three-to-one over the points of $\pi(D^3(L))$, one-to-one over the points of $\pi(D^2(L,(2)))$, two-to-one over the points of $\pi(D^2(L) \setminus (D^2(L,(2)) \cup D^3(L)))$, and one-to-one over the points of $\pi(N \setminus D^2(L))$. It follows that equation

$$C_0^{W(L)} = \alpha C_0^L + \beta C_0^{D^2(L)} + \gamma C_0^{D^2(L,(2))} + \delta C_0^{D^3(L)}$$

is equivalent to the system of linear equations

$$\begin{pmatrix} 1 \\ 1 \\ 1 \\ 1 \end{pmatrix} = \begin{pmatrix} 1\,0\,0\,0 \\ 2\,2\,0\,0 \\ 1\,1\,1\,0 \\ 3\,3\,0\,3 \end{pmatrix} \begin{pmatrix} \alpha \\ \beta \\ \gamma \\ \delta \end{pmatrix}.$$

Solving the above linear system gives $\alpha = 1, \beta = -1/2, \gamma = 1/2$ and $\delta = -1/6$. Thus, we have

$$\chi(W(L)) = \chi(L) - \frac{1}{2}\chi(D^2(L)) + \frac{1}{2}\chi(D^2(L,(2))) - \frac{1}{6}\chi(D^3(L)). \quad (10.3)$$

We have, by definition, $\chi(D^2(L,(2)) = S(L)$ and $\chi(D^3(L)) = 3T(L)$. Since $D^2(L)$ is the union of closed curves on the surface L with $3T(L)$ crossings, we can triangulate it with $(3T(L) + n)$ 0-cells and $(6T(L) + n)$ 1-cells, where n is the number of circles in $D^2(L)$. We get the desired result by substituting these in (10.3). □

Remark 10.1. We have a similar formula to that in Theorem 10.2 when considering cross-caps or cuspidal cross-caps instead of swallowtails [Izumiya and Marar (1993)]. The reason is that these points are locally homeomorphic to a swallowtail. Further generalisations of Theorem 10.2 can be found in [Sedykh (2012); Nuño Ballesteros and Saeki (2001); Houston (1999); Kossowski (2007)].

We can apply Theorem 10.2 to closed surfaces in $\mathbb{R}^3$. For an orientable surface $M \subset \mathbb{R}^3$, we have the global Gauss map $N : M \to S^2$ and the corresponding cylindrical pedal surface given by the image of the map $\mathrm{CP}_e : M \to S^2 \times \mathbb{R}$, with $\mathrm{CP}_e(p) = (N(p), \langle p, N(p)\rangle)$. In Chapter 5, we have shown that the cylindrical pedal $\mathrm{CP}_e(M)$ is the wavefront of a certain Legendrian immersion. We have also shown in Chapter 6 that the swallowtail singularities of CP_e are exactly the cusps of the Gauss map. We have the geometric characterisation of the cusps of the Gauss map (Theorem 6.3). The triple points of $\mathrm{CP}_e(M)$ are the tri-tangent planes of M. From the Gauss-Bonnet theorem we get the following corollary of Theorem 10.2.

Corollary 10.3. *For a generic closed and orientable surface M in $\mathbb{R}^3$, we have*

$$\chi(\mathrm{CP}_e(M)) = \frac{1}{2\pi}\int_M K\, d\mathfrak{v}_M + T^t(M) + \frac{1}{2}C^g(M),$$

where $T^t(M)$ is the number of tri-tangent planes of M and $C^g(M)$ is the number of the cusps of the Gauss.

As with the classical Gauss-Bonnet formula, all invariants of the right hand side of the formula in Corollary 10.3 are differential geometric invariants and the left hand side is a topological invariant of the surface. Other applications of Theorem 10.2 can be found in [Izumiya and Marar (1993); Izumiya and Romero Fuster (2006); Izumiya (2009); Houston and van Manen (2009)].

10.1.3 *Surfaces in* $\mathbb{R}^4$

Inflection points of a surface M in $\mathbb{R}^4$ are the singular points of the asymptotic curves on M (see Chapter 7). For a generic surface, the fields of asymptotic directions have index $\pm\frac{1}{2}$ at inflection points of imaginary type.

The inverse of the stereographic projection from $\mathbb{R}^3$ to S^3 maps the lines of principal curvature of a surface in $\mathbb{R}^3$ to the asymptotic curves of the spherical image of the surface (which we consider as a surface in $\mathbb{R}^4$). It follows that a closed surface in $\mathbb{R}^4$ which lies in the unit sphere S^3 and is height function generic has two orthogonal asymptotic directions at each point except maybe at a finite number of inflection points of imaginary type. This property holds for a more general class of surfaces in $\mathbb{R}^4$.

Definition 10.4. We say that a surface M in $\mathbb{R}^4$ is *locally convex* if it has locally a support hyperplane at each point.

In fact, any surface in S^3 is locally convex. The orthogonal direction to a support hyperplane at a point $p \in M$ determines a height function that has either a (local) minimum or a (local) maximum at p.

A height function generic surface M in $\mathbb{R}^4$ is locally convex if and only if it consists of hyperbolic and (isolated) inflection points ([Mochida, Romero Fuster and Ruas (1995)]). It follows that generic locally convex surfaces have two globally defined asymptotic directions whose critical points are isolated imaginary inflection points. We can now apply the generalised Poincaré-Hopf formula ([Pugh (1968)] for surfaces with boundary to get the following result.

Theorem 10.3. *For a height function generic closed and locally convex surface M in $\mathbb{R}^4$ we have*

$$2|\chi(M)| \leq \sharp\{\textit{inflection points}\}.$$

Remark 10.2. A similar inequality to that in Theorem 10.3 is given in [Garcia, Mochida, Romero Fuster and Ruas (2000)] for the Euler characteristic of the closure of the hyperbolic region $cl(M_h)$ of a generic non-locally convex surface M. Also in [Bruce and Tari (2002)] are obtained formulae for $\chi(M)$ in the locally convex case and for $\chi(cl(M_h))$ in the general case in terms of the number and type of the inflection points of M.

A consequence of Theorem 10.3 is the following.

Corollary 10.4. *Any height function generic closed and locally convex surface in $\mathbb{R}^4$ with non-vanishing Euler characteristic has at least 4 inflection points.*

Using the stereographic projection we recover the following result from [Feldman (1967)] for the number of umbilics of a generic closed surface in $\mathbb{R}^3$.

Corollary 10.5. *Any 2-sphere generically immersed into $\mathbb{R}^3$ has at least 4 umbilic points. Here generic means that the surface is distance-squared function generic.*

The result in Corollary 10.5 is the *generic* version of the following conjecture.

Carathéodory conjecture: *any closed, convex and sufficiently smooth surface in three dimensional Euclidean space has at least two umbilic points.*

A possible approach for solving the conjecture is by using Poincaré-Hopf formula and investigating the possible values for the index of an umbilic point. It is known, for instance, how to construct examples of local immersions of surfaces with umbilics of any index ≤ 1. This leads to the following conjecture which can be considered as a local version of the Carathéodory conjecture.

Loewner conjecture: *the index of the principal directions field at an umbilic point of a sufficiently smooth surface in three dimensional Euclidean space is at most* 1.

A review on the state of Carathéodory's conjecture can be found in [Gutierrez and Sotomayor (1998)]. In view of Corollary 10.4, it is reasonable to propose the following conjecture for surfaces in $\mathbb{R}^4$.

Carathéodory conjecture for surfaces in $\mathbb{R}^4$: *every closed, convex and sufficiently smooth surface in four dimensional Euclidean space which is homeomorphic to a 2-sphere has at least two inflection points.*

The above conjecture is shown to be true in some particular cases in [Gutiérrez and Ruas (2003); Nuño Ballestero (2006)].

Remark 10.3. It is shown in [Gutierrez and Sánchez Bringas (1998)] that, given any $n \in \mathbb{Z}$, there is an analytic immersion $f : \mathbb{R}^2 \to \mathbb{R}^4$ having a normal field ν and a ν-umbilic point p with index $\frac{n}{2}$. This implies that the Loewner conjecture does not hold for principal configurations associated to arbitrary normal fields on submanifolds in 4-space. The Carathéodory conjecture for surfaces in $\mathbb{R}^4$ is only about the singular points of the asymptotic curves.

10.1.4 *Semiumbilicity*

Recall from Definition 7.2 that a point p on $M \subset \mathbb{R}^4$ is semiumbilic if the curvature ellipse at p is a line segment that contains p. It can be shown that the asymptotic directions of M are orthogonal at semiumbilic points. Also, a point p is semiumbilic if and only if it is a ν-umbilic point of some normal field on M defined locally at p.

Let $N(p)$ be the curvature of the normal bundle of M at p (see §7.1). In fact, the area of the curvature ellipse of M at p is equal to $\frac{1}{2}|N(p)|$ ([Little (1969)]). We have the following characterisation of semiumbilic points.

Theorem 10.4. *Let M be a smooth surface in $\mathbb{R}^4$ and let p be a point on M. Then the following assertions are equivalent*

(i) *p is semiumbilic.*
(ii) $N(p) = 0$.
(iii) *p is ν-umbilic for some normal field ν on M.*
(iv) *There are two orthogonal asymptotic directions at p.*

It is shown in [Banchoff and Farris (1993)] that a surface in $\mathbb{R}^4$ is orientable if and only if it admits some globally defined normal field. As a consequence, and using the Poincaré-Hopf formula, we obtain the following corollary.

Corollary 10.6. *Any embedded orientable closed surface in $\mathbb{R}^4$ with non-vanishing Euler characteristic has semiumbilic points.*

Little proved in [Little (1969)] that any embedding of the torus in $\mathbb{R}^4$ has semiumbilic points. We have thus the following.

Corollary 10.7. *Any embedded orientable closed surface in $\mathbb{R}^4$ has semi-umbilic points.*

By Theorem 10.4, Corollary 10.7 is equivalent to stating that there does not exist any closed orientable surface embedded in $\mathbb{R}^4$ with never vanishing normal curvature.

A surface whose points are all semiumbilic is said to be a *totally semi-umbilic surface*. An example of a totally semiumbilic surface is a surface lying in a 3-sphere in $\mathbb{R}^4$. It is natural to search for sufficient conditions on a totally semiumbilic surface to be hyperspherical, i.e., to lie in a 3-sphere.

A totally semiumbilic surface M has two binormal fields $\mathbf{b}_1$ and $\mathbf{b}_2$ which are globally defined. Their singular points are the inflection points and the umbilic points considered as a particular case of inflection points of M. The binormal fields are characterised by the fact that one of their two principal curvatures vanishes identically on M; the other principal curvature we call *binormal curvature* of M. We denote by κ_1 and κ_2 the two binormal curvatures associated to $\mathbf{b}_1$ and $\mathbf{b}_2$ respectively of a totally semiumbilical surface M. The following result gives necessary and sufficient conditions on a totally semiumbilical surface to lie in a 3-sphere.

Theorem 10.5 ([Romero Fuster and Sánchez-Bringas (2002)]). *Suppose that M is a surface with isolated inflection points in $\mathbb{R}^4$. Then M is hyperspherical if and only if its asymptotic curves are globally defined and orthogonal and its binormal curvatures κ_1, κ_2 satisfy the following relation*

$$\left(\frac{\kappa_1}{\kappa_2} + \frac{\kappa_2}{\kappa_1} + 2\cos\Omega\right)E = constant,$$

where Ω is the angle between the two binormals at each point and E is the coefficient of the first fundamental form of M in isothermal coordinates.

10.2 Spacelike submanifolds of Minkowski space-time

We introduced new invariants in Chapter 9 as applications of singularity theory to differential geometry of submanifolds of Minkowski space-time. We prove here a Gauss-Bonnet and Chern-Lashof type theorems for closed spacelike surfaces M in Minkowski space-time $\mathbb{R}^4_1$ using those invariants.

Consider the canonical projection π from $\mathbb{R}^4_1$ to the Euclidean space $\mathbb{R}^3_0$ given by

$$\pi(x_0, x_1, x_2, x_3) = (0, x_1, x_2, x_3),$$

and denote by $\overline{M}$ the image of M by π. Since M is spacelike, $\pi|M : M \to \mathbb{R}^3_0$ is an immersion. If M is orientable, then so is $\overline{M}$. In this case we have a globally defined Gauss map $N : \overline{M} \to S^2$ on $\overline{M}$.

Recall from Definition 9.3, that we have the normalised lightcone Gauss map $\widetilde{\mathbb{L}} : U \to S^2_+$ and the normalised lightcone Gauss-Kronecker curvature $\widetilde{K}_\ell$ associated to $\widetilde{\mathbb{L}}$.

Lemma 10.1. *Under a suitable choice of direction of N, $\pi \circ \widetilde{\mathbb{L}}$ and N are homotopic.*

Proof. By Lemma 9.2, we have $\pi \circ \widetilde{\mathbb{L}}(p) \notin d(\pi|M)_p(T_pM) \subset \mathbb{R}^3_0$. Therefore $\langle \pi \circ \widetilde{\mathbb{L}}(p), N(p)\rangle \neq 0$ at any $p \in M$. We choose the direction of N that makes $\langle \pi \circ \widetilde{\mathbb{L}}(p), N(p)\rangle > 0$ and construct a homotopy $F : M \times [0,1] \to S^2$ between $\pi \circ \widetilde{\mathbb{L}}$ and N as follows:

$$F(p,t) = \frac{tN(p) + (1-t)\pi \circ \widetilde{\mathbb{L}}(p)}{||tN(p) + (1-t)\pi \circ \widetilde{\mathbb{L}}(p)||},$$

where $\|\cdot\|$ is the Euclidean norm in $\mathbb{R}^3_0$. Suppose that there exists $t' \in (0,1)$ and $p' \in M$ such that

$$t'N(p') + (1-t')\pi \circ \widetilde{\mathbb{L}}(p') = 0.$$

Then we would have $N(p') = -((1-t')/t')\pi \circ \widetilde{\mathbb{L}}(p')$. But this contradicts the assumption that $\langle \pi \circ \mathbb{L}(p), N(p)\rangle > 0$. Therefore, F is a continuous mapping satisfying $F(p,0) = \pi \circ \widetilde{\mathbb{L}}(p)$ and $F(p,1) = N(p)$ at any point p on M. □

The degree $\deg N$ of the map N is a homotopy invariant and satisfies $\deg N = (1/2)\chi(M)$ (see for example [Guillemin and Pollack (1974)]). A consequence of this and of Lemma 10.1 is the following result.

Corollary 10.8. *Let M be a closed orientable spacelike surface in $\mathbb{R}^4_1$. Then*

$$\deg \widetilde{\mathbb{L}} = \frac{1}{2}\chi(M).$$

Proposition 10.1. *Denote by $d\mathfrak{v}_M$ (respectively, $d\mathfrak{v}_{S^2_+}$) the volume form of M (respectively, S^2_+). Then the following relation holds:*

$$\widetilde{K}_\ell d\mathfrak{v}_M = \widetilde{\mathbb{L}}^* d\mathfrak{v}_{S^2_+},$$

where $\widetilde{\mathbb{L}}^* d\mathfrak{v}_{S^2_+}$ is the pull back by $\widetilde{\mathbb{L}}$ of the differential form $d\mathfrak{v}_{S^2_+}$.

Proof. Take a local parametrisation $\mathbf{x} : U \to \mathbb{R}^4_1$ of M, where U is an open subset of $\mathbb{R}^2$ with coordinates (u_1, u_2). Assume first that the normalised lightcone Gauss map $\widetilde{\mathbb{L}}$ is not singular at $p = \mathbf{x}(u)$. In this case, there exists an open neighbourhood $W \subset U$ containing p such that $\widetilde{\mathbb{L}} : W \to S^2_+$ is an embedding. Therefore, $\{\widetilde{\mathbb{L}}_{u_1}, \widetilde{\mathbb{L}}_{u_2}\}$ is a basis of $T_z S^2_+$ at any point $z \in V = \widetilde{\mathbb{L}}(W)$. We denote by $\widetilde{g}_{ij}$ the Riemannian metric on V and by $g_{\alpha\beta}$ the Riemannian metric on W given by the restriction of the Minkowski metric. Since $\mathbb{L} = \ell_0 \widetilde{\mathbb{L}}$ (see Chapter 9 for notation), $\ell_0 \widetilde{\mathbb{L}}_{u_i} = \mathbb{L}_{u_i} - \ell_{0u_i}\widetilde{\mathbb{L}}$. By Proposition 9.2, we have

$$
\begin{aligned}
\widetilde{g}_{ij} &= \langle \widetilde{\mathbb{L}}_{u_i}, \widetilde{\mathbb{L}}_{u_j} \rangle_1 \\
&= \frac{1}{\ell_0^2} \langle \mathbb{L}_{u_i}, \mathbb{L}_{u_j} \rangle_1 \\
&= \frac{1}{\ell_0^2} \langle -\sum_{\alpha=1}^{2} h_i^\alpha \mathbf{x}_{u_\alpha}, -\sum_{\beta=1}^{2} h_i^\beta \mathbf{x}_{u_\beta} \rangle_1 \\
&= \frac{1}{\ell_0^2} \sum_{\alpha,\beta=1}^{2} h_i^\alpha h_j^\beta \langle \mathbf{x}_{u_\alpha}, \mathbf{x}_{u_\beta} \rangle_1 \\
&= \frac{1}{\ell_0^2} \sum_{\alpha,\beta=1}^{2} h_i^\alpha h_j^\beta g_{\alpha\beta}.
\end{aligned}
$$

From the definition of $\widetilde{K}_\ell$ and the proof of Corollary 9.2 we get $\widetilde{K}_\ell = (1/\ell_0)^2 \det(h_j^i)$ and therefore

$$
\det(\widetilde{g}_{ij}) = \widetilde{K}_\ell^2 \det(g_{\alpha\beta}).
$$

Denote by $(\widetilde{u}_1, \widetilde{u}_2)$ the local coordinates on V via the embedding $\widetilde{\mathbb{L}}$. This means that the pull-back $\widetilde{\mathbb{L}}^*(d\widetilde{u}_1 \wedge d\widetilde{u}_2)$ of the volume form $d\widetilde{u}_1 \wedge d\widetilde{u}_2$ in V by $\widetilde{\mathbb{L}}$ satisfies

$$
\widetilde{\mathbb{L}}^*(d\widetilde{u}_1 \wedge d\widetilde{u}_2) = \begin{cases} du_1 \wedge du_2 & \text{if } \widetilde{K}_\ell(u) > 0, \\ -du_1 \wedge du_2 & \text{if } \widetilde{K}_\ell(u) < 0. \end{cases}
$$

Here (u_1, u_2) is a local parameter of $W \subset M$ and $(\widetilde{u}_1, \widetilde{u}_2)$ is a local parameter of $V \subset S_+^2$ via $\widetilde{\mathbb{L}} : W \to V$, that is, $\widetilde{\mathbb{L}}(u_1, u_2) = (\widetilde{u}_1, \widetilde{u}_2)$. If $\widetilde{\mathbb{L}}$ is orientation preserving, $\widetilde{\mathbb{L}}^*(d\widetilde{u}_1 \wedge d\widetilde{u}_2) = du_1 \wedge du_2$, otherwise $\widetilde{\mathbb{L}}^*(d\widetilde{u}_1 \wedge d\widetilde{u}_2) = -du_1 \wedge du_2$. By definition of the volume form, we have

$$
d\mathfrak{v}_V = \sqrt{\det(\widetilde{g}_{ij})} d\widetilde{u}_1 \wedge d\widetilde{u}_2, \; d\mathfrak{v}_W = \sqrt{\det(g_{\alpha\beta})} du_1 \wedge du_2,
$$

so that

$$
\widetilde{K}_\ell d\mathfrak{v}_W = \widetilde{\mathbb{L}}^* d\mathfrak{v}_V.
$$

If p is a singular point of $\widetilde{\mathbb{L}}$, then both sides of the equality are zero. This completes the proof. □

We have now all the ingredients to prove the following theorem.

Theorem 10.6 (The Lightcone Gauss-Bonnet Theorem). *Let M be a closed orientable spacelike surface in Minkowski space-time $\mathbb{R}_1^4$. Then*

$$
\int_M \widetilde{K}_\ell d\mathfrak{v}_M = 2\pi\chi(M).
$$

Proof. By Proposition 10.1 and Corollary 10.8 we have

$$\int_M \widetilde{K}_\ell d\mathfrak{v}_M = \int_M \widetilde{\mathbb{L}}^* d\mathfrak{v}_{S^2_+} = \deg(\widetilde{\mathbb{L}}) \int_{S^2_+} d\mathfrak{v}_{S^2_+} = \deg(\widetilde{\mathbb{L}}) 4\pi = 2\pi\chi(M).$$

□

An immediate consequence of the Lightcone Gauss-Bonnet Theorem is that the total lightcone Gauss-Kronecker curvature is a topological invariant. However, the lightcone Gauss-Kronecker curvature is not invariant under the Lorentz group. In fact, it is an SO(3)-invariant where SO(3) acts canonically on $\mathbb{R}^4_1$ as a subgroup of $SO_0(1,3)$.

If we consider the total absolute lightcone curvature, we obtain the following Chern-Lashof type inequality.

Theorem 10.7 (The Lightcone Chern-Lashof Type Theorem). *The following inequality holds for any closed orientable spacelike surface M in Minkowski space-time $\mathbb{R}^4_1$:*

$$\int_M |\widetilde{K}_\ell| d\mathfrak{v}_M \geq 2\pi(4 - \chi(M)).$$

Proof. We define the two subsets $M^+ = \{p \in M \mid \widetilde{K}_\ell(p) > 0\}$ and $M^- = \{p \in M \mid \widetilde{K}_\ell(p) < 0\}$ of M. Then we can write

$$\int_M |\widetilde{K}_\ell| d\mathfrak{v}_M = \int_{M^+} \widetilde{K}_\ell d\mathfrak{v}_M - \int_{M^-} \widetilde{K}_\ell d\mathfrak{v}_M$$

and

$$\int_M \widetilde{K}_\ell d\mathfrak{v}_M = \int_{M^+} \widetilde{K}_\ell d\mathfrak{v}_M + \int_{M^-} \widetilde{K}_\ell d\mathfrak{v}_M.$$

By Theorem 10.6 and the above equalities, we get

$$\int_M |\widetilde{K}_\ell| d\mathfrak{v}_M = 2\int_{M^+} \widetilde{K}_\ell d\mathfrak{v}_M - 2\pi\chi(M).$$

Thus, it is enough to show that

$$\int_{M^+} \widetilde{K}_\ell d\mathfrak{v}_M \geq 4\pi.$$

Let M_0, M_1, M_2, M_2^+ be the subsets of M defined by $M_0 = (\widetilde{K}_\ell)^{-1}(0)$, $M_1 = \{p \in M \setminus M_0 \mid \exists q \in M_0 \text{ with } \widetilde{\mathbb{L}}(q) = \widetilde{\mathbb{L}}(p)\}$, $M_2 = M \setminus (M_0 \cup M_1)$ and $M_2^+ = M^+ \cap M_2$. Since M_0 is the singular set of $\widetilde{\mathbb{L}}$, we have by Sard's Theorem that $\widetilde{\mathbb{L}}(M_0)$ and $\widetilde{\mathbb{L}}(M_0) \cup \widetilde{\mathbb{L}}(M_1)$ are measure zero sets in S^2_+. For any $\mathbf{v} \in S^2_+ \setminus (\widetilde{\mathbb{L}}(M_0) \cup \widetilde{\mathbb{L}}(M_1))$, the lightcone height function $h_\mathbf{v}$ has at least two critical points: a maximum and a minimum. We know that

$$\widetilde{K}_\ell(p) = \frac{\det \operatorname{Hess}(h_\mathbf{v}(p))}{\det(g_{ij}(p))},$$

for $\mathbf{v} = \widetilde{\mathbb{L}}(p)$. Since $\mathbf{v}$ is a regular value of $\widetilde{\mathbb{L}}$, $h_{\mathbf{v}}$ has a Morse-type singular point of index 0 or 2 at the minimum and maximum points. The light-cone Gauss-Kronecker curvarture $\widetilde{K}_\ell$ is positive at such points, so $\widetilde{\mathbb{L}}|M^+$ is surjective. Since the area of the unit sphere is 4π, we have the desired inequality by Proposition 10.1. □

Remark 10.4. The classical Gauss-Bonnet Theorem (see for example [do Carmo (1976)]) and the Chern-Lashof Theorem ([Chern and Lashof (1957)]) for surfaces in Euclidean space can be obtained as particular cases of Theorems 10.6 and 10.7 by taking $\mathbf{n}^T$ defining $\mathbb{L}$ as a constant timelike unit vector.

10.3 Notes

Curves. We have not touched on the subject of plane curves in this book. However, there are some global properties of plane and space curves that are worth mentioning here. We start with plane curves and the celebrated 4-Vertex Theorem which states that any smooth closed simple curve in Euclidean plane has at least 4 vertices. An analogous result for curves in the Minkowski plane $\mathbb{R}^2_1$ is proved in [Tabachnikov (1997)]. Using stereographic projections $H^2_+(-1) \to \mathbb{R}^2$ and $S^2_1 \to \mathbb{R}^2$, one can show that an analogous result of the 4-vertex Theorem is true for curves in $H^2_+(-1)$ and in S^2_1.

Results of global nature on space curves obtained by similar methods to those explored in this book concerning the number of tri-tangent planes, bi-tangent osculating planes and torsion zero points can be found in [Banchoff, Gaffney and McCrory (1985); Freedman (1980); Nuño Ballesteros and Romero Fuster (1992); Sedykh (1989); Ozawa (1985)].

It is worth observing that the case of spacelike embeddings of a circle (i.e., spacelike knots) in $\mathbb{R}^3_1$ is different from ordinary knots $\mathbb{R}^3$. There are, for instance, many cases of closed spacelike curves which are un-knotted in the ordinary sense ([Izumiya, Kikuchi and Takahashi (2006)]).

Umbilics on surfaces in $\mathbb{R}^3_1$ *and* $H^3_+(-1)$. It is shown in [Tari (2013)] that any closed and convex surface in Minkowski 3-space $\mathbb{R}^3_1$ of class C^3 has at least two umbilic points. For ovaloids (i.e., strictly convex surfaces) of class C^3, the umbilic points all lie in the Riemannian part of the surface and there are at least two of them.

Using the stereographic projection and Feldman result for generic surfaces in $\mathbb{R}^3$ ([Feldman (1967)]), it can be shown that the number of lightcone umbilic points (singularities of the lightcone lines of curvature) of a closed surface M generically immersed in $H^3_+(-1)$ is greater or equal than $2|\chi(M)|$. Furthermore, if M is homeomorphic to a 2-sphere, then it has at least 4 lightcone umbilic points.

Characterisation of metric spheres in Minkowski space-time. Morse theory played a principal role in differential topology. A Morse function on $f : M \to \mathbb{R}$ induces a decomposition of M as a CW-complex whose cell types are determined by the distribution and the indices of critical points (see for instance [Hirsch (1973); Milnor (1963); Matsumoto (1997)]). A well known consequence of this decomposition is the theorem of Reeb ([Reeb (1946)]), which says that any closed (compact without boundary) manifold that admits a Morse function with only two critical points is homeomorphic to a sphere. A natural question related to Reeb's theorem is to characterise the submanifolds in Euclidean or Lorentz-Minkowski spaces which are metric spheres. This problem is studied by Nomizu and Rodríguez in the case of submanifolds of Euclidean space. Let $f : M^n \to \mathbb{R}^n$ be an immersion. They proved in [Nomizu and Rodríguez (1972)] that if M^n is connected and complete in the induced Riemannian metric, and almost every distance squared function on M^n has index 0 or n at each of its critical points, then f embeds M^n either as an Euclidean n-sphere or as a flat affine subspace.

The analogous problem for submanifolds of Hyperbolic 3-space was treated by Cecil and Ryan. Given a surface M in $\mathbb{R}^3_1$, Cecil ([Cecil (1974)]) proved that a connected compact surface M immersed in $H^3_+(-1)$ is a metric sphere if and only if every Morse function of type timelike has exactly two singularities. Later, Cecil and Ryan ([Cecil and Ryan (1979a)]) considered the three families of height functions describing respectively the contact of a surface with spacelike, timelike and lightlike hyperplanes in $\mathbb{R}^3_1$ and proved that a connected complete surface M of H^3_+ is either a sphere, a horosphere, or an equidistant plane if and only if every Morse function of one of the above 3 types on M has index 0 or 2 at all its critical points. These results are extended in [Izumiya, Nuño Ballesteros and Romero Fuster (2010)] for closed surfaces in $\mathbb{R}^4_1$ that admit a globally defined non-degenerate parallel normal field.

Gauss-Bonnet and Chern-Lashof Theorems. For surfaces in the hyperbolic space, we have the Horospherical Gauss-Bonnet Theorem ([Izumiya and Romero Fuster (2006)]) and the Horospherical Chern-Lashof Type Theorem ([Buosi, Izumiya and Ruas (2010)]). The Lightcone Gauss-Bonnet Theorem holds for codimension two closed and orientable spacelike submanifolds in Minkowski space-time $\mathbb{R}^n_1$, $n \geq 4$ ([Izumiya and Romero Fuster (2007)]). The Lightcone Chern-Lashof Type Theorem for spacelike submanifolds of higher codimension in $\mathbb{R}^n_1$, $n \geq 4$, holds ([Izumiya (2014)]). We point out that this is a generalisation of both the Chern-Lashof type theorem for horospherical curvatures in hyperbolic space ([Buosi, Izumiya and Ruas (2010)]) and the original Chern-Lashof theorem for Euclidean space ([Chern and Lashof (1957)]). An interesting consequence is the introduction of the notion of lightcone tightness which generalises the concepts of horotightness in hyperbolic space ([Buosi, Izumiya and Ruas (2011); Cecil and Ryan (1979b)]) and tightness in Euclidean space ([Cecil and Ryan (1985)]).

kth-regular immersions. The concept of kth-regular immersions of a submanifold in Euclidean space was introduced in terms of maps between osculating bundles in [Pohl (1962)] and [Feldman (1965)]. In the case of curves, the k-regularity condition means that the first k derivatives are linearly independent. The existence of 3-regular embedded closed space curves was investigated in [Costa (1990)]. Corollary 10.2 and its generalisations show that convexity is an obstruction for the 3-regularity of embedded closed space curves.

A point p of a surface M in $\mathbb{R}^n$ is 2-regular if and only if, in some coordinates near p, the subset of vectors determined by the first and second derivatives of the immersion at p has maximal rank. Otherwise, p is 2-singular. Feldman ([Feldman (1965)]) proved that the set of 2-regular immersions of any closed surface M in $\mathbb{R}^n$ is open and dense for $n = 3$ and $n \geq 7$. The 2-singular points of surfaces in $\mathbb{R}^4$ coincide with the inflection points.

The 2-regular points of a surface in $\mathbb{R}^5$ coincide with the M_2 points (Proposition 8.5). Examples of 2-regular immersions of closed orientable surfaces with non-zero genus are not known so far. On the other hand, it was shown in [Romero Fuster (2007)] that there are no 2-regular embeddings of orientable closed surfaces immersed in S^4.

Thom polynomials and Chern classes. We pointed out in the Notes of Chapter 3 that there is a rich branch of research on global topological invariants using Thom polynomials, see for example Ohmoto lecture notes [Ohmoto (2013)] and [Ando (1996); Kazarian (2006, 2001)].

Topological approach. Since Morse functions are stable mappings whose target manifold is $\mathbb{R}$, it was natural to seek to extend Morse theory to stable mappings with higher dimensional target manifolds, see for example [Saeki

(1996, 2004); Lippner and Szucs (2010); Szabó, Szucs and Terpai, (2010)].

Other approaches to global problems using Vassiliev type invariants and h-principal can be found, for example, in [Ando (1985, 2007a,b); Eliashberg and Mishachev (2002); du Plessis (1976a,b); Gromov (1986); Goryunov (1998); Yamamoto (2006); Ohmoto and Aicardi (2006)].

Bibliography

Abraham, R. (1963), Transversality in manifolds of mappings. *Bull. Amer. Math. Soc.* **69**, pp. 470-474.

Ando, Y. (1985), On the elimination of Morin singularities. *J. Math. Soc. Japan* **37**, pp. 471–487.

Ando, Y. (1996), On Thom polynomials of the singularities D_k and E_k. *J. Math. Soc. Japan* **48**, pp. 593–606.

Ando, Y. (2007a), A homotopy principle for maps with prescribed Thom-Boardman singularities. *Trans. Amer. Math. Soc.* **359**, pp. 489–515.

Ando, Y. (2007b), The homotopy principle for maps with singularities of given K-invariant class. *J. Math. Soc. Japan* **59**, pp. 557–582.

Arbarello, E., Cornalba, M., Griffiths, P. A. and Harris, J. (1985), *Geometry of algebraic curves. Vol. I.*, Grundlehren der Mathematischen Wissenschaften [Fundamental Principles of Mathematical Sciences], **267**.

Arnol'd, V. I. (1979), Indices of singular points of 1-forms on a manifold with boundary, convolution of invariants of reflection groups, and singularities of projections of smooth surfaces. *Uspekhi Mat. Nauk* **34:2**, pp. 3-38 (*Russian Math. Surveys* **34:2**, pp. 1-42).

Arnol'd, V. I. (1981), Lagrangian manifolds with singularities, asymptotic rays, and the open swallowtail. *Funct. Anal. Appl.* **15-4**, pp. 235–246.

Arnol'd, V. I. (1983), Singularities of systems of rays. *Russian Math. Surveys* **38**, pp. 87–176.

Arnol'd, V. I. (1990), *Singularities of caustics and wave fronts.* Mathematics and its Applications (Soviet Series), **62**. Kluwer Academic Publishers Group.

Arnol'd,V. I., Guseĭn-Zade S. M. and Varchenko A. N. (1985), *Singularities of differentiable maps. Vol. I. The classification of critical points, caustics and wave fronts.* Monographs in Mathematics, **82**. Birkhäuser.

Atique, R. G. W. (2000), On the classification of multi-germs of maps from $\mathbb{C}^2$ to $\mathbb{C}^3$ under $\mathcal{A}$-equivalence. *Chapman & Hall/CRC Res. Notes Math.* **412**, pp. 119–133. Chapman & Hall/CRC, Boca Raton, FL.

Banchoff, T. and Farris, F. (1993), Tangential and normal Euler numbers, complex points, and singularities of projections for oriented surfaces in four-space. *Pacific J. Math.* **161**, pp. 1-24.

Banchoff, T., Gaffney, T. and McCrory, C. (1982), *Cusps of Gauss mappings.* Research Notes in Mathematics **55**. Pitman (Advanced Publishing Program), Boston, Mass.-London.

Banchoff, T., Gaffney, T. and McCrory, C. (1985), Counting tritangent planes of space curves. *Topology* **24**, pp. 15–23.

Banchoff, T. and Thom, R. (1980), Erratum et compléments: "Sur les points paraboliques des surfaces" by Y. L. Kergosien and Thom. *C. R. Acad. Sci. Paris Sér. A-B* **291**, A503–A505.

Basto-Gonçalves, J. (2013), Local geometry of surfaces in $\mathbb{R}^4$. Preprint, *arXiv* : 1304.2242.

Binotto, R. (2008), Projetivos de curvatura. Doctoral Thesis, Instituto de Matemática, Estatística e Computação Científica, UNICAMP (Brasil).

Binotto, R, Costa, S. I. R. and Romero Fuster, M. C. (2015), Geometry of 3-manifolds in Euclidean space. To appear in *RIMS Kôkyûroku Bessatsu.*

Birbrair, L. (2007), Metric theory of singularities. Lipschitz geometry of singular spaces. *Singularities in geometry and topology*, pp. 223–233. (World Sci. Publ., Hackensack, NJ).

Blair, D. E. (1976), *Contact manifolds in Riemannian geometry*, Lecture Notes in Mathematics, **509**, Springer-Verlag, Berlin.

Bleeker, D. and Wilson, L. (1978), Stablility of Gauss maps, *Illinois J. Math.* **22**, pp. 279–289.

Brieskorn, E. and Knörrer, H (1986), *Plane algebraic curves.* Birkhäuser.

Bröcker, Th. (1975), *Differentiable germs and catastrophes.* London Mathematical Societry Lecture Notes **17** (Cambridge University Press).

Bruce, J. W. (1981), The duals of generic hypersurfaces; *Math. Scand.* **49**, pp. 36–60.

Bruce, J. W. (1984), Generic reflections and projections; *Math. Scand.* **54**, pp. 262–278.

Bruce, J. W. (1986), On transversality. *Proc. Edinburgh Math. Soc.* **29**, pp. 115-123.

Bruce, J. W. (1994a), Generic geometry, transversality and projections. *J. London Math. Soc.* **49**, pp. 183-194

Bruce, J. W. (1994b), Lines, circles, focal and symmetry sets. *Math. Proc. Cambridge Philos. Soc.* **118**, pp. 411–436.

Bruce, J. W. and Fidal, D. (1989), On binary differential equations and umbilics. *Proc. Royal Soc. Edinburgh,***111A**, pp. 147–168.

Bruce, J. W., Fletcher, G. J. and Tari, F. (2004), Zero curves of families of curve congruences. *Contemp. Math.,***354**, Amer. Math. Soc. Amer. Providence, RI, pp. 1–18.

Bruce, J. W. and Giblin, P. J. (1990), Projections of surfaces with boundary. *Proc. London Math. Soc.* **60**, pp. 392–416.

Bruce, J. W. and Giblin, P. J. (1992), *Curves and Singularities.* Cambridge University Press.

Bruce, J. W., Giblin, P. J. and Gibson, C. G. (1985), Symmetry sets. *Proc. Roy. Soc. Edinburgh Sect. A* **101**, pp. 163–186.

Bruce, J. W., Giblin, P. J. and Tari, F. (1995), Families of surfaces: height functions, Gauss maps and duals. Real and complex singularities (São Carlos, 1994). *Pitman Res. Notes Math. Ser. (Longman, Harlow)* **333**, pp. 148–178.

Bruce, J. W., Giblin, P. J. and Tari, F. (1998), Families of surfaces: height functions and projections to plane. *Math. Scand.* **82**, pp. 165–185.

Bruce, J. W., Giblin, P. J. and Tari, F. (1999), Families of surfaces: focal sets, ridges and umbilics. *Math. Proc. Camb. Phil. Soc* **125**, pp. 243–268.

Bruce, J. W., Kirk N. P. and du Plessis A. A. (1997), Complete transversals and the classification of singularities. *Nonlinearity* **10**, pp. 253-275.

Bruce, J. W. and Nogueira, A. C. (1998), Surfaces in $\mathbb{R}^4$ and duality. *Quart. J. Math. Oxford Ser.* **49**, pp. 433–443.

Bruce, J. W., du Plessis, A. A. and Wall, C. T. C. (1987), Determinacy and unipotency. *Invent. Math.* **88**, pp. 521–554.

Bruce, J. W. and Romero Fuster, M. C. (1991), Duality and orthogonal projections of curves and surfaces in Euclidean 3-space. *Quart. J. Math.* **42**, pp. 433–441.

Bruce, J. W. and Tari, F. (1995), On binary differential equations. *Nonlinearity* **8**, pp. 255–271.

Bruce, J. W. and Tari, F. (2000), Duality and implicit differential equations. *Nonlinearity* **13**, pp. 791–811.

Bruce, J. W. and Tari, F. (2002), Families of surfaces in $\mathbb{R}^4$. *Proc. Edinb. Math. Soc.* **45**, pp. 181–203.

Bruce, J. W. and Tari, F. (2005), Dupin indicatrices and families of curve congruences. *Trans. Amer. Math. Soc.* **357**, pp. 267–285.

Bruce, J. W. and West, J. M. (1998), Functions on a crosscap. *Math. Proc. Cambridge Philos. Soc.* **123**, pp. 19–39.

Bruce, J. W. and Wilkinson, T. C. (1991), Folding maps and focal sets. Proceedings of Warwick Symposium on Singularities, Springer Lecture Notes in Math., Vol. 1462, pp. 63–72, Springer-Verlag, Berlin and New York.

Bryant, R. L. (1987), Surfaces of mean curvature one in hyperbolic space. in Théorie des variétés minimales et applications (Palaiseau, 1983–1984), *Astérisque* No. 154–155, 12, pp. 321–347, 353 (1988).

Buchner, M.A. (1974), *Stability of the cut locus.* Doctoral thesis, Harvard University.

Buosi, M., Izumiya, S. and Ruas, M. A. S. (2010), Total absolute horospherical curvature of submanifolds in hyperbolic space. *Adv. Geom.* **10**, pp. 603–620.

Buosi, M., Izumiya, S. and Ruas, M. A. S. (2011), Horo-tight spheres in hyperbolic space. *Geom. Dedicata* **154**, pp, 9–26.

do Carmo, M. P. (1976), *Differential geometry of curves and surfaces.* Prentice-Hall, 1976.

do Carmo, M. P. (1992), *Riemannian geometry.* Translated from the second Portuguese edition by Francis Flaherty Mathematics: Theory & Applications. Birkhuser Boston, Inc., Boston, MA.

Cecil, Th. E. (1974), A characterization of metric spheres in hyperbolic space by Morse theory. *Tohoku Math. J.* **26**, pp. 341–351.

Cecil, Th. E. (1980), Lie Sphere Geometry with Applications to Submanifolds. Universitext, Springer-Verlag, New York Berlin.

Cecil, Th. E. and Ryan, P. J. (1979a), Distance functions and umbilic submanifolds of hyperbolic space. *Nagoya Math. J.* **74**, pp. 67–75.

Cecil, Th. E. and Ryan, P. J. (1979b), Tight ant taut immersions into hyperbolic space. *J. Lond. Math. Soc.* **19**, pp. 561–572.

Cecil, Th. E. and Ryan, P. J. (1985), *Tight and taut immersions of manifolds.* Research Notes in Mathematics, **107**. Pitman (Advanced Publishing Program), Boston, MA.

Chen, L. and Izumiya, S. (2009), A mandala of Legendrian dualities for pseudo-spheres in semi-Euclidean space *Proc. Japan Acad.* **85**, pp. 49–54.

Cheng, S. -Y. and Yau, S.-T. (1973), Maximal space-like hypersurfaces in the Lonrentz-Minkowski spaces. *Ann. of Math.* **104**, pp. 407–419.

Chern, S. and Lashof, R. K. (1957), On the total curvature of immersed manifolds. *Amer. J. Math.* **79**, pp. 306–318.

Cleave, J. P. (1980), The form of the tangent developable at points of zero torsion on space curves. *Math. Proc. Camb. Phil.* **88**, pp. 403–407.

Costa, S. I. R. (1990), On closed twisted curves. *Proc. Amer. Math. Soc.* **109**, pp. 205–214.

Costa, S. I. R. and Romero Fuster, M. C. (1997), Nowhere vanishing torsion closed curves always hide twice. *Geom. Dedicata* **66**, pp. 1–17.

Costa, S. I. R., Moraes, S. M. and Romero Fuster, M. C. (2009), Geometric contacts of surfaces immersed in $\mathbb{R}^n, n \geq 5$. *Differential Geom. Appl.* **27**, pp. 442–454.

Damon J. N. (1984), The unfolding and determinacy theorems for subgroups of $\mathcal{A}$ and $\mathcal{K}$. *Mem. Amer. Math. Soc.* **50**, No. 306.

Damon J. N. (1988), Topological triviality and versality for subgroups of $\mathcal{A}$ and $\mathcal{K}$. *Mem. Amer. Math. Soc.* **75**, No. 389.

Damon J. N. (1992), Topological triviality and versality for subgroups of $\mathcal{A}$ and $\mathcal{K}$. II. Sufficient conditions and applications. *Nonlinearity*, **5**, pp. 373–412.

Damon J. N. (2003), Smoothness and geometry of boundaries associated to skeletal structures. I. Sufficient conditions for smoothness. *Ann. Inst. Fourier* (Grenoble) **53**, pp. 1941–1985.

Damon J. N. (2004), Smoothness and geometry of boundaries associated to skeletal structures. II. Geometry in the Blum case. *Compos. Math.* **140**, pp. 1657–1674.

Damon J. N. (2006), The global medial structure of regions in $\mathbb{R}^3$. *Geom. Topol.* **10**, pp. 2385–2429.

Dara, L. (1975), Singularités génériques des équations differentielles multiformes. *Bol. Soc. Brasil Math.* **6**, pp. 95–128.

David, J. M. S. (1983), Projection-generic curves. *J. London Math. Soc.* **27**, pp. 552–562.

Davydov, A. A. (1994), *Qualitative control theory.* Translations of Mathematical Monographs, **142**. AMS, Providence, RI.

Dias, F. S. and Nuño Ballesteros, J. J. (2008), Plane curve diagrams and geometrical applications. *Q. J. Math.* **59**, pp. 287–310.

Dreibelbis, D. (2001), Bitangencies on surfaces in four dimensions. *Quart. J. Math.* **52**, pp. 137–160

Dreibelbis, D. (2004), Invariance of the diagonal contribution in a bitangency formula. *Real and complex singularities*, pp. 45–56, Contemp. Math., **354**, Amer. Math. Soc., Providence, RI.

Dreibelbis, D. (2006), Birth of bitangencies in a family of surfaces in $\mathbb{R}^4$. *Differential Geom. Appl.* **24**, pp. 321–331.

Dreibelbis, D. (2007), The geometry of flecnodal pairs. *Real and complex singularities,* pp. 113-126, Trends Math., Birkhäuser, Basel.

Dreibelbis, D. (2012), Self-conjugate vectors of immersed 3-manifolds in $\mathbb{R}^6$. *Topology Appl.* **159** (2012), pp. 450–456.

Eliashberg, Y. and Mishachev, N. (2002), *Introduction to the h-principle.* Graduate Studies in Mathematics, **48**. American Mathematical Society, Providence,RI.

Epstein, C. L. (1986), The hyperbolic Gauss map and quasiconformal reflections. *J. Reine Angew. Math.* **372**, pp. 96–135

Feldman, E. A. (1965), Geometry of immersions I. *Trans. AMS* **120**, pp. 185–224.

Feldman, E. A. (1967), On parabolic and umbilic points of immersed hypersurfaces. *Trans. Amer. Math. Soc.* **127**, pp. 1–28.

Fletcher, G. J. (1996), *Geometrical problems in computer vision.* Ph.D thesis, Liverpool University.

Freedman, M. H. (1980), Planes triply tangent to curves with non-vanishing torsion. *Topology* **19**, pp. 1–8.

Gaffney, T. (1983), The structure of the $T\mathcal{A}(f)$, classification and application to differential geometry. *Singularities, Part 1 (Arcata, Calif., 1981)*, pp. 409–427, Proc. Sympos. Pure Math. **40**, Amer. Math. Soc., Providence, RI.

Gaffney, T. and Massey, D. (1999), Trends in equisingularity theory. Singularity theory (Liverpool, 1996), xix–xx, pp. 207–248, *London Math. Soc. Lecture Note Ser.*, **263**. (Cambridge Univ. Press, Cambridge).

Gaffney, T. and Ruas, M. A. S. (1979), Singularities of mappings and orthogonal projections of spaces. *Notices Am. Math. Soc.* **26**, no.1, January 1979, 763-53-23.

Garcia, R., Mello, L. F. and Sotomayor, J. (2005), Principal mean curvature foliations on surfaces immersed in $\mathbb{R}^4$. *EQUADIFF 2003*, pp. 939-950, World Sci. Publ., Hackensack, NJ.

Garcia, R., Mochida, D. K. H., Romero Fuster, M. C. and Ruas, M. A. S. (2000), Inflection points and topology of surfaces in 4-space. *Trans. Amer. Math. Soc.* **352**, pp. 3029-3043.

Garcia, R. and Sotomayor, J. (1997), Structural stability of parabolic points and periodic asymptotic lines. *Mat. Contemp.* **12**, pp. 83–102.

Garcia, R. and Sotomayor, J. (2000), Lines of axial curvature on surfaces immersed in $\mathbb{R}^4$. *Differential Geom. Appl.* **12**, pp. 253–269.

Genin, D., Khesin, B. and Tabachnikov, S. (2007), Geodesics on an ellipsoid in Minkowski space. *Enseign. Math.* **53** , pp. 307–331.

Giblin, P. J. and Brassett, S. A. (1985), Local symmetry of plane curves. *Amer. Math. Monthly* **92**, pp. 689–707.

Giblin, P. J. and Holtom, P. (1999), The centre symmetry set. *Banach Center*

Publ., Polish Acad. Sci. Inst. Math., Warsaw **50**, pp. 91–105.

Giblin, P. J. and Janeczko, S. (2012), Geometry of curves and surfaces through the contact map. *Topology Appl.* **159**, pp. 466–475.

Giblin, P. J. and Weiss, R. (1987), Reconstruction of surfaces from profiles. Proc. First Internat. Conf. on Computer Vision; Computer Society of the IEEE, pp. 136–144.

Giblin, P. J. and Zakalyukin, V. M. (2005), Singularities of centre symmetry sets. *Proc. London Math. Soc.* **90**, pp. 132–166.

Gibson, C. G. (1979), *Singular points of smooth mappings.* Research Notes in Mathematics, **25**. Pitman (Advanced Publishing Program), Boston, Mass.-London.

Gibson, C. G., Hawes, W. and Hobbs, C. A. (1994), Local pictures for general two-parameter planar motions. Advances in robot kinematics and computational geometry (Ljubljana), *Kluwer Acad. Publ.*, Dordrecht, pp. 49–58.

Gibson, C. G., Wirthmüller, K., du Plessis, A. A. and Looijenga, E. J. N. (1976), *Topological stability of smooth mappings.* Lecture Notes in Mathematics, **552**. Springer-Verlag, Berlin-New York,

Golubitsky, M. and Guillemin, V. (1973), *Stable mappings and their singularities* GTM, **14**. Springer-Verlag, New York.

Golubitsky, M. and Guillemin, V. (1975), Contact equivalence for Lagrangian manifold, *Adv. Math.* **15**, pp. 375–387.

Golubitsky, M. and Schaeffer, D. G. (1985), *Singularities and groups in bifurcation theory. Vol. I.* Applied Mathematical Sciences, **51**. Springer-Verlag, New York.

Gonçalves, R. A., Martínez Alfaro, J. A., Montesinos Amilibia, A. and Romero Fuster, M. C. (2007), Relative mean curvature configurations for surfaces in $\mathbb{R}^n, n \geq 5$. *Bull. Braz. Math. Soc. New Series* **38**(2), pp. 1–22.

Goryunov, V. V. (1981a), *Surface projection singularities.* Ph. D. thesis, Moscow's Lomonosov State University.

Goryunov, V. V. (1981b), Geometry of bifurcation diagrams of simple projections on a line. *Funktsional. Anal. i Prilozhen* **15**, pp. 1–8, 96.

Goryunov, V. V. (1990), Projections of Generic Surfaces with Boundaries. *Adv. Soviet Math.* **1**, pp. 157–200.

Goryunov, V. V. (1998), Vassiliev type invariants in Arnold's J^+-theory of plane curves without direct self-tangencies. *Topology* **37**, pp. 603-620

Gromov, M. (1986), *Partial differential relations.* Ergebnisse der Mathematik und ihrer Grenzgebiete (3) [Results in Mathematics and Related Areas (3)], 9. Springer-Verlag, Berlin.

Guillemin, V. and Pollack, A. (1974), *Differential topology.* Prentice-Hall, Inc., Englewood Cliffs, N.J.

Gutierrez, C., Guadalupe, I., Tribuzy, R. and Guíñez V. (1997), Lines of curvature on surfaces immersed in $\mathbb{R}^4$. *Bol. Soc. Brasil. Mat.* (N.S.) **28**, pp. 233–251.

Gutierrez, C., Guadalupe, I., Tribuzy, R. and Guíñez, V. (2001), A differential equation for lines of curvature on surfaces immersed in $\mathbb{R}^4$. *Qual. Theory Dyn. Syst.*, pp. 207–220.

Gutierrez, C. and Guíñez, V. (2003), Simple umbilic points on surfaces immersed in $\mathbb{R}^4$. *Discrete Contin. Dyn. Syst.* **9**, pp. 877–900.

Gutiérrez, C. and Ruas, M. A. S. (2003), Indices of Newton non-degenerate vector fields and a conjecture of Loewner for surfaces in $\mathbb{R}^4$. Real and complex singularities, pp. 245–253, *Lecture Notes in Pure and Appl. Math.*, **232**, Dekker, New York.

Gutierrez, C. and Sánchez-Bringas, F. (1998), On a Loewner umbilic-index conjecture for surfaces immersed in $\mathbb{R}^4$. *J. Dynam. Control Systems* **4**, pp. 127–136.

Gutierrez, C. and Sotomayor, J. (1998), Lines of curvature, umbilic points and Carathéodory conjecture. *Resenhas* **3**, pp. 291–322.

Hilbert, D. and Cohen-Vossen, S. (1932), *Anschauliche Geometrie.* Julius Springer, Berlin. *Geometry and the Imagination.* Chelsea Publishing Company, NY, 1952.

Hirsch, M. W. (1976), *Differential topology.* Graduate Texts in Mathematics, **33**. Springer-Verlag, New York-Heidelberg.

Hörmander, L. (1971), Fourier integral operaators I. *Acta. Math.* **127**, pp. 71–183.

Houston, K. (1999), Images of finite maps with one-dimensional double point set. *Topology Appl.* **91**, pp. 197–219.

Houston, K. and van Manen, M. A. (2009), Bose type formula for the internal medial axis of an embedded manifold. *Differential Geom. Appl.* **27**, pp. 320–328.

Ishikawa, G. (1993), Determinacy of envelope of the osculating hyperplanes to a curve. *Bull. London Math.Soc.*, **25**, pp. 787–798.

Ishikawa, G. (1995), Developable of a curve and determinacy relative to osculation-type. *Quart. J. Math.* **46**, pp. 437–451.

Izumiya, S. (1993), Perestroikas of optical wave fronts and graphlike Legendrian unfoldings. *J. Differential Geom.* **38**, pp. 485–500.

Izumiya, S. (1995), Completely integrable holonomic systems of first-order differential equations. *Proc. Royal Soc. Edinburgh* **125A**, pp. 567–586.

Izumiya, S. (2009), Legendrian dualities and spacelike hypersurfaces in the lightcone *Moscow Math. J.* **9**, pp. 325–357.

Izumiya, S. (2014), Total lightcone curvatures of spacelike submanifolds in Lorentz-Minkowski space. *Differential Geom. Appl.* **34**, pp. 103–127.

Izumiya, S. and Janeczko, S. (2003), A sympletic framework for multiplane gravitational lensing. *J. Math. Physics* **44**, pp. 2077–2093.

Izumiya, S. and Kasedou, M. (2014) Lightlike flat geometry of spacelike submanifolds in Lorentz-Minkowski space. *Int. J. Geom. Methods Mod. Phys.* **11**, pp. 1450049–1–35.

Izumiya, S., Kikuchi, M. and Takahashi, M. (2006), Global properties of spacelike curves in Minkowski 3-space. *J. Knot Theory Ramifications* **15**, pp. 869–881.

Izumiya, S. and Kossioris, G. (1995), Semi-local Classification of Geometric Singularities for Hamilton-Jacobi Equations. *J. Differential Equations* **118**, pp. 166–193.

Izumiya, S. and Kossioris, G. (1997a), Bifurcations of shock waves for viscosity solutions of Hamilton-Jacobi equations of one space variable. *Bull. Sci. math.* **121**, pp. 166–193.

Izumiya, S. and Kossioris, G. (1997b), Geometric Singularities for Solutions of Single Coservation Laws. *Arch. Rational Mech. Anal.* **139**, pp. 255–290.

Izumiya, S., Kossioris, G. and Makrakis, G. (2001), Multivalued solutions to the eikonal equation in srtratified media. *Quart. Applied Math.* **LIX**, pp. 365–390.

Izumiya, S. and Marar, W. L. (1993), The Euler characteristic of a generic wavefront in a 3-manifold. *Proc. Amer. Math. Soc.* **118**, pp. 1347–1350.

Izumiya, S. and Marar, W. L. (1995), On topologically stable singular surfaces in a 3-manifold. *J. Geom.* **52**, pp. 108–119.

Izumiya, S., Nuño Ballesteros, J. J. and Romero Fuster, M. C. (2010), Global properties of codimension two spacelike submanifolds in Minkowski space. *Advances in Geometry* **10**, pp. 51–75.

Izumiya, S. and Otani, S. (2015) Flat approximations of surfaces along curves. *Demonstratio Math.* XLVIII, pp. 217–241.

Izumiya, S., Pei, D. and Romero Fuster, M. C. (2006), Horospherical geometry of surfaces in Hyperbolic 4-space. *Israel J. Math.* **154**, pp. 361–379.

Izumiya, S., Pei, D., Romero Fuster, M. C. and Takahashi, M. (2005), The horospherical geometry of submanifolds in hyperbolic space. *J. London Math. Soc.* **71**, pp. 779–800.

Izumiya, S., Pei, D. and Sano, T. (2003), Singularities of Hyperbolic Gauss maps. *Proc. London Math. Soc.* **86**, pp. 485–512.

Izumiya, S. and Romero Fuster, M. C. (2006), The horospherical Gauss-Bonnet type theorem in hyperbolic space. *J. Math. Soc. Japan* **58**, pp. 965-984.

Izumiya, S. and Romero Fuster, M. C. (2007), The lightlike flat geometry on spacelike submanifolds of codimension two in Minkowski space. *Sel. Math. New ser.* **13**, pp. 23–55.

Izumiya, S, Romero Fuster, M.C. and Saji, K(2009), Flat lightlike hypersurfaces in Lorentz-Minkowski 4-space. *J. Geom. Phys.* **59**, pp. 1528–1546.

Izumiya, S. and Saji, K (2010), The mandala of Legendrian dualities for pseudo-sphers in Lorentz-Minkowski space and "flat"spacelike surfaces. *J. Sing.* **2**, pp. 92–127.

Izumiya, S. and Takahashi, M. (2011), Pedal foliations and Gauss maps of hypersurfaces in Euclidean space. *J. Sing.* **6**, pp. 84–97.

Izumiya, S. and Takeuchi, N. (2001), Singularities of ruled surfaces in $\mathbb{R}^3$. *Math. Proc. Cambridge Philos. Soc.* **130**, pp. 1–11.

Izumiya, S. and Takeuchi, N. (2003), Geometry of Ruled Surfaces, in *Applicable Mathematics in the Golden Age* (Edited by J. C. Misra), Narosa Publishing House, New Delhi, India, pp. 305–338.

Izumiya, S. and Tari, F. (2008), Projections of hypersurfaces in the hyperbolic space to hyperhorospheres and hyperplanes. *Rev. Mat. Iberoam.* **24**, pp. 895–920.

Izumiya, S. and Tari, F. (2010a), Projections of timelike surfaces in the de Sitter space. Real and complex singularities, pp. 190–210, *London Math. Soc. Lecture Note Ser.* **380**, Cambridge Univ. Press, Cambridge.

Izumiya, S. and Tari, F. (2010b), Projections of surfaces in the hyperbolic space along horocycles. *Proc. Roy. Soc. Edinburgh Sect. A* **140**, pp. 399–418.

Izumiya, S. and Tari, F. (2010c), Self-adjoint operators on surfaces with a singular metric. *J. Dyn. Control Syst.* **16**, pp. 329–353.

Izumiya, S. and Tari, F. (2013), Projections of surfaces in the Minkowski 3-space along lightlike directions. *Nonlinearity* **26**, pp. 911–932.

Izumiya, S. and Yıldırım, H. (2011), Slant geometry of spacelike hypersurfaces in lightcone. *J. Math. Soc. Japan* **63**, pp. 715–752

Kasedou, M. (2009), Singularities of lightcone Gauss images of spacelike hypersurfaces in de Sitter space. *J. Geometry* **94**, pp. 107–121.

Kasedou, M (2010a), Spacelike submanifolds of codimension at most two in de Sitter space. in Real and complex singularities. *London Math. Soc. Lecture Note Ser.* **380**, pp. 211–228.

Kasedou, M (2010b), Spacelike submanifolds in de Sitter space. *Demonstratio Math.* **43**, pp. 401–418.

Kazarian, M. (2001), Classifying spaces of singularities and Thom polynomials. New Developments in Singularity Theory (Cambridge 2000), NATO Sci. Ser. II Math. Phys. Chem, 21, Kluwer Acad. Publ., Dordrecht, pp. 117–134.

Kazarian, M. (2006), Thom polynomials. Singularity theory and its applications, *Adv. Stud. Pure Math.* **43**, Math. Soc. Japan, Tokyo, pp. 85–135.

Kergosien, Y. L. (1981), La famille des projections orthogonales d'une surface et ses singularités. *C. R. Acad. Sci. Paris Ser. I Math.* **292**, pp. 929–932.

Kergosien, Y. L. and Thom, R. (1980), Sur les points paraboliques des surfaces. *C. R. Acad. Sci. Paris Sér. A-B* **290**, A705–A710.

Kirk, N. P. (2000), Computational aspects of classifying singularities. *LMS J. Comput. Math.* **3**, pp. 207–228.

Klein, F. (1974), Le programme d'Erlangen. Considérations comparatives sur les recherches géométriques modernes. *Collection "Discours de la Méthode".* Gauthier-Villars Éditeur, Paris-Brussels-Montreal, Que., 1974.

Klotz, C., Pop, O. and Rieger, J. H. (2007), Real double-points of deformations of $\mathcal{A}$-simple map-germs from $\mathbb{R}^n$ to $\mathbb{R}^{2n}$. *Math. Proc. Camb. Phil. Soc.* **142**, pp. 341–363.

Kobayashi, O. (1983), Maximal surfaces in the 3-Dimensional Minkowski space. *Tokyo J. Math.* **6**, pp. 297–309.

Kobayashi, T. (1986), Asymptotic behaviour of the null variety for a convex domain in a non-positively curved space form. *J. Fac. Sci. Univ. Tokyo* **36**, pp. 389–478.

Koenderink, J. J. (1984), What does the occluding contour tell us about solid shape? *Perception* **13**, pp. 321–330.

Koenderink, J. J. (1990), *Solide Shape.* (MIT Press, Cambridge, MA).

Koenderink, J. and van Doorn, A. J. (1976), The singularities of the visual mapping. *Biological Cybernetics* **24**, pp. 51–59.

Kokubu, M., Rossman, W., Saji, K., Umehara, M. and Yamada, K. (2005), Singularities of flat fronts in Hyperbolic space. *Pacific J. Math.* **221**, pp. 303–351.

Kommerell, K. (1905), Riemannsche Flächen im ebenen Raum von vier Dimensionen. *Math. Ann.***60** no. 4, pp. 548–596.

Kossowski, M. (1987), Pseudo-Riemannian metrics singularities and the extendability of parallel transport. *Proc. Amer. Math. Soc.* **99**, pp. 147–154.

Kossowski, M. (1989), The intrinsic conformal structure and Gauss map of light-like hypersurface in Minkowski space. *Trans. Amer. Math. Soc.* **316**, pp. 369–383.

Kossowski, M. (2007), Extrinsic Euler characteristic of a non-immersed hypersurface (focal surfaces, pedal images, and extrinsic Morse indices). *Differential Geom. Appl.* **25**, pp. 44–55.

Landis, E. E. (1981), Tangential singularities. *Funktsional. Anal. i Prilozhen* **15** (1981), pp. 36–49.

Lê Dung Tráng and Ramanujam, C. P. (1976), The invariance of Milnor's number implies the invariance of the topological type. *Amer. J. Math.* **98**, pp. 67–78.

Lippner, G. and Szucs, A. (2010), Multiplicative properties of Morin maps. *Algebr. Geom. Topol.* **10** , pp. 1437–1454.

Little, J. A. (1969), On singularities of submanifolds of higher dimensional Euclidean spaces. *Ann. Mat. Pura Appl.* **83** (4), pp. 261–335.

Looijenga, E. J. N. (1974), *Structural stability of smooth families of C^∞ functions.* Ph. D. thesis, University of Amsterdan.

Lyashko, O. V. (1979), Geometry of bifurcation diagrams. *Uspekhi Mat. Nauk* **34**:3 (1979), 205–206. (Russian Math. Surveys **34**:3 (1979), pp. 209–210.)

Mac Lane, S. and Birkhoff, G. (1967), *Algebra.* The Macmillan Co., New York; Collier-Macmillan Ltd., London.

Martinet, J. N. (1982), *Singularities of smooth functions and maps.* London Mathematical Society Lecture Notes **58**. Cambridge University Press.

Martins, R. and Nuño-Ballesteros, J. J. (2009), Finitely determined singularities of ruled surfaces in $\mathbb{R}^3$. *Math. Proc. Cambridge Philos. Soc.* **3**, pp. 701–733.

Massey, D. B. (1995), *L cycles and hypersurface singularities.* Lecture Notes in Mathematics, 1615. Springer-Verlag, Berlin.

Mather, J. N. (1968), Stability of C^∞ mappings, I. The division theorem. *Ann. of Math.* **87**, pp. 89–104.

Mather, J. N. (1969a), C^∞ mappings, II. Infinitesimal stability implies stability. *Ann. of Math.* **89**, pp. 254–291.

Mather, J. N. (1969b), Stability of C^∞ mappings, III. Finitely determined map-germs. *Publ. Math., IHES* **35**, pp. 279–308.

Mather, J. N. (1969c), Stability of C^∞ mappings, IV. Classification of stable germs by $\mathcal{R}$-algebras. *Publ. Math., IHES* **37**, pp. 223–248.

Mather, J. N. (1970), Stability of C^∞ mappings, V. Transversality. *Advances in Math.* **4**, pp. 301–336.

Mather, J. N. (1971), Stability of C^∞ mappings, VI. The nice dimensions. *Proc. Liverpool Singularities Symp. I,* LNM **192**, Springer, pp. 207–253.

Matsumoto, Y. (1997), *An introduction to Morse theory.* Translated from the 1997 Japanese original by Kiki Hudson and Masahico Saito. Translations of Mathematical Monographs, **208**. Iwanami Series in Modern Mathematics. American Mathematical Society, Providence, RI, 2002. xiv+219 pp.

McDuff, D. and Salamon, D. (1995), *Introduction to symplectic topology.* (Oxford

University Press, New York.)

Mello, L. F. (2003), Mean directionally curved lines on surfaces immersed in $\mathbb{R}^4$. *Publ. Mat.* **47**, pp. 415–440.

Milnor, J. (1963), *Morse Theory.* Annals of Mathematics Studies, No. 51 Princeton University Press, Princeton, N.J.

Milnor, J. (1968), *Singular points of complex hypersurfaces.* Annals of Mathematics Studies, No. 61 Princeton University Press, Princeton, N.J.

Mochida, D.K.H. (1993), *Geometria generica de subvariedades em codimensao maior que 1.* Ph.D. thesis, University of São Paulo.

Mochida, D.K.H., Romero Fuster, M.C. and Ruas, M. A. S. (1995), The geometry of surfaces in 4-space from a contact viewpoint. *Geom. Dedicata* **54**, pp. 323–332.

Mochida, D. K. H., Romero Fuster, M. C. and Ruas, M. A. S. (1999), Osculating hyperplanes and asymptotic directions of codimension two submanifolds of Euclidean spaces. *Geom. Dedicata* **77**, pp. 305–315.

Mochida, D. K. H., Romero Fuster, M. C. and Ruas, M. A. S. (2001), Singularities and duality in the flat geometry of submanifolds of Euclidean spaces. *Beiträge Algebra Geom.* **42**, pp. 137–148.

Mochida, D. K. H., Romero Fuster, M. C. and Ruas, M. A. S. (2003), Inflection points and nonsingular embeddings of surfaces in $\mathbb{R}^5$. *Rocky Mountain J. Math.* **33**, pp. 995–1009.

Mond, D. M. Q. (1982), *The Classification of Germs of Maps from Surfaces to* 3-*space, with Applications to the Differential Geometry of Immersions.* Ph.D. thesis, University of Liverpool.

Mond, D. M. Q. (1985), On the classification of germs of maps from $\mathbb{R}^2$ to $\mathbb{R}^3$. *Proc. London Math. Soc.* **50**, pp. 333–369.

Mond, D. M. Q. (1989), Singularities of the tangent developable surface of a space curve. *Quart. J. Math.* **40**, pp. 79–91.

Montaldi, J. A. (1983), *Contact with applications to submanifolds.* Ph.D. thesis, University of Liverpool.

Montaldi, J. A. (1986a), On contact between submanifolds. *Michigan Math. J.* **33**, pp. 81–85.

Montaldi, J. A. (1986b), Surfaces in 3-space and their contact with circles. *J. Differential. Geom.* **23**, pp. 109–126.

Montaldi, J. A. (1991), On generic composites of mappins. *Bull. London Math. Soc.* **23**, pp. 81–85.

Moraes, S. and Romero Fuster, M.C. (2005), Semiumbilic and 2-regular immersions of surfaces in Euclidean spaces. *Rocky Mountain J. Math.* **35**, pp. 1327–1345.

Morris, R. J. (1996), The sub-parabolic lines of a surface. The mathematics of surfaces, VI (Uxbridge, 1994), pp. 79–102, Inst. Math. Appl. Conf. Ser. New Ser., 58, Oxford Univ. Press.

Mostowski, T. (1985), Lipschitz equisingularity. *Dissertationes Math. (Rozprawy Mat.)* **243**, 46 pp.

Nabarro, A. C. (2003), Duality and contact of hypersurfaces in $\mathbb{R}^4$ with hyperplanes and lines. *Proc. Edinb. Math. Soc.* **46**, pp. 637–648.

Nabarro, A. C. and Romero Fuster, M. C. (2009), 3-manifolds in Euclidean space from a contact viewpoint. *Comm. Anal. Geom.* **17**, pp. 755–776.

Nabarro, A. C. and Tari, F. (2009), Families of surfaces and conjugate curve congruences. *Adv. Geom.* **9**, pp. 279–309.

Nabarro, A. C. and Tari, F. (2011), Families of curve congruences on Lorentzian surfaces and pencils of quadratic forms. *Proc. Roy. Soc. Edinburgh Sect. A* **141**, pp. 655–672.

Nagai, T. (2012), The Gauss map of a hypersurface in Euclidean sphere and the spherical Legendrian duality. *Tology Appl.* **159**, pp. 545–554.

Nogueira, C. A. (1998), *Superfícies em $\mathbb{R}^4$ e dualidade.* Ph.D. thesis, Univesity of São Paulo.

Nomizu, K. and Rodríguez, L. (1972), Umbilical submanifolds and Morse functions. *Nagoya Math. J.* **48**, pp. 197–201.

Nuño Ballesteros, J. J. (2006), Submanifolds with a non-degenerate parallel normal vector field in euclidean spaces. Singularity Theory and Its Applications. *Adv. Stud. Pure Math.*, **43**, pp. 311–332.

Nuño Ballesteros, J. J. and Romero Fuster, M. C. (1992), Global bitangency properties of generic closed space curves. *Math. Proc. Cambridge Philos. Soc.* **112**, pp. 519–526.

Nuño Ballesteros, J. J. and Romero Fuster, M. C. (1993), A four vertex theorem for strictly convex space curves. *J. Geom.* **46**, pp. 119–126.

Nuño Ballesteros, J. J. and Saeki, O. (2001), Euler characteristic formulas for simplicial maps. *Math. Proc. Cambridge Philos. Soc.* **130**, pp. 307–331.

Nuño Ballesteros, J. J. and Tari, F. (2007), Surfaces in $\mathbb{R}^4$ and their projections to 3-spaces. *Proc. Roy. Soc. Edinburgh* Sect. A, **137**, pp. 1313-1328.

Ohmoto, T. (2013), *Singularities and characteristic classes for differentiable maps.* Lecture Notes of a mini-course in the International Workshop on Real and Complex Singularities, ICMC-USP (São Carlos, Brazil), 2012. arXiv:1309.0661v3

Ohmoto, T. and Aicardi, F. (2006), First order local invariants of apparent contours. *Topology* **45**, pp. 27–45.

Oliver, J. M. (2010), *Pairs of geometric foliations of regular and singular surfaces.* Ph.D. thesis, Durham University. http://etheses.dur.ac.uk/280/.

Oliver, J. M. (2011), On the characteristic curves on a smooth surface. *J. London. Math. Soc.* **83**, pp. 755–767.

Olver, P. J. (1999), *Classical Invariant Theory.* London Math. Soc. Stud. Texts, vol. **44**, Cambridge University Press,.

O'Neill, B. (1983), *Semi-Riemannian Geometry. With applications to relativity.* Pure and Applied Mathematics, **103**, Academic Press.

Oset Sinha, R., Ruas, M. A. S. and Atique, R. G. W. (2015), On the simplicity of multigerms. To appear in *Math. Scand.*

Oset Sinha, R. and Tari, F. (2010), Projections of space curves and duality. *Quart. J. Math.* **64**, pp. 281–302.

Oset Sinha, R. and Tari, F. (2015), Projections of surfaces in $\mathbb{R}^4$ to $\mathbb{R}^3$ and the geometry of their singular images. *Rev. Mat. Iberoamericana* **31**, pp. 33–51.

Ozawa, T. (1985), The numbers of triple tangencies of smooth space curves.

Topology **24**, pp. 1–13.

Pelletier, F. (1995), Quelques proprietes geomtriques des varietes pseudo-riemanniennes singulieres. *Ann. Fac. Sci. Toulouse Math.* **4**, pp. 87–199.

Penrose, R. (1965), Gravitational collapse and space-time singularities. *Phys. Rev. Lett.* **14**, pp. 57–59.

Platonova, O. A. (1984), Projection of smooth surfaces. *Trudy Sem. Petrovsk*, **238**, pp. 135–149.

du Plessis, A. (1976a), Homotopy classification of regular sections. *Compositio Math.* **32**, pp. 301–333.

du Plessis, A. (1976b), Contact-Invariant regularity conditions. *Singularités d'applications différentiables*, pp. 205–236 . Lecture Notes in Math. Vol. **535**, Springer, Berlin.

du Plessis, A. A. and Wall, C. T. C. (1995), *The geometry of topological stability.* London Mathematical Society Monographs. New Series, **9**. Oxford Science Publications. Oxford University Press.

Pohl, W. F. (1962), Differential geometry of higher order. *Topology* **1**, pp. 169–211.

Porteous, I. R. (1983a), The normal singularities of surfaces in $\mathbb{R}^3$. Singularities, Part 2 (Arcata, Calif., 1981), pp. 379–393, *Proc. Sympos. Pure Math.* **40**, Amer. Math. Soc., Providence, RI, 1983.

Porteous, I. R. (1983b), Probing singularities. Singularities, Part 2 (Arcata, Calif., 1981), pp. 395–406, *Proc. Sympos. Pure Math.* **40**, Amer. Math. Soc. Providence, R.I., 1983.

Porteous, I. R. (1987), Ridges and umbilics of surfaces. The mathematics of surfaces, II (Cardiff, 1986), pp. 447–458, *Inst. Math. Appl. Conf. Ser. New Ser.* **11**, Oxford Univ. Press, New York, 1987.

Porteous, I. R. (2001), *Geometric differentiation. For the intelligence of curves and surfaces.* Cambridge University Press, Cambridge.

Poston, T. and Stewart, I. (1996), *Catastrophy theory and its applications.* New Yerk: Dover.

Pugh, C. (1968), A generalized Poincaré index formula. *Topology* **7**, pp. 217-226.

Ramírez-Galarza, A. I. and Sánchez-Bringas, F. (1995), Lines of curvature near umbilical points on surfaces immersed in $\mathbb{R}^4$. *Ann. Global Anal. Geom.* **13**, pp. 129–140.

Reeb, G. (1946), Sur les points singuliers d'une forme de Pfaff completement integrable ou d'une fonction numérique. *C. R. Acad. Sci. Paris* **222**, pp. 847–849.

Remizov, A. O. (2009), Geodesics on 2-surfaces with pseudo-Riemannian metric: singularities of changes of signature. *Mat. Sb.* **200:3**, pp. 75–94.

Rieger, J. H. (1987), Families of maps from the plane to the plane. *J. London Math. Soc.* **2**, pp. 351–369.

Rieger, J. H. and Ruas, M. A. S. (1991), Classification of $\mathcal{A}$-simple germs from $\mathbb{K}^n$ to $\mathbb{K}^2$. *Compositio Math.* **79**, pp. 99–108.

Rieger, J. H. and Ruas, M. A. S. (2005), M-deformations of $\mathcal{A}$-simple Σ_{n-p+1}-germs from $\mathbb{R}^n$ to $\mathbb{R}^p$, $n \geq p$. *Math. Proc. Camb. Phil. Soc.* 139, pp. 333–349.

Romero Fuster, M. C. (1981), *The convex hull of an immersion.* Ph.D. thesis,

University of Southampton.

Romero Fuster, M. C. (1983), Sphere stratifications and the Gauss map. *Proc. Edinburgh Math. Soc.* **95**, pp. 115–136.

Romero Fuster, M. C. (1988), Convexly generic curves in $\mathbb{R}^3$. *Geom. Dedicata* **28**, pp. 7–29.

Romero Fuster, M. C. (1997), Stereographic projections and geometric singularities. Workshop on Real and Complex Singularities (São Carlos, 1996). *Mat. Contemp.* **12**, pp. 167–182.

Romero Fuster, M.C. (2004), Semiumbilics and geometrical dynamics on surfaces in 4-spaces. Real and complex singularities, pp. 259–276, Contemp. Math., 354, *Amer. Math. Soc.*, Providence, RI.

Romero Fuster, M. C. (2007), Geometric contacts and 2-regularity of surfaces in Euclidean space. Singularity theory, pp. 307–325, World Sci. Publ., Hackensack, NJ, 2007.

Romero Fuster, M. C., Ruas, M. A. S. and Tari, F. (2008), Asymptotic curves on surfaces in $\mathbb{R}^5$, *Communications in Contemporary Maths.* **10**, pp. 1–27.

Romero Fuster, M. C. and Sanabria Codesal, E. (2002), On the flat ridges of submanifolds of codimension 2 in $\mathbb{R}^n$. *Proc. Roy. Soc. Edinburgh Sect. A* **132**, pp. 975–984.

Romero Fuster, M. C. and Sanabria-Codesal, E. (2004), Lines of curvature, ridges and conformal invariants of hypersurfaces. *Biträge Algebra Geom.* **45**, pp. 615–635.

Romero Fuster, M. C. and Sanabria Codesal, E. (2008), Conformal invariants interpreted in de Sitter space. *Mat. Contemp.* **35**, pp. 205–220.

Romero Fuster, M. C. and Sanabria Codesal, E. (2013), Conformal invariants and spherical contacts of surfaces in $\mathbb{R}^4$. *Rev. Mat. Complut.* **26**, pp. 215–240.

Romero Fuster, M. C. and Sánchez-Bringas F. (2002), Umbilicity of surfaces with orthogonal asymptotic lines in $\mathbb{R}^4$. *Differential Geom. and Appl.* **16**, pp. 213–224.

Romero Fuster, M. C. and Sedykh, V. D.(1995), On the number of singularities, zero curvature points and vertices of a simple convex space curve. *J. Geom.* **52**, pp. 168–172.

Romero Fuster, M. C. and Sedykh, V. D. (1997), A lower estimate for the number of zero-torsion points of a space curve. *Biträge Algebra Geom.* **38**, pp. 183–192.

Ruas, M. A. S. and Tari, F. (2012), A note on binary quintic forms and lines of principal curvature on surfaces in $\mathbb{R}^5$. *Topology Appl.* **159**, pp. 562–567.

Saeki, O. (1996), Simple stable maps of 3-manifolds into surfaces. *Topology* **35**, pp. 671–698.

Saeki, O. (2004), *Topology of singular fibers of differentiable maps.* Lecture Notes in Mathematics, 1854. (Springer-Verlag, Berlin).

Saji, K. (2010), Criteria for singularities of smooth maps from the plane into the plane and their applications. *Hiroshima Math. J.* **40**, pp. 229–239.

Saji, K., Umehara, M. and Yamada, K. (2009), The geometry of fronts. *Ann. of Math.* **169**, pp. 491–529.

Saloom, A. and Tari, F. (2012), Curves in the Minkowski plane and their contact with pseudo-circles. *Geometria Dedicata* **159**, pp. 109–124.

Seade, J. (2007), On Milnor's fibration theorem for real and complex singularities. *Singularities in geometry and topology*, pp. 127–158, World Sci. Publ., Hackensack, NJ, 2007.

Sedykh, V. D. (1989), Double tangent planes to a space curve. (Russian) *Sibirsk. Mat. Zh.* **30**, pp. 209–211; translation in *Siberian Math. J.* **30**, pp. 161–162.

Sedykh, V. D. (1992), The four-vertex theorem of a convex space curve. *Funct. Anal. Appl.* **26**, pp. 28–32.

Sedykh, V. D. (2012), On Euler characteristics of manifolds of singularities of wave fronts. (Russian) *Funktsional. Anal. i Prilozhen.* **46**, pp. 92–96; translation in *Funct. Anal. Appl.* **46**, pp. 77–80.

Shcherbak, O. P. (1986), Projectively dual space curve and Legendre singularities. *Sel. Math. Sov.* **5**, pp. 391–421.

Siddiqi, K. and Pizer, S. (Eds.) (2008), *Medial Representations. Mathematics, Algorithms and Applications.* Computational Imaging and Vision, Vol. 37. (Springer).

Sotomayor, J. (2004), Historical Comments on Monge's Ellipsoid and the Configuration of Lines of Curvature on Surfaces Immersed in $\mathbb{R}^3$. *ArXiv*: math/0411403v1.

Sotomayor, J. and Gutierrez, C. (1982), Structurally stable configurations of lines of principal curvature. Bifurcation, ergodic theory and applications (Dijon, 1981), pp. 195–215, *Astérisque*, **98-99**, Soc. Math. France, Paris.

Steller, M. (2006), A Gauss-Bonnet formula for metrics with varying signature. *Z. Anal. Anwend.* **25**, pp. 143–162.

Szabó, E., Szucs, A. and Terpai, T. (2010), On bordism and cobordism groups of Morin maps. *J. Singul.* **1**, pp. 134–145.

Tabachnikov, S. (1997), Parametrized plane curves, Minkowski caustics, Minkowski vertices and conservative line fields. *Enseign. Math.* **43**, pp. 3–26.

Tari, F. (1991), Projections of piecewise-smooth surfaces. *J. London Math. Soc.* **44**, pp. 155–172.

Tari, F. (2009), Self-adjoint operators on surfaces in $\mathbb{R}^n$. *Differential Geom. Appl.* **27**, pp. 296-306.

Tari, F. (2010), Pairs of foliations on surfaces. Real and complex singularities, pp. 305–337, London Math. Soc. Lecture Note Ser., 380, Cambridge Univ. Press, Cambridge.

Tari, F. (2012), Caustics of surfaces in the Minkowski 3-space. *Quart. J. Math.* **63**, pp. 189–209.

Tari, F. (2013), Umbilics of surfaces in the Minkowski 3-space. *J. Math. Soc. Japan*, **65**, pp. 723–731.

Thom, R. (1956), Les singularités des applications différentiables. *Ann. Inst. Fourier, Grenoble* **6**, pp. 43–87.

Thom R. (1972), Sur les équations différentielles multiformes et leurs intégrales singulières. *Bol. Soc. Brasil. Mat.* **3** , pp. 1–11.

Thom, R. (1976) *Structural stability and morphogenesis. An outline of a general theory of models.* Translated from the French by D. H. Fowler. With a foreword by C. H. Waddington. Second printing. W. A. Benjamin, Inc., Reading, Mass.-London-Amsterdam.

Thom, R. (1983), *Mathematical models of morphogenesis.* Halsted Press [John Wiley & Sons, Inc.], New York.

Uribe-Vargas, R. (2001), *Singularités symplectiques et de contact en géomtrie différentielle des courbes et des surfaces.* Ph. D. thesis, Université Paris 7.

Uribe-Vargas, R. (2006), A projective invariant for swallowtails and godrons and global theorems on the flecnodal curve. *Mosc. Math. J.* **6**, pp. 731–768.

Vaisman, I. (1984) *A first course in Differential Geometry.* Marcel Dekker.

Wall, C. T. C. (1977), Geometric properties of generic differentiable manifolds. *Geometry and Topology*, Rio de Janeiro, 1976, Springer Lecture Notes in Math. **597**, pp. 707–774.

Wall, C. T. C. (1981), Finite determinacy of smooth map-germs. *Bull. London Math. Soc.* **13**, pp. 481–539.

Wall, C. T. C. (2004), *Singular points of plane curves.* (London Mathematical Society Student Texts 63, Cambridge University Press, Cambridge).

West, J. (1995), *The differential geometry of the cross-cap.* Ph.D. thesis, University of Liverpool.

Whitney, H. (1944), The singularities of a smooth n-manifold in $(2n-1)$-space. *Ann. of Math.* **45**, pp. 247–293.

Whitney, H. (1955), On singularities of mappings of euclidean spaces. I. Mappings of the plane into the plane. *Ann. of Math.* **62**, pp. 374–410.

Wilkinson, T. C. (1991), *The geometry of folding maps.* Ph.D. thesis, University of Newcastle-upon-Tyne.

Wong, Y. C. (1952), A new curvature theory for surfaces in a Euclidean 4-space. *Comment. Math. Helv.* **26**, pp. 152–170.

Yamamoto, M. (2006), First order semi-local invariants of stable maps of 3-manifolds into the plane. *Proc. London Math. Soc.* **92**, pp. 471–504.

Zakalyukin, V. M. (1976), Lagrangian and Legendrian singularities. *Funct. Anal. Appl.* **10**, pp. 23–31.

Zakalyukin, V. M. (1984), Reconstructions of fronts and caustics depending on a parameter and versality of mappings. *J. Soviet Math.* **27**, pp. 2713–2735.

Zakalyukin, V. M. (1995), Envelopes of families of wave fronts and Control theory. *Proc. Steklov Institute of Math.* **209**, pp. 114–123.

Index

P-$\mathcal{R}^+$-equivalence, 62, 64
P-$\mathcal{K}$-equivalent, 63
ν-umbilic point, 40
ν-flat umbilic point, 41
ν-principal curvature, 38
ν-shape operator
  submanifolds in $\mathbb{R}^{n+r}$, 37–39
  surfaces in $\mathbb{R}^4$, 216
h-principal, 346
k-jet space, 47
ν-parabolic point, 41, 43
ν-principal direction, 38
4-Vertex Theorem, 332, 343

kth-regular immersion, 345

apparent contour, 159, 160, 163, 164
asymptotic curve
  surfaces in $\mathbb{R}^3$, 144, 157, 186, 187
  surfaces in $\mathbb{R}^4$, 213–217, 339
  surfaces in $\mathbb{R}^5$, 264, 266, 269
asymptotic direction
  surfaces in $\mathbb{R}^3$, 144, 148, 158, 160
  surfaces in $\mathbb{R}^4$, 213, 214, 223, 235, 240, 241
  surfaces in $\mathbb{R}^5$, 260–262, 265, 268–270, 272

bi-Lipschitz equivalence, 71
bifurcation set, 9, 64, 101
binormal curvature (surfaces in $\mathbb{R}^4$), 338, 339
binormal direction
  surfaces in $\mathbb{R}^4$, 222, 223, 236, 238, 241
  surfaces in $\mathbb{R}^5$, 259, 260, 265
binormal vector, 14

canal
  cylindrical pedal, 134
  hypersurface, 42, 43, 134, 225
  surface, 43
canonical
  1-form, 98
  contact structure, 106
  symplectic structure, 99
Carathéodory conjecture, 337
catastrophe
  map, 9, 102
  map-germ, 64
  set, 9, 101
caustic, 9
Christoffel symbols, 322
classification, 59
codimension of an extended orbit, 54
codimension of an orbit, 54
complex curve, 219
computer vision, 139
conjugate curve congruence, 145
conjugate direction, 144, 163
contact, 7
  between submanifolds, 75
  form, 105
  group, 75

manifold, 105
map-germ, 75
structure, 105
type, 76
with a foliation, 80
contactomorphic, 107
contactomorphism, 107, 108
contour generator, 159, 160
convex, 330, 336
convex hull, 330
cosmic censorship, 292
cotangent bundle, 9, 98
criminant, 48
critical set, 48
cross-cap, 13, 18, 50, 234, 235, 239–241
elliptic, 239
hyperbolic, 239
parabolic, 239
curvature
de Sitter Gauss-Kronecker, 321
de Sitter mean, 321
Gauss-Kronecker, 26–28, 33, 34, 43
geodesic, 142, 158
hyperbolic Gauss-Kronecker, 321
hyperbolic mean, 321
Lipschitz-Killing, 38, 39, 43
mean, 26
normal, 142, 143
plane curve, 2, 3
principal, 26, 27, 29–31, 164
space curve, 14
curvature ellipse
surfaces in $\mathbb{R}^4$, 205, 208, 210, 213
surfaces in $\mathbb{R}^5$, 252, 257, 258
cusp, 4, 6, 49
cusp of Gauss, 155–157, 185, 188, 332, 335
cuspidal cross-cap, 20
cuspidal edge, 20, 154
cylindrical pedal, 32, 153
singularities of, 154

Damon, 69
Darboux Theorem, 99
de Sitter 3-space, 284
de Sitter horosphere, 305
degree of a map, 340
developable surface, 19
discriminant, 10, 48, 64, 68
distance squared function, 7
distance squared function generic
surface in $\mathbb{R}^3$, 179, 191
surface in $\mathbb{R}^4$, 247
surfaces in $\mathbb{R}^5$, 273
dual
coordinates, 99
hypersurface, 32
dual surface, 154
Dupin foliation
spherical, 133, 183
tangential, 132, 153
Dupin indicatix
tangent, 152

elliptic point
surfaces in $\mathbb{R}^3$, 12, 33
surfaces in $\mathbb{R}^4$, 207, 212, 213, 215, 223, 239
equidistant surface, 307
equisingularity, 71
Euler characteristic of
a cylindrical pedal, 335
a locally convex surface in $\mathbb{R}^4$, 336
a wavefront, 333
evolute, 3, 4, 30
extended family of
distance squared functions, 90, 178
height functions, 88, 119, 147
lightcone height functions, 300
extended tangent space $L_e\mathcal{G} \cdot f$, 54
external point, 330
extremal of a principal curvature, 190, 196
extrinsic properties, 141

family of
distance squared functions, 8, 9, 34, 90, 120, 178, 246
height functions, 33, 88, 119, 129, 147, 221, 224, 226, 236

lightcone height functions, 293, 319
Lorentzian distance squared functions, 309
orthogonal projections, 93, 159
finite determinacy, 55
first fundamental form, 24, 25, 36, 140
flat rib, 230
flat ridge, 230, 232, 265
flecnodal
curve, 187
point, 165, 166, 187
set, 166, 188
focal set, 13, 30, 182, 195
fold, 49
future
directed, 284
direction, 284

Gauss map, 12, 25–27, 32, 43, 44, 140
Gauss map with respect to a normal vector field, 37
Gauss-Bolyai-Lobachevski, 284
Gauss-Bonnet Theorem, 335
Gaussian curvature, 12, 141, 163, 204, 239
generalised Gauss map, 226, 227
generating family, 10, 11, 102, 109, 116
generic
embedding, 81
immersion, 81, 85–87
property, 81
geodesic inflection, 142, 186, 187, 197
geometric subgroups, 69
germ, 46
graph-like
Legendrian
immersion, 113
submanifold, 113
Morse family of hypersurfaces, 115
wavefront, 116
gravitational collapse, 292

height function generic
surfaces in $\mathbb{R}^3$, 147, 155, 156, 184
surfaces in $\mathbb{R}^4$, 226, 227, 229, 231, 232
surfaces in $\mathbb{R}^5$, 259
Hessian matrix, 34, 41
horosphere, 307
horospherical
Chern-Lashof Type Theorem, 344
Gauss-Bonnet Theorem, 344
Horospherical Geometry, 281, 325
hyperbolic 3-space, 284
hyperbolic Gauss map, 307
hyperbolic point
surfaces in $\mathbb{R}^3$, 12, 33
surfaces in $\mathbb{R}^4$, 207, 212, 215, 223, 229, 238, 239
hypocycloid, 193, 198

inflection, 3, 164, 186
inflection point
surfaces in $\mathbb{R}^4$, 206, 210, 212, 214, 215, 220, 223, 228, 231, 234, 238, 335, 336
internal points, 330

Koenderink Theorem, 163

Lagrangian
diffeomorphism, 100
equivalent, 101
fibration, 100
immersion, 9
map, 9, 101
stable, 102
submanifolds, 100
surface, 9
left group, 51
left-right group, 51
Legendrian
diffeomorphism, 108
equivalent, 108, 109
fibration, 107
immersion, 10
lift, 108
map, 10, 108
stable, 109

submanifold, 107
surface, 10
Lie group, 52
lightcone
Chern-Lashof Type Theorem, 342
dual surface, 318
Gauss map, 287
Gauss-Bonnet Theorem, 341
Gauss-Kronecker curvature, 288
mean curvature, 288
normal vector, 318
parabolic point, 288, 291
pedal, 300
pedal surface, 300
principal curvature, 288
second fundamental form, 288
shape operator, 319
Theorema Egregium, 320
tightness, 345
umbilic, 288
Weingarten formula, 289
lightlike
focal set, 312
hyperplane, 283
hypersurface, 311
plane, 283
tangent hyperplane, 298
vector, 283
Lightlike Geometry, 281
line of principal curvature, 141
Liouville form, 98
Little, 201, 204, 205, 338
local ring, 112
Loewner conjecture, 337
Lorentz-Minkowski space, 281

mandala, 317
map-germ, 46
marginally trapped, 292
strongly, 293
Mather, 45
Mather's groups, 51
Maurer-Cartan structural equations, 203
maximal spacelike surface, 293
mean curvature, 141
mean curvature vector
surfaces in $\mathbb{R}^4$, 205
surfaces in $\mathbb{R}^4_1$, 292
surfaces in $\mathbb{R}^5$, 253
Metric properties, 71
Milnor, 70
Milnor Fibration Theorem, 70
Milnor number, 70
minimal surface, 293
miniversal unfolding, 57, 58, 62, 63, 65
Minkowski space-time, 281, 283
Monge form, 146
Morse family of
functions, 102
hypersurfaces, 108
multi-germ, 47
multi-jet space, 47

nice dimensions, 86
normal curvature, 204
normal form, 59
normal frame
future directed, 287
past directed, 287
normal section, 143, 163
normal vector, 14
normalised lightcone
Gauss map, 291
Gauss-Kronecker curvature, 291
mean curvature, 291
principal curvatures, 291
shape operator, 291

open lightcone, 284
osculating
circle, 73
family of hyperspheres, 132
hyperplane, 222, 249, 260
hypersphere, 73, 247

parabolic point
hypersurfaces in $\mathbb{R}^{n+1}$, 33, 34
surfaces in $\mathbb{R}^3$, 12, 148

surfaces in $\mathbb{R}^4$, 207, 212, 215, 223, 231, 238
parabolic set
hypersurfaces in $\mathbb{R}^{n+1}$, 33
surfaces in $\mathbb{R}^3$, 44, 184
surfaces in $\mathbb{R}^4$, 215
parallel, 5, 6, 10
past directed, 284
pencils of quadratic forms, 212, 216
Penrose inequality, 292
Plücker conoid, 18
pleat, 49
Poincaré, 284
Poincaré-Hopf formula, 329, 336
principal
curvature, 12, 141
direction, 12, 26
Principal Axis Theorem, 25
profile, 159
projection P-generic
surfaces in $\mathbb{R}^4$, 233, 235, 236
surfaces in $\mathbb{R}^5$, 267
projection generic surface in $\mathbb{R}^3$, 165
projective cotangent bundle, 10, 106

regular point, 48
rib, 274
rib of order k, 247
ridge, 189–192
ridge of order k (surfaces in $\mathbb{R}^4$), 248
ridge point, 189
right-group, 51
robust features, 139, 184
ruled surface, 15, 18

second fundamental form
hypersurfaces in $\mathbb{R}^{n+1}$, 27
submanifold in $\mathbb{R}^{n+1}$, 36
surfaces in $\mathbb{R}^3$, 140
surfaces in $\mathbb{R}^4$, 203, 204, 209, 211, 219
surfaces in $\mathbb{R}^5$, 252, 254
semiumbilic point, 206, 248, 249, 337, 338
Serret-Frenet, 15
shape operator, 12, 25, 28, 29, 43
shape recognition, 139
simple germ, 60
singular fibre, 70
singular map-germ, 48
singular point, 47
space-time singularities, 292
spacelike
hyperplane, 283
plane, 283
surface, 285
vector, 283
spacelike knot, 343
sphere (in $H^3_+(-1)$), 307
standard contact structure, 105
stereographic projection, 336
strata, 330
bifurcation, 331
conflict, 331
Morse, 331
stratification, 329, 330
Maxwell, 331
stratified set, 330
sub-parabolic curve, 195–197
swallowtail, 11, 20, 154
swallowtail point, 333
Symmetry Lemma, 75
Symmetry Set, 21
symplectic
form, 98
manifold, 98
structure, 98
symplectomorphism, 99

tangent
affine hyperplane, 129
indicatrix, 130
tangent developable, 19
tangent space $L\mathcal{G} \cdot f$, 54
The contact Darboux Theorem, 107
The Morse Lemma, 60
Theorema Eugregium of Gauss, 12
Thom, 95
Thom's seven elementary catastrophes, 63
Thom's splitting lemma, 61

Thom's transversality theorem, 81
Thom-Bordman symbols, 51, 227
time-orientable, 284
time-orientation, 285
timelike
  hyperplane, 283
  plane, 283
  vector, 283
torsion, 15
torsion zero point, 332
totally ν-umbilic submanifold, 40
totally semiumbilic surface, 338
totally umbilic hypersurface, 29
transversality, 81
tri-tangent plane, 335
tri-tangent support planes, 332
triple point, 333

umbilic point, 12, 29, 141, 193, 194, 336
umbilical curvature (surfaces in $\mathbb{R}^5$), 276
umbilical focal hypersurface (surfaces in $\mathbb{R}^5$), 273
umbilical focus (surfaces in $\mathbb{R}^5$), 273

Vassiliev type invariant, 346
versal deformation, 56, 62
versal unfolding, 56–58
vertex, 3

wavefront, 5, 108
Weingarten
  formula, 28
  map, 25, 140
Weingarten map, 12
Whitney, 45
Whitney C^∞-topology, 80
Whitney umbrella, 50

Printed in the United States
By Bookmasters